UK Health (

Recollections and Reflections

The British Computer Society

The BCS is the leading professional body for the IT industry. With members in over 100 countries, the BCS is the professional and learned Society in the field of computers and information systems.

The BCS is responsible for setting standards for the IT profession. It is also leading the change in public perception and appreciation of the economic and social importance of professionally managed IT projects and programmes. In this capacity, the Society advises, informs and persuades industry and government on successful IT implementation.

IT is affecting every part of our lives and that is why the BCS is determined to promote IT as the profession of the 21st century.

Joining the BCS

BCS qualifications, products and services are designed with your career plans in mind. We not only provide essential recognition through professional qualifications but also offer many other useful benefits to our members at every level.

BCS membership demonstrates your commitment to professional development. It helps to set you apart from other IT practitioners and provides industry recognition of your skills and experience. Employers and customers increasingly require proof of professional qualifications and competence. Professional membership confirms your competence and integrity and sets an independent standard that people can trust. Professional Membership (MBCS) is the pathway to Chartered IT Professional (CITP) status.

www.bcs.org/membership

Further Information

Further information about the BCS can be obtained from:
The British Computer Society, First Floor, Block D,
North Star House, North Star Avenue, Swindon SN2 1FA, UK.

Telephone: 0845 300 4417 (UK only) or +44 (0)1793 417 424 (overseas)
Contact: www.bcs.org/contact

UK Health Computing

Recollections and Reflections

EDITED BY

Glyn Hayes and Denise Barnett

 BCS

© 2008 The British Computer Society

The right of the authors to be identified as authors of this work has been asserted by them in accordance with sections 77 and 78 of the Copyright, Designs and Patents Act, 1988.

The British Computer Society,
First floor, Block D,
North Star House,
North Star Avenue,
Swindon, Wiltshire SN2 1FA,
UK
www.bcs.org

ISBN 978–1–902505–99–2

British Cataloguing in Publication Data.
A CIP catalogue record for this book is available at the British Library.

Edited and typeset by: Standard Eight Limited, www.std8.com
Printed and bound by Antony Rowe Ltd, Chippenham, Wiltshire.

Contents

Contents

Contents

Abbreviations

A&E	Accident and Emergency
ABHI	Association of British Healthcare Industries
AGNIS	Awareness among GPs of the Need for Information Security / A Gnomic Nursing Information System
AHA	Area Health Authority
AI	Artificial Intelligence
AIM	Advanced Informatics in Medicine
ALICE	Algorithms through Interaction with Clinical Experts
API	Application Program Interface
ART	Algebraic Reconstruction Technique
BAMM	British Association of Medical Managers
BARD	British Database of Research into Aids for the Disabled
BCS	British Computer Society
BJHC	*British Journal of Healthcare Computing*
BJHC&IM	*British Journal of Healthcare Computing & Information Management*
BMA	British Medical Association
BSI	British Standards Institution
CAER	Computer-based Accident and Emergency Records
CAL	Computer-Assisted Learning
CAMAC	Computer Automated Measurement and Control
CANIS	Computer-Assisted Nursing Information System
CANO	Computer-Assisted Nursing Orders
CAS	Comparative Analysis Service
CASPE	Clinical Accountability, Service Planning and Evaluation
CAT	Computer-Assisted Testing / Computer-Assisted Tomography
CBS	Common Basic Specification
CBT	Computer-Based Training
CCPS	Computerized Care Planning System
CCTA	Central Computer and Telecommunications Agency
CDS	Clinical Data System
CDSS	Clinical Decision Support System
CEN	Comité Européen de Normalisation
CHI	Community Health Index
CHIME	Centre for Health Informatics and Multiprofessional Education
CIS	Clinical Information System
CISP	Clinical Information Systems Project
CIT	Centre for Information Technology
CMMS	Casemix Management System
CNN	Case Notes Number

CNS	Community Nursing System
CNSS	Computerized Nurse Scheduling System
CPC	Computer Policy Committee
CPD	Continuing Professional Development
CPHVA	Community Practitioners' and Health Visitors' Association
CPNG	Computer Projects Nursing Group
CPR	Computer-based Patient Record
CRAMM	CCTA Risk Analysis and Management Methodology
CRISP	Computer-Recorded Information System for Psychiatry
CRS	Care Records Service
CSA	Common Services Agency (Scotland)
CT	Computerized Tomography
CTG	Cardiotocograph
DBMS	Database Management System
DBO	Database Object
DEB	Dental Estimates Board
DEC	Digital Equipment Corporation
DHA	District Health Authority
DHSIS	Directorate of Health Service Information Systems
DHSS	Department of Health and Social Security
DIANE	Dental Institute Area Network
DIN	Doctors' Independent Network
DoH	Department of Health
DRG	Diagnosis Related Group
DSM	Digital Standard MUMPS
DSS	Decision-Support System
DTI	Department of Trade and Industry
EASE	The European Association of Science Editors
ECG	Electrocardiogram
EDI	Electronic Data Interchange
EDP	Electronic Data Processing
EDTA	European Dialysis and Transplant Association
EEG	Electroencephalogram
EFMI	European Federation for Medical Informatics
EHR	Electronic Health Record
EPR	Electronic Patient Record
ERDIP	Electronic Record Development and Implementation Programme
ESPRIT	European Strategic Programme for Research and Development in Information Technology
FBP	Filtered Back-Projection
FECG	Fetal Electrocardiogram
FHSA	Family Health Service Authority
FIP	Financial Information Project
FPC	Family Practitioner Committee

FPS	Family Practitioner Service
GDEP	GPASS Data Evaluation Project
GEIS	General Electric Information System
GIPL	Generalized Inter-Processor Link
GIS	Geographical Information System
GMSC	General Medical Services Committee
GP	General Practice/Practitioner
GPASS	General Practice Administration System for Scotland
GPRD	General Practice Research Database
GUI	Graphical User Interface
HA	Health Authority
HIF	Health Informatics Forum
HIS	Health Information System
HISS	Hospital Information Support Systems
HMC	Hospital Management Committee
HPA	Hospital Physicists Association
HPG	Homogeneous Patient Group
HRG	Healthcare Resource Group
ICD	International Classification of Disease
ICPC	International Classification of Primary Care
ICT	Information and Communications Technology
ICU	Intensive Care Unit
ICWS	Integrated Clinical Workstation
IDPS	IBM Desktop Pilot System
IFIP	International Federation for Information Processing
IHSM	Institute of Health Services Management
IMG	Information Management Group
IMIA	International Medical Informatics Association
IM&T	Information Management and Technology
IPL	Inter-Processor Link
IPMI	*Information Processing in Medical Imaging* (conference)
IRC	Inter-Regional Collaboration
IRIA	Institut de Recherche en Informatique et en Automatique
ISHTAR	Implementing Secure Healthcare Telematics Applications in euRope
ISO	International Organization for Standardization
ITIN	*Information Technology in Nursing*
ITU	Intensive Therapy Unit
IVI	Interactive Video Instruction
JANET	Joint Academic Network
JEC	John Ellicott Centre
KCHMRP	King's College Hospital Medical Records Project
LAN	Local Area Network
LEO	Lyons Electronic Office
LIMS	Laboratory Information Management System

LMSG	London Medical Specialist Group
LP	Linear Programming
LSP	Local Service Provider
MAAG	Medical Audit Advisory Group
MCA	Medicines Control Agency
MDIS	McDonnell Douglas Information Systems
MIE	Medical Informatics Europe
MOH	Medical Officer of Health
MRC	Medical Research Council
MRI	Magnetic Resonance Imaging
MSG	Medical Specialist Groups
MTS	Michigan Terminal System
MUGAL	MUMPS User Group Application Library
MUMMIES	Maternity Users – Medical, Midwifery and Management, Information and Expert Systems
MUMPS	Massachusetts General Hospital Utility Multi-Programming System
NAO	National Audit Office
NCASP	National Clinical Audit Support Programme
NCC	National Computing Centre
ND	Norsk Data
NHS	National Health Service
NHS CFH	NHS *Connecting for Health*
NHSCR	NHS Central Registry
NHS IA	NHS Information Authority
NHSME	NHS Management Executive
NHSTA	NHS Training Authority
NIS	Nursing Information System
NMSG	Northern Medical Specialist Group
NOMDS	National Organ Matching and Distribution Service
NPfIT	National Programme for Information Technology
NPHT	National Provincial Hospitals Trust
NPL	National Physical Laboratory
NSG	Nursing Specialist Group
NUMINE	Network of Users of Microcomputers in Nurse Education
OCHP	Oxford Community Health Project
OMR	Optical Mark Reader
OR	Operational Research
OSL	Open Software Library
OT	Occupational Therapy
OU	The Open University
PAC	Parliamentary Committee on Accounts
PACS	Picture Archiving and Communication Systems
PACT	Prescribing Analysis and Cost
PAM	Professions Allied to Medicine
PAS	Patient Administration System

PC	Personal Computer
PCCIU	Primary Care Clinical Informatics Unit
PCG	Primary Care Group
PCIU	Primary Care Information Unit
PCS	Pathology Communications Server
PCT	Primary Care Trust
PETA	Programme for Enabling Technology Assessment
PFI	Private Finance Initiative
PHCSG	Primary Health Care Specialist Group
PI	Performance Indicator / Programmed Instruction
PIU	Programmed Investigation Unit
PMI	Patient Master Index
PPS	Prospective Payment System
PRIMIS	Primary Care Information Services
QEH	The Queen Elizabeth Hospital, Birmingham
RCC	Regional Computer Centre
RCGP	Royal College of General Practitioners
RFA	Requirements for Accreditation
RHA	Regional Health Authority
RHB	Regional Hospital Board
RLI	Royal Lancaster Infirmary
RM	Resource Management
RMI	Resource Management Initiative
RXDS	Rank Xerox Data Systems
S&DP	Security and Data Protection
SAGNIS	Strategic Advisory Group for Nursing Information Systems
SCRAP	System of Computing in Radiotherapy and Treatment Planning
SEISMED	Secure Environment for Information Systems in MEDicine
SHHD	Scottish Home and Health Department
SMIS	Standard Maternity Information Service
SMMIS	St Mary's Maternity Information System
SNOMED®	Systematized Nomenclature of Medicine
SPSS	Statistical Processing for Social Sciences
SQL	Structured Query Language
SSADM	Structured Systems Analysis and Design Method
UCH	University College Hospital
UKCHIP	UK Council for Health Informatics Professions
UKTS	UK Transplant Service
UMIST	University of Manchester Institute of Science and Technology
VDU	Visual Display Unit
WIMP	Windows, Icons, Menus, Pointing device
WIMS	Works Information Management System
XML	eXtensible Mark-up Language

List of tables and figures

Recollections and reflections

I.T.

Far,
and long,
have we travelled,
to make knowledge
so easy to convey –
even the knowledge
of how it is conveyed
voluminous enough
to justify at least
a Bachelor's Degree:

> "the wireless" first,
> then "television", (each image
> "worth a thousand words");
> and, bursting late on the scene,
> the "personal computer",
> with "word processor", and
> "databases" accessed at
> button's touch – the whole of
> the Encyclopaedia Britannica
> on a single "CD-ROM".
> And – latest of all – "emailing",
> with "surfing the World Wide Web" –
> and "Google" the Open Sesame
> to a Global Knowledge Store.

So –
what more?

> Midst such welter of
> technology, have we, in fact,
> forgotten the simplest (but
> most profound) technique
> of all, whereby is acquired
> the deepest of all
> knowledge? –

> "Be *still* –
> and know
> that I am
> God."

Lloyd A W Kemp 2005, Pioneering Medical Physicist (age 93)

Recollections and reflections

Preface

The introduction of information technologies into health care has not followed a clear, smooth, well-marked path. Rather it has been like cutting routes through very tall grass, sometimes well over head-height, without a compass or a guide. Even now there may be times when it still feels a bit like that, especially when health care and information professionals first set off into health care informatics. From above it may be easy to spot the simple crop circles, the complex maze or the straight path that got the explorers to their intended destination. When deep in the middle it is less easy to assess the situation.

Membership of the health information groups within the British Computer Society (BCS) offers the opportunity to join with colleagues in this exploration and to start along the tracks made by others. To celebrate the 25th anniversary this guide sets out to record some of the main paths and to suggest how current and new members of the health informatics groups might contribute.

In the early days, like-minded people came together to share ideas about the direction of the paths and the methods of marking them out. They came with different professional backgrounds but with shared interests in getting the new technological developments to work for patients and to help staff support those patients. This book sets out to record some of the routes already created and thus to explain some of the possible reasons why some projects have yet to reach a satisfactory destination. The past and current environments and the tools made available and developed are outlined to help explain some of the problems encountered in the past and that may still be recognized as present today.

The BCS Health Informatics Forum (HIF) provided the opportunity to ask the early pioneers how they started out and to recall some of their routes and decisions. This should help new members to understand how their interest groups have contributed and where it has taken them. There are suggestions for ensuring ideas continue to be shared and clinical influence brought into projects so that patients benefit. In these days of matrix management, where people with a wide range of skills need to work together to develop any health care project, this book provides a way to understand the perspectives of other professional groups. It also provides a lasting memento for the pioneers who set off first and bravely led the way.

This is a book written by BCS HIF members for members, and is about the many BCS members and others who helped.

Health care computing in the UK has come a long way since its early origins in the 1960s. Many lessons have been learnt, many mistakes made.

At the time of writing, in 2006, there are major initiatives taking place to expand the use of IT in health care. Have these new steps taken account of the experience over 50 years? Have the hard won lessons been learnt? Are we in danger of reinventing wheels?

Many of the early pioneers are no longer with us. The BCS HIF felt that we should try to document the early days. Partly to record what has been done and who did it, partly to document what has been learnt, but also to provide a resource for future generations.

This book contains some recollections and reflections on the development, in the UK, of health care computing, or medical computing as it used to be called. It is more like an autobiography than a history because some of the key players have already left the field – either through death, loss of memory or loss of papers and inclination. There will be other players who commit their recollections and reflections to paper and, indeed, we hope that this book will encourage them to do so. However, this is not intended as the stories of old men and women told round the fireside on a winter's evening about what might or might not have happened but rather is based on the contemporary evidence of papers and reports; thus the references are a key element of the book, which will allow researchers to follow the train of thought for themselves. The references, therefore, are as complete as the authors can make them to help the interested reader follow up a topic, but given the foregoing aims, it is inevitable that some details may have got lost over the years, and their completeness will vary between authors.

The period covered by this book is from the early starts up to the new millennium. The end date is variable because many activities spanned a long period of time and it would have been inappropriate to use an arbitrary cut-off date.

In Chapter 29, *Lessons for the future*, there is a paragraph that should be driven into the hearts of all those who, at whatever level, wish to conceive, develop or implement IT in health care.

> There is a wealth of advice on project management and systems development but the pathology of failed systems has a smell about it. The budgets and project milestones have been set by political rather than technical requirements. The clinical staff who will be using the systems have not been heavily involved in the design processes. The technical staff have not acquired an adequate understanding of the clinical requirements. The organization's management have not been fully involved in the project nor have they grasped the implications of project failure.

That lesson reverberates throughout the story that this book tells.

There are parts of the story missing, either because individuals were not available to tell them or did not have the time. We do not think these omissions detract from the story but do hope they will be told in other places at other times.

I am extremely indebted to the Editorial Board members who have given so much of their time freely for this project. I would particularly like to recognize the work done in the early stages of this work by Pat Bishop. Pat, a true pioneer of health care computing, sadly passed away during the writing of the book. He will be remembered with affection and respect by all who knew him.

The initial reviewers, Jean Roberts, Sheila Bullas and Nettie de Glanville provided a vital critique of our early work. John Bryden from Scotland, who was originally on the Editorial Board, withdrew for personal reasons. I am also indebted to those others who have provided parts of this story. They are acknowledged in the relevant sections. In particular we are all indebted to the work of Chris Mayes, the HIF manager at BCS headquarters, for keeping us on track and providing the very necessary administrative support.

Note: Over the period covered there were changes in the political organization of the NHS: Ministry of Health, Department of Health, Department of Health and Social Services – Home and Health Department in Scotland – as well as devolution of responsibilities to national assemblies. We have used the convention DoH throughout for simplicity.

Glyn Hayes, Chairman,
BCS Health Informatics Forum 2006

1 BCS involvement in health care informatics

DENISE BARNETT AND BARRY BARBER

The British Computer Society (BCS) was founded in 1957 to provide a national body to represent all aspects of the activities of the emerging computing profession. HRH the Duke of Kent became patron in 1976. The Society was awarded a royal charter in 1984 when it had over 30,000 members in various grades, ranging from the full professional grades of Member and Fellow through to the Affiliate grade.

Affiliates were individuals from other professions who were allowed to participate in BCS activities. Access to the professional grades was through a combination of academic and professional qualifications with supervised, structured experience. The Professional Board supervised the BCS examinations, the professional development scheme and the BCS code of conduct. This was one of the four main operating boards, each chaired by a vice-president, allocated resources by a Finance Board, and over seen by Council.

By 1985 the Branches' Board linked together the activities of 42 local branches, each running a local programme of events. Of greater relevance to the health professions was the Specialist Groups Board. This covered 40 groups, including nursing, medical and primary health care. Each group put on a programme of activity for its members. Any member of the BCS was also free to join any specialist group.

There was a Technical Board to oversee computer developments and to help BCS members keep up to date with issues such as privacy and confidentiality, and the security of the census, as well as the representation of the BCS on the specialist committee of the British Standards Institution.

The BCS maintained a joint library with the Institute of Electrical Engineers. The health section was relatively limited, but it provided a home for the proceedings of national and international conferences. Members of the specialist groups, as BCS affiliates, were able to use the material for reference on site, but not to borrow it. In the 1980s, the BCS published the *Computer Journal*, the *Bulletin* and a BCS Newsletter. The Society also provided a weekly page of BCS news and activities in the trade newspaper *Computing*.

The membership structure of the BCS has now changed, with more professional members joining through other routes. The expected professional and ethical standards have been clearly stated.

HEALTH SPECIALIST GROUPS

There were five groups. The first were the London Medical Specialist Group (LMSG), which was set up in 1967, the Northern Medical Specialist Group founded in 1968 and the Scottish Medical Specialist Group in 1974. The Nursing Specialist Group was established in 1983, but it had its antecedents in the DoH Computer Project Nurses Group, which had been formed in 1974. The Primary Care Specialist Group was established in 1982. The groups' titles might sound to 21st century ears to be uni-professional, but in fact there was a great deal of sharing and cross-fertilization of ideas between and within them. The only reason for developing additional groups was that the first two groups could not come round to exploring topics in Primary Care or Nursing sufficiently often to fulfil the growing need for discussion.

The individual groups

Founder and early members of the specialist groups recall the early days.

THE LONDON MEDICAL SPECIALIST GROUP

The BCS provided a useful forum for discussing medical computing possibilities as a specialist. In the early days the format was usually an evening meeting in central London with a speaker and a glass of sherry to follow – the latter was quite an attraction and a reward for meeting in 'our own time'. This after-hours format worked for quite a while in the early days because that was the only way that time could be found for informal discussion with colleagues, often working in London.

It is interesting in retrospect that after talks on the use of computers in hospitals, audience comments were often along these lines: 'We would like to do this sort of thing but we cannot find the cost of a computer.'

Over time the comments changed to: 'We can find the cost of a computer but we cannot find out the cost of computer maintenance,' and later to: 'We can find the cost of a computer and its maintenance but we cannot find out the cost of security.'

Among the early chairmen of the LMSG were Dr Malcolm Forsythe, Dr Mike Abrams, Dr Barry Barber, Professor John Anderson and Professor Richard Vincent. They were at the heart of the many medical informatics developments across the country. The detailed constitution, minutes and documents have been handed down from chairman to chairman and secretary to secretary but all appear to have been lost in the mists of time.

Doubtless fragments of these records will emerge among individuals' professional papers in the future. All that remains is the idiosyncratic listing of some committees and chairmen when conference proceedings were published or when copy was contributed to the *British Journal of Healthcare Computing* or the various versions of the *Kluwer Handbook*.[1, 2]

Curiously, despite the fact that the chairmen of the BCS Specialist Groups were always approved and recorded by the Technical Board or the Specialist Groups Board it appears that even these records have vanished.

The LMSG joined with the other groups to run national meetings in 1969 and 1971 and then provided the driving force behind the establishment of the annual *Current Perspectives and Healthcare Computing* conferences. The BCS encouraged the group to make the most of available opportunities in medical computing and the Society believed that it was better able to do this without any strong central direction. Senior officers provided the group with support at critical junctures which enabled it to:

- join IFIP TC4 – later to become the International Medical Informatics Association (IMIA);
- help establish the European Federation for Medical Informatics (EFMI);
- run the first EFMI conference in Cambridge in 1978 – the group was given a budget of £30,000 in the year when the library budget of £10,000 was cut;
- run the first International Conference on Nursing Informatics in London and Harrogate in 1982 – which enabled the IMIA to establish its Nursing Informatics Group;
- establish our annual conferences – sometimes the BCS had a conference department that they wanted us to use and sometimes it had not; and
- establish new specialist groups as required to meet our needs – including the Medical Specialist Groups Coordinating Committee which was the forerunner of the present Health Informatics Forum.

The annual conferences very successfully linked with the *British Journal of Healthcare Computing* (BJHC) in providing a combined BCS conference and BJHC Exhibition. Although this turned the event into a very large affair requiring substantial financial underpinning, its success over many years enabled the event to develop sufficient monies without recourse to central BCS funds. In times of financial crisis various BCS Treasurers have discovered how successful the group's enterprises had been. Throughout this time many BCS members continued to run conferences in conjunction with their own professional bodies and their hospitals, regions or the NHS or DoH, but the BCS provided a more permanent home for LMSG activities. However, for these reasons, much of the early medical informatics and medical operational research literature is scattered across many professional journals, organizations and national and international conferences.

BARRY BARBER

THE NORTHERN SPECIALIST GROUP

The origins of the Northern Specialist Group, or the Northern Medical Specialist Group (NMSG) as it was originally known and is still affectionately referred to, can be traced back to 1968. Following the example of the London

Group, the NMSG was founded by Dr David 'Nobby' Clark who was then a newly appointed research fellow in medical computing at Manchester University. He collaborated in this venture with Dr Peter Harvey, a consultant from Lancaster Royal Infirmary, Professor Bernard Richards and several others including Dr Gillian Cooper, Dr Peter Farrer, and Dr Brian Gill.

The NMSG met three or four times a year in the Computation Department at the University of Manchester Institute of Science and Technology (UMIST). One of the first speakers to be invited was a high court judge. Judge Sigismund, discussed some of the ethico-legal issues that were likely to arise when computers were used to store medical records. I joined Nobby's unit, and the Group, in 1970. I believe it was to one of these meetings, held in Lancaster in the early 1970s, that we gave what must have been one of the very first demonstrations of tele-computing in medicine. We showed that we could search records held at Manchester and see the results on a terminal 60 miles away. Unfortunately due to a programming error on my part the demonstration came to a halt as soon as we tried to enter a search term in the second half of the alphabet.

Jean Roberts joined the Group in 1973 and played a major role in several key developments. An important milestone was the introduction of a national meeting for health computing. In his role as NMSG chair Nobby took on the task of organizing the exhibition when the inaugural meeting was held in Lancaster in 1978, and was reprimanded – so he tells me – for overspending his budget on overseas telephone calls to potential exhibitors. The first EFMI meeting was held in 1978 in Cambridge, when many of the NMSG were heavily involved.

After the 1970s, and as Nobby and I moved into other areas of activity, the Group entered a fallow period. It was re-launched in 1996 with the resumption of active participation by Professor Bernard Richards and by me, under the leadership of Steve Kay. It has been extremely active ever since – organizing up to 12 meetings a year, mostly in Manchester.

TOM SHARPE

THE SCOTTISH MEDICAL SPECIALIST GROUP

Half a dozen or so of us from Scotland ended up at that seminal BCS Medical Specialist Group meeting held in Birmingham in 1969. There were three or four of us there from Glasgow and we had not even met before. Following that meeting internal chat suggested that we should have a Scottish group. I personally approached the BCS HQ and went through the hoops to have established the Medical (Scotland) Specialist Group.

We held our first meeting in Paisley with about 60 attendees, and established a committee and held meetings around Scotland about every six to eight months. Having a meaningful committee with the geography of Scotland was a challenge. We established the principle of a core group committee of secretary, treasurer and chairholder, based on Glasgow, and the rest of the committee coming from the Scottish city where we were holding the meetings in that particular year.

Sometime before 1981 Barry Barber and Bud Abbott set up a meeting with Mike Heasman – the then guru of Scottish health IT – and me to encourage the Scottish Group to become part of the Health Informatics Committee. Up until then we had been solely represented through the BCS Specialist Groups Committee.

The rest is by and large history. I was certainly too long as chairholder. Ray Jones succeeded me and then we went into the doldrums of early electronic mailing lists and deciding how everyone was to be kept posted. Following Ray's move south there was a problem period of a new chair and secretary far removed from rapport with the others.

Heather Strachan has now succeeded as an excellent chairholder and the group is beginning to thrive again.

JOHN BRYDEN

PRIMARY HEALTH CARE SPECIALIST GROUP

In 1979, Geoffrey Dove, a London GP set up a Medical Computer Club. Similar organizations were set up by Geoffrey Clayton, from Norwich, and Alastair Malcolm in the north-east of England. These organizations came together to form the Primary Health Care Specialist Group (PHCSG). The inaugural meeting was held at the ICI Conference Centre at Alderley Edge, Cheshire in 1981. Geoff Clayton was the first chairman and Geoff Dove the treasurer. Both were emphatic that medical computing should be open to all contributors, that meetings should be well lubricated, and that the responsibility of the treasurer was to ensure that this could be afforded.

From the start the Group's membership was an eclectic mixture of GPs, GP staff, academics, health authority and DoH staff, plus technical professionals working in primary care. This mixture of members and the independence of being a BCS Specialist Group gave it greater influence and wide-ranging activities.

Soon after its formation the Group set up a sub-group for the BBC Micro, which was being used by many GPs at the time. It also accepted an invitation to send a representative to the Joint Computer Policy Group of the British Medical Association and Royal College of General Practitioners. The groups started holding two major conferences each year. These became a major source of debate and academic publication of work in the field. The Group published a regular newsletter and was and is still a melting pot for technical and organizational strategy development. In the early years one of its main activities was spreading the word about the benefits of computers to the general practice population.

After five years in the post Geoff Clayton retired as chairman and his place was taken by Glyn Hayes. Under his leadership the group developed a more political role, applying pressure and offering advice on primary care IT to all levels of the NHS. It also started think-tanks on technical subjects, which played a major role in the development of GP computer systems.

In 1990, the PHCSG organized a workshop on International Primary Care Computing in Brighton under the auspices of the IMIA. This workshop produced a book that described the thinking on the subject and included a large bibliography.[3]

The newsletter grew and subsequently, under the editorship of Sheila Teasdale become the *Journal of Informatics in Primary Care*, an indexed scientific publication with an international exposure.

GLYN HAYES

NURSING SPECIALIST GROUP

The nursing profession owes a particular debt to the BCS in that when the experimental projects were being developed, the small band of nurses involved in these projects grouped together for mutual discussion and support. They then looked to their national professional body, the Royal College of Nursing, for further support but found little interest or enthusiasm for the subject of computing at that time. The group welcomed the possibility offered to them of creating a Nursing Specialist Group (NSG) within the BCS. Furthermore, it gave them easy links with the other health professionals, both nationally and internationally. The Group has kept a careful list of its officers as follows:

Claire Ashton, Chair 1983–87
Yvonne Bryant, Secretary 1983–90, Chair 1990–93
Brian Layzell, Treasurer 1983–95, Vice Chair (administration) 1996
Ron Hoy, Chair 1987–90
Deirdrie Gossington, Vice Chair 1989
Madeline Gillies, Vice Chair 1990
Heather Strachan, Secretary 1990–96, Vice Chair 1997
Keith Oswin, Treasurer 1995–2002
Derek Eaves, Secretary 1996–97, Membership Secretary 1991
Nick Hardiker, Vice Chair 1996
Graham Wright, Chair 1993–97
Peter Murray, Chair 1997–99
Paula Procter, Secretary 1997–2004
Carol Cooper, Chair 1999–2001, Membership Secretary 1996
Helen Sampson, Vice Chair 1999, Chair 2001–04
Richard Hayward, Chair 2004–
Janette Bennett, Treasurer 2002–

The group, like the PHCSG also, established their own conferences to address nursing issues in a multidisciplinary way. The themes of these conferences offer a good reflection of the concerns of the NSG.

1984	Microcomputers in Nursing
1990	IT Beyond the Theory
1991	Information Technology – Nursing and the Multidisciplines
1992	Informatics for the Nursing Professions: The What, Why, When, Where and Who of Information Use
1993	Managing Information for the Benefit of Patients
1994	Sharing Information – Focusing on the Patient
1995	Patient Privacy, Confidentiality and Data Security
1996	Making the Right Connections Electronic and Human Networks
1997	Clinical Informatics – A Health Alliance
1998	IM&T – The New Strategy
1999	Solving the Jigsaw – Supporting the Frontline Practitioner
2002	Information in a Multidisciplinary World

More detail about the BCS NSG is in Chapter 19.

In the 1990s, with encouragement from members of the NSG who were also members of the College, the Royal College of Nursing appointed a part-time and then full-time advisor. It was the first of the three professional nursing bodies to do this. It also provided an important resource in Anne Casey who did a lot of work on nursing and clinical terms, both nationally and internationally.

The advent of the Nursing Professions Information Group encouraged the Royal College of Midwives and the Community Practitioners' and Health Visitors' Association (CPHVA) section of the more general health staff union, Unison, to develop their advisory membership structures and to invest in staff time on informatics issues. Members of these organizations were often also members of the NSG.

The NSG, like its fellow BCS specialist groups, provided printed versions of the papers given at its conferences, first as supplements within its newsletters and journal, later as the INFOrmed Touch Series. These papers were priced to come within the purse of the average staff nurse and were later added to the NSG website from where they can be downloaded without charge. The NSG faced major logistics problems in communicating with the very large number of nurses, midwives and health visitors working within the NHS, independent hospitals, hospices, private nursing homes and charities.

DENISE BARNETT AND MAUREEN SCHOLES

Internet contact

The current BCS health informatics member groups are available on the internet to individuals who may join one or more of the groups. Visit www. bcs.org/forums or a group's individual website:

Health Informatics Interactive Care

www.hiicsg.bcs.org

London and South East

www.hilsesg.bcs.org

Health Informatics Northern

www.bcs-nmsg.org.uk

Health Informatics Nursing

www.nursing.bcs.org

Health Informatics Primary Health Care

www.phcsg.org.uk

Health Informatics Scotland

www.scotshi.bcs.org.uk

Health Informatics South West

(No website.)

EXPANSION OF THE HEALTH SPECIALIST GROUPS

For a period, Barry Barber was vice-president (specialist groups) and during this time it was recognized that the existing groups could not cover all the topics that were of interest to those concerned with computing in primary care or nursing. This was a difficult decision because it was important that the health specialist groups should not disintegrate into separate islands of knowledge when they were all emphasizing the need for systems and organizational integration.

The Medical Specialist Groups Coordinating Committee

The Medical Specialist Groups (MSG) Coordinating Committee grew out of the organizing committee that was assembled to run the first international nursing informatics conference. It included representatives from all the health specialist groups as well as from other professional groups that were known to be involved in health informatics. Sometimes this was linked to individuals being active members of other professional groups as well as their BCS specialist group. For example, Barry Barber chaired the first computer topic group of the Hospital Physicists Association (HPA) in 1965–66.

The MSG came together to organize the first national conference from 6–10 January 1969 at the University of Birmingham, which attracted around 400 delegates. The Organizing Committee was chaired by Malcolm Forsythe of the Wessex Regional Hospital Board, the secretary was Julian Bogod from ICL, the proceedings editor was Mike Abrams of Guy's Hospital; Conway Berners Lee of ICL and Barry Barber of The London Hospital were additional committee members. The proceedings were published as *Medical Computing – Progress and Problems* for the BCS.[4]

This was followed by a smaller meeting, held at the University of Bristol from 15–17 September 1971. The proceedings were again edited by Mike

Abrams as *Spectrum 71*.[5] The Minister of State, Lord Aberdare, gave the opening address. After this the serious health informatics effort was channelled into local meetings of the three specialist groups. There was a lull in big conferences while people got on with some serious work arising from the first flood of ideas and opportunities, as well as work coming out of the DoH Experimental Real-Time Computer Programme.

The MSG Coordinating Committee held a conference in 1984 when the HPA was represented by John Newell, and the MUMPS User Group by Jo Milan. The lessons of the success of the coordinated effort were not lost. The MSG was gradually developed into the Health Informatics Committee and then the Health Informatics Forum, to be as inclusive as was possible with other professional organizations concerned with health informatics and its development.

The BCS medical specialist groups return to Birmingham University in March 1984 initiated the present series of *Current Perspectives in Healthcare Computing*, the title given to the published proceedings of successive conferences.

1984   Birmingham University

1985   Sussex University

1986   UMIST, Manchester

1987   University College, Cardiff

1988   Brighton Metropole

1989   Harrogate, Conference Centre

1990   Brighton Metropole

After that, the conferences were held annually in Harrogate as this offered the best way of fulfilling the needs of the conference and the exhibition.

In the introduction to the 1984 proceedings, Barry Barber said:

For many years Health Computing has been an uphill struggle; a struggle for funds, a struggle for suitable facilities and a struggle for acceptance.

He went on to indicate that things had changed.[6]

An online conference planner was introduced to help each participant create and print a personal schedule for the event. This could be linked to obtaining an appropriate attendance certificate for continuing professional development credits.

When the conference was moved from a university to a modern hotel conference setting, the arrangements proved not only more expensive, but also more difficult for the exhibitors, who struggled to bring in their big trucks full of equipment. The move to an annual conference in Harrogate each March came about because it appeared to be the only site in the UK that provided a large hall with enough smaller conference rooms, as well as sufficient space for the large exhibition that is now associated with it. The media facilities at Harrogate, including a well-equipped press area, encouraged a wide dissemination of news from the conference, including video clips on local, and occasionally national, television news programmes.

The organizing committee found that the exhibition was at the end of a long series of tasks, such as sorting out the programme and the organizational arrangements. In 1986, it invited Jenny Alloway of the *British Journal of Healthcare Computing* to organize the exhibition. This very important step freed the BCS Organizing Committee from the worries of trying to develop an appropriate exhibition to partner the expanding conference. It also established a professional, year-round basis for that exhibition. Soon it was bringing large numbers of people to the Harrogate Conference Centre: some 3,000 visitors over a three-day period, as well as about 1,000 conference delegates. The arrangements had to be booked years in advance in order to ensure continuity and the appropriate financial planning. Nettie de Glanville has now overseen the exhibition arrangements for many years and it has grown into a major attraction for National Health Service (NHS) staff and for those providing computing and consulting services.

Over the years new activities and facilities have been added, such as a business centre; a large area for poster presentations, with guided tours; internet access; and a new technology zone for innovators to present cutting-edge medical devices or smart health care solutions.

The *Exhibition Guide*, produced by BJHC Limited, provides both details of new products and contact information for each exhibitor, while the conference guide is a useful source of biographical details of the plenary speakers. Award ceremonies have also been added to the programme, with the presentation of the Healthcare IT Effectiveness Awards, and latterly with the BCS HC Achievement Award to honour an individual's lifetime contribution to the field of health care informatics. The Leadership in Health Informatics Accolades acknowledge and reward good practice in the collection and management of data and information in the English NHS.

BCS Health Informatics Forum

The BCS Health Informatics Forum (BCS HIF) was formed in January 2005. It evolved from the BCS Health Informatics Committee to provide leadership in all aspects of informatics in support of health. It also acts as a source of professionally recognized expertise. Its main activities are to:

- co-host the annual *Health Computing* conference and exhibition;
- comment on all major reports and proposals from Government;
- meet with Government, ministers or departments where necessary;
- compile, via its Policy Task Force, responses to relevant Government documents from the home countries, the European Commission and internationally, and also prepare BCS HIF policy on other health informatics issues;
- publish work, fund projects and arrange workshops to develop the science of health informatics;

- appoint the UK representatives to the two relevant international bodies: IMIA and EFMI;
- work with the NHS, clinical royal colleges, NHS IT organizations and other bodies towards developing health informatics professionalism in the UK, through the UK Council for Health Informatics Professions; and
- host quarterly meetings that welcome liaison groups.

RADICAL STEPS

Radical Steps is the generic name for a series of think-tanks and consultations on issues surrounding major topics of concern in informatics to support health care in the UK since 2002 – when a phase of significant investment in informatics was launched. It was named 'Radical' because it makes possible plain speaking and observations on both the positive and negative aspects of the way development and implementation plans are being realized – without fear of personal retribution.

Radical Steps initially related to consultation about the NHS National Programme for IT (NPfIT) and has been repeated annually since then. The concept has also been replicated in scrutinies of open source use and educational content relevant to the health and social care domain to date, both nationally and internationally. These projects produce formal position papers after views are synthesized and validated across the market; these statements have proved to be of interest more widely than the domain of health in the UK. (Visit www.bcs.org/bcs/Forums/health for more details.)

JEAN ROBERTS

INTERNATIONAL INFORMATION EXCHANGE

In the early days of developing computers to help in health care, enthusiasts came together for mutual support and to share ideas. The international exchange of ideas relied on the telephone, on paper and on face-to-face meetings. Although the European and worldwide conferences and related publications developed at the same time, often involving the same people, for clarity the European activities are described first.

European Federation for Medical Informatics

The French organization, Institut de Recherche en Informatique et en Automatique (IRIA), had been running International Medical Informatics conferences (Journées d'Informatique Médicale) annually in Toulouse since 1964. John Anderson and Malcolm Forsythe from the UK were keen supporters of these events. The proceedings of the 1976 conference were edited by M. Laudet, John Anderson and F. Begon and published by Taylor and Francis of London.[7] However, there was a growing feeling that there should be a European organization to promote these activities. The advent of the successful *MEDCOMP* conference, held in Berlin as a result of an initiative by OnLine Conferences Ltd from Brunel University, helped to focus minds among the European medical informatics community.[8]

It was clear that the IFIP TC4 *MEDINFO* conferences, which were being planned on a three-year cycle, could not come back to Europe much more often than once in a decade and it was necessary that the European activity should be wider than the French Toulouse meetings. Accordingly, the BCS was invited to a meeting to explore these issues, to be held under the auspices of the World Health Organization (WHO) (European region) in Copenhagen on 10 and 11 September 1976. In addition to the UK there were participants from Belgium, Denmark, the Federal Republic of Germany, France, Finland, Italy, Norway, Sweden and The Netherlands, as well as representatives from the WHO.

Following discussions during the meeting, a Declaration of Intent was drawn up and signed as follows:

European Federation for Health/Medical Informatics

The Federation shall be constituted as a non-profit organization concerned with the Theory and Practice of Information Science and Technology within the Health and Health Sciences in a European context.

We declare that the ten delegates here today from ten national societies shall constitute the Preliminary Council of the Federation which hereby exists.

COPENHAGEN, 11 SEPTEMBER 1976

The declaration was signed by a participant from each country.

There were a number of key issues that had to be resolved. The first was whether the organization should be one of national societies or of individuals. This was resolved in favour of national societies in order to foster their development: one organization per country. In EFMI the voting would be on the basis of one vote per country represented. A second issue was the adoption of the term 'Medical Informatics' or 'Health Informatics' to describe the activities. There was no problem of scope, but rather that of its accurate description. The working language adopted was English.

The scope of the 'European' membership was based on the WHO's list of European countries, which explains why countries such as Israel were included.

The Copenhagen meeting elected Antoine Remond of France as the preliminary chairman, Barry Barber from the UK as the preliminary secretary and Peter Reichertz from the Federal Republic of Germany as the preliminary treasurer, together with three other members to carry through the agreed arrangements. The decisions taken in Copenhagen were gradually developed into detailed statutes, or a constitution. The first meeting of the Preliminary Council was scheduled to take place during the next Journées d'Informatique Médicale in Toulouse to be held 22–25 March 1977. There were many decisions regarding the details of the arrangements that were used later to develop the organization's constitution, but a clear three-year rotating period of office was established. This did not prevent the same individuals from serving several terms of office but it did ensure that the national society, on whose behalf they were acting, considered the matter and made an appropriate appointment at regular intervals.

In addition, it was decided that the membership fees should be kept low to encourage the widest spread of members: the initial fee was set at 100 Swiss francs. This, of course, was not the cost of membership because this would involve the travel and subsistence of a representative attending Council meetings once or twice a year: but a membership fee that included such expenditure would appear prohibitively expensive. In the early years it was possible to combine the work of EFMI Council with that of the Scientific Programme Committee for the conferences, thus reducing the costs of membership.

The statutes were approved at the second and final Preliminary Council meeting held on 23 March 1977 during the meeting in Toulouse. The statutes were deposited in Switzerland by Werner Schneider: he was then able to open the EFMI bank account. This groundwork cleared the way for the first EFMI Council meeting that was held on the same day.

The second EFMI Council meeting was held in London on 8 December 1977 to review the arrangements for the congress and the third took place in Cambridge on 3 September 1978 during *MIE 78*. Peter Reichertz was elected as the chairman, Barry Barber as the secretary and Antonio Fernandez Perez de Talens from Italy as the treasurer.

BCS runs the first EFMI Congress

It became clear that the EFMI would need to run European Conferences. The UK delegate was able to table two working papers that proposed holding the first EFMI Congress in Cambridge on 4–8 September 1978, together with a letter of support from the BCS. These plans were gratefully accepted and it was noted that the national societies should pay for the attendance of their representatives at the required two or three programme committee meetings, but that the congress fees should subsequently cover these expenses. It was also agreed that for this first congress, any excess of income over expenditure should be shared between the BCS and EFMI on a 60:40 basis in favour of the BCS.

The second Congress was allocated to the Federal Republic of Germany and was held – following a successful independent conference organized by Online Conferences Ltd, which had many BCS members on the programme committee – in a new conference hall in Berlin in 1979. It was agreed that EFMI should not attempt to compete with the world *MEDINFO* meetings and hence the next congress was arranged for 1981 at Toulouse in France.

The programme committee for the first EFMI Congress was chaired by John Anderson and met in London on 11 and 12 December 1976 to finalize the call for papers. It met in March 1977 in Toulouse to discuss session chairmen; and in London in December 1977 to finalize the programme and issue the delegate brochure. Barry Barber, then chairman of the LMSG chaired the BCS Organizing Committee. This included Ken Goulding, Peter Hammersley and Stan Sargent from the LMSG, David Clark and Peter Harvey from the Northern MSG and John Bryden from the Scottish MSG, in addition to the Programme Committee Chairman, John Anderson. Similar timescales and the separation of a programme committee from an organizing committee have continued to be used for subsequent EFMI congresses.

The 822-page proceedings of the conference were edited by John Anderson and published by Springer Verlag in Berlin as the first volume of a new series: *Lecture Notes in Medical Informatics*.[9] The series was edited by Donald Lindberg and Peter Reichertz.

The report of the organizing committee for the first EFMI Congress, *Medical Informatics Europe 78*, showed an attendance of 435 participants compared with an initial estimate of 405 and, of these, 394 were fee-paying participants. Fifty-six per cent of the delegates came from outside the UK.

The BCS were fortunate in having the right honourable Mr David Ennals, Secretary of State for the Social Services, to open the congress and speak at the congress dinner. This nearly did not happen because the Prime Minister Jim Callaghan had pencilled that day in as a possible general election date. Sir Douglas Black, President of the Royal College of Physicians, and Professor Maurice Wilkes gave keynote addresses in the opening session. Political and professional support have always been a matter of national pride for hosting nations.

The good news of an excess of income over expenditure of £4,673 was reported to the BCS Finance and General Purposes Committee. The BCS received its £2,804 share at about the same time as the organizers of the BCS national conference, *BCS 79*, reported an excess of expenditure over income of about £5,000 for the conference and a similar amount for the associated exhibition. The EFMI results were warmly received.

The second EFMI Congress to be held in the UK was held in Glasgow on 20–23 August 1990. The proceedings, *Medical Informatics Europe 90*, were edited by Rory O'Moore, Stellan Bengtsson, John Bryant and John Bryden.[10] This was number 40 in the series of lecture notes.

Barry Barber continued as an officer of EFMI, becoming vice-president, president and vice-president Europe (IMIA Vice President (Europe)) and eventually an honorary fellow. The UK has continued to support EFMI with John Bryant and John Bryden as officers, with numerous delegates, and the second UK congress, *MIE 90*. In 1984, Bud Abbott organized a coach-load of delegates to travel from the North East Thames Regional Management Services headquarters in Brentwood to the congress in Brussels. The BCS established a productive trend that has now been followed for over a quarter of a century.

The EFMI (Medical Informatics Europe – *MIE*) Conferences were:

Cambridge	1978
Berlin	1979
Toulouse	1981
Dublin	1982
Brussels	1984
Helsinki	1985
Rome	1987
Oslo	1988
Glasgow	1990
Vienna	1991
Jerusalem	1993
Lisbon	1994
Copenhagen	1996
Thessalonica	1997
Ljubljana	1999
Hanover	2000
Budapest	2002
Saint Malo	2003
Geneva	2005

The BCS continues to be represented on the EFMI Council and EFMI continues to develop its activities with main conferences in between the triennial *MEDINFO* conferences, special interest conferences and the activities of its 13 working groups (www.efmi.org/). It continues to bring Eastern European countries on board and medical informatics societies in 29 countries are currently EFMI members. The costs in terms of time and resources in travelling to EFMI activities are substantially less than is required to attend IMIA activities and they are often more closely focused on the needs of European countries.

European projects

The then European Economic Community launched successive 'Framework Programmes for Research and Development'. The second framework in 1989 included telematics applications in health care (AIM) and flexible and distance learning (DELTA). Work was done in the AIM projects on:

- decision support;
- image processing;
- multimedia patient records;
- cooperative working;
- telemedical applications;

- man–machine interfaces;
- signal handling; and
- information transfer and advanced data storage.

The DELTA programme included: co-authoring multimedia courses; two-way distance teaching using satellite links; and ISDN video-conferencing.

In 1990–94, the European Commission set up a series of research and technological development programmes to bring together people of different nationalities from the, then, 12 member states. The 1993 telematics programme had six areas: health care, along with flexible and distance learning provided research funding, contributing about 50 per cent of the costs for 36 projects, but competition for funds was intense.

As the reflections in the remainder of this book will demonstrate, health service staff and academics in the UK were involved in pioneering in these and other areas.

BSI support for CEN TC 251

National and international exchange of information is easier if there are common systems and terminology. The European Committee for Standardization (Comité Européen de Normalisation or CEN) was made up of the standardization bodies across Europe. In the UK the British Standards Institution (BSI) provides the national standards. The European informatics community used its technical committee structure to identify, discuss and agree its response to the proposed standards. Technical Committee 251 (TC 251) was shadowed in the UK through the BSI, so that proposals could be reviewed for application or for potential problems for the UK. BCS members were actively involved and were able to take back the relevant issues to their specialists groups for further discussion.

The CEN TC 251 working groups covered such areas as:

- medical records structure;
- vocabulary;
- confidentiality and personal data protection;
- information security;
- health care information frameworks;
- syntax for interchange;
- coding systems;
- semantics representation;
- medical image and related data interchange formats;
- interchange formats and protocols for electrocardiography;
- information content of patient data cards; and
- decision support.

Worldwide activities

International Medical Informatics Association (IMIA)

In 1969, the BCS received a letter from Professor François Gremy suggesting that the BCS might like to join a technical committee of the International Federation for Information Processing (IFIP). There was surprise at the thought that anyone would travel 'overseas' to go to a committee meeting, but nevertheless BCS joined, being represented by Malcolm Forsythe on Technical Committee 4. This was known by the obscure and unwieldy acronym of IFIP TC4. It developed into the world body linking medical informatics organizations. Malcolm was the first secretary of the group with Professor Gremy as the chairman.

The first worldwide IFIP TC4 conference was held in Stockholm as *MEDINFO 74*. John Anderson and Malcolm Forsythe from the UK edited the proceedings. BCS members from the UK have continued to play a part in these triennial conferences, which provide the health informatics community with a snapshot of what is going on across the world, as well as the chance to meet experts in every aspect of health informatics. Naturally, these events, like the Olympics, do not come to an individual continent or country very often:

Stockholm	1974
Toronto	1977
Tokyo	1980
Amsterdam	1983
Washington	1986
Singapore and Beijing	1989
Geneva	1992
Vancouver	1995
Seoul	1998
London	2001
San Franciso	2004
Brisbane	2007

Tallberg and her colleagues analysed the underlying themes of the papers given at *MEDINFO* as part of a comparison with those at *MIE* and international nursing informatics conferences.[11] Some topics have given way to new approaches as technological solutions have developed.

After the first ten years IFIP TC4 became more autonomous – as the International Medical Informatics Association – and this event was celebrated in Paris in 1979 with a long bus trip to Tours. (In the tradition of medical events there were various interesting stops on the way, like the visit to the royal abbey at Fontevraud and the wine cellars in Saumur.)

Following the untimely death of Peter Reichertz, Barry Barber edited the *MEDINFO 89* proceedings with Gustav Wagner. This *MEDINFO* had to be hastily transferred from Beijing to Singapore, but the proceedings reflect the conference that the programme committee intended to run rather than the actual situation of two rather different events.

MEDINFO IN LONDON

The BCS applied to host *MEDINFO* on two occasions before becoming successful at the IMIA General Assembly in Sydney in 1997 with its application to host *MEDINFO 2001*. A core team was established to bring the event to the UK from activists in the community, lead by Jean Roberts as local organizing committee chair. The proposed month of September and IMIA-preferred London location resulted in the use of the new ExCeL congress and exhibition centre in London Docklands, which at the time of the bid had not been built. The 24/7 webcam, which showed site progress, was visited very frequently by IMIA board members as the development came to fruition.

In addition to logistics, the UK team were involved in preparing the congress management, scientific programme selection, editorial production of the proceedings, a full tutorial programme, social programme and travel arrangements, exhibition, bursary scheme and all the marketing – by snail mail and web. No mean feat for people who all had day jobs to hold down.

The event was ultimately financially and scientifically successful, with an internationally refereed 12-stream programme with 270 formal papers, 28 workshops and 11 panels. The plenary session speakers were invited from seven countries and 160 posters complemented the verbal presentations. A parallel exhibition contained vendor, multinational project and Government agency information stands. Proceedings were produced on CD-Rom and in hard copy. At the end of the event, the UK team handed over to the next city, San Francisco.

JEAN ROBERTS

The first international nursing conference

The collaboration of the medical specialist groups around the first EFMI conference continued as the Medical Specialist Groups Coordinating Committee. Their experience enabled the BCS to run the first International Nursing Informatics Conference in 1982 following the interest in nursing informatics expressed at the Tokyo *MEDINFO* in 1980. The BCS put together an organizing committee that was able to handle the technical conference arrangements but also the participation in international medical informatics activities enabled it to bring together an international group of nurses involved in computing projects. In this sense it was required to act as scientific programme committee as well as organizing committee.

After *MEDINFO 80* soundings were taken within the BCS and elsewhere to explore the possibility of organizing and funding an international nursing informatics conference. When these soundings proved positive, an organizing committee was established under the chairmanship of Maureen Scholes who was at that time the chair of the Computer Project Nurses Group established by Dame Phyllis Friend, the Chief Nursing Officer at the then Department of Health and Social Security (DHSS). The committee comprised six nurses involved in various computing activities, three people from the DoH, one from the NHS training centre at Harrogate and two from the LMSG. Between them they were able to develop a two-centred international conference; the first part was a large, two-day, open gathering at Church House, Westminster. This helped to raise money for the second event, which was a closed workshop at The White Hart, the NHS Training Centre in Harrogate. There were 550 participants in the open conference and 59 invited to the closed workshop. The proceedings were published by North-Holland as *The Impact of Computers on Nursing*.[12]

The worldwide interest in this topic was reflected by the establishment of IMIA Working Group 8 on Nursing Informatics at the 1983 *MEDINFO* in Amsterdam, with Maureen Scholes as the first chairman. The group thought that there should be a series of such conferences every few years and that the chair should be rotated on a three-year basis to ensure worldwide coverage. The second conference in the series was held in Calgary, Canada in 1985. The BCS Nursing Specialist Group continued to play a large part on the international scene. The list of the initial conferences is as follows:

London and Harrogate	1982
Calgary	1985
Dublin and Killarney	1988
Melbourne and Sorrento	1991
San Antonio and Austin	1994
Stockholm and Lidingö	1997
Auckland	2000

These developments are outlined in *International Nursing Informatics*.[11]

Worldwide specialist groups

IFIP TC4/IMIA developed a number of working groups (WG) and they held working conferences on topics in their particular area. The ones that spring to mind most immediately are WG1 on education and WG4 on data protection and security. Both groups were active from a very early date and have continued this activity over many years. The BCS has been well represented at IMIA with Malcolm Forsythe being followed by Mike Abrams and Bud Abbott. Also the UK has been well represented in the working groups with David Kenny and Barry Barber following Gerd Griesser's initial chairmanship of WG4.

In the early days it was difficult to get programme committees to include whole sessions on nursing issues, even though *MEDINFO 74* had included papers from four British nurses. However, by the time of *MEDCOMP 77* in

19

Berlin there was a full nursing session of papers from three British nurses and by *MIE 79* there was a nursing session comprised of papers from four British nurses. As noted above, at the 1980 *MEDINFO* in Tokyo there was a strong interest in nursing informatics issues. It was during these discussions that the BCS thought that we might be able to host an international nursing informatics conference that would assess developments and the need for a nursing working group within IMIA. As well as nursing informatics activity in the BCS Health Informatics Specialist Groups, Maureen Scholes was the initial chair of the Nursing Working Group, WG8, which eventually became the IMIA Special Interest Group Nursing Informatics.

All the various IMIA Working Groups are accessible to BCS members but they all work in different ways and have different ways of pursuing their activities. The website www.imia.org shows the sort of activities that are in progress and the current BCS representatives can advise about the best ways of participating.

SPREADING THE WORD

In the early days of clinical computing members of the clinical professions with an interest in IT were thinly scattered across the UK. Finding out what was being done within one's own specialty was difficult, let alone in another specialist area with similar general problems. Finding the money and the annual leave to attend conferences was also difficult if not impossible, so the printed word was a vital source.

Although the BCS held its second conference on medical computing in Bristol in 1971 and the papers were published, most of the content was on the role of computers in management and in relieving staff of repetitive tasks in handling information about care or hotel-type services. Clinical issues were only addressed when the technology improved.

Another problem for the clinician was to finding a way through the many abbreviations and acronyms. There were three sets to learn: the jargon used by the engineering and software professionals in the NHS; that of European information technology projects; and the clinical jargon and NHS terms. It was not until 1995 that Hugh de Glanville and Adrian Stokes published their very useful *Abbreviary*.[13] Projects were known by their initials and some people twisted the title to produce a clinically related acronym.

Most of the papers from the UK conferences and those of the individual specialist groups were published, either as proceedings, series such as *INFOrmed Touch* from the nursing group, or as papers within professional journals. Sadly many of the key plenary papers, particularly from the *Healthcare Computing* conferences, were not submitted far enough ahead of their presentation to be included in the relevant *Current Perspectives in Healthcare Computing*. Most have evaporated into the ether.

In coming together to review the early development of health information systems some of the remaining pioneers were concerned that a large volume of potentially useful material had never been published in book form or as conference papers. The local reports and commercial systems descriptions and other 'grey literature' were not systematically collected in any central location. The frequent changes in the structure of the NHS, with relocations of offices and responsibilities, have not helped in retaining items. Nor have the acquisitions and mergers of commercial suppliers. Informal arrangements were made to collect items held by individuals coming up to retirement from their jobs and those older members who were moving house. There remains a major challenge to catalogue and conserve the material. In 2006, the University of Lancaster bravely took on this role. It will also hold the relevant health informatics publications.

International conference papers

The International Federation for Information Processing held the first world conference on Medical Informatics in Sweden in 1974. There were 18 themes and the resulting publication ran to nearly 1,000 pages.[14] Similar fat volumes appeared every third year. The education of staff in computing techniques and the use of computers to help in medical education were discussed. Anderson – of King's College Medical School – and two French colleagues reported the first international survey by questionnaire for medical professionals, nurses and health administrators' health workers in Western Europe.[15] Sayers, of Imperial College, London, gave an inaugural plenary paper on the analysis of biological signals.[16]

The First Congress of the European Federation of Medical Informatics was held in Cambridge, England, in 1978 and was recorded as the first of the *Lecture Notes in Medical Informatics* series. From the start there was an editorial board and the approach was scientific. Associated industry presentations on hardware and software were not included, so some of this 'soft' data were lost within the individual company archives, at least until mergers and acquisitions slimmed the initial bulge of small firms. There were, and still are, pre-conference teaching sessions for clinical and technical staff who are just getting involved in clinical informatics, most recently in the middle-European countries, with UK professionals acting as experts. Again their contributions have been lost to general view within personal collections. Photographic slides, overhead projection acetates and later the ubiquitous software presentations are no longer available.

Reports of international meetings also have a bias towards those able to attract the funding to attend, including university professors and clinicians involved in nationally funded or multi-site European projects. There was a core of 'familiar faces' at every European and International Medical Informatics Association conference. This was reflected in the first authors of the papers, because that was the person expected to present the paper. Getting the money to attend did not mean the first author had done the bulk of the work on the clinical application. There was also a bias introduced by the location of the conference, with nationals offering more papers, such as at the ninth congress in Glasgow, in 1990. The total number of papers submitted exceeded the number of available 'slots' so good clinical work had to compete with computer science papers.

BCS members contribute to 'blind' reviewing and scoring potential papers for consideration by the scientific committee of a conference. This role was made easier when the internet became more widely available. More recent conferences have been able to provide papers on CD-Rom to reduce the bulk to be carried home.

Journals

In the early days, when clinicians and technicians worked together on computing problems, publication of the technical issues and possible solutions might be in journals such as *Medical & Biological Engineering & Computing* or the *Journal of Medical Engineering & Technology*. One example was an article on a microprocessor-controlled signal generator for the functional testing of electrocardiographs. In 1984, with the fourth issue, the *British Journal of Healthcare Computing* began to publish a bibliography of articles on health care computing from such journals. The bibliography was compiled by the South West Thames RHA's library service. Clinicians faced the challenges of finding a library that stocked the journal to obtain a copy of the referenced article, and then of making sense of the technical details familiar to another discipline. Early computer developments could be hard intellectual slog.

British Journal of Healthcare Computing

In 1983, Jeny Alloway carried out a market survey about the content of a magazine that could carry papers and news about medical computing activities. This research developed into *The British Journal of Healthcare Computing* (BJHC).

It was thanks to Hugh and Nettie de Glanville that this magazine made it from the drawing board into practical reality in 1984. The venture was supported from the first by the various BCS health specialist groups and their members (who initially received free copies of the BJHC). The early issues included regular reports from, and contact details of, the BCS specialist groups, as well as other groups with a particular interest in a clinical specialism. However, as the BJHC grew and flourished, it developed in a variety of directions. It has been a keen supporter of BCS activities in health care, as well as being a sharp commentator on the growing field of health care computing. The journal of the Nursing Specialist Group, *Information Technology in Nursing* (ITIN), was initially typeset and published through the BJHC, with the staff providing practical help and advice to the novice editor. Hugh de Glanville, as a member of the European Association of Science Editors (EASE), ensured rigorous standards for both journals. (For more information about ITIN see Chapter 19.)

The BJHC became *The British Journal of Healthcare Computing & Information Management* (BJHC&IM) and ran a number of conferences in this field as well as handling the exhibition accompanying the annual *Health Computing* conferences. The BJHC&IM has also printed conference news, with many of the proceedings of the conferences printed for the BCS through BJHC Ltd, Weybridge.

Medical informatics and the internet in medicine

The *Medical Informatics and the Internet in Medicine* journal was started in 1975 and is published by Taylor and Francis in London. It was edited by John Anderson for many years, before John Newell, John Bryant and then Steve Kay took over as editor. For a long time it was one of the few places where academic work in medical informatics could be published apart from in *Methods of Information in Medicine* and at world (IMIA) and European (EFMI) conferences. Throughout, this journal continues to provide a valuable place for refereed papers in health informatics to be published worldwide.

Books

Initially, the majority of books were published by authors from North America. Over time the publications by UK authors increased, particularly in relation to clinical computing topics. The BCS also had an influential publishing committee that comprised the editors of its journals and series. North-Holland (Elsevier) published *MEDINFO* and IMIA working group conference proceedings from the beginning. Springer Verlag published *MIE* conference proceedings in their series of *Lecture Notes in Medical Informatics* and IOS Press from Amsterdam published the results of many Advanced Informatics in Medicine projects, and later projects in their series of *Studies in Health Technology & Informatics*.

Handbook of Information Technology in Health Care

Discussions with the Institute of Health Services Management (IHSM) led to a decision to publish a loose-leaf, updateable book on information technology in health care. At that time the BCS did not wish to undertake such a project, even though almost all the authors were going to be BCS members or colleagues of members. The initial publisher was Kluwer Publishing Ltd and then the Longman Group. The initial publication was in 1986 and issue 13 came out in 1991.[2] The driving force for this work came from Bud Abbott and he ensured existing authors updated their material and he found new authors as new topics emerged or old authors got worn out. The list of contributors and their short bibliographies provide details of many of the UK's most influential national and international pioneers.

In due course the IHSM decided not to continue the project and the BCS Health Informatics Specialist Group bought the rights to the material. It published the updated material in four volumes in 1996 as a one-off publication.[1] There were four handbooks: *Handbook A: Introductory*

Themes; *Handbook B: Aspects of Informatics*; *Handbook C: Primary Care*; and *Handbook D: Hospital Systems.*

Other resources

Pamphlets and leaflets

The close involvement of the DoH in funding many of the major hospital developments, and some of the general practice systems, led to regular publications to support and explain developments. The Information Management Group was, for a while, a very active participant at the annual exhibition associated with the *Healthcare Computing* conference, taking a large stand and providing its own seminars, demonstrations and presentations. Many of the IMG staff were also active members of the BCS, Stan Lajka was a regular presence at conferences, able to pull together all the resources in a distant location. Financial support through the IMG was also given to support the publication of relevant papers from BCS groups.

System suppliers and some of the large consulting groups distributed copies of research papers and descriptions of the development of their systems. These ranged from explicitly promotional material to robust scientific papers. Again the annual *Healthcare Computing* exhibition was a favoured outlet.

Open Software Library

The Open Software Library was based in the Education Centre at Warrington District School of Nursing. It was organized by members of the Warrington Computer Group with help from lecturers from North Cheshire College, Manchester Polytechnic and tutors from a number of schools of nursing. The main function was to collect computer programs dealing with health care that might not be generally available and to disseminate them on cassette tapes or floppy disks. The story of its development is given in Chapters 19 and 21.

Videos, teaching materials and websites

The development of video programmes on specific subjects can be found in later chapters. BCS members were also very active in creating teaching materials, websites and CD-Roms. Some of these were developed as part of their 'day jobs' for their employers, others were activities undertaken by a BCS specialist group or for commercial publishers.

REFERENCES

1 Abbott, W., Bryant, J.R. and Barber, B. (1996) *Information Management in Health Care: Handbook A: Introductory Themes.* BCS, London. ISBN 0 901865 85 0; *Handbook B: Aspects of Informatics.* BCS, London. ISBN 0 901865 90 7; *Handbook C: Primary Care.* BCS, London. ISBN 0 901865 95 8; *Handbook D: Hospital Systems.* BCS, London. ISBN 0 901865 96 6.

2   Abbott, W., Barber, B. and Peel, V.J. (1986–93) *Information Technology in Health Care.* Kluwer Publishing for the Institute of Health Service Management. Updatable work – Issue 1: September 1986 to Issue 20: August 1993.

3   Hayes, G. (1990) International primary care computing. In Hayes, G.M. and Robinson, M. (eds) *Proceedings of the IMIA Workshop on Primary Care Computing,* Brighton, 5 April 1990. North-Holland, Amsterdam.

4   Abrams, M. (ed.) (1969) *Medical Computing – Progress and Problems.* Chatto and Windus for BCS, London.

5   Abrams, M. (ed.) (1972) *Spectrum 71: BCS Conference on Medical Computing.* Butterworths, London.

6   Barber, B. (1984) Introduction. In Kostrewski, B. (ed.) *Current Perspectives in Health Computing.* Cambridge University Press for BCS, Cambridge. ISBN 0 521267 05 6.

7   Laudet, M., Anderson, J. and Begon, F. (eds) (1976) *Proceedings of Medical Data Processing,* Toulouse. Taylor and Francis, London.

8   Online Conferences Ltd (eds) (1977) Proceedings of *MEDCOMP 77.* ISBN 0 903796 16 X.

9   Anderson, J. (ed.) (1978) *MIE, Cambridge 78: Lecture Notes 1.* Springer Verlag, Berlin. ISBN 3 540089 16 0.

10 O'Moore, R., Bengtsson, S., Bryant, J. and Bryden, J. (eds) (1990) *MIE, Glasgow 90: Lecture Notes 40.* Springer Verlag, Berlin. ISBN 3 540529 36 5.

11 Scholes, M., Tallberg, M. and Pluyter-Wenting, E. (2000) *International Nursing Informatics: A History of the First Forty Years 1960–2000.* BCS, Swindon. ISBN 0 953542 72 6.

12 Scholes, M., Bryant, Y. and Barber, B. (1983) *The Impact of Computers on Nursing: An International Review.* North-Holland, Amsterdam. ISBN 0 444866 82 5.

13 de Glanville, H. and Stokes, A.V. (1995) *The BJHC Abbreviary.* BJHC Books, Weybridge. ISBN 0 948198 21 4.

14 Anderson, J. and Forsythe, J.M. (eds) (1974) *MEDINFO 74,* Stockholm. North-Holland, Amsterdam. ISBN 0 444107 71 1.

15 Anderson, J., Gremy, F. and Pages, J.C. (1974) (Title unknown.) In Anderson, J. and Forsythe, J.M. (eds) *MEDINFO 74,* Stockholm, pp. 207–11. North-Holland, Amsterdam. ISBN 0 444107 71 1.

16 Sayers, B.McA. (1974) The analysis of biological signals. In Anderson, J. and Forsythe, J.M. (eds) (1974) *MEDINFO 74,* Stockholm, pp. 13–20. North-Holland, Amsterdam. ISBN 0 444107 71 1.

PART 1

The background to early health informatics

INTRODUCTION

The early development of health informatics was unorganized and idiosyncratic. Developments depended on individuals who saw opportunities for exploration and who found ways in which their organizations could benefit from these opportunities. Universities set up large computing systems that could be used in batch-processing mode and then subsequently as remote entry job-shops. The finance departments of some hospitals and hospital boards saw opportunities for the replacement of their punched card and accounting machines. In the commercial sector, Lyons developed the LEO computer to support office automation and at a more affordable level Elliott Automation promoted the use of their Elliott 803 as a job-shop that anyone could use. In a more specifically health informatics context, Elliott Automation set up Elliott Medical Automation Ltd and they established a 'medical job-shop' within the precincts of University College Hospital (UCH) under the direction of Dr L. C. Payne. Apart from running appreciation sessions, the intention was to open up the market for the use of computers in health care and further E A Medical installations followed at the Edinburgh Royal Infirmary and the Hammersmith Hospital in London.

The first two computers at UCH were Elliott 803 computers and the third was a more advanced and powerful Elliott 4120. The application that was of most interest at UCH involved the biochemistry laboratory automation, under Dr Freddie Flynn, using auto-analysers that generated paper-tape output from which reports could be generated and which could be analysed for effective machine quality control and performance. Additionally, the medical physicists under John Clifton were greatly interested in the use of the machine for radiation treatment planning. This latter application moved fairly rapidly onto more powerful computers in order to obtain results fast enough for the clinical decision-making required to start patients on treatment. Both of these were key applications that fed into the development of a line of special purpose computers designed to address these issues.

The Hospital Physicists Association established a Computer Topic Group that was initially chaired by Dr Barry Barber and the Association of Clinical Biochemists also found itself heavily involved with these computer developments. It was always understood that the arrangement at UCH involved free processing facilities in lieu of rent for the accommodation. This appeared to lead to great care in the use of the computing services, whereas the purchase at The Royal London Hospital meant that the greater the use of the machine, the more experience would be obtained.

The case for the Elliott 803 at The Royal London Hospital was based on the savings to be derived from handling the finance system in all its complexity. The Cooper Brothers Report noted that the hospital would:

> . . . *have available, at no extra cost, a computer for research, medical and teaching work to the extent of about 80 hours a month on the first shift, with the potentiality of using the second and third shifts if required.*[1]

The Board of Governors was very clear that the machine was to be made fully available for scientific and medical applications providing this did not prejudice the financial systems. The House Governor's computer meetings allocated priorities for the major activities but, thereafter, great efforts were made to explore opportunities across the hospital and the medical college for the use of these facilities using either standard analytical or survey programs or programs specially written by individual enthusiasts. It is interesting to note that the system was soon being used into the second and third shifts operated by those self-same enthusiasts. It is also worth noting that the finance system did not have top priority on the machine – that honour was accorded to the Kidney Matching Program because the clinical requirement was paramount. The time available was used extensively for various medical surveys, such as Aubrey Sheiham's Nigerian Dental Survey; the work of the obstetric department; and the study of the emergency anaesthetic service. It included the computerization of a wide variety of hospital statistical returns. Together these provided the concept that the next step was the integration of various administrative aspects of patient care that were described as the patient administration system, the basic administrative infrastructure that underpins the delivery of care.

During these early days, meetings were held under the auspices of many different organizations: the British Computer Society; the Hospital Physicists Association; Association of Clinical Biochemists; the Institute for Health Services Management; and the Operational Research Society, to name just a few, as well as within hospitals, hospital boards and the DoH. However, many of these organizations specialized in their own specific fields and it was the BCS that provided the enabling facilities to allow the exploration of health informatics in a wider context with the setting up of appropriate specialist groups as they were required.

When the DoH had got over the shock of finding a teaching hospital that decided to purchase its own computer – a shock that it had tried to prevent because it was trying to establish a computer system for the financial needs of all the London Teaching Hospitals – it set about exploring the opportunities for the use of computers in the NHS and that led to the extraordinary

Experimental Real-Time Computer Programme. This programme sought to exploit the opportunities for the use of real-time facilities within the NHS, instead of relying on the batch-processing facilities that had previously been available. This was a quantum leap and at that time the only comparisons were the airline seat-booking systems, which required many more resources that the NHS had available. This initial exploration was accompanied with a review of all computer applications, an examination of the need for training, paying and retaining specialist computer staff and for the development of standard programs that might be exploited across the NHS.

The DoH kept a close watch on developments. For many years it ran a useful forum in which the participants in the Experimental Real-Time Programme met to exchange ideas, problems and technical solutions. Slowly, the concept of real-time computing developed both in small, dedicated systems and in larger, multiprocessing systems – although the Ockenden and Bodenham study suggested that this difficult and expensive activity should be left to the English.[2] Similarly, the concept of Patient Administration Systems was accepted, and eventually developed further with the Resource Management and the Hospital Information Support Systems Initiatives. The 'Holy Grail' was always the complete medical record but that was, inevitably, unobtainable until the necessary infrastructure had been developed.

Developments in health informatics also depended on the speed with which the computer technology could keep up with the requirements of the potential users and the changes in political expectations. The early steps relied on the personal interests and skills of the pioneers, who often adapted hardware and ideas from outside the health service. The finance departments were early adopters and in some health organizations this remained the location for management of the local computing resources well into the late 1990s.

The computers available in the 1960s were physically large, generated a lot of heat and needed to be housed in large, air-conditioned rooms. Thus they were expensive. They also needed careful physical maintenance and with, by today's standards, tiny amounts of memory, management of the data and computing load. Investment by the Government through the DoH was crucial if hospitals and general practitioners were to be able to develop systems for support services and for clinical use. There was also a desire from the Department of Trade for support to create a British commercial sector in computer hardware and software. The national initiatives were therefore key to these developments.

In his foreword to *Walk Don't Run*, a collection of essays in honour of Mrs Edith Körner, Robert Maxwell of The King's Fund commented on Patrick Jenkins' decision to set up a long overdue joint review of health services information by the DoH and NHS.

> *That seemed appropriate to the task in hand, and was in keeping with the Ministerial mood of that time, reflected in* Patients First. *The emphasis was on greater decentralisation of management in the service, more influence by the*

periphery on the centre. A major study of information systems at all levels from local to national appeared a 'natural' for a new experiment in joint enterprise, going much beyond traditional patterns of NHS representation on Departmental working parties.

REFERENCE 3

The dependence on central funding continued and, with further changes in the organization of health care, those in the service often felt that central control through standards, accountability and inspection made the provision of accurate information both a help and a hindrance. It could have a direct effect on financial resources. Some of the tensions were illustrated by the misinformation associated with waiting lists.[4]

BARRY BARBER AND DENISE BARNETT

REFERENCES

1 Cooper Bros (1964) *Electronic Data Processing*. A report prepared for The London Hospital, 4 February.

2 Ockenden, J.M. and Bodenham, K.E. (1970) *Focus on Medical Computing*. Scicon Report commissioned by the Scottish Home and Health Department. Oxford University Press for NPHT, Oxford.

3 Maxwell, R. (1985) Foreword. In Mason, A. and Morrison, V. (eds) *Walk, Don't Run: A Collection of Essays on Information Issues*. Published to honour Mrs Edith Körner CBE, Chairman of the NHS/DHSS Health Services Information Steering Group 1980–84. King Edward's Hospital Fund for London, London. ISBN 0 197246 31 1.

4 National Audit Office (2001) *In-patient and Outpatient Waiting in the NHS*. The Stationery Office. ISBN 0 102911 08 8.

2 Finance

BUD ABBOTT

By the early 1950s information technology was already present in the NHS in the form of punched cards and accounting machines and it was heavily oriented toward finance. Payroll, cost accounting, budgeting, committal accounting, etc. were all demanding extra resources for the financial systems and these forms of information technology were perceived to be the answer. Generally the punched-card facilities were expensive both in terms of capital and revenue and several regional hospital boards (RHBs) became the providers of a central bureau service for their hospital management committees (HMCs), particularly in response to the requirements of the first national standard costing system introduced in 1956.

At the hospital level, the accounting machines were much less expensive and these were used to meet the needs of the finance systems. There was no even pattern – some HMCs relied exclusively on the RHB bureau, others used accounting machines, still others used both. The boards of governors of the teaching hospitals used accounting machines and small, punched-card facilities depending on their resources. The use of the facilities was also being extended into areas other than finance – hospital statistics were a natural use of punched cards, from works departments, catering and even the clinical areas. So the stage was set for the introduction of computers.

The dream began for me at The London Hospital in 1958 when I was in the finance office, fairly junior and responsible for the punched card and accounting machines of that era. I had a protracted spell as an in-patient and the hospital librarian offered me a book on computers – it started with Babbage and ended with the Lyons Electronic Office (LEO). Over the next few months I picked up some more information – not that there was much about finance or business applications – and I wrote a paper for my chief. I set out how I thought that computers could be the perfect answer to all the system requirements of The London – finance, stores, works, medical records and so on. The treasurer was not convinced and did not commission me to go and purchase one of these marvels but he did encourage me to do more, particularly if I could find some more immediate answer to some of his more pressing problems. So I continued to explore this computer concept – binary arithmetic, the accumulator, B lining, machine code, nickel delay lines, magnetic film and so on.

During the late 1950s and early 1960s there was little general awareness of the successes of the work at Bletchley Park in respect of the development of computing technology and of the spin-off that these developments

31

might have in the wider world. Indeed, while the Americans were exploiting these developments commercially, it appeared to have been Government policy to continue the secrecy that was necessary during the Second World War – possibly because Churchill remembered the damage done by his revelations about code-breaking during the First World War. It was only with the publication of Winterbottom's book *The Ultra Secret* that it slowly became clear that some of the technology that had appeared on the market was derived from the computing breakthroughs at Bletchley Park.[1] The fast paper-tape readers and printers turned out to have been the practical implementation of some aspects of the Turing Machine as theoretically described in Turing's famous 1937 paper *On Computable Numbers with an Application to the Entscheidungsproblem*.[2]

These readers and printers became available on the Elliott 803 and 503 computers and they provided a cheaper, easier to use, computer facility than much of the punched-card equipment used in the larger machines that derived more obviously from the Jacquard loom and the carpet industry. The large commercial systems that were developed in the 1950s and 1960s required major investments that could be made only by major industrial or Government organizations. The first commercial machine was the Lyons Electronic Office, LEO.[3] It was still based on thermionic valve technology and mercury delay lines, but it was much more expensive than the later transistor-based technology of the Elliott machines.

GOVERNMENT INTEREST

In 1956, a DoH working party was set up to consider the use of automatic data processing by the RHBs as an extension to the punched-card bureaux then being developed. This initiative was concerned with the advancing finance systems. The thinking that developed in the late 1950s and early 1960s led to the establishment of major computer systems at the RHBs or the Regional Health Authorities (RHAs) as they later became. It was 1961 before Manchester got an IBM 1401, 1963 when Bristol got an ICT 1301, and 1964 before Birmingham got an ICT 1500. These installations were all developing payroll applications but progress was slow.

At this stage the London teaching hospitals were going rather faster. Inevitably it was finance that led the way and the first trials were on the first generation valve machines. In 1958, Jack Rowlandson, the treasurer of St Bartholomew's Hospital, wrote a payroll program to run on the Elliott 405 and invited other London teaching hospitals to join him in experimental runs. Some did – King's College, Middlesex, St. Mary's, Royal Free and The London (as I remember) – most did not. I was nominated from The London and, after I had become acquainted with some of the mysteries of programming and operating the equipment, The London ran a full duplicate payroll for student nurses in 1961–62. I had expected (naively) that there would be a fairly rapid progress to a full payroll service for the London teaching hospitals but there were the now familiar problems of disagreements over costs, purposes, controls, etc. and there were delays. The fundamental systems questions, which – as always – needed to be resolved before the next steps could be taken, were:

- What should be the scope of the system?
- Who is in and who is out?
- What is in and what is out?
- What parts of the finance system should be included?
- How does one secure agreement to these decisions?
- What size of machine is appropriate to the final scope agreed?
- What organization is required to operate the machine?

Almost inevitably, tidy-minded administrators opted for more and more inclusive and ambitious arrangements that, necessarily, involved the development of larger and more complex systems with correspondingly greater problems for the creation of the necessary consensus and for the technical systems themselves.

The London was a very strong supporter of the teaching hospitals computer project. However, it became clear that these developments would not become available before it was imperative to re-equip the finance department with replacements for its ageing 40-column punched-card equipment and accounting machines, which were becoming increasingly unreliable. The situation was further complicated because the equipment was also used to provide the monthly hospital statistical returns for the DoH. These requirements, equally inevitably, led the hospital toward the acquisition of a small computer system. The London simply could not wait.

> The payroll subsystems went live at the beginning of the new financial year in April 1965, and by May things were going well enough for Bud Abbott to take a holiday and recover from the stress and long hours. Just before he went, at the computer meeting with the House Governor, the question arose as to who would supervise the finance systems in his absence. The House Governor had assumed that he would deal with this, until I explained that he knew nothing about the finance system. It was at this time that the 'penny dropped' that the details of the whole finance system were in Bud Abbott's head. The hospital took out an insurance policy on his life and asked him to get writing as soon as he got back from holiday!
>
> **BARRY BARBER**

LOCAL DEVELOPMENT

In 1962, Barry Barber, who was then a principal physicist at The London, and I put our heads together and, at the instigation of the treasurer, wrote a paper for the House Governor of The London Hospital. In it we proposed the purchase of a second-generation computer using transistors for The London. A comprehensive set of financial and scientific applications was outlined, plus a range of statistical programs for the research needs of the medical staff, as well as indicating the further potential of the equipment. The Board of Governors called in Cooper Brothers as consultants. The consultants reviewed the proposal and rewrote our paper, raising all the numbers (costs, staffing, timescales, etc.) and most importantly said it was a good idea. The case for installing The London Hospital computer system, set out by the consultants' report, rested on the immediate needs of the finance systems.[4] Also included were the opportunities that it gave for dealing with some Medical Records department activities – the diagnostic index, DoH statistical returns and enquiries, EEG records, VD returns and special analyses. The scientific needs of the Physics department and the wider opportunities in operational research, medical statistics and medical research were not overlooked.

Following discussions at the DoH, in the era when the teaching hospitals were not subject to an RHA, the Board of Governors accepted the report in 1963 and placed the order for an Elliott 803, which was finally delivered in October 1964.

It should be noted that the Ministry were, at that time, not enthusiastic about the proposal but the Board of Governors was persuaded of the financial and technical merits of the scheme.

The Elliott machines could be purchased by smaller organizations, such as The London Hospital, and were made available at University College Hospital and the Edinburgh Royal Infirmary by Elliott Medical Automation Ltd, initially on a service cum rental contractual basis. Although these machines were slow by contemporary standards they provided hitherto unavailable opportunities to complement the way in which the large machines were, also, slowly becoming available with batch-processing facilities and then with remote terminal access via teletype terminals.

The Elliott 803 ordered by The London had a 8,192 word store, each of 39 bits capable of containing two instructions of six bits and an address of 13 bits covering any location in the memory. The cycle time was 256 microseconds and the word also contained a 'B line modifier' allowing addresses to be incremented following the execution of an instruction and thus facilitating the establishment of powerful computing loops. The backing store was provided by 35 mm magnetic film handlers. As a portent of things to come, the Elliott 503 was 60 times faster than the Elliott 803 but only twice as expensive.[5] The actual installation is shown in Figure 2.1 with a young Bud Abbott and his punch room supervisor Beryl Baker examining the operation of one of the 35 mm magnetic film handlers.

It is significant to recognize the special position of The London Hospital in the structure of the NHS at that time. It was blessed with first-class managers, who had a certain level of independence as a major teaching hospital, the Board of Governors being responsible directly to the DoH. These managers were fully supported by their board, which included some

FIGURE 2.1 *The Elliott 803 computer installation at The London Hospital* (reprinted by kind permission from *The London Hospital Illustrated*, vol IV, No. 4, Spring 1969. © Royal London Hospital Archive. Courtesy of Royal London Hospital Archives)

of the best financial brains in the City and were anxious that the hospital should match the best in its care for its patients and its staff. There was a ready acceptance of new ideas, special encouragement for junior staff with training support and acceptance of the need for constructive change. The focus was still on the paramount needs of the hospital to serve the local community of the East End where there was a tradition of support for the hospital going back many years. The computer project fitted the ethos of the hospital.

> As with many computers at the time, the Elliott 803 room was provided with a 'viewing window', so that visitors could inspect the marvel without interrupting the actual work. Hence while we were working, from time to time, various faces would appear at the window and then vanish. I remember one time while we were waiting for the engineers to come and repair the system, I suddenly saw the House Governor, the Hon. John L.C. Scarlett, at the window with a visitor. I was really glad when they went on their way without coming in for a discussion because the room was totally silent and I was acutely aware that this great and expensive beast was not working!
>
> **BARRY BARBER**

Once the computer was ordered, adaptation of some available space enabled the data preparation room and the computer room to be built. The equipment was delivered and the installation went live in summer 1964. The finance system had been comprehensively flow-charted in 1962, so that programming could begin as soon as the installation was ordered. Applications were batch processed. That is, the program was put in the main store, the previous file on film was loaded onto the handler, a new film to receive the updated file was loaded onto another handler and the prepared input on paper-tape was set in the paper-tape input reader. The program was run through for all the records, taking account of the previous file and the input tape and produced an updated file and the paper-tape output for printing offline.

The basic case for the computer was based around the finance systems and the development and running of these program suites took general priority in the use of the system. However, during the first year of operation, substantial demand for the use of the system arose from many parts of the hospital and the medical college that had not been evident before the computer facilities became accessible.

The demands from users involved in medical and scientific work proved complementary to the work of developing the financial systems. Full single-shift working was required from the beginning and double-shift working was already being considered as the service entered its second year of operation. At Christmas 1964, when the computer had only been installed two months, the treasurer then asked for the original implementation plan – commencing with the stores system – to be held over and to begin instead with implementation of the various payroll programs at the beginning of the financial year. This was instead of the finance systems that had been due to go live in April 1965. To make matters more complicated, the only trainee programmer left at the point where he might have been considered to have finished his training in machine code programming.

It was with some difficulty that the payroll applications started in April 1965. Everyone concerned with the computer installation had a great deal to learn about the different jobs that had to be tackled successfully to ensure that the new computer systems worked correctly and produced the right results. The information had to be punched onto paper-tape – usually 'punched and verified' to ensure accuracy – the computer programs had to be operated on the machine in the correct order and with the right data and film, there had to be effective operations management and job control – quite apart from the original systems analysis, systems design, programming and program testing. It was a very new world and it involved a very steep learning curve for everyone.

By April 1966 the full financial system was in place – it was tailor-made to the requirements of the hospital. There were five major sections: the salaries and wages subsystem; the goods inwards subsystem; the stores issues subsystem; the miscellaneous subsystem; and the ledgers subsystem. It had been intended that we would start with the goods inwards subsystem but, due to increasing problems with the accounting machines, the payroll subsystem was implemented first. Barry Barber and Bud Abbot described all the subsystems in their book *Computing and Operational Research at The London Hospital*.[6]

As stated, this finance system had been tailor-made for The London Hospital. Indeed, the principle had been adopted that it was the duty of the Computer Unit to provide a service for the hospital and, as a consequence, the system had been made as flexible as possible. An illustration of this was that the unit was organized to receive the last of the clock cards from the Finance department on Wednesday morning and yet payrolls were available to be paid to members of the staff at Wednesday midday. While this approach necessarily entailed a certain loss of efficiency in the utilization of the Computer Unit's resources, it nevertheless provided the greatest efficiency in the use of the combined resources of the Computer Unit and the Finance department.

Many lessons were learnt from the establishment of the Computer Unit at The London Hospital, not least that a complete hospital finance system is a very complex structure and one requiring a good deal of effort to create and perhaps even more effort to maintain. The embryonic patient administration system (PAS), although not then central to the operation of the hospital, had become the next major system for development. The London participated in the experimental programme to explore the application of real-time computing across the hospital. A new facility based on the UNIVAC 418-III with terminals in all the wards and departments was installed in 1971 and the PAS went live in 1972, followed by other developments of patient systems. The finance system on the Elliott 803 needed to be replaced and in 1973 a second 418-III installation was purchased to provide for the finance system and a back-up for the real-time system. During the 1970s the finance system was improved and transferred to the new installation, still as a batch-processing operation.

Meanwhile, the DoH began to take more interest in computing and the Central Organization and Methods (O & M) Unit published a table showing the computers installed in the NHS in 1966. In 1968, the department established its computer branch under Don White and computer development in the NHS gained a most important supporter. The London teaching hospitals, building on the technical success of the Rowlandson experiment, gradually finalized the requirements for a bureau service in London. With support from the King's Fund and the DoH, the Hospital Computer Centre for London was opened in 1969. The Centre was equipped with an ICL 1904E, with priority being given to the implementation of finance systems, and in 1974 the Centre was transferred to the South East Thames Regional Health Authority.

At the first reorganization of the NHS in April 1974 the South West Metropolitan Regional Hospital Board was split in two and South West Thames and Wessex Regional Health Authorities were formed. The board had a computer centre in Winnal Valley, Winchester for payroll accounting and hospital activity analysis. When the RHAs were formed the centre continued to operate on behalf of both regions. However, almost the first resolution of the new Wessex RHA was to thank

South West Thames for its fine computer centre and politely invited it to find another source for its computing. Hence I was recruited to build a new computer centre to serve the South West Thames region.

Manchester region had been commissioned by the DoH to write a new payroll system to become a standard application for all regions. The system was due to become available within the period that Wessex agreed to continue supplying South West Thames. It was decided to adopt the new payroll system as soon as it was available for transfer.

However, there were no firm plans then for a national accounting system. A team from SW Thames spent several months at Winnal trying to produce adequate documentation to enable them to support that finance system. They had found the inevitable backs of envelopes and fag packet documents but nothing sufficient to make it practical to take on running the software at another site. It was also clear that the finance officers in the area, district health authorities and the hospitals all had different ideas on what a finance system should provide. At that time Oxford region were writing new software and with some difficulty the regions' finance officers agreed to accept that specification. This was a time when the National Computing Centre (NCC) was promoting systems documentation standards. These were adopted by several regions that began to make feasible the sharing of systems.

When the South West Thames new Computer Centre was opened in 1977 the Manchester payroll was thought to be ready to make available to other regions. That was when our troubles really began. Unfortunately none in the NHS had sufficient experience of what was needed to support software being run at a remote site; certainly nothing as time critical as payroll. There was a long and painful learning curve to be traced by all concerned. But, eventually, lessons were learned and payroll settled as a solid working system. A national user group monitored performance and agreed changes on behalf of users. This was a time when Department and Treasury officials were getting a taste of the power of computers and their demands for information, complex pay agreements and deadlines of 'yesterday not being soon enough' put a great deal of pressure on the Manchester software team. It took a long time for realism to take hold.

This was all done at a time when money was very tight. Any spending on computers could easily be presented as frivolous and denying funds for patient care. It was thanks to the vision of civil servants such as Don White and certain RHA officers that any progress was possible.

Running the 'Oxford' accounting system was fine for a while, but finance officers are never to be satisfied. If you put three in a room they will come up with four different specifications. South West Thames eventually had to produce its own system. At this time we started experimenting with a package to extract information from the finance files on demand using Commodore PET computers – precursors to today's PCs – as remote terminals. At the same time, by purchasing American software for hospital patient administration run on Digital Equipment Corporation (DEC) VAX computers, we had begun to move away from the Government's ICL straightjacket. Computing power and storage capacity was beginning to take off and after a while PCs were becoming affordable in quantity and replaced the PETs. Demand for access mushroomed and we were fortunate to be able to continue developing systems to provide it.

In the meantime other national systems were emerging for supplies, manpower, hospital activity and vaccination and immunization. A national computer policy committee (CPC) was formed to determine NHS computing needs rather than have the Department dictate everything. This was chaired by Sir Gordon Roberts, chairman of Oxford RHA. Five regional officers were appointed to serve on this committee representing the five disciplines comprising each regional management team. A management services officer and a regional computing services officer (myself for the first two years) were invited to attend to represent and consult with their respective regional colleagues. They appointed Pat Bishop from the West Midlands Regional Computer Centre (RCC) to be the first secretary. The CPC took over the direction of the national programme of standard systems and began the slow process of educating colleagues in the RHAs of the potential for using computers in the NHS. They recognized the need for standards, including common data definitions and began measures to define them and see that they were understood through an education programme.

After a while Mike Fairey was appointed head of the DoH's computer branch (immediately preceding Ray Rogers) and on the retirement of Sir Gordon Roberts, he took over the chair of the CPC. Work on standard systems continued but the promulgation of a national accounting system never bore fruit. Almost without exception it was impossible to get regional finance officers to accept one. Their ideas of what they wanted and their failure to agree a system of coding meant that no region wanted to take it.

Later the first NHS trusts were created with new freedom to find their own computer services. There had always been a tension between regions and hospitals. Hospitals resented 'their money' being top sliced to provide regional services and computing was not exempt from this. Economies of scale, concentration of expertise and the critical mass effect counted for little. Sharing became a dirty word and competition was the thing. It seemed that with every appointment of a new trust director of finance the existing accounting systems had to be replaced by what the new incumbent was used to. The RCCs were all sold off or privatized in various ways leading to a dispersal of NHS in-house expertise. It has taken a long time to recover and now RCCs are being reinvented.

NEVILLE VINCENT

REGIONAL DEVELOPMENTS

By 1969 it was clear that the RHBs were establishing bureau computer services primarily for the finance systems of their region. Ten of the RHBs had a computer by 1969, mostly ICL 1902s, plus English Electric 4/30s and 4/40s and NCR 315s, all primarily concerned with payroll with some other finance applications. The standardization programme began in 1971 with the promulgation by the DoH of a minimum hardware configuration for each RHB based on the ICL 1903T/1904. Subsequently, a programme of applications for establishing standard programs to be used by all NHS installations was drawn up, and centres of responsibility were established for each application. The range of finance systems largely comparable to the original London subsystems was included in this programme. However, while the payroll and the vaccination and immunization applications were widely adopted there was less enthusiasm for the others. The policy was reviewed many times, most significantly in 1978–79 when it was accepted that the NHS had not taken the standard systems universally because each region's investment in computing was on widely differing scales and timings and with very different priorities. Payroll was still used as a standard application with the centre of responsibility at Manchester RHA and, with rather less acceptance, the accounts payable and ledgers standard applications were supported by Birmingham RHA.

A view by the Regional Computing Services Officer at South West Thames RHA of the situation at this time and subsequent developments of the RHA computing services is covered earlier.

As indicated above, the late 1970s were a period of consolidation by each of the regional computer services who upgraded their facilities, improved their operations, and chose their own combination of available systems for the major routine requirements of the various aspects of the finance system, and other applications. Many systems were written or adapted to run on minicomputers that were usually used as stand-alone facilities (e.g. stores) or, sometimes, as front ends to the mainframes (e.g. budgetary control). Overall the spread of applications grew steadily – except for PAS, which was not adopted too widely outside the experimental project sites. By the end of the 1970s the standard applications were reduced from 11 down to three, and these were financed by a consortium of user-authorities. Some finance data collection over dedicated telephone networking also came into use during this period.

At this stage the RHAs had established their individual finance systems based on batch-processing techniques and mainframe computers. Finance systems were the economic drivers for the early computer installations both at the hospital and regional levels. In consequence, although computer units were not usually part of the finance departments of these organizations, operationally the use of the computer was heavily oriented toward ensuring the finance systems were kept to schedule. Also, as the basic elements of the finance systems were established so the finance departments found other needs for financial information could be met and required more time from the batch-processing process.

The advent of minicomputers and especially microcomputers spread the availability of computer power across the various organizations. No longer was it necessary for computer requirements to be slotted into the mainframe schedules largely set to meet finance time frames. Contracting out to private suppliers for computer services for finance systems soon followed and the regional centres that were an integral part of the NHS finance systems were gradually disbanded.

REFERENCES

1   Winterbottom, F.W. (1974) *The Ultra Secret*. Weidenfeld & Nicolson, London.

2   Turing, A.M. (1937) *On Computable Numbers with an Application to the Entscheidungsproblem*. Proceedings of the London Mathematical Society, 2, 42.

3   Ferry, G. (2003) *A Computer Called Leo*. Fourth Estate, London.

4   Cooper Bros. (1964) *Electronic Data Processing*. A report prepared for The London Hospital, 4 February.

5   Barber, B. (1963) Memorandum on a conversation with various members of the Elliott's sales staff regarding the advantages of the 503 over the 803. Medical Physics Dept Report No. 994, 5 November.

6   Barber, B. and Abbott, W.A. (1972) *Computing and Operational Research at The London Hospital*; 'Computers in Medicine' series. Butterworths, London. ISBN 0 407517 00 6.

3 Early innovations

BARRY BARBER

There was a period during the 1960s and 1970s when there was a growing interest in the development of computing systems in the NHS but when there was no easy route to securing resources to carry out the necessary developments. In many cases the Medical Physics departments contained the only potentially usable computing resource. It was expected that a significant amount of graduate time should be spent on the research and development aspects of their work in order to keep abreast of opportunities in health care as distinct from the routine support of radiation equipment and radiation-emitting isotopes used for therapeutic and diagnostic purposes.

The Medical Physics department at The London Hospital had been looking for support for its radiation treatment calculations to complement the work that had been done in precision dosimetry by Dr Lloyd Kemp with and automatic isodose plotter[1] and his ionization comparator.[2] He had found deficiencies in the implementation of the unit of dose, the roentgen, at the National Physical Laboratory (NPL)[3] and had developed a new approach to the design of a cavity ionization chamber.[4] The exploration of these possibilities required further experimental work with the ionization current comparator as well as access to improved facilities for carrying out complex calculations. Even though the department had purchased an Archimedes-Diehl electric calculator[5,6] (which had a keyboard of nine digits and an accumulator of 18 digits and, most usefully, a single storage register into which the results of calculations could be summed) in 1957, the initial design of such ionization chambers was explored experimentally.[7,8]

Later in that year consideration was given to the use of a Ferranti Pegasus machine in order to analyse the behaviour of the new design of cavity ionization chamber. However, at that time the programming of the necessary calculations was going to be very complex and it was with some relief an analytical solution was derived to the problem, which simply involved modified Bessel functions of order zero and one that allowed the calculations to be carried out on the Archimedes-Diehl calculator.[9,10]

As a result of discrepancies between the experimental work and the calculations, additional work was undertaken on an analogue computer at the GEC laboratories at Wembley that had been developed for the design of cylindrical electron tubes. This work showed up an error of a factor of two in the textbook solution of the problem.[11] Further work led to the finalization of the design curves for the chamber[12] and the practical comparison of The

London Hospital's kilocurie cobalt unit, as determined through a dosimeter calibrated by the NPL, and the absolute measurements using the guarded-field technology.[13]

The needs of the Medical Physics department for mathematical support in its dosimetry had been well established by the time the computer arrived.[14] The plans involved an examination of an existing program for radiation treatment planning for external beams, already available from Elliott Medical Automation Ltd, as well as working on new treatment plans, handling the corrections for body inhomogeneities and moving beam analysis. There was also a program for handling the radiation fields from sealed radiation sources that was expected to provide a useful research tool. Beyond these obvious areas of activity the computer provided opportunities for handling ad hoc physical and statistical calculations as well as providing a tool for dealing with isotopic analyses and the developing field of medical ultrasonics. The work outlined was seen as tentative and dependent on the pressures and priorities of the whole work programme, but it was noted that:

> At the moment there is no medical computing centre in Great Britain and no one knows what benefits it will provide or how useful those benefits will be . . . . Nevertheless, it is likely that the most substantial rewards will accrue from the chance that this will give for research staff to make a start on some of their problems with modern methods and in the long run this should be of decisive importance in the standing of the hospital and the quality of the staff that it can attract.

REFERENCE 14

Work had already been undertaken in the design of a non-standard type of radium needle[15] that could automatically fulfil the requirements of the Paterson-Parker rules and it had been made available from the Radiochemical Centre at Amersham. The scope of the work foreseen by Elliott Medical Automation Ltd was outlined by Payne.[16] It went beyond the immediate issues of radiation treatment planning to medical diagnostics and medical records, envisaging extensive operational research studies of hospital procedures. The paper ended:

> The noble profession will be no less noble for being business-like in respect of those many aspects of its work where business methods have something to offer. None of this will interfere with the sanctity of the doctor–patient, nurse–patient relationships and all parties will benefit from the improved salaries, conditions and further amenities which technological innovation can wrest from £1B annual expenditure.

I had already gone on a two-day Elliott Autocode course at their Borehamwood centre because it was intended that the scientific work would be handled in the 'high-level' language, rather than in 803 machine code, which was necessary at that time for the efficient programming of the finance systems and required three months' training. During the course, we were all set an exercise to see how much we had understood. This involved the calculation of the circumference, the area, the surface area of the equivalent sphere and the volume of the sphere, from knowledge of the radius. Most of the class read in a number and then did the calculation rather rapidly. In my case, I did not know where the radius was to come from so I set up a loop from 0 to 10 by steps of 0.1 together with the normal line spacing used in printing tables. The program was allowed to run for about five minutes before it was stopped, but it produced a credible tabulation that still reminds me of my first effort at programming.

At a later stage I had to attend the much higher level ALGOL course in order to use the Elliott simulation package designed for the Elliott 503. Unfortunately, it ran so slowly on our Elliott 803 that run-times for some clinic simulations were used overnight, over weekends and over the Christmas holidays. An indication of the detailed work being undertaken in the Medical Physics department at that time was given in a computer progress report.[17] At this stage, the Medical Physics and Clinical Chemistry departments were the only source of computing expertise available in many hospitals. In some cases the computing facilities could be made available to help more widely within the hospital. These ideas were first noted in the Medical Physics department's Annual Reports for the year 1961–62.[18]

EXPLORATION OF OPERATIONAL RESEARCH TECHNIQUES

In 1960, the Nuffield Provincial Hospitals Trust (NPHT) held a meeting at Oxford on the application of operational research (OR) in the health services. It was concluded that there were 'widespread opportunities for the useful applications of operational research . . . at various levels of administration . . . and the number of studies should be increased as soon as possible'. These conclusions from the NPHT's Fifth Report were noted in a departmental brief report[19] and were amplified in the context of the hospital,[20–22] again after Dr Kemp and I had attended a course on OR techniques at Borough Technical College. During this time a series of reports were produced concerning various OR techniques and hospital applications.[23–25]

About this time we took the opportunity to use the computer to tabulate information about some of the simple queues: with random arrivals and random service times, with different numbers of servers and different ratios of the average service time to the average arrival time. This work was extended to explore the situation where the service times arose from the more complex Erlang family of distributions. This led to some tables in *Operational Research Quarterly*[26] within a month or so of similar tables being published in the *Journal of the Royal Statistical Society*. Fortunately, the numerical figures agreed where the tables overlapped.

In addition, Dr Kemp was able to visit the Johns Hopkins Hospital Operational Research Division while he was in the USA on a visit to the *International Congress of Radiology* in Montreal. He brought back a lot of ideas from the work being undertaken in Charles Flagle's OR group at that hospital,[27] and these ideas matured into more detailed ideas about the work that might be done at The London Hospital.[28]

The NPHT took their own advice and published work on an overall view of what might be achieved with OR in the NHS,[29] as well as work on casualty services,[30] the organization of X-ray departments[31] and queuing in outpatient departments.[32] The 1965 outpatient queuing studies followed work by J.D. Welch and N.T.J. Bailey that the NPHT had supported back in 1952.[33, 34] M.J. Blanco White and M.C. Pike had, also, been interested in queuing issues, but from an architectural point of view in respect of queuing space required in clinics.[35] It was interesting that, by 1965, the NPHT had found that the overall situation had not changed much since the original work of Welch and Bailey over a decade before. Preliminary work at The London Hospital suggested that things were the same there.

The encouragement of the NPHT in attempting to support the use of OR techniques within the NHS was significant and it is a pity that their encouragement was so frequently ignored. A great deal of effort was put into the exploration of some of these issues at The London Hospital and a flurry of reports emanated from the Medical Physics department concerning work carried out in various areas at this time.

By 1963 Elliott Automation was providing computing facilities on an Open Shop basis together with short programming courses in Autocode. The first scientific work carried out using these facilities involved the analysis of observations on outpatient clinics, and a number of small projects both from within the Medical Physics department and from the hospital at large. This initial work in OR, together with the other projects, was sufficiently useful that it provided the basis for other computing requirements, as well as the requirements of the Medical Physics services and medical survey analyses. About this time Elliott Automation established EA Medical Automation Ltd and reached agreement with University College Hospital for the installation of an Elliott 803 computer, which would be available on an Open Shop basis to explore the medical applications of computers in a health care setting as well as providing UCH with computing facilities. This was the first computer system installed in a UK hospital for medical applications under a commercial arrangement. It was followed about a year or so later by an Elliott 803 at The Royal Infirmary, Edinburgh, and the Elliott 4120 at the Hammersmith Post Graduate Medical School. In addition, an IBM 1440 was installed at the United Birmingham Hospital group supported by the NPHT. The Elliott 803 installed in 1964 at The London Hospital was the first computer installed by a UK hospital at its own expense.

Even at this early stage Elliott Automation was examining issues of diagnostic support, calculations for matters such as radiation treatment and medical record keeping.[16] Indeed, the history of medical computing has been that of slowly transforming the medical and technical vision into routinely usable systems. For this we all had to wait for much more reliable and powerful hardware to become available. We also had to learn how much effort was needed to provide 'user-friendly' access to the computer systems and to implement and integrate these across the health care environment. Much as with the motor car, it has taken many years for a mass market to develop and for computers to become conveniently usable throughout the community.

Although some activities come unexpectedly, 'out of the blue', more often than not they come as developments built on existing activities. In this sense, the developments in the handling of the finance activities at The London Hospital with accounting machines, punched-card facilities, as well as system centralization, all paved the way for the implementation of the system on a computer. Similarly, the struggles in the Medical Physics department to improve radiation treatment planning using precision measurement techniques, an optical simulator, mechanical calculating facilities and the exploration of OR techniques all set the scene for the scientific use of a computer system when it became available. Correspondingly, the work that was subsequently carried out on hospital statistics and patient surveys, using the Elliott 803, laid the foundation for the development of patient administration activities and patient record systems.

THE OPERATIONAL RESEARCH UNIT

In 1966, I was acting head of the Medical Physics department for a period and was invited to establish the Operational Research Unit as a companion unit to the Computer Unit. My brief was to concentrate on the opportunities that had been established for the use of OR in the hospital. In these early days a wide variety of projects were tackled, including:

- handling of emergency admissions – information and communication problems;
- continued work on queuing in outpatient clinics;
- student nurse duty rotas;
- handling of medical statistics;
- emergency anaesthetic work;
- theatre usage;
- communications with general practitioners; and
- flexibility in the timing of patients' meals.

Some of these studies developed into major pieces of published work while others provided part of the background of decision-making within the hospital. The work ebbed and flowed between the Computer Unit and the Operational Research Unit activities according to the needs of the moment and, indeed, systems analysis and operations analysis (to give it the then current American name) provided twin tools to address problems.[36] In some cases it was possible to make progress by implementing some computer system, while in other cases an OR analysis helped to explore the underlying issues and to find useful improvements without installing a computer system. Subsequently, the earlier OR work provided suggestions for the installation of more effective computer systems as a result of the greater insight into the underlying problems.

Curiously, at that time it was often easier to persuade colleagues to install a computer system – with all its complications and expense – than to persuade them to allow staff to undertake an OR survey and some mathematical analysis. Often improvements could be had by some basic reorganization rather than by massive data collection. Some of this early work was outlined together with information about the development of the Elliott 803 work and the initial plans for the Real-Time System in *Computing and Operational Research at The London Hospital*.[37]

Work on queues

The work on queuing had prompted a substantial amount of data collection and monitoring, as well as some complicated calculations on particular mathematical types of queues,[26] but the real problem turned out to be that of basic queue management. The problems could readily be resolved if those concerned ensured that patients were booked at sensible intervals, the doctors were available at the agreed times, and any pre- or post-consultation activities had been programmed into the system. A typical outpatient clinic queue simulation shown in Figure 3.1 indicates how the size of the queue changes when the clinic is over-booked, correctly booked and under-booked.

The study of the anaesthetic emergency work was pursued in conjunction with ex-London anaesthetists at the Northampton General Hospital, and at Alder Hey Children's Hospital, Liverpool, in order to give a wider spread of relevance to the results. The anaesthetists categorized their work according to the time within which the patient required anaesthetic help, as follows:

- immediate – within three minutes
- urgent – within 15 minutes
- emergencies – next available anaesthetist but no definite time limit
- other emergencies – could wait several hours if necessary
- routine work carried out in emergency time by duty anaesthetist.

This study resulted in a much clearer understanding of the requirements of this service and of the manpower needed to provide it, as well as the best ways of organizing the available manpower. The use of queuing theory helped to illuminate the staffing issues. Figure 3.2 shows how the emergency anaesthetic workload – expressed in terms of the average emergency anaesthetic time required per unit of elapsed time – and the average length of time required to handle the work affects the number of emergency anaesthetists required to achieve the specified response times.

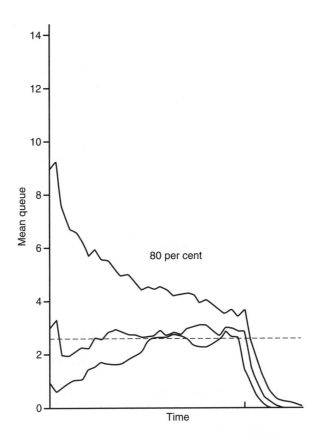

FIGURE 3.1 *Typical outpatient clinic simulation showing the effect of the initial patient queue.*[37] © Elsevier

It was noted at one stage that, while I was spending five minutes a point making calculations on the 803 computer, Alan Jennings was getting books out of the local library and using his slide rule to produce faster results. The work led to a series of five papers in the *British Journal of Anaesthesia*,[38] as well as other papers by Alan Jennings.[39] The work continued when the Operational Research Unit moved to the North East Thames RHA.[40, 41] The results were used extensively by the anaesthetists in discussions with the DoH.

Steve Farrow, David Fisher and David Johnson explored a number of issues in respect of the programme for handling chronic renal failure. This work used Markov chains as a means of modelling the situation. It helped to throw a very practical light on the development and budgetary costs of the Dialysis and Transplant programme at The London Hospital.[42] Figures 3.3 and 3.4 outline the Markov chain model used and the expected growth of patients on the programme. The authors continued to look at the national picture and the use of Markov chain modelling in a wider context.[43, 44] The calculations were based on the work of the Transplantation Immunology department at The London Hospital.[45]

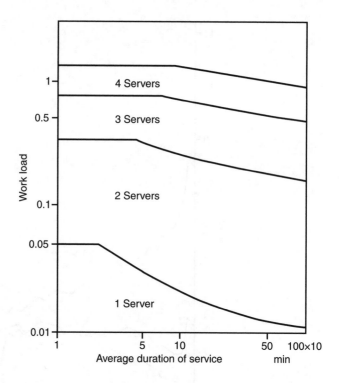

FIGURE 3.2 *Emergency anaesthetic service staffing requirement.*[37] © Elsevier

Work on bed usage

While we were examining the medical statistics for the hospital, we found a way of plotting the key bed-usage parameters on a single graph: using a graph of length of stay against turnover interval. The lines of equal percentage bed occupancy appeared as radial lines through the origin, while the lines of equal numbers of discharges per available bed showed as lines with a slope of –1. The resulting 'Barber–Johnson diagram'[46] had the advantage of allowing all the parameters to be displayed at once, instead of having the usual fragmented discussions about all the various parameters separately. One further advantage was that the calculations were based on available beds and it was immediately clear where beds were being borrowed from other consultants or consultant groups. This advantage was missed by the DoH because, after publication, they explained that they had modified the calculations to avoid having beds that were occupied by more than 100 per cent – thus losing the information that the statistics might have conveyed. The paper was adopted by Professor Alberto Franci of the University of Urbino in Italy, and he had data coming off his computers well before this had been achieved in England.[47]

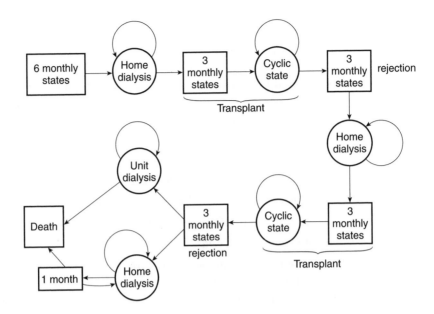

FIGURE 3.3 *Absorbing Markov chain model of a combined haemodialysis/transplantation programme.*[37] © Elsevier

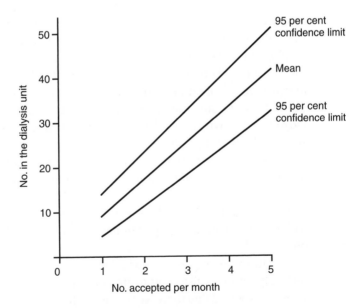

FIGURE 3.4 *Five-year growth of patients in a combined haemodialysis/transplantation programme according to the rate of acceptance of patients.*[37] © Elsevier

Admissions and discharges

Another major effort was made in the multidisciplinary Admission and Discharge Study[47–49] in which a group explored the various problems encountered by a hospital running at a very high bed occupancy, and the ways in which these might be alleviated. The key opportunities for dealing with the problem were:

- locating new admissions in the 'wrong' ward ('lodging');
- moving patients who were nearly ready for discharge into a 'wrong' ward for a period before discharge ('pre-discharge lodging');
- refusing emergency admissions ('closing the hospital');
- discharging patients before their intended discharge time;
- putting up extra beds.

There was a widespread perception that 'the system had broken down'. The project decided to analyse the admissions and discharge information for a three-year period, including an assessment of the number of avoidable patient transfers. At the same time two attitude surveys were conducted to assess the views of the various groups of staff affected by the admission and discharge system. The attitude surveys were also intended to explore suggestions for improving the system. This work was undertaken in the context of the Revans' Hospital Internal Communications project.[50, 51]

One of the key issues was that all these various devices had unsatisfactory side effects on patient care. The final perception was that the high occupancy inevitably led to problems but that many of these coping techniques could be employed for short periods. But the problems needed to be monitored and the side effects kept to a minimum; lodging should be kept to an acceptably low proportion of admissions. Patients who had been lodged once should not be lodged again. Care should be taken to ensure that patients who were lodged in the wrong ward were not overlooked by the relevant medical staff. Early discharge should not be considered unless the patient's home circumstances made safe continuing care feasible. Extra beds were almost never used because of the difficulties of caring for patients without the basic necessities.

The OR Unit developed a measure of system strain that correlated well with bed occupancy, to help in understanding the underlying mathematical processes.[52]

The attitude surveys showed that the organization was taking these issues seriously and was listening to practical suggestions from those involved. They also showed that there were widely different perceptions of the problem among the various disciplines. The practical work and the attitude surveys helped the staff to achieve a common understanding of the problems and the ways in which these problems might be alleviated.

SHARING INFORMATION

During this period the Computer and Operational Research Units were invited to speak about the various developments at an increasing number of BCS and NHS meetings and to publish their finding.[53-56] The last of these papers set out for the first time how the patient administration systems had developed out of the processing of statistics and medical surveys and outlined the scope of the patient administration systems of the future. The developing ideas about the application of OR techniques within the hospital environment and their benefits were also set out.[57] In addition, Michael Fairey developed his ideas about information systems in hospital administration. Thinking from the complexities of an industrial production line he began to outline the sort of information that might help in the process of managing patients through their period of care; the 'Fairey Tube' (Figure 3.5) was born.[58] It continued to be of interest as plans were developed later to use the functioning real-time PAS to provide this type of data about the practical operation of the hospital.[59] Another proposal that was followed up some years later and was never funded.

The London Hospital was able to bring together a critical mass of individuals who knew the hospital's organization and people, and who were respected in their own specialist fields. They were able to deploy computer and OR expertise in searching for solutions to the hospital's problems, and they had experience of exploring successfully many different aspects of the hospital's organization. The work on the admission and discharge study provided a good starting point when it came to implementing a real-time computer system to handle these processes. Attitude surveys also provided useful suggestions from and interaction with the various groups of staff.

The management of change in a complex multidisciplinary organization like a hospital is a serious and complex affair that demands as much effort as is required to determine the technical changes that are needed.[60]

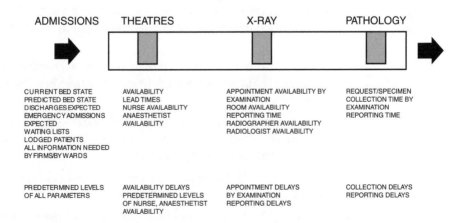

FIGURE 3.5 *Schematic representation of in-patient progress.* (Fig 3 (p 387) Ref 59 Reprinted by permission of The Random House Group Ltd.)

REFERENCES

1   Kemp, L.A.W. (1946) Exploration of X-ray dose distribution: automatic method. *Br. J. Radiol.*, 19, 488–501.

2   Kemp, L.A.W. (1954) A review of the theory, calibration techniques and applications of an ionization current comparator with an investigation of its capabilities as a precision instrument. *Amer. J. Roentgenology, Radium Therapy and Nuclear Medicine*, 71, 853–63.

3   Kemp, L.A.W. (1956) The roentgen: deficiencies in present-day free-air chamber standards. *Nature*, 178, 1250.

4   Kemp, L.A.W. (1956) Guarded field ionization chambers: a new principle applied to both free-air and thimble versions. *Nature*, 178, 1250–1.

5   Barber, B. (1957) *Memo on the Archimedes-Diehl High Speed Calculating Machine with a Double Product Register*. Medical Physics Dept, The London Hospital, 8 March.

6   Barber, B. (1957) *Report on the Desk Calculating Machines Demonstrated and Left on Trial in the Physics Department During June & July 1957*. Medical Physics Dept, The London Hospital, Report No. 615, 14 June.

7   Kemp, L.A.W. and Barber, B. (1957) Guarded-field thimble ionization chambers. *Nature*, 180, 1, 116–17.

8   Kemp, L.A.W. and Barber, B. (1958) The construction and use of a guarded-field cavity ionization chamber: a precision ionometric investigation. *Phys. Med. Biol.*, 3, 123–37.

9   Barber, B. (1959) The basic design data of a guarded-field thimble ionization chamber: a theoretical investigation. *Phys. Med. Biol.*, 4, 1–9.

10 Barber, B. (1960) *An Investigation Into the Design and Characteristics of a Guarded-field Ionization Chamber*. London University PhD Thesis, The London Hospital and the Middlesex Medical Colleges.

11 Barber, B. (1961) The basic design data of a guarded-field thimble ionization chamber: an analogue investigation. *Phys. Med. Biol.*, 6, 259–70.

12 Barber, B. (1962) The basic design data of a guarded-field thimble ionization chamber: further analytical investigations. *Phys. Med. Biol.*, 6, 389–400.

13 Kemp, L.A.W. and Barber, B. (1963) The basic design data of a guarded-field thimble ionization chamber for the measurement of supervoltage radiation. *Phys. Med. Biol.*, 8, 149–60.

14 Barber, B. (1963) *Computer Applications*. Medical Physics Dept, The London Hospital, Report No. 988, October.

15 Barber, B. (1963) A new type of radium needle for use on radium surface applicators. *Br. J. Radiol.*, 36, 774–5.

16 Payne, L.C. (1963) Towards medical automation. *World Medical Electronics*, 2, 1. Hanover Press, London.

17 The London Hospital (No date)*Computer Applications in Hand or Completed*. Medical Physics Dept, The London Hospital, Report No. 1042.

18 The London Hospital (1962) *Report on the work of the Physics department*. The London Hospital, April 1961–April 1962.

19 The London Hospital (1961) *Conclusions of the 1960 Oxford Conference on the Medical Applications of Operational Research*. Medical Physics Dept, Report No. 824, 14 June.

20 The London Hospital (1961) *An Appreciation of the Possibilities Offered by the Techniques of Operational Research*. Medical Physics Dept, Report No. 809, 18 May.

21 The London Hospital (1961) *The Application of Queueing Theory to Hospital Departments*. Medical Physics Dept, Report No. 825, 15 September.

22 The London Hospital (1962) *Further Discussion of the Possible Applications of Operational Research in The London Hospital*. Medical Physics Dept, Report No. 876, 13 April.

23 The London Hospital (1961) *Games and Decisions: A Summary of the Methods Given by JD Williams in The Compleat Strategyst*. Medical Physics Dept, Report No. 810, 15 September.

24 The London Hospital (1962) *A Preliminary Report on the Behaviour of Queues with a Single Consultant*. Medical Physics Dept, Report No. 861, 16 January.

25 The London Hospital (1962) *A Preliminary Report on the Theory of Multi-server Queues*. Medical Physics Dept, Report No 865.

26 Barber. B. (1964) Some numerical data for single-server queues involving deterministic input arrangements. *Operational Research Quarterly*, 15, 107–15.

27 Kemp, L.A.W. (1962) *Report on the Visit to the Operational Research Division at the Johns Hopkins Hospital, 11 September 1962*. Medical Physics Dept, The London Hospital, Report No. 907, 31 October. (Includes appendices on the publications and activities of the division.)

28 The London Hospital (1963) *Operational Research*. Medical Physics Dept, Report No. 982, September.

29 Nuffield Provincial Hospitals Trust (1962) *Towards a Measure of Medical Care: Operational Research in the Health Service*. Oxford University Press for NPHT, Oxford.

30 Nuffield Provincial Hospitals Trust (1960) *Casualty Services and Their Setting: A Study in Medical Care*. Oxford University Press for NPHT, Oxford.

31 Nuffield Provincial Hospitals Trust (1962) *Towards a Clearer View: The Organisation of Diagnostic X-ray Departments*. Oxford University Press for NPHT, Oxford.

32 Nuffield Provincial Hospitals Trust (1965) *Waiting in Outpatients Departments*. Oxford University Press for NPHT, Oxford.

33 Welch, J.D. and Bailey, N.T.J. (1952) Appointment systems in hospital outpatient departments. *The Lancet*, 1, 1105.

34 Bailey, N.T.J. (1952) A study of queues and appointment systems in hospital outpatient departments, with special reference to waiting times. *J. Royal Stat. Soc.* (B), 14, 185–99.

35 White, M.J.B. and Pike, M.C. (1964) Appointment systems in outpatients' clinics and the effect of patients' unpunctuality, *Med Care*, 2, 3, 133–45.

36 Barber, B. (1971) The interaction of operations analysis and systems analysis. *Sperry-UNIVAC Medical Computing Conference*, Rome, 28–30 September.

37 Barber, B. and Abbott, W.A. (1972) *Computing and Operational Research at The London Hospital*; 'Computers in Medicine' series. Butterworths, London. ISBN 0 407517 00 6.

38 Taylor, T.H., Jennings, A.M.C., Nightingale, D.A., Barber, B., Lievers, D., Styles, M. and Magner, J. (1969) A study of anaesthetic emergency work. *Br. J. Anaesth*. Paper I: The method of study and introduction to queuing theory, 40, 70–5; Paper II: The workload of the three hospitals, 41, 76–83; Paper III: The application of queuing theory to anaesthetic emergency work, 41, 167–75; Paper IV: The practical application of queuing theory to staffing arrangements, 41, 357–62; Paper V: The recorded service failures and general conclusions, 41, 362–70.

39 Jennings, A.M.C. (1969) A statistical study of anaesthetic emergency work. In Abrams, M. (ed.) *Medical Computing: Progress and Problems*, pp. 321–31. Chatto & Windus for BCS, London.

40 Bates, T. (1982) Deployment of emergency anaesthetists. In O'Moore, R.R., Barber, B., Reichertz, P.L. and Roger, F. (eds) *MIE, Dublin 82: Lecture Notes 16*, pp. 874–80. Springer Verlag, Berlin.

41 Sharratt, M., Jennings, A.M.C. and Barber, B. (1976) Staffing levels for emergency anaesthetic services. In Barber, B. (ed.) *Selected Papers on OR in the Health Services*, pp. 195–221. OR Society.

42 Farrow, S.C., Fisher, D.J.H. and Johnson, D.B. (1971) Statistical approach to planning an integrated haemodialysis transplantation programme. *BMJ*, 2, 671–6, June.

43 Farrow, S.C., Fisher, D.J.H. and Johnson, D.B. (1972) Dialysis and transplantation – the national picture. *BMJ*, 3, 686–90, September.

44 Davies, R., Johnson, D.B. and Farrow, S.C. (1975) Planning patient care with a Markov model. *Operational Research Quarterly*, 26, 3, 599–607.

45 Festenstein, H., *et al.* (1969) (Title unknown.) *The Lancet*, 2, 389.

46 Barber, B. and Johnson, D. B. (1973) The presentation of acute hospital in-patient statistics. *Hospital and Heatlth Services Journal*, January, 11–14.

47 Franci, A. and Tatti, P. (1977) *Riforma Sanitaria: Rapporto tra cost-effiicienza dei sanitari e utilizzo ottimale delle risorse esistenti.* University of Urbino.

48 Fairey, M.J., Barber, B. and Webster, B. (in association with Arie, T. and Tooley, P.) (1970) *A Study of Admission and Discharge Procedures 1966–1968.* The London Hospital, February.

49 Fairey, M.J., Webster, B. and Barber, B. (1971) Admission procedures at a London teaching hospital: part 1. *Nursing Times*, 29 April, 65–8.

50 Fairey, M.J. (1971) Admission procedures at a London teaching hospital: parts 2 & 3. *Nursing Times*, 29 April, 69–74.

51 Revans, R.W. (ed.) (1972) *Hospitals: Communication, Choice & Change: Hospital Internal Communications Project Seen from Within.* Tavistock Press, London.

52 Wieland, G.F. and Leigh, H. (eds) (1971) *Changing Hospitals: Hospital Internal Communications Project.* Tavistock Press, London.

53 Operational Research Unit (1972) *A Measure of System Strain.* Report 23 November.

54 Abbott, W. (1966) A year's experience of a hospital computer. *Brit. Hospital J. & Social Service Review*, 20 May, 913–14.

55 Barber, B. and Abbott, W. (1966) Computers in the hospital service. *The Hospital*, November.

56 Abbott, W. (1969) The computer in hospitals. *The Practitioner*, 203, 306–12, September.

57 Barber, B. and Abbott, W. (1968) Approaching the computerised hospital. *The Hospital*, March, 83–6.

58 Barber, B. (1969) Operational research at a hospital level. In Abrams, M. (ed.) *Medical Computing: Progress and Problems*, pp. 332–44. Chatto & Windus for BCS, London.

59 Fairey, M.J. (1969) Information systems in hospital administration. In Abrams, M. (ed.) *Medical Computing: Progress and Problems*, pp. 384–89. Chatto & Windus for BCS, London.

60 Barber, B. (1983) *Data analysis from The London Hospital Computer System – The Fairey Tube Revisited.* Management Services Division, NE Thames RHA, 7 June.

61 Fairey, M.J. (1976) The management of change. In Barber, B. (ed.) *Selected Papers on Operational Research in the Health Service*, pp. 53–65. Operational Research Society, Health & Welfare Study Group.

4 Towards patient administration systems

BARRY BARBER

The previous chapters on finance and the early innovations in scientific computing have outlined the way in which these activities developed at The London Hospital. They have also shown how much scientific computing and medical statistics developed from a desire to explore how the computer might best be used across the spectrum of activities of a hospital and medical college. This chapter charts the way in which a specific patient administration system (PAS) developed – almost of its own accord.

THE ELLIOTT 803 AT THE LONDON HOSPITAL

Since everyone was in a new situation, there was a lot to learn: how to run the computer effectively; what jobs were needed to do that work; how to meet the key deadlines; what the computer could do and what was best done in other ways. There was a flood of patient survey work from college and hospital staff and it was necessary to set out what facilities were available and how they could be best used.[1] Standard programs could be used from the library, the necessary analysis could be programmed by the research worker, or the Computer Executive could be convinced that the project was of sufficiently high priority that the Computer or Operational Research Units should do the work with their limited resources.

The following activities were recorded in the first year the hospital had the computer:[2]

- Finance
  - wages
  - goods inwards
  - remittance advice
- Patient statistics
  - clinic attendance
  - in-patient stays
  - bed occupancy
  - SH3 statistical returns

- Patient surveys
  - obstetric department
  - EEG department
  - physiotherapy department
  - development of general survey program
- Operational research
  - major survey of diagnostic X-ray departments
  - laundry and linen room survey
- Physics department
  - radiotherapy calculations
  - exposure times for Kilocurie Cobalt Unit
  - decay factors for various radioactive isotopes
  - theoretical study of multichannel collimators
  - initial work on the Elliott treatment planning program
  - exponential curve fitting
- Sundry statistics for various staff
- Medical college and other research users
  - Nigerian dental survey
  - pharmacological department
  - MRC social medicine unit
  - MRC neonatal unit.

This first year laid the foundations for much that was to come. The initial survey work led to the creation of a general purpose machine code program to write survey data onto magnetic film for analysis in Autocode. The implementation of the hospital's statistical returns led to a concern for increased computer education to take full advantage of the new technology. At the close of the first year's operations it was observed that:

> *The experience gained in this phase will enable us to specify the real machine needs of the next phase leading to systems capable of doing the bulk of the hospital's information recording and providing immediate access to information of clinical and administrative interest.*

REFERENCE 2

In the second year's report,[3] the term patient administration was already entrenched. It was coined to include the administrative information associated with a patient's care as distinct from the direct medical record, on which many others were working. We strongly believed that it was important to get a good grasp of the new technology and its limitations before putting patients' care and lives at risk. A full set of 33 pages of flowcharts for a batch-processing PAS accompanied the third year's report.[4] However, it was already evident that a real-time computing approach would be required if information were to be captured as a result of normal hospital activities and an application had already been made to the DoH to take these ideas forward.

The third year report showed that the average weekly use of the computer had climbed from an initial 50.7 hours per week to 74.4 and then 104.6 hours per week (Table 4.1).[4] The Elliott 803 soon became so heavily used that a second one was purchased in March 1969 to support the finance and other operational systems. The term 'Patient Administration System' appeared for the first time in print in the Financial Times.[6]

It is clear from Table 4.1 how much machine time came to be used by the PAS – the demands of operational research also continued to escalate. The EEG department work continued to build up even after they purchased a departmental microcomputer (a Digico 16). Douglas Maynard and Pam Prior initiated a major line of research in the development of a Cerebral Function Monitor.[7, 8] Some departments' work grew until they obtained access to other facilities and, of course, much depended on the research interests of individual members of staff.

These usage figures immediately show the rapid growth of the PAS activity, overtaking the time used by the finance systems. They also outline the wide variety of other uses that arose as the system became available for other activities. It is worth remembering that all this work was undertaken

TABLE 4.1 *Computer usage at The London Hospital (in hours per year)*

	1964–65	1965–66	1966–67
Systems			
Financial system	592.9	1,261.5	1597.1
Patient administration system	272.5	599.1	1753.1
Other	0.0	2.5	0.0
Research and other uses			
Operational research	118.9	439.2	363.3
Patient surveys	313.9	164.1	128.0
EEG department	14.9	271.1	550.9
Sundry statistics	12.2	18.4	45.0
Mathematical analysis (inlcuding Physics department)	112.2	59.5	248.7
Department of Pharmacology	145.9	184.9	17.9
MRC units	143.0	253.8	171.1
Miscellaneous	595.8	218.4	78.0
Computer Unit	363.5	399.2	484.0
Total	2,685.7	3,871.7	5,437.1
Average weekly running time	50.7	74.4	104.6

on a computer with an immediate access store of 8,192 words of 39 bits and a cycle time of 256 microseconds. The secondary store comprised reels of 35 mm magnetic film.

Among the patient surveys, the Obstetric Survey involved significant effort and provided a useful training ground for such work. Most of the information about a woman's obstetric history was formulated into binary (yes/no) responses to specific questions about her history. This allowed for very efficient storage of this information in the computer records. In addition, there were a series of numerical results and some textual information. The concept was that these computer coding sheets would be filled in for all the deliveries in the department and then it would be possible to produce the information for the department's annual report by direct calculation from the computer.

This system was somewhat of a nuisance for the registrars who had to fill in the coding sheets but it provided a considerable insight into the problems of handling medical data. The programs worked, the data input were generally satisfactory, but it was difficult to use the results directly for tabulating in the department's annual report. In practice, there were substantial delays in the system between the patient's discharge and the availability of the input information in punched-tape format for processing. Even worse, there was a tendency for the 'medically interesting cases' to linger on the consultants' desks, so that the computer results were neither a complete set of information for the annual report, nor were they a statistically valid sample of the workload. These were salutary lessons to be learned. In due course these problems were ironed out by Christine Tuck, Alaric Cundy and their colleagues.[8]

Following the experience of the obstetric survey a special purpose medical survey program was developed that allowed for rapid processing of binary information as well as allowing a certain amount of numerical and textual information to be processed. This arrangement allowed a great deal of research work to be undertaken on a 'Do It Yourself' basis. Aubery Sheiham brought back a great deal of information from his periodontal studies of Western Nigerian Children and this was analysed on the Elliott 803.[9] In contrast to that, some studies required small amounts of statistical work.[10]

The PAS developed out of the work on medical statistics for the DoH returns. The original SH3 returns simply involved counting various types of attendance, but the developing Hospital Activity Analysis system, which was a development of the ten per cent Hospital In-Patient Enquiry, was much more demanding. This approach was initiated by Bernard Benjamin and then was institutionalized by the DoH.[11, 12]

The concept of the PAS was that the administrative framework of health care should be developed as part of the process of delivering care and should automatically provide the management information for the hospital and for the DoH as a by-product of those processes. This approach foreshadowed the one taken by Edith Körner's steering group by nearly a decade.[13, 14]

After the initial work on the Elliott computer system it was possible to consider undertaking some other major activity. The obvious possibilities were:

- using the experience of Professor Flynn in the Biochemistry Laboratory at UCH,[15–19] where he was using equipment to produce paper-tape output from his auto-analysers, which were then processed on the Elliott 803, to provide the printed results;

- using the nurse allocation program developed by Dr Alex Barr of the Oxford RHB to improve the processes of nurse allocation and to follow up his work on nursing care;[20, 21] or

- introducing some computer monitoring of the outpatient clinics to focus work on those most in need of attention.

In the event, the first opportunity was researched using the statistics available from the Medical Records department. These data, when plotted on log–linear paper showed an exponential increase in test requests with a doubling time of about three years. However, at that time, the computer approach was rejected by the Director of the Laboratories.

The issue of nurse allocation was explored with Matron's Office staff in some depth. During this preliminary work it became evident that the key issue was not going to be the running of Alex Barr's allocation program, but the collection of the detailed information about all the learner and qualified nurses and their clinical experience. With some 1,250 nurses this would be a major data collection exercise in its own right and, from both a computing and an OR point of view, it was clear that this was where such work should start.[22–24] This was too big an undertaking to be considered initially, however, these early investigations led to a stream of work on nurse staffing and record keeping over the next decade.[25–30]

Work had been progressing on outpatient queueing issues and it was thought worthwhile to attempt to develop an outpatient monitoring system that might help focus on the clinics that needed special attention.[31] In many ways, this allowed the organization to begin to absorb the concepts that were being developed without there being a great impact on the more clinical issues, where major changes would have significant organizational effects. This monitoring of outpatient clinics was the result of the previous work on clinic organization and the concepts and detailed approach are described elsewhere.[32] The monitoring was sufficiently successful that it was moved from the Elliott 803 onto the UNIVAC machine when it became available as part of the patient administration suite of programs.[33]

THE DOH EXPERIMENTAL REAL-TIME COMPUTER PROGRAMME

Before the elaboration of this programme, almost all the computing activity in health care had been handled using batch-processing techniques. Information was captured, turned into a form that could be read by the computer, usually on punched paper-tape or on punched cards, 'punched and verified', and then processed in a convenient, daily, weekly or monthly cycle. The process provided a record of what had happened at a certain time, but not a record of the current situation.

Scientists in universities had already begun to use large, time-sharing systems that could handle batch input and output thus giving access to many users on a 'time-sharing' basis. Batch systems were a major development from the previous manual systems and remote access, time-sharing systems provided great opportunities for university research. However, the state of computing had already changed enough to allow 'real-time' processing to take place.

The most obvious, non-military, systems that brought about this change were the airline booking systems that were already beginning to cope with an increasing volume of air passenger traffic. Their approach to computer systems could be achieved by having a small computer dedicated to a single set of tasks, or by having larger systems that could handle the processing initiated at multiple terminals. It was thought that there were many tasks in the health care setting that could be automated in this way to considerable advantage but there were, also, considerable problems. For computers to be of real service, it was necessary to find:

- ways in which non-computer specialists could communicate rapidly and efficiently with the computers;
- hardware systems that could provide response times that were fast enough to allow the health care activities to proceed without interruption;
- ways of developing appropriate software to use the fast hardware effectively;
- arrangements for training staff to use the systems once they had been developed;
- ways of integrating these systems within the health care community.

Fundamentally, the question was whether it could all be made to work within the health care environment. Real-time systems were acknowledged as being very difficult to develop in comparison with the batch systems used by the earlier computers. Intrinsically, the batch-processing approach provided time in the systems cycle in which any problems with machines or programs could be sorted out. The real-time approach had to provide instant and accurate answers. If this could not be achieved, then some form of batch-type processing had to be brought into play, whether manual or computer-based, as the opportunity for completing the transaction had been missed. At the time that the DoH Experimental Programme was instituted, it was not clear that all these requirements could be achieved, nor where the NHS might benefit from such techniques. The original objectives of the programme were:

- the achievement of better patient care;
- improved clinical efficiency;

- improved administrative efficiency;
- improved research facilities.

However, as time went by and it became clear that the systems could be made to work, the last of these objectives was discarded and the DoH focused on various approaches to the evaluation of the systems. Each of these projects had to submit a feasibility study for the proposed project. If this was accepted the project proceeded to develop an 'Invitation to Tender', which could be issued to suppliers to enable them to quote a cost for the system.

Tom Maddison outlined the DoH view of NHS computing,[34] and the developing Policy and Objectives were set out by Dave Gedrych.[35] Later on the DoH pulled together its understanding of the ways in which computers could be used, to improve the health services, in a publication in 1972.[36] Mike Caddick and D.T. Lee combined the DoH views on the planning and control of their developing policy,[37] and John Handby developed ideas on the design of terminal-based systems.[38] Roger Harding-Smith had been involved in the purchase of The London Hospital's Elliott 803 computer and subsequently moved to UNIVAC in time to help with The London Hospital's next major purchase. He outlined his, private sector, ideas about the systems that could be used in hospitals and they were not far from those being developed at the DoH.[39]

The interest and excitement about the opportunities offered by real-time systems was not fully shared. The Scicon report for the Scottish Home & Health Department suggested that 'major investment is likely to be in the unglamorous but effective batch-processing area but with some encouragement of remote processing on a limited scale'.[40]

The term 'benefits realization' first came to notice in connection with information systems in Norwood's paper at *MEDINFO 74* but it has echoed down the years as people struggled to assess the value of their systems.[41] There were problems in understanding what were essentially infrastructure activities, that made other applications possible, and what were straight-forward applications that could be assessed on their own merits. Sharpe set out an early view of the DoH approach to evaluation at the same conference and Barber set out The London Hospital view about how the evaluation should be handled in terms of a 'before and after' scenario.[42, 43] This approach sought to collect a great deal of information about the functioning and modelling of the hospital before and after the installation of the computer, particularly in respect of such issues as:

- the transit times for laboratory results;
- a simulation of the laboratory office activities;
- the completeness of the laboratory requests;
- the probability of blood collection for tests;
- staff activity;

- all analysis of hospital performance;
- the quality of patient notes and hospital activity analysis returns;
- the time a patient spent in bed waiting for some activity (patient states);
- staff attitude surveys; and
- a simulation of the computer system.

These somewhat conflicting views of evaluation were synthesized subsequently in considering four aspects of evaluation: project evaluation, technical evaluation, application evaluation and programme evaluation according to which aspects of the activity were to be evaluated.[44] That paper was written when the programme had been expanded to include the projects at King's College Hospital, North Staffordshire Royal Infirmary, The London Hospital, the Queen Elizabeth Hospital in Birmingham, the Exeter Community Services project, and the four coordinated projects at Addenbrooke's Hospital in Cambridge, and Charing Cross, St Thomas' and University College hospitals in London. The DoH were especially interested in selecting applications for wide-scale introduction, which were in some sense more profitable. However, the DoH inevitably had to be interested in the success of the various individual projects, the technical success of the approaches that were adopted and, of course, the success of the whole expensive programme in generating approaches to information systems development that would be of benefit to the whole NHS.

These issues were further explored in the context of the whole programme but the key issue was the focus of the questions that needed answers.[45] Often a great deal of data collection simply addressed issues that were irrelevant or insignificant. The DoH produced an Interim Evaluation in 1977,[46] outlining what applications had been tackled, where and what benefits appeared to have been secured. This work continued and produced a large volume of data that disappeared, leaving very little trace, within one of the many major NHS reorganizations. The best account available of where this work in developing performance criteria went, together with 24 references, is given by Alaric Cundy and Julie Nock.[47]

During the early years of the experimental real-time projects, Don White organized a number of meetings at the DoH so that the various projects could share information about their practical problems and their progress. Attendees also heard from the DoH about the development of their ideas and from others with technical expertise to offer. At one such meeting Morris Collen attended and sat quietly listening as we described the work that we were all going to do on our systems. When Don White enquired, towards the close of the meeting, what Morris thought about our plans he said he wished that he could do with his computing facilities a fraction of what we were planning. This was a salutary reminder to us all that we were attempting quite difficult tasks because he had much greater computing facilities, at the Kaiser Foundation Research Institute and Permanente

Medical Group, than any of the projects were going to have available. Don White realized that there would be important issues to be addressed when such projects became successful and started to provide a key part of the NHS infrastructure.[48]

MEDICAL RECORDS PROJECT AT KING'S COLLEGE HOSPITAL

The first of these projects was the medical records project at King's College Hospital (KCHMRP) directed by Professor John Anderson. It was planned that the project would capture the medical and nursing records of all his patients. This was a very ambitious plan for the time, given the availability of hardware and software. John Anderson was an important pioneer in developing medical informatics in the UK and he published widely on the topics of computerized medical records and the value of such records in medical education. He was active in IFIP TC4 and its successor, the IMIA. He edited the proceedings of the first *MEDINFO* congress, held in Stockholm in 1974, and he chaired the first congress of the EFMI in Cambridge in 1978. In addition, he was active in the Education Working Group of TC4/IMIA, as well as being a keen supporter of the French medical informatics conferences (Journées d'Informatique Médicale de Toulouse), held in Toulouse from 1970 until 1980, when they were superceded by the EFMI congresses. He was also very involved with the BCS Medical Specialist Group.

Probably the best description of the KCHMRP is contained in a lengthy chapter in Morris Collen's book on hospital computer systems, published in 1974.[49] It outlines John Anderson's philosophy concerning medical records and sets out his perceptions of the requirements for a medical information system, its structure, the inclusion of the nursing record and the privacy requirements. He was quite clear that:

> ... the problems in medicine, especially those in relation to the medical record, were complex and difficult, although not beyond the range of the possible.

He also thought that there was a danger in:

> ... the rising tide of optimism about the use of computers in the Health Service in the late 1960s and early 1970s .... There was a general feeling that everything could be achieved with a great expenditure of resources and not much thinking, Utopia was just around the corner.

The project was not approached in any superficial fashion and a great deal of work had been done to try to sort out the 'tree-branching' display menus that were needed to enable doctors to enter their medical findings. Previous work on implementing a small punched-card system had indicated the difficulties of implementing medical record systems in particular clinics. It was clear that the records needed to be appropriately structured if they were to be useful. It was with these difficulties in mind that John Anderson undertook feasibility studies in conjunction with International Computers

Ltd (ICL) – the UK preferred national supplier. These studies were submitted in 1967 and 1968 and the project was authorized to go ahead in 1968.

John Anderson had developed a dictionary of medical terms to help this venture. The computer supplied was an ICL 1905E with a 48k store of 24-bit words, six magnetic tape drives, two line printers and eight visual display units (VDUs). Six of the VDUs were for the pilot area in the wards, while the other two were used for systems development and teaching. At one stage, some VDUs were trolley mounted so that they could be wheeled to the patient's bedside for convenient interaction between the doctors and the patient. This arrangement apparently caused sufficient alarm to one patient that he ran out of the ward. The objectives of the system were to provide a lifetime summary medical record and an in-depth episode record of care in hospital.

The system went live with 40 medical beds in early 1970 and it took three months before problems of training and software were sorted out. By June or July 1970 some 600 messages were entered per week, with some 400 created by the doctors and 100 by the nurses. By mid-1971 this had risen to about 1,500 messages per week. The project demonstrated that it was possible to create and use a medical record in the fashion suggested, but it did not deliver a workable solution for the hospital in which the beds were located. The system was only available for two shifts rather than on a 24-hour basis and the handling of medical data outside of this time presented great problems.

The nursing record had always been seen as an integral part of the medical record. In many ways the nursing system was a greater success than the medical record itself, because the latter involved a considerable amount of time and effort on the part of the junior medical staff in entering information into the system and some staff had criticized this process. Opit and Woodruffe had worked on the project for a period and they published two very critical papers after they left,[50] which had a very damaging effect on the perception of the project. They continued to criticize the real-time programme and these comments did not help when the project was called before the Parliamentary Accounts Committee.[51, 52] The issue of data input remained an important one. It is possible that it was not capable of being solved for complex medical records because of the limitations of the equipment available at the time.

Although the project was clearly an experimental one, rather than an operational part of the hospital, it demonstrated that it was possible to develop a patient's medical record in a computer using a tree-branching menu approach. In particular it was evident that the nursing record system, which required a much simpler record and data input structure than the medical record system, could be made to work very effectively and productively. However, after the project had been severely criticized by the Parliamentary Accounts Committee, funding was withdrawn and the project eventually died. It provided a number of very useful lessons for other experimental projects. The BBC also made an interesting film, narrated by

Fyfe Robertson, which illustrated the problems with manual case notes and the hopes for more complete and effective computer-based case notes.

In the process of developing the system, which never really got beyond the experimental stage, some 4,000 VDU display frames were developed for the menu selection approach to the input of the medical record. Special purpose middleware was written to ease the use of tree-branching menu selection from the various information frames. The medical record system had not been further developed between 1971 and mid-1972 and there was obvious tension between the medical record project and the management. At one stage it was hoped that the system might be developed into a full-scale hospital communication system, but it became clear that the facilities were hopelessly inadequate for that purpose. Professor Anderson resigned as Project Director and much that might have been achieved was allowed to waste. One exception was that it was understood that the augmented operating system, that had been developed for the project, was used very successfully in the banking sector in Germany.

John Anderson's fundamental interest was in medical education. As one might expect, his interest in medical record systems lay in the possibilities that they offered to enable clinicians to understand more about the investigation and treatment of their patients than was possible from rather patchy handwritten case notes. These interests are clearly evident from examples of his publications.[53–56]

THE LONDON HOSPITAL COMPUTER PROJECT

The London Hospital Computer Project was the third of the DoH's experimental real-time projects. It was initiated when the hospital applied to the DoH to develop a real-time PAS based on the work illustrated by the proposal to develop a batch-processing PAS, which had been outlined in 1967 in the third annual report of the Computer and Operational Research Units.[4] This report included appendices outlining a proposed PAS to replace the existing Elliott 803 systems in April 1968. It might have been attractive to pursue an approach to handling medical records because work had already been undertaken in respect of using Kardveyor's to ease access to patients' case notes. However, it was believed that the complications of handling a large real-time system were sufficiently great to warrant careful testing and consideration before moving into clinical areas where patients' safety might be at issue. It had even been decided to avoid carrying simple 'Medicalert' information in case an error in that information might prejudice a patient's care. The PAS's infrastructure would cause quite enough problems without attempting to take on the whole clinical record system for a very large hospital.

The initial consultations with the DoH indicated the need for a feasibility study and the hospital immediately put work in hand to explore the necessary issues. In anticipation of this work, the hospital had already set up a Computer Executive to oversee the hospital's computer activities and which reported directly to the House Governor (Chief Executive). The executive consisted initially of five members:

- The Deputy House Governor, Michael Fairey – Chairman and *primus inter pares.*
- Professor Robert Cohen – Consultant Physician.
- Maureen Scholes – Senior Assistant Matron.
- William (Bud) Abbott – Director of the Computer Unit.
- Barry Barber – Director of the Operational Research Unit.

This group were well known within the hospital community and had participated extensively in a wide variety of projects over the years that were relevant to an understanding of how the hospital functioned and to the development of a successful computer system. They had between them over 100 years of experience at The London Hospital and knew a lot about how it worked in personal terms. It was the function of the doctor, nurse and administrator to establish priorities within their own professional areas and to advise their colleagues on the Executive of those requirements. In addition, it was their responsibility to interpret the development of the computer project to their professional colleagues within the hospital and to the computer team leaders as they developed the various aspects of the new system. The belief was that these professional responsibilities could not be delegated to junior doctors, nurses or administrators but were the vital province of top management; a view that was unfashionable at the time but which would not be challenged now. However, it did impose considerable burdens on these senior staff because there was no diminution of their other professional activities.

Initially, the Computer Executive was responsible for formulating the general strategic approach to developing a real-time PAS for the hospital and for carrying out the necessary detailed consultations and tests to ensure that the strategy was acceptable and feasible. As the project developed, the Executive interacted with the computer team leaders as they submitted reports on the following key stages in the implementation of various computer applications:

- Initial study of existing application.
- Outline of proposed computer application.
- Detail of proposed computer application.
- Implementation of computer application.

The Executive had to satisfy itself that the description of the existing application was correct and accurate and that the proposals were feasible and acceptable and then, finally, that the application had been implemented successfully and was running smoothly. There was, additionally, a small representative medical committee to support Professor Cohen in providing medical advice to the Computer Executive in respect of any contentious issues.

In order to develop the feasibility study a number of critical decisions had to be addressed. Some of which were highlighted by the experience of the KCHMRP.[49] The key issues that were identified were those of project scope, application priorities, availability of the system and the basic user interface.

Pilot studies versus operational systems

It was clear that no amount of work on pilot studies would enable the replacement of any of the existing hospital systems. There were thus no obvious financial savings. It was also thought that successful pilot studies on one set of wards did not necessarily imply that similar success would be achieved on another set of wards. However, it was clear that it was not possible to tackle all wards and all applications at one time, even if the computing resources were unlimited.

Decisions in this area quite clearly affected the amount of hardware that was required, as well as the nature of the software that would have to be written. These concepts were explored at length and were thought of in graphical terms as the 'horizontal development' of a few applications in all wards, versus the 'vertical development' of a wide variety of applications in a few wards. Unlike the two-ward initial development of the KCHMRP, The London Hospital team decided to attempt a development in all wards. This immediately established a requirement for at least 47 VDUs as compared with the eight deployed at KCHMRP. The cost of these terminals immediately became a critical concern and it sharpened the issues of computer response time that had already emerged as a problem at KCHMRP.

This, in turn, focused attention on the fundamental speed of the computer, as well as on the ability of the software to take advantage of that speed. The concept was that each application would be layered onto previous applications because the earlier applications had bedded down and were functioning properly. Using this modular approach it was important that agreed applications should be frozen and not be continually modified in the way that had been possible with the batch systems on the Elliott 803. In principle, each application could affect all the other applications and extensive testing was required before each application was brought on stream. This effort could be wasted if other changes were being continually introduced that might have the effect of bringing the whole system down.

Priorities

The initial objective and equipment was for:

- Model 1(a): handling in-patient administration; clinical laboratories' investigations and reporting; blood transfusions and infection control; some specialist medical records; research.
- Model 1(b): booking theatre time; theatre and anaesthetic records; nursing administration and duty scheduling; Intensive Therapy Unit (ITU) records.

- Model 2: outpatient appointments; outpatient records.
- Model 3: further medical records as required clinically; additional departments such as Diagnostic X-ray and Pathology, Radioactive Isotope Laboratory, Physics, Radiotherapy, Cardiac, EEG; facilities for analogue information from electronic instruments and monitoring equipment.

Data input and access

The WIMP (windows, icons, menus and pointing device) approach has provided almost universal access to computer systems, so that it is difficult now to remember the problems that were encountered in seeking to get non-computer professionals to use computer systems to input their information and to access it subsequently. Navigational techniques, using pull-down menus, are now taken for granted and many professionals type information into their computers quite routinely. Even when they do not have touch-typing skills, their fingers can usually keep up with their thoughts as they write their papers and reports.

In 1968, this was all in the future. Typing was considered as a skill reserved for secretaries and beneath the attention of serious health professionals. The complexities of computer programming had been restricted to systems programmers and scientists who had begun to unleash the power that some high-level programming expertise provided for their scientific research and statistical studies.

It was quite evident from the experience with the Elliott 803 that most doctors and nurses could not be relied on to enter information in coded form, so it was vital that some other means was found and tested. The so-called tree-branching approach was similar to the pull-down menu approach but it was important to find out if it was acceptable to the staff and whether the note taking could be completed in an acceptable time. One possibility was a Plessey VDU incorporating wires that could be labelled by software and used for menu selection by touching with a finger tip. This process could fulfil the project's requirements. The concepts were tested using a Plessey VDU linked into a Micro 16 computer, which provided a fast response to touch-wire selections. We found that the interface could work with non-computer staff provided the system response was under a second. This was important because it was vital that the time to effect the transaction should not be longer than would be required for a manual system. Detailed examination of the transactions later showed that an average transaction would require eight VDU selections to complete.

The touch-wire system in use for air traffic control at West Drayton had great possibilities as these touch-wires could be labelled with different menu selections under software control. However, unexpectedly this approach failed in the hospital testing because some older members of staff had finger tremors which played havoc with the software.

BARRY BARBER

This finger tremor might have been resolved simply by asking staff to press the touch-wires firmly but in the event the Plessey touch-wire screens were not included in any of the final tenders. The workable solution was found to be fast menu selection, but we recognized that the minimum amount of typing required in some instances would normally have to be handled by specially trained clerical staff.

Twenty-four hour operation

In the transition from batch systems to real-time systems we recognized that the computer transaction had to be handled within the normal time span of the manual transaction if it were to be effective. If this could not be done, for any reason, then the transaction would have to be recorded on paper for subsequent input, processing or response to the doctor, nurse or patient. Real-time systems were definitely time critical. It was recognized that the system would have to be made available as close to twenty-four hours a day as was possible once it was put into serious use. The Executive was also aware that the systems and applications programmers needed access to the computer in order to test, debug and generally maintain their programs. Generally, this would require a duplex system, which was not on offer from the DoH. These access problems had already caused difficulties at KCHMRP.

Key decisions taken by the Executive

Taken together these logical, key decisions raised serious issues for the hardware and software requirements. The DoH accepted the feasibility study and work progressed on the development of the invitation to tender on which suppliers could quote. This was completed a year and a day after the feasibility study.[57, 58] This time was taken up in exploring the facilities available from various suppliers and in sorting out the estimated file sizes and likely peak transaction rates. After detailed examination it was estimated that the combined Model 1 system, when fully implemented, would be subjected to 1.44 transactions per second at peak hours. It would require 10.3 menu formats to be accessed per second and that this would require 2.27 main file accesses per second. These figures allowed for average peak-time transaction rates with an appropriate peak to mean ratio so that the system did not let staff down just when they were at their most busy. In modern terms, the Executive were trying to address the issue of scalability. If the project were successful, the hospital would take charge of a system that could continue to meet its needs as new applications were implemented.

During the process of examining possible systems and exploring the degree to which they could meet the hospital's needs, the Computer Executive indicated that the DoH funding would not be sufficient to meet the full demands of the Model 1 system. However, it believed that it would be a great opportunity for particular manufacturers to join forces with the team at The London Hospital in order to get in on the beginning of a major new market that was being developed. The Executive also made the point that their brief was to get the best equipment for the specified job and that they were not being restricted to a 'Buy British' policy.

In developing the calculations for the transaction rates and the formats required in the tree-branching menu process some care was given to ensuring that the facilities were fully adequate for the task. It was always expected that the hardware and software requirements would escalate during the programming phase and this would lead to disaster if the original estimates had been too tightly drawn.

The format of the Invitation to Tender was outlined by the DoH and, in particular, it divided the system requirements into Minimum Requirements, and Desirable Features (reproduced below). The hospital was helped by Scicon in translating its requirements into a verifiable specification.

THE INVITATION TO TENDER: MINIMUM SYSTEM REQUIREMENTS

- The total cost must be less than £400,000 installed.
- The system must be able to sustain at least 46 VDUs and eight teletypewriters in real-time operation at a peak rate of one enquiry a second for ten minutes. The system must be capable of processing the batch load which will always be run as background to the real-time work.
- The response time under peak loading should not be worse than 95 per cent of responses in under three seconds, from depression of the send key to display of the last character of the new frame at the terminal.
- The system must be capable of operating continuously, except for emergency repairs and scheduled maintenance. The real-time system must be available at least 98 per cent of the time during any month and maintenance requiring this system to be inoperative must be carried out between 23.00 and 07.00 hours.
- A multi-access operating system that will make efficient and economic use of core and computing power.
- Fast direct access storage of less than 20 milliseconds access time for systems residence, communications queuing and VDU formats – or some acceptable alternative.
- The following items of hardware must be included in the configuration offered:
  - Hardware memory protection
  - Direct access storage – at least 100 million characters
  - Real-time clock
  - 47 VDUs with full alphanumeric keyboard
  - Eight teletypewriters
  - Line printer
  - Three magnetic tape transports

- ◆ Interface for clinical laboratory computer
- ◆ Card and paper-tape input
- The company must guarantee maintenance of the equipment while it is being used at The London Hospital.

REFERENCE 58

© Royal London Hospital Archive. Courtesy of Royal London Hospital Archives.

THE INVITATION TO TENDER: DESIRABLE FEATURES

- The real-time system should not break down more than twice a month and in general the repair times should be within the following ranges:
  - ◆ 5 minutes in peak hours (09.00–12.00 and 14.00–17.00 Monday to Friday)
  - ◆ 10 minutes in basic working day (07.00–23.00, except peak hours)
  - ◆ 30 minutes in low-level and emergency period (23.00–07.00)
  - ◆ The system should never be inoperative continuously for more than 24 hours
- Response time to display standard formats should be better than one second. The response to confirm file updating can be rather slower – say, better than three seconds. The aim should be to provide this service 99 per cent of the time.
- A linked programmable processor for terminal handling.
- Back-up on electromechanical devices to enhance reliability of the real-time system.
- Additional equipment:
  - ◆ Extra magnetic tape transport
  - ◆ Fast printing equipment to replace two of the teletype terminals.
- On-site expansibility of the system for Models 2 and 3.
- Ability to link further satellite computers.
- Provision of supporting software:
  - ◆ Test and linkage re real-time program modules.

REFERENCE 58

© Royal London Hospital Archive. Courtesy of Royal London Hospital Archives.

The suppliers' responses were outlined in a detailed and standardized questionnaire covering all the matters thought relevant to the successful development of the project. In addition to any demonstration that the supplier thought helpful, the suppliers' responses detailed a simulation exercise that could help demonstrate how their system would work under various transaction loads and the build-up to saturation.

Evaluation of tenders

We had detailed discussions with a number of manufacturers to assess how best their systems might handle the hospital's requirements, after which we received quotations from:

- IBM for a 360/40;
- ICL for a 1905E running the George 3 operating system;
- UNIVAC for a 418/III with 4.25 ms fast access drums and Fastrand II;
- Honeywell for a Cluster of three minicomputers.

The first step was to ascertain that the systems offered met the minimum requirements set out in the Invitation to Tender. The Board of Governors had retained Scicon to help with the evaluation of tenders as well as to help prepare the tender documents. However, a hospital team under the Chief Systems Analyst and the Chief Programmer worked independently from the Scicon team on the evaluation process. It soon became clear that none of the four tenders received by the deadline met the minimum requirements, nor did the suppliers provide sufficient information or assurances to enable their tenders to be fully assessed.

After extensive discussions and visits to installations in the UK and the USA, some suppliers wished to modify their bids. All the suppliers were allowed to re-tender and it was these second bids that were finally evaluated.

When the Computer Executive began its own evaluation it had a considerable amount of information available from the DoH and the DTI, as well as information from different users of systems, similar to those being offered in the bids, from the UK and the USA. The Executive had to carry out cost comparisons on a uniform basis including software programming support, machine facilities and maintenance costs – which were aligned over a seven-year period.

The next step was the verification of the systems capacity to handle the work listed in terms of handling the minimum requirements. Unfortunately, no supplier met these requirements unequivocally and the considered view of the hospital's Computer Executive was that the answers to some of these requirements varied from a clear 'yes' through 'probable', 'possible', 'doubtful' to an outright 'no'. The desirable features were similarly assessed, but these produced more definite 'no's'.

Some suppliers had made special claims about the speed with which established programmers could learn to use their languages and operating systems. Before the tendering process had been completed, an experienced programmer was allocated to each supplier to explore how the tree-branching approach would work in each supplier's systems environment. These programming trials were very revealing as the hospital gained further knowledge of the various operating systems on offer. Also, one supplier was clearly unable to achieve the required response times with the operating system offered.

Three of the four suppliers carried out simulation runs on the basis of the exercise included in the Invitation to Tender. Again, these exercises proved valuable to the hospital staff, highlighting the functioning of the operating system in ways that would not otherwise have become apparent. In one case the supplier who had a bad programming trial had a bad simulation exercise and it was eventually proved that the facilities tendered could not run the hospital's system. As a result of all this work it was clear that two tenders could not meet the minimum requirements. After this stage, the questionnaires provided very little help in finalizing the decision between the remaining suppliers. Most of the additional information had to be extracted directly from the suppliers once its significance had been established.

Finally, the Scicon and the hospital teams allocated weights to the principal characteristics of the systems that made them important in implementing the real-time system that had been designed, and then assessed each tender in respect of these characteristics. The Computer Executive found very similar results from the two teams and then set about establishing its own judgement of the issues concerned, weighing the detailed arguments put forward for the weights and assessments. When set out in this fashion, one supplier appeared to have an advantage.

Following that, the system costs were combined with the technical assessment to provide a cost-effectiveness assessment and the leading supplier's advantage increased, particularly when the long-term costs were considered. A cost value assessment was developed using a technique described by Joslin.[59] The desirable features were costed and deducted from the system cost to deduce a notional cost for the system including the minimum requirements only. This exercise again showed the advantage of the leading supplier. By these various techniques, the Executive attempted to explore the stability of the choice of supplier and it became convinced that the UNIVAC 418/III was the best choice for implementing the system that had been designed.

The Executive informed the DoH of their decision and, after further discussions between the DoH, the DTI and the hospital, their choice was endorsed, thus providing the hospital with the best chance of making the system work within a live hospital environment. The system was eventually installed in March 1971.

The machine had been developed from the UNIVAC 418/II, which had been used for communications and message switching. The core was organized into three banks having separate processors with a total of 80k 18-bit words, with a 750 ns cycle time, and the system allowed for simultaneous processing between different core banks and the separate input/output and command/arithmetic processors. The UNIVAC Real Time Operating System (RTOS) was basic and it was necessary for the hospital to write a lot of special purpose middleware in order to run the system. UNIVAC provided programming support to help in this process and it was noted that programmers accustomed to the Elliott 803 machine code found the transition to the UNIVAC assembler language fairly easy. Additional software included COBOL and Fortran compilers as well as standard statistical and mathematical packages and a simulation package (GASP).

By comparison with the KCHMRP project, The London Hospital had a reasonable number of VDUs – even though they were monochrome, not colour screens – an adequate backing store in the Fastrand II and some very fast 'Floating Head' drums to handle the tree-branching. The computer was established in a newly built John Ellicott Centre. The Centre was named after John Ellicott, the chairman of the Hospital's House Committee of 1757, who was largely responsible for the petition leading to the grant of the royal charter. He was a member of The Royal Society and a renowned clockmaker, mathematician and astronomer.

The new configuration was quite a contrast to the original Elliott 803 installation, which was located in a basement next door to the sexually transmitted diseases clinics. Having obtained the facilities, the hard work was about to begin – it was now up to the team to 'do the business'. A second UNIVAC 418/III was purchased in January 1972 to take over the finance system from the, by then, aging Elliott computers. This provided back-up for the first machine and was as near to achieving a duplex system as was possible at the time.

THE LONDON HOSPITAL'S REAL-TIME PATIENT ADMINISTRATION SYSTEM

As described above, the first battle in implementing the outline PAS following the work on the Elliott 803 was to understand the different requirements of a real-time system. The second battle was to ensure that the facilities obtained were sufficient to deliver these requirements across the hospital as a whole. The final battle was to deliver the PAS as outlined by the Executive and by John Rowson.[60, 61] Throughout the work on this project the Computer Executive had solid support from the Board of Governors and the senior management. The chairman, Sir Harry Moore, consulted some of the computer staff at Hill Samuels where he was a director and provided the Executive with a copy of B.J. Brocks' paper on why computer projects fail.[62] Sir Harry advised us that he did not expect his hospital computer project to fail for any of the reasons that Brocks had identified in his paper. There may be other, previously unknown, reasons for failure but we should make sure that we addressed all the known problems in a serious fashion. It is interesting, re-reading that paper after some 30 years, to realize how many projects had failed in the intervening years as a result of problems that Brocks had so clearly identified.

The key components of the PAS, as they were listed in 1983 by The John Ellicott Centre at The London Hospital,[63] were as follows:

- Waiting list system
  - data entry
  - amendment

- ◆ deletion
- ◆ transfers
- ◆ inspection
- ◆ printouts
- ◆ statistical analyses
- ◆ system housekeeping
- In-patient system
  - ◆ admissions
  - ◆ discharges
  - ◆ transfers
  - ◆ inspections
  - ◆ amendments
  - ◆ printouts
  - ◆ entry reversals
- Booked list system
  - ◆ data entry
  - ◆ inspection
  - ◆ amendment
  - ◆ appointments
  - ◆ call from waiting list
  - ◆ deletion
  - ◆ cancellations
  - ◆ printouts
- Inter-ward transfers system
  - ◆ inspection – including avoidable transfers (lodging)
  - ◆ cumulative monthly summaries
  - ◆ overnight summarization
- Hospital activity analysis system
  - ◆ local and national analyses
- Bedstate confirmation system
- Patient index
  - ◆ inspection
  - ◆ amendment
  - ◆ update
- Patient registration (medical)
- Patient identification labels
- Dental services.

The Pathology system included the following:

- Laboratory requesting
  - requesting tests from manual forms
  - requesting from wards and lists of patients 'to come in'
  - printing request forms
- Laboratory procedures
  - check-in requests
  - logsheets of tests
  - worksheets by test categories
  - results entry – manual
  - results entry – automatic
  - quality control
  - results approval
- Test reporting
  - individual reports
  - cumulative reports
- Viewing results
  - from patient index
  - from work category lists
  - from wards via bedstate lists
- Record keeping
  - workload statistics
  - microfiche records in date order by patient
  - cross infection analysis
  - archiving for subsequent computer analysis
  - weekly summary printouts by destination as back-up.

The X-ray system included the following:

- Online X-ray requesting
- Patient appointments
- Patient arrivals
- Requests in progress
- X-ray reporting
  - dictating direct to secretary entering report into the system
  - dictating indirectly to cassette for subsequent system entry and radiologist approval
- X-ray report approval
- Report printing
- Statistics.

The Nursing Personnel System included the following:

- Creating nursing personnel record
- Record inspection
- Enter or change record items – recording details of changes
- Changing payroll or staff numbers
- Enter staff allocations – updating the nurses history
- Enter leave
- Make staff allocations via the 'due to move' list
- Enter learner set details – common pre-planned activities
- Delete set record
- Request printed output
- Request print of an individual's Kardex record
- Inspect or edit a location file
- Demonstration or training system.

A comparison with the Invitation to Tender shows that the major difference between the original intentions and the final outcome was that the Theatre Booking and Record System was not developed at that time.

At an early stage the Computer Executive provided a whole-day training and listening session in the Princess Alexandra School of Nursing for all consultants, senior nurses and heads of departments. Some 20 or so sessions were provided to enable staff to understand some of the key issues of computer hardware, computer software and its development, and system selection, as well as an outline of the proposed real-time system and the initial applications that would run on it.

The discussion periods enabled the Executive to answer questions and secure advice from their colleagues about the proposed system and the best way of proceeding.[64, 65] These discussions helped shape the project so that it would fit in most comfortably with the hospital's way of doing things. Professor Cohen's Medical Advisory Group also provided useful input. These were two-way processes in that when advice had been given and was followed by the Executive, our clinical and administrative colleagues would accept the results. The system designers could not keep going round and round in circles on key issues. Another part of this process was the development of appropriate training for the staff who would actually be using the systems. A general-purpose training course was devised and supplemented by a demonstration system that provided an introduction to the tree-branching approach to using the system.

Each VDU had a 'home display' that essentially outlined the services available at that terminal. It was always possible to return to the home display if a new user got lost among the branches of the tree. In general, the services available at a terminal were only those that were required at that location and staff could not gain access to patient information that they would not previously have had access to. For instance, each ward only had information about its own, and its paired ward's patients. The Executive had decided that the confidentiality of the system should not be worse than the previous manual arrangements but that the confidentiality arrangements should not make the use of the system unreasonably difficult.[66] The computer terminals were under the control of the ward or laboratory staff in the same way as the patients' case notes or laboratory equipment.

A great deal of thought was given to the best way of handling inter-ward transfers as it affected the change of access to the patient's records. In the end it was thought that the records were best 'pulled in' by the receiving ward rather than being 'pushed out' by the sending ward. It was felt that this would work best because there was a greater incentive for the receiving ward to take action whereas the other ward would only take action when it was convenient – especially as the delay provided them with a spare bed that was not listed! It was not long before some enterprising medical students found that they could 'pull in' records of their nursing friends in the Nurses Sick Room. However, at that time the information on the system was limited and the staff records were not available to be handled this way so no details could be obtained. Nevertheless, it presented the medical students with a problem because they could not 'push the records back again'! The traces of their efforts led to a serious warning about computer misuse being added on the system and the problem went away.

BARRY BARBER

In each case, following the approval of the Computer Executive for the development of any aspect of the computer system, a senior systems analyst would study the existing systems that were to be computerized and develop a substantial systems report describing how that system was functioning. Each report would be carefully reviewed by the Executive to ensure that the analyst had fully understood the situation and that the report fully reflected this understanding. This process of examination and questioning could take some time while the analyst went back to carry out further investigations. Having experienced, or participated in many studies of various aspects of the hospital's operation, the Executive were quite knowledgeable about how the various hospital systems functioned. However, it appeared from time to time that things had changed since a particular member of the Executive had last been involved in some of these activities. At one stage it was discovered that nurses were requesting some laboratory tests. In some cases because this had been requested by a doctor who was busy and at other times because the senior nurse knew what was required. This was a surprise to our clinical colleagues but the computer teams needed to know who should be trained in the new requesting system. The Executive did not wish to stop a useful process nor did it wish to countenance a wholesale transfer of duties from one group of staff to another. The final compromise was for the consultants on each ward to establish protocols for what routine tests were required for what types of condition and to authorize trained nurses and third-year nursing students to order tests within the context of the established protocols.

After a systems report had been accepted, the analyst developed a systems proposal for the computer-based approach to carry out the same functions with suggested additions that might make the computer system more useful. This report would, again, be examined carefully by the Computer Executive in conjunction with appropriate colleagues to ensure that any misunderstandings were first located and then eradicated. After the system had been agreed, the system was 'frozen' and no further changes and modifications would be allowed – unless some dangerous emergency surfaced that demanded immediate remedial action. This was important because throughout the process of systems implementation, new ideas were always emerging that would make the computer system more useful but any attempt to keep changing and modifying a real-time system would lead to instability and chaos. The new ideas had to be held over until the system had been implemented and stabilized – and there were resources to enable the modifications to be made and tested. The system was then developed along the lines that the Executive had agreed and it was tested section by section and finally was ready for live running. The staff were trained in the various facilities of the applications that were being introduced and then it was implemented with the minimum of duplicate running – a process known as 'sudden death'.

Following the implementation, and while the new system was bedded in, the analyst produced an Implementation Report which would highlight:

- remedial action that had had to be taken;
- changes that needed to be made to refine the system when the next modifications were being considered;
- suggestions for future developments of that application; as well as
- any lessons that had been learned for the future development of the system as a whole.

> An anaesthetist decided to 'improve' the new computer system by inventing inaccurate and potentially dangerous information in order to see if anyone noticed. Not only was it noticed, but the computer log showed the exact time it was entered and it was traced to him. We had always feared medical student pranks but not this shame-faced consultant.
>
> **BARRY BARBER**

The Computer Executive made decisions about modifications to existing computer applications and the introduction of new applications on the basis of these Implementation Reports and on the responses of and informal contacts with other hospital staff.

The Executive had a firm policy of not publishing intentions but only publishing accomplishments. Hence, it was not until about the time of the 1974 *MEDINFO* in Sweden that some of this material became publicly available.[43, 64, 67–69]

The first paper outlined how the hospital intended to evaluate the benefits of its new systems and two years later it set out an examination of the key decisions taken by the Executive at an IRIA conference in Toulouse.[70] It was clear that:

- the implementation of a real-time system was more formidable than anticipated and transcription errors were not automatically eliminated, but that such systems did have enormous advantages:

  - The speed and access to information, restricted only by the Medical Advisory Committee's advice on confidentiality, had substantially improved – 77 per cent of reports within two seconds and 90 per cent within three seconds.

  - The information system was more complete than the previous manual system – waiting list 22 per cent increase in record completeness and 44 per cent increase in the number of items per patient. The information on incomplete laboratory request forms was weighted according to importance, and the system improved from 7.5 per cent to 0.65 per cent.

  - The information was legible and hence many tasks no longer had to be handled by clerical and technical staff who were specialized in the range of activities carried out in a specific department. After a little training the computer system could be handled by those who could read English, thus allowing greater flexibility in the deployment of clerical staff.

  - The crude transcription error rate in the Waiting List System was 2.23 per cent, the missed entry rate was 0.58 per cent and the number of forms with at least one error or missing item was 33 per cent. Although these figures were unexpectedly high, it was believed that they were a considerable improvement on the previous manual system and most of the errors were not of clinical significance.

- although the hospital was anxious to have the system operational as close to 24 hours per day as possible and the computer had been operational continuously since 1972, it was not possible initially to have the real-time hospital system operational for longer than 8 am to midnight. The remaining time was required for preventive maintenance, housekeeping, file protection (back-up), batch processing on real-time files and system set-up for the next day. At that time about 17 per cent of the systems activity occurred outside the traditional prime shift of 9 am to 5 pm. The peak hourly transaction rates then were 225, compared with 45 between 7 pm and 10 pm, and 30 between 10 pm and midnight. These figures clearly showed that the system was being effectively used outside of the prime shift. The problem with having the system offline at any time was that users reverted to some form of manual system, the results of which might not be subsequently entered into the real-time system. The hospital had originally stated that the real-time system required full duplex facilities but the DoH were not able to fund this requirement. However, the hospital moved in this direction when it bought a second UNIVAC 418/III to carry out its batch-processing needs on the financial system. Unfortunately, it was not possible to justify the cost of installing a fully duplex system, so there were severe limitations on some of the applications testing that could be undertaken on the batch machine. Before the implementation of the real-time system, the rate of progress was limited to the available staff time to develop the system. However, once the system had become operational, the amount of testing time available on the computer plummeted and staff time was taken up in the maintenance of the real-time application. These twin effects had a substantial impact on the speed with which other applications could be developed and implemented;

- the success of the project showed that doctors, nurses and clerks were perfectly capable of handling the system with its menu and tree-branching approach and that the facilities were well accepted. This approach has become standard but it was not then known how acceptable it would become, however, no one had believed dedicated computer operators would have been feasible or desirable;

- the educational programme required to achieve success was considered a potential problem but in fact the approach adopted worked well. The staff training programmes included dummy patient records that could be used for visitor presentations as well. The programme included:

  - user involvement at the design stage
  - a computer induction programme for all
  - specific education for those using particular applications
  - the availabilty of a demonstration of the proposed system
  - specialist supervision at the time of implementation
  - a user handbook[71]
  - a follow-up programme to ensure user familiarity with the system
  - adequate user system support by telephone and otherwise – computer ward rounds were useful when new systems were implemented and
  - a staff induction programme to ensure the proper training of staff new to an application area;

- the selection of the applications for implementation had worked well, they were acceptable to the users and they provided the infrastructure for future development. The key decisions taken by the Executive provided the basis for this achievement; the system was implemented hospital wide and it was used by the ordinary hospital staff as just another innovation in a health service that was always innovating in terms of patient care.

The hospital was frequently asked to show visitors how the system was working and this was always most convincing when it was in practical use on the wards, but it was difficult to find times and places where the inconvenience to staff was kept to a minimum. It was perhaps a key evaluation moment when a group of visitors, who were being shown the system, were asked by a house officer to move away from the computer terminal because he had patients to care for.

Unfortunately, we never managed to persuade the DoH that the system should be replicated elsewhere and thus spread the development costs across many sites outside Tower Hamlets District. By comparison the team led by Professor Albert Bakker succeeded in establishing a good system at the Academic Hospital in Leiden,[72–74] and then in developing a non-profit organization, BAZIS. This was owned by the hospitals in The Netherlands and was able to install systems in many hospitals. In 1979, BAZIS had systems in six hospitals and later they had over 60 per cent of the market in The Netherlands.[75–77] This was definitely an opportunity that the DoH Experimental Real-Time Computer Programme and the DoH totally failed to grasp. The DoH had undertaken an important experiment at an important juncture in computing systems development, but it failed fully to capitalize on its success.

None of the special approaches by the Computer Executive to take the hospital staff along with the Executive would have been of the slightest interest if Bud Abbott had not been capable of understanding and carrying through the technical work that was required to make it all function as planned. He displayed the highest talents as a leader: he was able to choose capable colleagues and then inspire them with his understanding of the work to be done and the problems to be overcome. He picked the key team leaders in 1968: they stayed with the hospital developing the systems and keeping them running until they were pushed out in the late 1990s, just as the NHS was beginning to recognize the importance of information management. Throughout the previous three decades, treasurers across the NHS had been worried about recruiting computer specialists because their expertise was nationally in short supply. All of the computer staff at The London Hospital could have secured much more highly paid jobs in industry, but they cared enough to help the hospital develop a pioneering system and to keep it running and developing. The senior staff were:

John Rowson – Director following Bud Abbott's move to the North East Thames RHA

Tony Wills – Chief Systems Analyst following Peter Locke's move to the North East Thames RHA

Theo Brueton – Chief Software Programmer

Deidre Gillon – Senior Analyst/Programmer who had been recruited even earlier for work on the Elliott 803.

What happened to everyone from the Computer Executive?

Mike Fairey moved to the Region, and then to the DoH, then back to The Royal London and the Medical College.

Professor Robert D Cohen undertook serious medical studies after the computer abberation.

Maureen Scholes continued her interest until she retired.

William (Bud) Abbott established the Regional Computer Centre and much else.

Barry Barber moved to the Region to develop more operational research, and then moved to data protection and security with the NHS Executive.

PATIENT ADMINISTRATION IN OTHER EXPERIMENTAL REAL-TIME SYSTEMS

Despite concerns raised at intervals by the Parliamentary Accounts Committee, it is interesting to note that Don White, the Head of the DoH team spearheading the experimental programme, was considering the issues of unionization and the growing dependency of hospitals on their computer systems. This was as early as 1982, showing clearly that in many ways the programme had fulfilled its task of kick-starting health care computing in this important area of real-time systems development. Don White's concern foreshadowed information warfare concerns some two decades later, when the dependency of hospitals on their computer systems had become even more evident. His initiatives in health care computing enabled the NHS to explore techniques, opportunities and issues long before they became critical to the delivery of care. However, a great opportunity was lost to the NHS because so little of this work and expertise was transferred directly into non-profit NHS systems development.

When the time came for PASs to be implemented across the NHS, recourse was frequently to large US suppliers – the ICL–Stoke PAS was perhaps the most notable exception. The DoH initiated a major effort in systems evaluation in order to select the most useful applications for the technology, but the ground had already moved beneath their feet as it became clear that appropriate systems were becoming essential to the functioning of biochemical chemistry laboratories, Medical Physics departments and, in due course, to the functioning of hospitals. The evaluate, develop and implement approach had been overtaken by the speed requirements of the NHS.

Patient administration became the hallmark of many of the DoH Experimental Computer Projects because it provided the basic infrastructure on which other systems could be built. The second DoH project was located at North Staffordshire Hospital, Stoke-on-Trent, and the ICL System 4/50 was installed in September 1971, just six months after the UNIVAC at The London Hospital. The focus of the Stoke Project was on its large outpatient clinics, giving a useful development of the PAS into this area. In the summer of 1967, the feasibility study for the project was assigned to the hospital services section of English Electric Computers, which eventually became the medical area of ICL. This followed an earlier design study in the outpatient department. This work was outlined by Lawton from the ICL team,[78] and it gave a clear idea of the issues facing a hospital group that had no previous computing experience. The study involved a 'conventional' initial systems analysis in the three key areas identified by the hospital group as problems: communications; scheduling of resources; and the use of the clinical record.

In addition, they undertook an extensive series of discussions with senior staff on the potential benefits of computer systems. These discussions were extended to many other staff and were developed into a structured education programme at three levels with: appreciation level sessions to enable hospital staff to understand what was going on and to be able to help when they encounter computer staff; preparation level sessions with staff who would be affected by computer applications, and detailed level sessions with senior staff who would evaluate the detailed proposals.

The four main initial systems were the outpatient system, the in-patient system, the medical records system and the service department system. The major advance that the project was making was the development of the PAS into the outpatient area. The medical records system was concerned with the management of the medical case notes rather than the replacement of them. Throughout, the emphasis was on the team work needed between the outside computer staff and the hospital staff. The former had to learn about the operation of the hospital group and the latter had to learn about computers and how they could be made to help with those operations. Like The London Hospital Project, the objective was to set up an operational system, not simply to pilot what might be done. The Stoke system went on to be the basis of the ICL PAS that was adopted by a number of regions following the Körner Reports.

The suppliers were beginning to examine how their products could be used in this field and to see how they could develop their systems to address the health care needs that were being established.[79–81] In this context Ted Coles was one of the first clinicians to obtain a computing qualification to help in these exciting developments.

ICL was the prime contractor for much of the initial interest in health care computing, following the amalgamation of ICT and English Electric – which gave the company the problem of reconciling the System 4 computers and the 1900 series computers.

The Government's 'Buy British' policy gave ICL an important edge in tendering processes. It encouraged the company to invest seriously in resources in order to understand the nature of the health care market and to address the problems that were being encountered there. However, the tendering process was open and some of the experimental projects were quite clear about the difficulty of the tasks being undertaken and the need to secure the most appropriate equipment to address those tasks. They were also clear that the resources being made available were insufficient to secure 24-hour operation and to address the scale of development required. Suppliers were urged to include as much equipment and software support as possible, to provide a 'showcase' for their systems and thus secure more orders.

The ICL systems at King's College Hospital, the North Staffordshire Royal Infirmary at Stoke, and the Exeter Project linked to a Health Centre at Ottery St Mary, were followed by UNIVAC systems at The Royal London Hospital and the Queen Elizabeth Hospital in Birmingham. The four coordinated hospitals' project: University College Hospital, St Thomas' Hospital and Charing Cross Hospital in London and Addenbrooke's Hospital, Cambridge, (nick-named The Four Horsemen) purchased Xerox systems. In the latter case, considerable effort was devoted to the issues of transferability and to utilization of the same or similar software.[82-88] As has been mentioned previously, representatives from the various computer projects met regularly to share experiences at the DoH under the chairmanship of Don White as the NHS Computing Services Group.

Other agencies and individuals became interested in exploring some of the issues that affected developments in this field, for instance, concerning the efficient use of computing resources, the problems of terminals to be used by patients, the syntax and handling of medical information and the software requirements. Alex D'Agapeyeff had a particular interest in the complex software requirements for handing real-time systems and was especially concerned that the NHS should recruit and retain professional programmers and analysts.[89] There were limits to what could be achieved by retraining existing staff and using students from the emerging computer degree courses. Handling complex real-time systems was a very demanding professional activity. At that time some hospital managers appeared to think that complex computing systems could be designed by students and amateurs, when it would never occur to them to ask a medical student to design a new hospital or to undertake the most complex surgery.

The next successful project was the system developed at The Queen Elizabeth Hospital (QEH), Birmingham, based on a UNIVAC 418/III. The computer was installed in March 1972 and there was useful collaboration in the early days with the team at The London Hospital. Again, the system was developed on the, by now standard, patient administration infrastructure,[90-92] but QEH moved quickly onto the clinical laboratory systems, with facilities for laboratory management as well as result reporting to the wards and other departments. These arrangements were made possible because one professor, Tom Whitehead, was already heavily involved in the automation of his laboratory and he already had computer support there.[93-95] In addition, the Birmingham team were developing nursing systems at a very early stage,[96-97] and had been using operational research techniques to explore issues of hospital resource usage and nursing dependency.[98]

The Exeter Community Health Project installed an ICL 1904A computer in July 1972. This was an exceedingly innovative project because it was based initially on a general practice group at Ottery St Mary and it focused on 'care in the community' rather than on the running of a major hospital. It provided the link between community care and hospital care. The project provided a comprehensive approach to community care that was at least a decade ahead of DoH thinking. It had links from general practice information to outpatient, in-patient, service department information and nursing information as a composite record. It was based on work carried out by Professor Ashford[99] and it was developed over many years under the leadership of Jack Sparrow and Hugh Fisher.[100, 101] The approach was based on their famous 'Hexagon' diagram, which sought to synthesize the requirements of hospital health information systems with the systems required in general practice and the community. From outside this project it always appeared that it was a great pity that it was not exploited much more extensively.

The approach to the development of the nursing systems in the Exeter Project was outlined by Alison Head.[102] Clearly, the issues of confidentiality were much more serious for the Exeter Project than for the other hospital projects because the information was no longer under the control of a single organization. Hugh Fisher addressed these problems seriously and head on;[103] these issues have arisen even more forcibly in the context of the DoH's current 'Computing for Health' programme.

The next project involved exploring the issue of how a small hospital might take advantage of the developing and increasingly powerful technology. The North East Thames RHA established a project in Southend Health District to address this approach,[104, 105] and other hospitals were also examining ways in which they might profit from the expertise that was being accumulated.[106, 107]

A number of projects were not supported by the DoH and have dropped out of view, but the preparatory work that was done nevertheless frequently contributed to a better understanding of the local health care organizations and resulted in subsequent initiatives within their regions. For instance, The Guy's–Essex Project provided useful input.[108] There was continuing evidence of the activity of the various projects as they worked their way through the problems of handling novel real-time systems and novel health care applications.

David Clark was heavily involved in medical education at Manchester and at one stage was still using an old IBM 7090 in the medical college. It was hoped that work in Manchester would lead to a DoH experimental project but, although this did not materialize, a lot of other productive activities did take place. Among other things work took place at the Monsall Isolation Hospital in Manchester[109] and at St Mary's Maternity Hospital in Manchester.

The DoH continued to develop their policies in respect of health care computing, regional computing services and standard systems.[110–112] Professor Robert Cohen examined the way that computer evaluation was being handled,[113] and Professor John Anderson continued developing his ideas about medical education and Computer Record Systems, despite the closure of the King's College Hospital Medical Records Project.[114]

A number of UK-trained staff helped in developments elsewhere, and Alan Thomas' work in New Zealand is an example of the way that this happened.[115] But many others were involved in the development of health information systems, either within the health care organization, or as part of the private sector: the Middle East was a favourite destination for a period.

THE IMPACT OF THE KÖRNER COMMITTEES IN DEVELOPING PATIENT ADMINISTRATION SYSTEMS

The next significant initiative involved the NHS/DoH Health Services Information Steering Group (Körner committees), which were concerned with the assessment of the information required to manage the various layers of the NHS. The policy was that, as far as possible, these data should be obtained as a by-product of normal operations rather than as a separate information collection exercise. These reports have been listed previously[13] but the conclusions were set out concisely by Alastair Mason and his colleagues.[116] The period during which the Körner work was going on was a very exciting one for collaborative activity, even though the DoH failed to capitalize on this work and subsequently established multitudinous reporting systems to address all sorts of targets. The Körner work provided a major impetus for the installation of PASs across the NHS to collect the minimum data-sets that had been established as the required information. Each of the NHS RHAs adopted its own standard PAS and Stan Sargeant reported on the implementation of the systems across the West Midlands RHA.[117] The South Western RHA review of the available systems was carried out by staff of the Exeter Project and it was an illuminating survey of the systems that were then available for implementation. It noted that:

> *Although many of the systems investigated originated in the NHS it was most disappointing to learn first hand just how little had been done in coordinating the activities of the various computing groups within the NHS. Overall the lack of cooperation and wasteful effort is the biggest lesson to be learnt by the NHS. Any future efforts devoted to patient care computing in the Region should attempt to remedy this situation.*

REFERENCE 118

A PAS based on the ICL system at Stoke-on-Trent became a major competitor in the negotiations for regional standardization, as well as systems based on the DEC–MUMPS combination used at University College Hospital in the North East Thames Region. It was unfortunate that, at the time that these systems were needed, the system at The London Hospital had been transferred from UNIVAC 418/III computers onto DEC VAX computers and it was under-powered and not acceptable for implementation across the

North East Thames RHA. A year later all the problems had been dealt with, but it was too late to change.

American systems were offered by British Medical Data Systems – a subsidiary of Shared Medical Systems – and IBM, but there were often problems in finding a low-cost entry point for districts or hospitals who were initially only interested in a low level of statistical service, rather than a full PAS. The Exeter Project was a clear contender for the development of community-based PASs, and it was a great pity that the DoH did not properly use the systems and expertise available there.

The DoH Experimental Real-Time Computer Programme was a far-sighted scheme and the expertise fed into numerous parts of the NHS but the DoH/NHS was not able to capitalize on the successes that had been so hard won. Some of the NHS expertise in these areas surfaced in a joint venture with the Institute of Health Service Management in developing a loose-leaf book on the management of health information.[119]

REFERENCES

1 The London Hospital (circa 1965) *Computer Facilities at The London Hospital*. Internal document, later updated.

2 Abbott, W. and Barber, B. (1966) *The London Hospital Computer Unit: Progress Report 1965*. Internal document: The London Hospital, January.

3 Abbott, W. and Barber, B. (1967) *The London Hospital Computer & Operational Research Units: Progress Report 1966*. Internal document: The London Hospital, January.

4 Abbott, W. and Barber, B. (1968) *The London Hospital Computer & Operational Research Units: Progress Report 1967*. Internal document: The London Hospital, April.

5 Green, W. Computers XIV: uses in medicine. *The Financial Times*, Monday 11 December 1967.

6 Maynard, D.E., Prior, P. and Scott, D.F. (1969) Device for continuous monitoring of cerebral activity in resuscitated patients. *BMJ*, 11, 545–6.

7 Prior, P.F. and Maynard, D.E. (1986) *Monitoring Cerebral Function: Long-term Monitoring of E.E.G and Evoked Potentials*. Elsevier, Amsterdam.

8 Tuck, C., Cundy, A.D., Wagman, M.C., Usherwood, M. and Thomas, M. (1976) The use of a computer in an obstetrics department. *British Journal of Obstetrics and Gynaecology*, 83.2 (February), 97–104.

9 Sheiham, A. (1968) The epidemiology of chronic peridontal disease in Western Nigerian school children. *J. Peridont. Res.*, 3, 257.

10 Seekington, I.M., Huntsman, R.G. and Jenkins, G.C. (1967) The serum folic acid levels of grass-fed and stabled horses. *Vet. Rec.*, 81, 158.

11 Benjamin, B. (1965) Hospital activity analysis. *The Hospital London*, 61, 221.

12   DoH (1969) *Hospital Activity Analysis.* NHS memorandum HM(69) 79.

13   Körner Committee (1982–85) Reports from the NHS/DoH Health Services Information Steering Group. First: *The collection and use of information about hospital clinical activity* (1982). Second: *The collection and use of information about patient transport services* (1984). Third: *The collection and use of information about health services manpower* (1984). Fourth: *The collection and use of information about hospital clinical activity in hospitals and the community* (1984). Fifth: *The collection and use of information about services for and in the community* (1984). Sixth: *The collection and use of information about health services finance* (1984). Supplement to the first and fourth reports: *The collection and use of information about maternity services* (1985). A report from the Confidentiality Working Group: *The protection and maintenance of confidentiality of patient and employee data* (1984). The interim report of the Dental Working Group: *The collection and use of information about dental services* (1985).

14   Mason, A. and Morrison, V. (eds) (1985) *Walk, Don't Run: A Collection of Essays on Information Issues.* Published to honour Mrs Edith Körner CBE, Chairman of the NHS/DHSS Health Services Information Steering Group 1980–84. King Edward's Hospital Fund for London, London. ISBN 0 197246 31 1.

15   Flynn, F.V. and Vernon, J. (1965) Cumulative reporting of chemical pathology. *J. Clin. Pathol.*, 5 Sept., 18(5), 678–83.

16   Flynn, F.V. (1966) Use of a computer by a clinical chemistry service. *Proc. R. Soc. Med.*, Aug, 59(8), 779–82.

17   Flynn, F.V., Piper, K.A. and Roberts, P.K. (1966) Equipment for linking the auto-analyser to an off-line computer. *J. Clin. Pathol.*, Nov, 9(6), 633–9.

18   Flynn, F.V. (1969) Problems and benefits of using a computer for laboratory data processing. *J. Clin. Pathol. Suppl. (R .Coll. Pathol.)*, 3, 62–73.

19   Flynn, F.V. (1971) Computer assistance in the chemical pathology laboratory. *Biochem. J.*, Jan, 121(1), 2P.

20   Barr, A. (1965) *Scheduling of Student Nurses with the Aid of a Computer.* Operational Research Unit, Report 7, Oxford Regional Hospital Board.

21   Barr, A. (1967) *Measurement of Nursing Care.* Operational Research Unit, Report 9, Oxford Regional Hospital Board.

22   (Authors unknown) (1966) *Matters arising from discussions on student nurses duty allocation.* Operational Research Unit Report 2. Internal document: The London Hospital, 17 March.

23   (Authors unknown) (1966) *Notes on the weekend spent in matron's office 26–27 February 1966.* Operational Research Unit Report 3. Internal document: The London Hospital, 17 March.

24 (Authors unknown) (date unknown) *Confidential memorandum following a discussion of the strategic problems facing the nursing administration.* Operational Research Unit Report. Internal document: The London Hospital.

25 (Authors unknown) (date unknown) Nurse staffing at unsocial hours. Operational Research Unit Report. Internal document: The London Hospital.

26 (Authors unknown) (1973) *The use of jacket microfiche for holding the records of former members of the nursing staff.* Internal document: The London Hospital, December.

27 (Authors unknown) (1975) Initial proposals for a computer-based nurse record system for learners. Internal document: The London Hospital.

28 Shah, A. (1976) *Nurse manpower planning.* Internal document: The London Hospital, November.

29 Shah, A. and Farrow, S.C. (1976) Pre-registration house appointments: A computer-aided allocation scheme. *Medical Education*, 10, 474–9.

30 Shah, A. (1979) A computer-aided interactive procedure to improve nurse-training programmes. *Medical Informatics*, 4, 4, 209–18.

31 The London Hospital (1965) *The Collection of Basic Data from Outpatient Clinics.* Medical Physics Dept, Report 1135. Internal document, December.

32 Barber, B. and Abbott, W. (1968) Computer monitoring of hospital outpatient clinics. *The Hospital*, October, 64, 345–9.

33 Sharratt, M. and Yare, D. (1972) Computer monitoring of hospital outpatient clinics: an assessment. *Hospital & Health Services Review*, November, 398–402.

34 Maddison, T.J. (1969) NHS computer equipment. In Abrams, M. (ed.) *Medical Computing: Progress and Problems*, pp. 195–9. Chatto & Windus for BCS, London.

35 Gedrych, D.A. (1969) The Department's policy and objectives. In Abrams, M. (ed.) *Medical Computing: Progress and Problems*, pp. 345–52. Chatto & Windus for BCS, London.

36 DHSS (1972) *Using Computers to Improve Health Services.* DHSS, London, November.

37 Caddick, M.T. and Lee, D.T. (1975) The planning and control of a health care computing policy. In Anderson, J. and Forsythe, J.M. (eds) *MEDINFO 74*, Stockholm, pp. 39–44. North-Holland, Amsterdam. ISBN 0 444107 71 1.

38 Handby, J.G. (1975) Successful design management of integrated terminal-based medical systems. In Anderson, J. and Forsythe, J.M. (eds) *MEDINFO 74*, Stockholm, pp. 79–83. North-Holland, Amsterdam. ISBN 0 444107 71 1.

39 Harding-Smith, R.H. (1975) Priorities for health service computing development. In Anderson, J. and Forsythe, J.M. (eds) *MEDINFO 74*, Stockholm, pp. 33–38. North-Holland, Amsterdam. ISBN 0 444107 71 1.

40 Ockenden, J.M. and Bodenham, K.E. (1970) *Focus on Medical Computing*. Scicon report commissioned by the Scottish Home & Health Dept. Oxford University Press for NPHT, Oxford.

41 Norwood, D. (1975) Economic evaluation of total hospital information systems. In Anderson, J. and Forsythe, J.M. (eds) *MEDINFO 74*, Stockholm, pp. 149–54. North-Holland, Amsterdam. ISBN 0 444107 71 1.

42 Sharpe, J. (1975) Towards a methodology for evaluating new uses for computers. In Anderson, J. and Forsythe, J.M. (eds) *MEDINFO 74*, Stockholm, pp. 137–43. North-Holland, Amsterdam. ISBN 0 444107 71 1.

43 Barber, B. (1975) The approach to the evaluation of The London Hospital computer project. In Anderson, J. and Forsythe, J.M. (eds) *MEDINFO 74*, Stockholm, pp. 155–165 and pp. 1011–12. North-Holland, Amsterdam. ISBN 0 444107 71 1.

44 Barber, B., Abbott, W. and Cundy, A. (1977) An approach to the evaluation of the Experimental Computer Programme in England. In Shires, D.B. and Wolf, H. (eds) *MEDINFO 77*, Toronto, pp. 913–16. North-Holland, Amsterdam.

45 Barber, B. and Cundy, A. (1980) Evaluating evaluation: some thoughts on a decade of evaluation in the NHS. In Lindberg, D.A.B. and Kaihara, S. (eds) *MEDINFO 80*, Tokyo, pp. 602–6. North-Holland, Amsterdam. ISBN 0 444860 29 0.

46 DoH (1977) *Interim Report on the Evaluation of the NHS Experimental Computer Programme*. DoH, London.

47 Cundy, A. and Nock. J.D. (1979) An assessment of the use of performance criteria in the evaluation of the NHS Experimental Computer Programme. In Barber, B., Grémy, F., Überla, K. and Wagner, G. (eds) *MIE, Berlin 79: Lecture Notes 5*, pp. 117–130. Springer Verlag, Berlin. ISBN 3 540095 47 7.

48 White, D. (1982) Medical computing and industrial relations. In O'Moore, R.R., Barber, B., Reichertz, P.L. and Roger, F. (eds) *MIE, Dublin 82: Lecture Notes 16*, pp. 692–9. Springer Verlag, Berlin. ISBN 0 387112 08 1.

49 Anderson, J. (1974) King's College Hospital computer system. In Collen, M.F. (ed.) *Hospital Computer Systems*, pp. 457–516. Wiley Bio-medical Health Publications, John Wiley & Sons, New York.

50 Opit, L.J. and Woodruffe, F.J. (1970) Computer held clinical record system. *BMJ*, 4, 76–9 and 80–2.

51 Opit, L.J. and Woodruffe, F.J. (1971) Hospital based real-time computing systems. *BMJ*, (issue unknown).

52 Opit, L.J. (1972) Long-range planning for what? – a strategy to find out. In Abrams, M.E. (ed.) *Spectrum 71: BCS Conference on Medical Computing*, pp. 173–81. Butterworths for BCS, London.

53 Anderson, J. (1972) A user's view of an experimental computer project at King's College Hospital. In Abrams, M.E. (ed.) *Spectrum 71: BCS Conference on Medical Computing*, pp. 47–56. Butterworths for BCS, London.

54 Anderson, J., Grémy, F. and Pagès, J.C. (1975) Educational requirements for medical informatics. In Anderson, J. and Forsythe, J.M. (eds) *MEDINFO 74*, Stockholm, pp. 207–11. North-Holland, Amsterdam. ISBN 0 444107 71 1.

55 Anderson, J. (1977) Education of health staff in information processing techniques. In Shires, D.B. and Wolf, H. (eds) *MEDINFO 77*, Toronto, pp. 975–7. North-Holland, Amsterdam.

56 Anderson, J. (1981) The systematic doctor. In Parslow, R.D. (ed.) *Information Technology for the Eighties, BCS 81*, pp. 159–70. Heyden & Sons for BCS.

57 The London Hospital Computer Executive. (1968) *Preliminary Study for the Installation of a New Computer*. Internal document: The London Hospital, 9 October.

58 The London Hospital Computer Executive (1969) *Invitation to Tender: The Board of Governors of The London Hospital Invite Tenders for a Real-Time Computer System, Installation of a New Computer*. Internal document: The London Hospital, 10 October.

59 Joslin, E.O. (1968) *Computer Selection*. Addison Wesley, Reading MA, USA.

60 Fairey, M.J., Scholes, M., Abbott, W., Kenny, D.J., Cohen, R.D. and Barber, B. (1974) *A Case Study in the Implementation of a Major Computer System*. Conferences at The London Hospital Medical College, 27 November 1973 & 24 April 1974, London. ISBN 0 444867 49 X.

61 Rowson, J.E.M. (1983) Implementation of a hospital information system. In van Bemmel, J.H.V., Ball, M.J. and Wigertz, O. (eds) *MEDINFO 83*, Amsterdam, pp. 94–101. North-Holland, Amsterdam.

62 Brocks, B.J. (1969) What went wrong? An analysis of mistakes in data processing management. *Accountancy*, September, 666–75.

63 The John Ellicott Centre (1983) *Patient Administration System*. The London Hospital, internal document.

64 Scholes, M. (1975) Education of health staff in computing. In Anderson, J. and Forsythe, J.M. (eds) *MEDINFO 74*, Stockholm, pp. 213–15. North-Holland, Amsterdam.

65 Scholes, M., Forster, K.V. and Gregg, T. (1977) Continuing education of health service staff in computing. In Online Conferences Ltd (eds) *MEDCOMP 77*, pp. 639–48. ISBN 0 903796 16 X.

66 Barber, B., Cohen, R.D., Kenny, D.J., Rowson, J.E.M. and Scholes, M. (1976) Some problems in confidentiality in medical computing. *J. Med. Ethics.*, 2, 71–3.

67 Kenny, D.J. (1975) Management tactics for the introduction of computers into health care units: the experience of The London Hospital. In

Anderson, J. and Forsythe, J.M. (eds). *MEDINFO 74*, Stockholm, pp. 127–31. North-Holland, Amsterdam. ISBN 0 444107 71 1.

68   Barber, B. (1975) Radiotherapy computer applications. In Anderson, J. and Forsythe, J.M. (eds) *MEDINFO 74*, Stockholm, pp. 801–5. North-Holland, Amsterdam. ISBN 0 444107 71 1.

69   Mace, D. (1978) A common approach to a variety of clinical laboratories. In Anderson, J. (ed.) *MIE, Cambridge 78: Lecture Notes 1*, pp. 509–19. Springer Verlag, Berlin. ISBN 3 540089 16 0.

70   Barber, B., Cohen, R.D. and Scholes, M. (1976) A review of The London Hospital computer project. In Laudet, M., Anderson, J. and Begon, E. (eds) *Proceedings of Medical Data Processing*, Toulouse, pp. 327–38. Taylor & Francis, London. ISBN 0 850661 06 4.

71   The London Hospital (1972) *A Guide to The London Hospital Computer System*. Unpublished.

72   Bakker, A.R. and Costers, L. (1977) Experiences with an implementation approach for a hospital information system. In Online Conferences Ltd (eds) *MEDCOMP 77*, pp. 177–86. ISBN 0 903796 16 X.

73   Bakker, A.R. (1977) Centralized versus decentralized hospital information systems. In Shires, D.B. and Wolf, H. (eds) *MEDINFO 77*, Toronto, pp. 895–9. North-Holland, Amsterdam.

74   Bakker, A.R. (1977) Implementation approach and evaluation of the use of Leyden University Hospital information system. In Shires, D.B. and Wolf, H. (eds) *MEDINFO 77*, Toronto, pp. 943–7. North-Holland, Amsterdam.

75   Bongers, A. and Kouwenberg, J.M.L. (1978) A large database on a minicomputer. In Anderson, J. (ed.) *MIE, Cambridge 78: Lecture Notes 1*, pp. 431–42. Springer Verlag, Berlin. ISBN 3 540089 16 0.

76   Bongers, A., Kouwenberg, J.M.L. and Bakker, A.R. (1979) Storage structure in a large database and an approach to multi-organization usage. In Barber, B., Grémy, F., Überla, K. and Wagner, G. (eds) *MIE, Berlin 79*, pp. 590–601. Springer Verlag, Berlin. ISBN 3 540095 49 7.

77   Bakker, A.R. (1982) Organization of a cooperation for further development and implementation of an integrated hospital information system. In O'Moore, R.R., Barber, B., Reichertz, P.L. and Roger, F. (eds) *MIE, Dublin 82: Lecture Notes 16*, pp. 14–20. Springer Verlag, Berlin.

78   Lawton, M.D. (1970) Systems design for a management-oriented hospital information system. In Abrams, M.E. (ed.) *Medical Computing: Progress and Problems*, pp. 358–73. Chatto & Windus for BCS, London.

79   Davis, A.E. (1969) Computers – an evolutionary force in the development of medical records. In Abrams, M.E. (ed.) *Medical Computing: Progress and Problems*, pp. 112–19. Chatto & Windus for BCS, London.

80   Coles, E.C. (1969) A computer system for maintaining on-going medical records for chronic conditions. In Abrams, M.E. (ed.) *Medical Computing: Progress and Problems*, pp. 102–11. Chatto & Windus for BCS, London.

81 Harding-Smith, R.H. (1975) Priorities for health service computing development. In Anderson, J. and Forsythe, J.M. (eds) *MEDINFO 74*, Stockholm, pp. 33–8. North-Holland, Amsterdam. ISBN 0 444107 71 1.

82 Hammersley, P., Roach, M.E., Robertson, V.S., Campbell, T. and Terry, H. (1977) Planning for transferability. In Online Conferences Ltd (eds) *MEDCOMP 77*, pp. 259–70. ISBN 0 903796 16 X.

83 Hammersley, P. (1972) Principles of control in patient administration. In (eds unknown) *Proceedings of Toulouse Journées d'Informatique Medicale Conference*, pp. 319–35. IRIA.

84 Baker, G.J., Gardner, S.W. and Gradwell, D.J.L. (1974) A database for four hospitals in the UK. In Anderson, J. and Forsythe, J.M. (eds) *MEDINFO 74*, Stockholm, pp. 323–7. North-Holland, Amsterdam.

85 Campbell, T., Terry, H. and Roach, M.E. (1976) Development and implementation of medical applications as part of a co-ordinated project. In Laudet, M., Anderson, J. and Begon, E. (eds) *Proceedings of Medical Data Processing*, Toulouse, pp. 371–80. Taylor & Francis, London. ISBN 0 850661 06 4.

86 Bryant, Y.M. and Bryant, J.R. (1977) Towards an integrated system for nursing administration. In Online Conferences Ltd (eds) *MEDCOMP 77*, pp. 439–51. ISBN 0 903796 16 X.

87 Hammersley, P. (1977) A review of hospital auxiliary systems. In Shires, D.B. and Wolf, H. (eds) *MEDINFO 77*, Toronto, pp. 171–8. North-Holland, Amsterdam.

88 Hammersley, P., Roach, M.E., Robertson, V.S., Campbell, T. and Terry, H. Planning for transferability. In Online Conferences Ltd (eds) *MEDCOMP 77*, pp. 259–70. ISBN 0 903796 16 X.

89 D'Agapeyeff, A. and Hunter, D.G.N. (1969) Software in hospitals. In Abrams, M.E. (ed.) *Medical Computing: Progress and Problems*, pp. 203–10. Chatto & Windus for BCS, London.

90 Hills, P.M. (1969) The objectives and design philosophy of the Real-Time Computer Project at The Queen Elizabeth Medical Centre. In Abrams, M.E. (ed.) *Medical Computing: Progress and Problems*, pp. 375–83. Chatto & Windus for BCS, London.

91 Hills, P.M. (1978) The objectives and design philosophy of the Real-Time Computer Project at The Queen Elizabeth Medical Centre. In Anderson, J. (ed.) *MIE, Cambridge 78: Lecture Notes 1*, pp. 385–91. Springer Verlag, Berlin. ISBN 3 540089 16 0.

92 Sergeant, S. (1978) Clinical laboratory systems. In Anderson, J. (ed.) *MIE, Cambridge 78: Lecture Notes 1*, pp. 497–507. Springer Verlag, Berlin. ISBN 3 540089 16 0.

93 Whitehead, T.P. (1965) *Progress in Medical Computing*. Elliott Medical Automation, pp. 52–66.

94   Whitehead, T.P., Becker, J.F. and Peters, M. (1968) (title unknown) In McLachlan, G. and Shegog, R.A. (eds) *Computers in the Service of Medicine vol. 1*, pp.115–? Oxford University Press for the Nuffield Provincial Hospitals Trust, Oxford.

95   Whitehead, T.P. (1969) The computer in the laboratory. *Practitioner*, September; 203, 215, 294–505.

96   Ashton, C.C. (1976) A developing nursing record system. *Sperry-UNIVAC Symposium on Computer Applications in the Field of Medicine*, Rome, 19–21 October.

97   Ashton, C.C. and Bryant, Y.M. (1979) A review of nursing systems in the UK. In Barber, B., Grémy, F., Überla, K. and Wagner, G. (eds) *MIE, Berlin 79: Lecture Notes 5*, pp. 207–18. Springer Verlag, Berlin.

98   Rhys Hearn, C. (1971) The more fruitful uses of hospital resources. In Abrams, M.E. (ed.) *Spectrum 71: BCS Conference on Medical Computing*, pp. 57–74. Butterworths for BCS, London.

99   Ashford, J.R. (1969) The patient record in a community-based medical information system. In Abrams, M.E. (ed.) *Medical Computing: Progress and Problems*, pp. 86–93. Chatto & Windus for BCS, London.

100   Clarke, D.J., Fisher, R.H. and Ling, G. (1979) Computer-held patient record: the Exeter Project. *Information Privacy*, vol. 1, March, 64–71.

101   Fisher, H. (1984) The role of FPS computing with regard to the problems of problems of patient care as a whole – and the role of the Exeter FPS in FPS computing. In Kostrewski, B. (ed.) *Current Perspectives in Health Computing*, pp. 221–9. Cambridge University Press for BCS, Cambridge.

102   Head, A. (1977) Maintaining the nursing record with the aid of a computer. In Online Conferences Ltd (eds) *MEDCOMP 77*, pp. 469–83. ISBN 0 903796 16 X.

103   Fisher, H. (1978) Rules for confidentiality – practical experience in a community-based real-time system. In Anderson, J. (ed.) *MIE, Cambridge 78: Lecture Notes 1*, pp. 363–71. Springer Verlag, Berlin. ISBN 3 540089 16 0.

104   Kennedy, T.C.S. and Moss, N.B. (1975) A computer-assisted clerical system for management of hospital waiting lists. In Anderson, J. and Forsythe, J.M. (eds) *MEDINFO 74*, Stockholm, pp. 497–502. North-Holland, Amsterdam. ISBN 0 444107 71 0.

105   Bell, P.C. and Moss, N.B. (1979) Lessons from six years of using an interpretive language on a minicomputer to run a hospital in-patient management system. In Barber, B., Grémy, F., Überla, K. and Wagner, G. (eds) *MIE, Berlin 79: Lecture Notes 5*, pp. 907–23. Springer Verlag, Berlin.

106   Farrer, J.A., Harvey, P.W. and Roberts, J. (1977) Problems arising from the classification in ICD on non-fatal and minor conditions. In Shires, D.B. and Wolf, H. (eds) *MEDINFO 77*, Toronto, pp. 289–92. North-Holland, Amsterdam.

107 Roberts, J., Brook, S. and Broadey, K. (1982) The development of integrated computer facilities within a District Health Authority. In O'Moore, R.R., Barber, B., Reichertz, P.L. and Roger, F. (eds) *MIE, Dublin 82: Lecture Notes 16*, pp. 387–93. Springer Verlag, Berlin.

108 Bowden, K.F. and MacCallum, I.R. (1969) A computer-based approach towards an integrated patient record. In Abrams, M.E. (ed.) *Medical Computing: Progress and Problems*, pp. 94–101. Chatto & Windus for BCS, London.

109 Richards, B. and Heyworth, B. (1970) *Monsall Hospital Annual Report: 1969*. UMIST, Manchester.

110 Bishop, P.J. (1984) Review of current NHS computing policy. In Kostrewski, B. (ed.) *Current Perspectives in Health Computing*, pp. 7–15. Cambridge University Press for BCS, Cambridge.

111 Abbott, W.C. (1977) The Regional Computing Service subsequent to the NHS re-organisation. In Online Conferences Ltd (eds) *MEDCOMP 77*, pp. 279–87. ISBN 0 903796 16 X.

112 Maddison, T.J. (1977) The development of standard National Health Service systems in England. In Online Conferences Ltd (eds) *MEDCOMP 77*, pp. 247–58. ISBN 0 903796 16 X.

113 Cohen, R.D. (1979) Evaluation of computer systems in medicine. In Barber, B., Grémy, F., Überla, K. and Wagner, G. (eds) *MIE, Berlin 79: Lecture Notes 5*, pp. 931–7. Springer Verlag, Berlin. ISBN 3 540095 4 97.

114 Anderson, J. (1984) Clinical records system: an overview. In Kostrewski, B. (ed.) *Current Perspectives in Health Computing*, pp. 107–114. Cambridge University Press for BCS, Cambridge.

115 Thomas, A.W. (1975) Experience with an on-line Patient Information System. In Anderson, J. and Forsythe, J.M. (eds) *MEDINFO 74, Stockholm*, pp. 391–7. North-Holland, Amsterdam. ISBN 0 444107 71 1.

116 Mason, A., Annesley, P., Ashley, J., Cottrell, K. and Wainwright, L. (1982) Information about patients in hospital: English recommendations for a national minimum data set. In O'Moore, R.R., Barber, B., Reichertz, P.L. and Roger, F. (eds) *MIE, Dublin 82: Lecture Notes 16*, pp. 51–5. Springer Verlag, Berlin.

117 Sargeant, S. (1984) The implementation of PAS by the West Midlands RHA. In Kostrewski, B. (ed.) *Current Perspectives in Health Computing*, pp. 23–31. Cambridge University Press for BCS, Cambridge.

118 Fisher, R.H., Greenslade, G.R., Kumpel, Z., Eades, H.D. and Rafferty, J.A. (1981) *Patient Administration Systems Available to the NHS*. South Western RHA.

119 Abbott, W., Barber, B. and Peel, V. (eds) (1986–93) *Information Management in Health Care*. Issue 20. Longmans (initially Kluwer) for IHSM, London. Updatable text.

5 National initiatives

MIKE FAIREY

This chapter deals with the emergence of computing into the field of health (initially that of health management) in the early 1960s, and the reaction of central government to that, for the time, startling innovation. It goes on to chart the national initiatives that emerged in England and Wales in the decades that followed, and the way in which those initiatives interacted with current, frequently changing, views on how the NHS should be managed. It does not deal with developments in Scotland during the same period. Many developments mirrored their English counterparts but Scotland did not, in the first instance, favour real-time computing.

EARLY REACTIONS

The initial reaction of the DoH to computing efforts within the NHS was one of considerable caution: this was an untrodden path, one to be negotiated with great care. It was felt that there should be a conscious attempt to avoid duplication of effort and the applications pursued should, if at all possible, be in areas that had been tried and tested elsewhere. The Manchester Regional Hospital Board's (RHB) proposal to build a payroll system for all staff within the region, and to offer that system for use by any other RHB that chose to use it admirably fitted the Ministry's cautious approach. It thus became the first truly national approach, that of the RHB-based large standard system. Starting in 1961, by 1971:

> ... most RHBs (had) successfully developed computer systems for a large number of central administrative tasks that (were) particularly amenable to solution through electronic data processing techniques. Extensive coverage (had) been achieved in stores accounts, suppliers accounts, and financial and cost accounts .... Of the various statistical applications, Hospital Activity Analysis of course stands out, since it covers all regions.
>
> REFERENCE 1

In 1970–71, there were 11 RHB computer installations, employing 561 staff, at a total cost of £1.25 million.[1]

There were of course drawbacks to this approach. Users, in most cases with justification, complained about the length of time that it took to amend the standard programs or to design extensions to them. Senior RHB staff resented the influence or, as they saw it, the control exercised by what had now become the Department of Health and Social Security. The

Computer Policy Standardization Committee, chaired and organized by the department, was able to exercise a considerable influence on the process, not least because of the Department's potential to oversee all NHS-funded activity. The resentment that this sometimes caused was a theme that was to rumble on and was not to be resolved until a decade later.

DOH COMPUTING

The DoH itself had responsibility for a number of countrywide functions, all involving the manipulation of very large quantities of data. The caution that it exhibited, in relation to the NHS, was mirrored in its approach to these functions. Until 1978, the Prescription Pricing Agency, responsible for reimbursing pharmacists and appliance contractors for their services, relied on manual methods: in that year, 2,400 staff were employed to price 240 million prescriptions a year. The decision to computerize the process had dramatic results: by 1996, 1,200 staff were processing 480 million prescriptions from 13,000 dispensing contractors, and by 2003 the number of prescriptions handled exceeded 600 million. As an integral by-product of this process it has been possible to produce valuable data for the better management of prescribing.

For the Dental Estimates Board (DEB, now the Dental Practices Board) the task, though smaller in size, was very similar. From handling some five million claims in 1949, the number had risen to some 20 million in 1968 and in excess of 30 million in 1978. At this point the DEB turned to computerization and, with the ensuing rise in productivity, was able by 1991 to process in excess of 47 million claims.

THE EXPERIMENTAL COMPUTER PROGRAMME

It was against this background – cautious, centralist, in some respects authoritarian – that in 1968 the DoH surprisingly embarked on a far-sighted experimental programme:

> ... *with the objectives of determining what role computers should play in the future in the NHS, aimed at:*

> 1 *discovering whether how and where computers can help to:*
>
>   *(a) improve patient care;*
>
>   *(b) improve clinical and administrative efficiency;*
>
>   *(c) provide facilities for management and research, and*
>
> 2 *giving some NHS staff practical experience with computers.*

REFERENCE 1

The programme, in some ways astonishing in its breadth, covered 14 major projects:[1]

King's College Hospital

Stoke (Birmingham RHB)

The London Hospital

United Oxford Hospitals

United Birmingham Hospitals

Exeter (South Western RHB)

United Liverpool Hospitals

United Manchester Hospitals

Guy's Hospital/Essex University

Hammersmith/Kensington (South West Metropolitan RHB)

Coordinated Project

> Charing Cross Hospital
>
> United Cambridge Hospitals
>
> St Thomas' Hospital
>
> Univerity College Hospital

Eighteen smaller projects were funded partly or completely by the department:[1]

Bristol Royal Infirmary

Hammersmith Hospital

King's College Hospital

Poole General Hospital

Queen Elizabeth Medical Centre , Birmingham

Sheffield Royal Infirmary

South Warwickshire Group Pathology Laboratories

St Stephen's Hospital

University College Hospital (×2)

Leicester Royal Infirmary

Bristol Radiotherapy Centre

Newcastle General Hospital

Royal Marsden Hospital

St Bartholomew's Hospital (×2)

Westminster Hospital

Thurrock Hospital, Essex

The larger projects, costing between £100,000 and £280,000 a year, employed a total of 163 people in March 1971, and planned to employ 366 by 1974–75. The total cost of the programme in 1970–71 was £670,000: the planned cost for 1974–75 was to be £2.5 million. Surprisingly, there were no false illusions:

> *... the process of determining just what role computing should play in health care will continue to be lengthy and probably very expensive ... some people think that, even in 10 or 15 years, computing will benefit health care only marginally. Experience to date, here and abroad, has shown a large number of failures, and very few successes ... and outnumbering both ... are projects on which trial must continue for a considerable time before they can be judged as successes or failures.*

REFERENCE 1

Looking back from 30 years on, the breadth of the Experimental Computer Programme seems remarkable. At the time it was positively visionary, a vision for which Don White, the assistant secretary in charge of the project, was largely responsible. The 14 major projects ranged from an attempt to computerize the medical record (King's College Hospital), through pioneering versions of whole hospital information systems including order communications and reporting (The London Hospital, Queen Elizabeth's Hospital, Birmingham), a prototype patient information system that later became widely used nationally (Stoke), to a brave attempt to achieve standardization through multi-centre cooperation in hardware and software (St Thomas', Addenbrooke's, Charing Cross, University College Hospital).

As perhaps was to be expected, in a programme as broad as the Experimental Computer Programme, the individual projects achieved varying degrees of success. Despite a considerable body of work at project level, no formal evaluation of the overall programme was ever attempted, although the DoH published an interim report in 1977.[2] It noted that projects had not been implemented as quickly as had been expected, and that evaluation, from its innovatory state, was proving difficult. Despite that:

> *... many of those problems [of implementation] ... were understandable in view of the innovatory nature of the programme, and most projects now have operational systems playing an important part in the day-to-day activities of patient care .... It has been established that computer systems perform various information handling activities competently, quickly and accurately .... Better information is available ... but the value of this information in its effect on NHS management decisions, and in the management of individual patients, or groups of patients has not been fully determined .... Enough work has however been done to confirm that if large complex computer systems are to offer positive economic returns, these must come from significant changes in organization made possible by the availability of up-to-date accurate and relevant information to managers.*

REFERENCE 2

This balanced and cautious view was not echoed by the Parliamentary Committee on Accounts (PAC). In their examination of the Experimental Computer Programme, conducted in the 1976–77 session, they castigated the DoH for the programme's apparent lack of results, and for the fact that, despite its avowedly experimental nature, not every project had succeeded. This early reaction of the PAC to health computing was a harbinger of much that was to come later, not always to the benefit of the NHS.

THE 1974 REORGANIZATION AND THE THREE CHAIRMEN

In the turmoil that followed the 1974 NHS reorganization, there was initially a considerable uneasiness in the relationship between the Department (particularly ministers) and Regional Health Authorities (RHAs). As one of the measures to improve matters, ministers bravely agreed to allow regional chairmen to review how the Department worked. The resulting report proposed a number of measures that were accepted,[3] although its most radical – the creation of the NHS Management Executive (NHSME) – had to wait until 1984 for implementation. Among the recommendations that were accepted rather earlier was the proposal to review NHS information, its collection and its usage:

> ... *DHSS and NHS should study what data is collected for DHSS by the Service: the use to which it is put: and what can be altogether abandoned.*
>
> REFERENCE 3

The NHS/DoH Steering Group on Health Services Information (more easily recognizable as the Körner Committee) began work in February 1980. By March 1985 it had published five reports analysing data items collected nationally across the service: it proposed that data to be collected should relate to district needs, and should only be derived from operational data; and that as far as possible data required nationally should also be drawn from the same source. Both the NHS and ministers accepted these proposals, and agreed that the new data-sets, the Körner Minimum Data Sets, should supercede existing data systems throughout the NHS from April 1987.

The resulting transition was not entirely easy, though it was reasonably complete by early 1989, and was to prove a valuable base for work to come in the 1991 reorganization of the NHS.

One further sign of the improving relationship between RHAs and the DoH, and one also deriving from the overall approach of the Three Chairmen's Review, was the decision in 1980 to reduce the Department's responsibility for standard systems and to pass that responsibility to regional chairmen. That work was undertaken on behalf of the regions by the Computer Policy Committee, chaired by a Regional Chairman, with an entirely NHS secretariat. It started a systematic attempt both to extend the role of standard systems, and, of much greater importance, to improve the relationship between those running and those using standard systems. Ironically, the transition from Departmental to regional responsibility for these important systems did not entirely allay the suspicions of users at operational level, thus demonstrating one of the less happy features of NHS culture.

THE IMPACT OF GENERAL MANAGEMENT

As the work of the Körner Committee drew to a close, the NHS initiated the introduction of general management, a move which at national level involved the creation of the NHSME. An executive-level Director, drawn from the NHS, was given responsibility for NHS computing and planning. The Körner Committee and the Computer Policy Committee were first combined and then replaced by an Information Advisory Group that reported to the Director of Planning and Information Technology, and to the NHSME. The branches of the DoH dealing with the Hospital and Community Health Services computing and information, the NHS Centre for Information Technology and the NHS corporate Data Administration, which had supported the Computer Policy Committee, were drawn together in the fruitful alliance of the Information Management Group (IMG), a cheerful and practical hybrid to which, over the years, gravitated a number of computing talents from across the NHS.

The first practical achievement of the IMG was the publication in 1986 of the *National Strategic Framework for Information Management in the Hospital and Community Health Services.*[4] Drawing on existing plans, and based on widespread consultation, the framework set out clearly the management framework of the NHS, and the management systems necessary to operate it.

> *The key to success of these management systems lies in the more effective use of improved information ... information systems must meet two equally important needs; the provision of support to clinicians, nurses and other staff in their day-to-day work; and the supply of valid and flexible management information, wherever possible as a by-product of these systems. Plans and strategies which address information issues must form part of, and underpin, plans and strategies for providing health care. In short, information must be managed.*
>
> REFERENCE 4

This was the first unambiguous statement of the critical role of information in the management of the NHS, a theme which became more and more insistent as time progressed. The framework set out what ministers expected of the NHS, the precise objectives and the agreed plans for tasks to be carried out centrally (together with a timetable), the scope and format for regional information strategies, and the range of agreed common standards:

> *There will be no reversion to the 'standard systems' route for one, nationally prescribed, technical solution for each application area.*
>
> REFERENCE 4

From this solid base of clearly stated policy, the IMG was able to begin a range of projects that were to have a profound impact on the future expansion of information technology and management in the NHS. The Hospital Information Support Systems project aimed to show how such systems could bring about significant changes in hospital operation, particularly when linked with concepts such as resource management. The Common Basic Specification project set out to provide an easily accessible model of health care provision that authorities across the country could draw on when designing their own computing systems, thus offering the potential for greatly reduced design times. The need for a standardized coding structure led to the purchase for the NHS of the award-winning Read Coding structure, thus sparking considerable interest in the process across a wide range of specialties.

Running in parallel with these developments, the NHSME Finance Directorate embarked on the Resource Management Initiative (see Chapter 6), an ambitious and necessary project designed to draw doctors and nurses into the overall management process. Spending £300 million over the period 1987–91, it aimed to draw together existing sources, and where necessary create new sources, to create a Casemix package and thus provide clinicians with practical management information on which to plan and monitor their performance. The concept was an admirable complement to the work of the Information Directorate and was of considerable significance for the changes ahead. One of the most successful projects, led by Frank Burns, was that at the Wirral, which went on to become one of the most advanced sites in the country.

Spreading the impact of this wide range of projects across the NHS would in any circumstances be a major undertaking, needing a period of organizational stability. This, however, was not to be. The Conservative administration was determined to bring market forces to bear on the NHS, and initiated a radical restructuring that aimed at the separation of commissioners and providers of health care. The achievement of such a goal, however, required a major rethink of data flows within the NHS, to support the political concept of funds following the patient. The resulting work not only showed how the successful operation of the NHS now depended on accurate and speedy information flows, but also laid out the conceptual foundations for the future.

FRAMEWORK FOR INFORMATION SYSTEMS

The Next Steps set out in detail the measures to be taken by field authorities and by the centre in order to meet the demanding timetable for the 1991 reorganization.[5] In addition, however, it also set out the longer and wider picture into which the short-term measures fitted. 1992 saw the publication of *An Information Management and Technology Strategy for England* that set out the work needed to achieve the wider picture.[6] The strategy was based on five key principles:

- Information would be person-based.
- Systems would be integrated.
- Information would be derived from operational systems.

- Information would be secure and confidential.
- Information would be shared across the NHS.

From those principles was derived the national work programme – the creation of a new NHS numbering system uniquely to identify individuals and to enable the move towards electronic patient records, the development of NHS administrative registers, the development of NHS-wide networking, particularly to create a secure electronic network, and the establishment of GP and Health Authority links for the exchange of useful administrative information. Coupled with these projects, an Information Management and Technology (IM&T) training programme for local implementation was to be developed.

The change to a Socialist administration in 1997 saw the end of a formal attempt to run the NHS on market lines, but the distinction between provider and commissioner – with its attendant demands on the creation and flow of information – remained and, if anything, became more marked. Setting out its stall in *The New NHS: Modern and Dependable*,[7] the new administration began a lengthy series of organizational changes designed to achieve those ends. Amongst those changes was the abolition of the IMG (which had quite undeservedly achieved bogey status amongst some Socialist MPs) and its replacement by a central Information Policy Unit (in effect, the commissioner) and a NHS Information Authority (the provider).

Despite the organizational turmoil, the overall vision set out in *The Next Steps*, and *The IM&T Strategy for England*, together with their associated work programmes held good, and indeed became even more vital. *Information for Health: An Information Strategy for the Modern NHS 1998–2005: A National Strategy for Local Implementation* built on the 1992 strategy retaining many of its major elements,[8] for example, the five key principles. One of its cornerstones was to propose the creation of the electronic health record, combining individual electronic patient records from all providers, and which:

> . . . is eventually universally accessible and which records the health care of individuals throughout their life.

REFERENCE 8

It also proposed connecting computerized general practices to NHSnet, implementing the NHS Tracing Service, and opening a National Electronic Library for Health. Most importantly, the strategy recognized the uneven base of IT achievement across the country, and proposed a step-by-step approach to the targets that it contained.

The publication of *The NHS Plan* in 2000[9] emphasized just how important the role of information and its secure transferability was to be in delivering health care for the future. Indeed, without it, the plan was virtually unachievable. It was therefore not surprising that the political spotlight should now fall, not as in the past in endless and usually uninformed criticism of the programme, but on a serious and determined effort to make it happen.

> *There is much to do now in bridging the gap between the NHS now and a service shaped around the needs and preferences of individuals. There is an urgency to put workable and person centred systems and solutions throughout the NHS and to enable links with social services.*

REFERENCE 10

One major explanation previously offered by health authorities for the slow take-up of proven IT systems was the lack of funding. From 1999 onwards, considerable and increasing sums of money were made available centrally to meet this problem. At first those sums were not hypothecated, and the demands of other Government targets led many authorities to use the funds for other purposes. Faced with this reluctance at authority level to take on IT expansion, the Government took a firmer line. In 2002, *Delivering 21st Century IT Support for the NHS* took a much more centralist approach.[11] There would be increased funding for IT, targeted on critical national services to be progressed nationally across the NHS; improved central direction and performance management for IT; streamlined procurement; closer working partnership with NHS IT suppliers; and a more corporate approach that was to include clear national standards and specifications for IT functionality.

COMMON THEMES

From the brief review above of national policies and initiatives relating to information technology and management over nearly 40 years, a number of themes emerge. Perhaps the most striking is the way in which the central attitude to computing has varied. Regarded at first as a potentially expensive activity that needed to be strictly controlled, a brief moment of enlightenment allowed the Experimental Computer Programme to go ahead. A number of substantial gains emerged from that programme, as did a number of salutary lessons. Though the champions of that scheme were under no illusion as to the difficulty, and potentially lengthy time span, that might be involved, the Parliamentary Committee on Accounts was unimpressed. Central policy turned once again to control – a control it should be noted that related to process, rather than to purpose. The end of the 1970s, and with an easing of relationships between the department and RHAs, saw another relaxation with RHA chairmen in control of the operation of standard systems.

The introduction of general management in 1984 and, in particular, the creation of the NHSME saw a major change in approach. Management systems and information systems were firmly linked and acknowledged as the major element in the running of the NHS. Devolution of operational responsibility in the context of agreed plans, and within an agreed structure of national policies, was the basis on which the NHS was to operate. This major advance on philosophies that had gone before was one to which both the field and the centre took time to adapt, but with increasingly beneficial results. Politicians, however, were impatient, and the 1991 changes brought to an untimely end a genuinely enlightened attempt to manage the NHS in a sensibly devolved manner. Ironically, the introduction of the market concept brought even more dramatically to the fore both the need for accurate information and its critical dependence on information technology. The information plan for 1991 laid the foundation for everything that was to come thereafter.

A major change, such as that in 1991, could not but reassert an increased element of central control. So too, in the decade that followed, did the pronouncements of the Parliamentary Committee on Accounts on investigations into the procurement of the Wessex RHA's regional system, and the department's purchase of the Read Coding system. The resulting tightening of procurement procedures not only produced frustration at operational level but, of even greater importance, dramatically slowed the rate of IT growth across the NHS.

The 1997 change of administration, and a new NHS plan, once more emphasized the absolute dependence of the plan, for its achievement, on successful information systems. For possibly the first time, it became politically apparent that information systems across the NHS had to grow and to grow rapidly. When the injection of sizeable (but unhypothecated) additional funds proved to be ineffective, it became politically necessary to move to a further, and current, stage of central control over the introduction of essential systems.

A further feature of the events that have been described is the degree to which the need for a sound national infrastructure has been slowly realized and – on occasions grudgingly – funded. In political terms, and thus in the annual hunt for central funding, infrastructure is boring, dull, probably inexplicable to the public, and above all lengthy in achieving discernible results. It is enormously to the credit of the architects of the 1992 and 1998 Information Strategies that they enabled the foundation on which present strategies, which are far more visible, are able to operate.

The period that this chapter has reviewed has seen advances in technology that have made possible many of the projects about which the pioneers of medical computing could only dream. At the same time, although only too often lagging behind those advances, information technology has moved into a commanding position in both the operation and the management of the NHS. No longer a recondite pursuit for the cognoscenti, information systems lie at the heart of the NHS as it moves into the new century.

There are many problems still to be faced. Politicians are not alone in failing to understand both the power and, more particularly, the necessity of information systems if the NHS is to continue its record of service to the country. Many in the NHS have yet to learn the same lessons. The years immediately ahead have the potential for enormous advance – but only if carefully handled.

REFERENCES

1   DHSS (1971) *Using Computers to Improve Health Services: A Review for the NHS*. DHSS, London.

2   DHSS (1977) *Interim Report on the Evaluation of the National Health Service Experimental Computer Programme*. DHSS, London.

3   DHSS (1976) *Regional Chairmen's Enquiry into the Working of the DHSS in Relation to Regional Health Authorities*. DHSS, London.

4   DHSS (1986) *National Strategic Framework for Information Management in the Hospital and Community Health Services*. DHSS, London.

5   DoH (1990) *Framework for Information Systems: The Next Steps*. DoH, London.

6   DoH (1992) *An Information Management and Technology Strategy for England*. DoH, London.

7   DoH (1997) *The New NHS: Modern and Dependable*. DoH, London.

8   DoH (1998) *Information for Health: An Information Strategy for the Modern NHS 1998–2005: A National Strategy for Local Implementation*. DoH, London.

9   DoH (2000) *The NHS Plan*. DoH, London.

10 DoH (2001) *Building the Information Core: Implementing the NHS Plan*. DoH, London.

11 DoH (2002) *Delivering 21st Century IT Support for the NHS*. DoH, London.

6 Resource management

SHEILA BULLAS

The Resource Management Initiative (RMI), launched in 1986, was primarily an organizational development programme aimed at introducing clinicians formally into the management structures of acute hospitals. One of those structures, the Clinical Directorate, endures successfully into the twenty-first century. The RMI was a national initiative funded and developed by the DoH, drawing in expertise from NHS hospitals and from other public sector and commercial organizations. The concept emerged from hospitals and details were developed with six leading hospitals. RMI only applied to England although Wales, Scotland and Northern Ireland were doing similar things and experiences were shared.

It was believed that involvement of clinicians within devolved structures would only succeed if those involved had access to appropriate information to support them in their role. The Casemix Management System was designed principally to meet the needs of clinicians in devolved management roles.

The only reason for documenting a past programme is if there are lessons to be learnt for the present or future. The objectives for which the RMI was launched are still relevant today in an updated form. Clinical directorates and similar structures still need the information, as does the Payment by Results scheme. Diagnosis Related Groups (DRGs) have been replaced by Healthcare Resource Groups (HRGs), and service plans have been replaced by integrated service improvement plans, but in essence the information support is the same. And now, 20 years later, the technology has developed to the point where that support is achieved more easily.

The people involved in resource management (RM) are too numerous to mention by name. This does not diminish the valuable contribution they all made.

THE CONCEPT

The RMI crystallized, allegedly, during a discussion that Ian Mills, finance director of the NHS Management Executive and Professor (now Sir) Cyril Chantler had at a function at Guy's Hospital, where Professor Chantler had been involved in developing the medical directorate structure. The origins of the initiative were firmly in both the clinical and finance camps.

The objectives of the RMI were formally set out in a Health Notice in 1986.[1] The principal objective was to introduce a new approach to the management of resources and to demonstrate whether or not this resulted in measurable improvements in patient care. A concern at the time was the variable cost of treating similar patients in different parts of the country – there was often a fourfold difference. Despite the fragility of costing systems at the time, it was felt that this range of variability indicated widespread waste of scarce resources. It was also felt that such waste compromised the clinical care given, hence the view emerged that a reduction in this variability could lead to improvements in patient care. There was also much criticism at the time of poor management decisions in the allocation of resources; if only clinicians were more involved in these decisions, then the allocation of resources would be better balanced to achieve these improvements.

This gave rise to a number of subsidiary objectives, which were to:

- identify areas of waste and inefficiency;
- improve the management of resources by encouraging clinical group discussion and review;
- highlight areas that could most benefit from the input of resources;
- identify and expose the health care consequences of given financial policies and constraints; and
- understand the comparative costs of future health care options and hold informed debates about such options.

The hospitals themselves were encouraged to add local objectives to the programme. Support for the introduction of clinical audit programmes and improving clinical outcomes was a major addition by some sites.

The initiative involved the development of management structures led by clinicians (normally senior consultants) and supported by business managers (normally from the finance departments), and information support (normally from the local information department or team). There were various models, the most enduring of which has been the Clinical Directorate model. The teams were at the level of specialty (e.g. general surgery) or group of specialties.

In practice, the initiative required information that would allow clinicians to understand the resources (nursing, theatre, medications, diagnostics, bed days, etc.) used by individual and similar patients and to understand the cost, again at individual and group level.

The devolution of budgets for pathology tests and radiological investigations to these clinical teams was intended to encourage the financial decisions on resource use to be managed at the same level as clinicians. Ensuring clinical involvement in those decisions was designed to ensure that clinical priorities would be at the fore when taking resource and financial decisions.

BENEFITS

The key benefits obtainable by the development of clinical management units supported by information included:[2]

- For clinical and general managers:
  - informed decisions on the priorities for allocation of resources to maximize patient care and outcomes;
  - identification of variation in clinical practice as the basis for clinical improvements;
  - supporting the development and monitoring of operational plans;
  - predicting patient flow and resource demand;
  - earlier and more explicit warning of major deviations from planned patterns of care;
  - maximising income.
- For finance:
  - greater control over budgets and financial outcome.
- For information staff:
  - a move away from 'number crunching' to interpretation of information.

Many of these benefits are still sought today and are provided in some hospitals by data warehouses that are very similar to those specified by the RMI.

Pilot sites

Six acute hospitals were selected to pilot the *Doctors in Management* initiative and the information systems to support it. These were the sites that would develop the concept and make it a reality. The criteria were that they should provide models, big and small, have a good geographical distribution and, of course, have people with the enthusiasm, commitment and ability to do the job. The pilots were:

- Guy's Hospital, with Professor Cyril Chantler as the lead;
- The Royal Hampshire County Hospital, Winchester;
- Huddersfield Royal Infirmary, with Mr Peter Jackson, consultant obstetrician and gynaecologist as the lead;
- Freeman Hospital, Newcastle upon Tyne;
- Arrowe Park Hospital, Wirral, with Frank Burns in the lead; and
- Pilgrim Hospital, Boston, Lincolnshire.

These were true pilot sites and not the 'early adopters' now favoured. There were no preconceived ideas as to what the management structures should be or the information systems to support them. We had models for both from earlier work undertaken in the USA, New Zealand and Europe and the sites developed these to suit the circumstances that prevailed in the UK at the time.

Personal involvement

In 1988, I was working at the South West Thames Regional Computer Centre implementing patient administration systems and being frustrated by the endless releases of software with faults that made it impossible to deploy them operationally. The rounds of testing over a period of nine months ensured improvements, but were tedious in the extreme. As someone with an entry on her CV that reads 'gets bored easily', I was ready for a new challenge. It was against this background that I went to the *Health Computing* conference after my manager, Bill Hedges, broke his ankle and was unable to attend. It was at the conference in Brighton that I met Tim Scott while a group of us were putting the NHS to rights till late into the night. With a varied background as pathology technician, statistician, service planner and business systems analyst, our conversation led to an opportunity for me to participate in an exciting and innovative project and I joined the RM team later that year to lead on information systems.

Quality improvement

The work of Brent James at Intermountain Healthcare[3] in Utah informed the use of Casemix groupings to support clinical quality improvements. Brent and his team were one of the first to apply statistical process control to health care. Some research was undertaken as part of the RMI but lapsed when key stakeholders moved on.

This work challenged the perception of the time that improving clinical quality would inevitably result in cost increases. There was much evidence to show that 'getting it right first time' and avoiding clinical error, particularly in prescribing, resulted in increases in both the effectiveness and efficiency of clinical care.

It is encouraging to see that these ideas are being investigated again within the NHS.

DOCTORS IN MANAGEMENT

The Griffiths Report of 1983 expressed the view that clinicians should be 'involved more closely in the management process consistent with clinical freedom for clinical practice'.[4]

A lot of interest was shown in the medical directorate model of organizational design developed at Guy's Hospital. This model was medically led: the directorate had a medical leader. The membership of the directorate was predominantly medical with the medical head of each division as a member. The divisional director could be either appointed or elected, but major criteria to emerge were the need for the confidence and support of their peers and for this person to continue to undertake clinical sessions.

The RMI resulted in the set up of the British Association of Medical Managers (BAMM). BAMM exists to help clinical professionals to develop their leadership and management skills to improve care for patients. BAMM's Chief Executive is Dr Jenny Simpson OBE who championed the RMI at the Sheffield Children's Hospital.

CASEMIX MEASURES

The concept of Casemix underpinned the information systems. To understand the variation in the resources consumed by patients with similar conditions, we needed to allocate similar patients together into a manageable number of groups to compare them. We needed to be able to group them on the basis of data items that were readily available and easily collected. There are about 500 groups for in-patients, based on similar clinical responses and costs. Initial work focussed on hospital in-patients but groups were available for outpatients, A&E patients and others.

Robert (Bob) B. Fetter and his team in the USA had developed a grouping method (DRGs) in the 1960s as a means of categorizing the output of a hospital and measuring cost. Bob Fetter had freely distributed the definitions of his groups, both as hard copy and on disk, to anyone who was interested. As a result, DRGs were adopted across the world as a standard grouping method and were used for a variety of purposes by different professional groups.

Interest in DRGs in the UK originated from work on in-patient costs and the development of specialty costs by Feldstein.[5] This led to the use of average specialty costs to fund patient flows under the resource allocation (RAWP) formula.[6]

In the early 1980s, Dr Hugh Sanderson and Dr Graham Winyard were investigating the use of DRGs at the London School of Hygiene. Dr Winyard moved to Lewisham and North Southwark Health Authority, as district medical officer, where he met Tim Scott who was working as a financial planner in the Treasurers Department. Tim instigated a DRG user group that included people from Lewisham and North Southwark HA, London School of Hygiene, South East Thames RHA, CASPE (Clinical Accountability, Service Planning and Evaluation) and that had links to Bob Fetter's team at Yale. This quickly extended to people from St Thomas' Hospital, North West Thames statistics and information department, and the DoH Operational Research Service. Tim Scott moved to the DoH to head the RMI under director of finance, Ian Mills. Dr Hugh Sanderson was an obvious choice to lead the development of Casemix measures required for RM and he became head of the Casemix office.

In 1983, the US Congress passed the Social Security Amendments, which established the Prospective Payment System for Medicare In-patient services (Public Law 98–21). This necessitated the introduction of patient-based information systems where patient records were completely and accurately coded for diagnoses and operative procedures and could be grouped into DRGs to ensure payment.

DRGs were the obvious Casemix measure to use for RM: the algorithms were readily and freely available. They could be constructed from the diagnosis and operation codes used in the UK and other recorded data items, and there was experience in the UK in their use.

There were a few disadvantages: some of the codes did not work well for US practice and some were not coherent in the UK because of differences in practices. The greatest disadvantage, however, was made clear by the leading clinicians from the pilot sites at a meeting at the DoH in 1989. Their view was that their acceptance of DRGs would lead to an acceptance of a prospective payments system (PPS) similar to that adopted in the USA. Since they had no intention of accepting PPS, the clinicians said they would withdraw their support for RM if DRGs were adopted. There was no intention at that time to introduce any such a method of payment, but it highlights the suspicion that leading clinicians had of the DoH.

Work began to search for improvements to DRGs that would be acceptable to clinicians and would involve clinicians in their development. The outcome of this work led to the development of the Healthcare Resource Groups (HRGs).

While there were well-developed groups for outpatients, A&E patients and others, it was the in-patient groups that were widely used for RM. This remains the same some 15 years later with the use of HRGs for Payment by Results.

INFORMATION SYSTEMS

Casemix Management System

The Casemix Management System was defined with the following objectives:

- It was to be particularly targeted to the Clinical Directorate and similar structures.
- Data would be interpreted by managers with professional clinical backgrounds, supported by business managers.
- It would aim to understand how clinical resource decision related to financial budgets, in order to support budgets fully devolved to the main clinical specialties.

The leading hospitals contracted with external companies to define the requirements of the information systems and to determine whether there were appropriate systems in existence that could be procured to meet the requirements. Systems in use in the USA were investigated, but the conclusion at the time was that there were no systems appropriate to meet the objectives of the RMI. Three companies embarked on the development of systems to meet the requirements of the pilot sites:

- KPMG Peat, Marwick, McLintock and IBM with Huddersfield Royal Infirmary and Arrowe Park Hospital, Wirral.

- ISTEL (then AT&T ISTEL and now McKesson) with Freeman Hospital, Newcastle upon Tyne, Guy's Hospital and Royal Hampshire County Hospital, Winchester.

- ICL with Pilgrim Hospital, Boston, Lincolnshire.

HUDDERSFIELD COMPLETES RM SYSTEM

Peter Wood, chairman of Huddersfield Health Authority, announced the completion of the development of its clinical information system (CIS) recently.

Bob Steele, from the health service management unit at Manchester University, took two years to create CIS at the Huddersfield Royal Infirmary. Mr Steele said that CIS cost about £850,000 to install. Depending on the size of the site, the cost of implementing CIS in another hospital would be about £50,000, he said.

BJHC, MARCH 1989

In September 1988, the central RM team undertook a review of the requirements. It was found that none of the systems being developed in the pilot hospitals could fully meet the objectives of the programme although, between them, they had all the necessary elements. One system, Transition, which had previously been implemented in New Zealand was found to be the closest to meeting requirements with little modification being required.

The key elements of the system were:

- a patient record containing the resources consumed by each patient: stay days, pathology tests, etc;

- DRG or HRG grouper for grouping together records having a similar clinical response;

- a care profile, describing the resources that might be expected to be consumed by a patient in a particular group against which resources consumed by individual patients could be compared; and

- standard costs for each of the resources consumed, calculated by a 'top-down' process of apportionment.

These were the basis for costing the care of individual patients from the 'bottom up': from the sum of resources used.

We called this system the Casemix Management System (CMMS) as similar US systems had been named: a regrettable choice as it gave the misleading impression that the purpose of the system was to 'manage' or in some way change the Casemix of the hospital, when it was in fact to understand and manage the resources of the hospital. The name made it all the more difficult to put the concept of RM across. (Huddersfield Royal Infirmary adopted the name Clinical Information System (CIS). While equally a misnomer, it was more acceptable and left a more favourable perception, which must have influenced its use by clinical management teams.)

The review of existing systems from around the world led to the publication of minimum requirements in the CMMS Core Specification.[7] The system provided a common management information database to both clinicians and managers. The basis of the system was the individual patient record, including details of all events during an episode of treatment: nursing care, operations, medications, investigations, etc. Diagnoses and operative procedures were clinically coded and patient records could be grouped into DRGs, HRGs or other grouping measures. This grouping of patients highlights the degree of variation in resource usage attributable to similar patients. Information at a more detailed level was available for clinical research, quality control and clinical audit.

To receive central funds, a hospital had to purchase a system from a supplier who could demonstrate that their system met this minimum core specification. The aim was to ensure value for money for the NHS by encouraging competition. While this approach had been made clear from the outset, there was much lobbying by some pilot site suppliers to restrict this development. On one occasion it was made clear to me by a supplier that they were in regular contact with the NHS director of finance and the minister and had made it clear to them that they did not appreciate the view we took. On a different occasion, 17 representatives from another supplier met to persuade the team otherwise. Fortunately objectivity prevailed and several new suppliers entered the market and competition ensured system choices at reasonable prices.

By 1993 some 12 commercial systems were available that met the minimum requirements and the majority of acute hospitals had procured CMMSs.

The information systems introduced into Arrowe Park Hospital, as part of the RMI and the Hospital Information Support Systems (HISS) initiative, were particularly successful, not least because of the personal involvement of the CEO, Frank Burns. Frank Burns subsequently headed up the development of the DoH's information strategy: *Information for Health*.[8] The lessons that Frank learned in his implementation were very similar to those learned some two decades before at The Royal London and doubtless are still being learned elsewhere by others today.

Supporting medical/clinical audit and contracting

During the course of any innovative and lengthy project, the 'goalposts' move. RM was not immune to this.

The information requirements for supporting both medical and clinical audit as well as the introduction of the internal market, and contracting between providers and purchasers of health care in the 1991 NHS reforms, were very close to those required to support clinical management structures. Sensibly, it was decided that meeting these requirements would be combined within the single-system solution. The minimum data-set was updated to include requirements arising from the NHS White Paper *Working for Patients*.[9]

However, many hospitals saw these as separate systems and procured accordingly. To support medical and clinical audit, many hospitals bought microcomputer systems, running a variety of different software for different specialties. In addition to the support of audit programmes, these systems often provided some operational support, such as the production of discharge letters and summaries or appointment scheduling, more appropriate to operational systems. Some of the CMMSs were focused on their support of contracting and lost sight of both clinical management and audit.

Clearly the RM team had failed to put across the message that a single system could support all three purposes, but it was not in suppliers' interests to develop their systems in this way. The direction that the systems took was dependent on who felt they had ownership of those systems at a local level.

NURSING, OTHER OPERATIONAL SYSTEMS AND INTERFACES

The CMMS was essentially a data warehouse with excellent analysis and presentation tools. It derived its data from the operational systems of the hospital. Most hospitals at that time had a patient administration system (PAS) which provided the main data-feed. (See Chapter 4.)

While the CMMS could provide practical support with a single feed of patient demographic and episode details from the PAS, it became much more useful if data regarding tests, investigations, medications, nursing hours and other resources could be linked to those episodes. At the time, pathology and radiology systems were commonplace in hospitals. Since nursing systems existed but were not implemented in many hospitals, and since 70 per cent of the cost of health care comprises people costs, and much of this is nursing, it was essential that nursing data be captured in the CMMS. Both staff rostering and care planning were supported by the nursing systems funded by the RMI.

Capturing data from PAS alone could provide valuable Casemix information, with each successive interfaced system refining the clinical and management use of the data. Much effort went into the difficult and expensive interfacing of whatever local operational systems existed. At that stage of development of hospital operational systems, this was ambitious, both technically and for data quality reasons. Pathology, radiology and other departmental systems had developed and been implemented in isolation: patient identification varied and data quality left much to be desired.

Nursing systems were widely implemented, supported by RMI funds. Some are still in operation today. However, in the future, nursing systems as such will no longer exist because care planning functions will be integrated into patient-focused operational systems as multidisciplinary care planning driven by care pathways. Nurse rostering is now provided as a function to support the rostering of many staff groups and is incorporated into personnel and pay systems.

Apart from the nursing system, it was expected that the related HISS programme would fund the procurement and implementation of operational systems. These systems would be interfaced to the CMMS as part of their implementation. (See Chapter 7.)

Clinical coding

The complete and accurate coding of diagnosis and operative procedures is essential for the allocation of a Casemix group and is central to the objective of comparing the resource use of similar patients. Given that coding in many hospitals at the time was poor, the RMI had a stream of work aimed at improving the quality of these data. This covered awareness of coding importance, a training programme for clinical coders and the support of software encoders to improve the capture of data. As a result, the importance of coding was highlighted and quality improved.

The introduction of *Payment by Results* again put the focus on accurate and comprehensive clinical coding.

Costing

Each individual event (bed day, investigation, etc.) was allocated a standard cost in order to provide the cost of an individual patient's care. The expected profile of care for a patient in each group was costed using the same standard costs. The comparison of a patient 'cost' against the relevant profile highlighted variation in the clinical care of that patient, since in a Casemix system, increased cost is directly related to the consumption of more resources: longer than expected length of stay, more investigations, etc. For the clinical community, 'cost' provided a common currency or way of comparing resources that were defined in different units. For the finance community, it reflected the actual cost of patient care.

The inclusion of costs within Casemix systems served two main purposes:

- Cost provided a common currency by which the consumption of all resources could be compared. As such they could be expressed in any suitable unit. This happened in many hospitals and evolved into the means of costing and monitoring contracts.

- The devolution of budgets for all direct resources (e.g. diagnostic tests and investigations) to the main clinical specialties (e.g. surgery, medicine). This level of devolution of budgets occurred in only a few hospitals and is still not common today.

The standard costs would be calculated using the 'top-down' methods that were developing within the finance profession. There was a lot of debate on the level of detail in the cost figures: finance treating it as a science rather than an art and seeking the 'true cost' of each resource. This is a 'holy grail' that is always just out of reach and was unnecessary. Much effort went into the refinement of costs that distracted from discussion of the interpretation of the data. In general, finance departments considered that the devolution of budgets to clinical teams, based on the standard costs, would lead to loss of financial control unless they were 'true costs'. This was an inaccurate perception that remains today. It may have been more appropriate to consider these as 'prices' rather than 'costs' as was the case in similar US systems. However, the term 'price' was less acceptable and the term 'cost' was adopted.

In April 1993, the director of finance and corporate information at the NHS Executive issued guidance on the principles and approach for cost allocation as it applied to the costing process.[10] Casemix costing has continued to develop and is now reflected in the tariffs used for *Payment by Results*.

RELEVANCE TO THE NHS TODAY

Is RM something best consigned to the past or is there any relevance to the NHS today?

Most implemented Casemix systems have long since been decommissioned. However, the concept and practice is entirely relevant to supporting the management structures within hospitals today. The system also provides the necessary information to support *Payment by Results*, in particular, an understanding of the cost of individual patients by a 'bottom-up method': a costing based on the resources consumed by individual patients. It provides the information needed to compare the cost of groups with national tariffs and the variation of cost of individual patients within a group. This is essential to ensure that clinical quality improvement drives cost improvement, rather than the other way round: the focus on driving down cost leading to reduced quality.

A formal evaluation of the RMI was produced.[11] The evaluation set out to assess to what extent the objectives for the programme were met and how effectively it was managed. The report concludes that overall the programme was successful in meeting its objectives and was managed successfully. With hindsight, a number of things would have been done differently. These are lessons that still apply today.

RM was perhaps the first initiative that aimed at major organizational change supported by the development of a management information system specifically designed to support it.

Budgets were only devolved as envisaged in a few hospitals. Most finance departments felt that such widespread devolution would result in a lack of financial control. There was also a view in finance circles that costing was a science and not an art: indeed there was much debate about calculating the 'true cost' or the 'real cost' of a resource unit when no such thing exists.

It must be remembered that, in most hospitals in the late 1980s, and to a lesser extent today, responsibility for information systems resided with finance directors. In those hospitals that took the plunge, budgetary control was enhanced. However, it was counter to the financial culture of absolute accuracy, rationality and comprehensiveness. It took a brave chief executive to pursue this approach. With hindsight, greater support should have been given to the finance community.

There was a search for another 'holy grail' when it came to CMMS: the interfacing of all available systems before operational use of the data – patient administration, pathology, radiology and nursing as a minimum. In fact a great deal could be achieved once data from the PAS became available, such as the 'cost' of an individual patient based on day-stay cost and specialty cost; comparisons of length of stay of patients within a group; and comparisons across hospitals. With the poor data quality of many departmental systems, and the difficulty of extracting data and matching them to PAS data, interfacing became a major limiting issue. The difficult task of interfacing the departmental systems and a prevailing view that Casemix systems were not useful until the fullest possible information became available, meant that some hospitals did not use Casemix systems for supporting clinical management.

With hindsight, the organizational development and information systems aspects of RM could have been better integrated. While the links were made at a national level, the organizational change and information systems implementation at a local level were often treated as separate projects, with different project teams. This was not helped by the organizational development team being based in Manchester, while the information team were based in London. Conference calls, email and video conferencing were not readily available at the time to keep closer links.

As a result, changes in structures, roles and responsibilities were frequently not linked directly to the information needed to support those responsibilities. In many sites, finance departments resisted the full devolution of budgets, based on standard costs, to clinical directorates. Information systems were directed to the needs of the finance departments to meet contracting needs, rather than being made widely available and used as intended. Since ownership of RM was often perceived to be a finance project and finance departments on the whole were also responsible for IT, this result should have been foreseen. Few hospitals developed the sense of ownership by clinical teams that was needed for widespread success.

CONCLUSION

The concept of RM and Casemix management was entirely right for the 1980s and 1990s. Clinical directorates were widely implemented and remain a model for clinical involvement in management today. The technology and competence of those who implemented it had not developed sufficiently to make the information systems an overwhelming success in more than a small number of hospitals.

RM laid the foundations for future developments. The focus on accurate clinical coding and the training programmes established are still in evidence today. Those hospitals that have implemented data warehouses to support internal and external management processes have built on these concepts.

There is still a lot to be learnt from RM, including the full devolution of budgets to the main clinical specialties, allowing clinicians to decide where resources are best focused in the interests of the patients. The introduction of standard costs applied to actual patient care and care profiles are elements that few have developed but one that will highlight variation in clinical practice and focus for improvement.

REFERENCES

1 NHS Management Board (1986) Health Notice HN(86)34. November.

2 Bullas, S. (1994) *Managing Quality and Cost: Using Patient-based Information.* Longman, London.

3 Webref (accessed Dec 2006) Intermountain Healthcare. www.inter-mountainhealthcare.org/xp/public/aboutihc/quality/

4 DHSS (1983) *Report of the NHS Management Inquiry.* Chairman Roy Griffiths. DHSS, London.

5 Feldstein, M.S. (1967) *Economic Analysis for Health Services Efficiency.* North-Holland, Amsterdam.

6 Newman, T. and Jenkins, L. (1991) *DRG Experience in England 1981–1991.* CASPE.

7 NHS Management Executive (1989) *Casemix Management System Core Specification.* NHS Management Executive, January.

8 Webref (accessed Dec 2006) DoH (1998) *Information for Health: An Information Strategy for the Modern NHS 1998–2005.* DoH, London. www.dh.gov.uk/PublicationsAndStatistics/Publications/Publicatio nsPolicyAndGuidance/PublicationsPolicyAndGuidanceArticle/fs/ en?CONTENT_ID=4007832&chk=KdSoKm

9 DoH (1989) *Working for Patients.* White Paper (Cm 555) HMSO, London.

10 NHS Management Executive (1993) *Costing for Contracting.* Report EL(93)26, April.

11 Webref (accessed Dec 2006) DoH (1996) *The Evaluation of the NHS Resource Management Programme in England: A Report Submitted to the NHS Executive.* DoH, London. www.dh.gov.uk/PublicationsAndSta tistics/Publications/PublicationsPolicyAndGuidance/PublicationsPoli cyAndGuidanceArticle/fs/en?CONTENT_ID=4008641&chk=H8/YvW

7 The HISS programme and the development of electronic health records

JOHN BRYANT

The electronic health record has been the 'holy grail' of health informatics for at least 40 years. The goal to computerize the medical record, including the nursing record, was the raison d'être for Professor John Anderson's Medical Records Project at King's College. It was also prominent in the thoughts of those involved in The London Hospital Computer Project in the late 1960s. These projects are described in Chapter 4. Although a very small number of early developments attempted to build functional clinical record and support systems, the pathway to the electronic health record was essentially based on the development of the various patient administration systems (PASs).

The Experimental Computer Programme dominated the 1970s (see Chapter 4). These projects formed the early pathway as they struggled to debug and harness the new information technologies associated with transaction processing and database management systems. At that time, these projects comprised the main NHS development resource with respect to online, real-time computer systems. The Regional Computer Centres concentrated on batch processing. As a result of Government pressure the regional centres were standardized on ICL hardware. However, when the regional health authorities (RHAs) finally got around to considering the needs of the acute hospitals they diverged, with six regions standardizing on ICL hardware, the others using DEC VAX hardware.

The 1980s saw the emergence of standardized approaches to patient administration and the gradual development of order communications systems; largely unknown outside The London Hospital. The standardization effort was led by the Inter-Regional Collaboration (IRC), which was established, in late 1983, as a loose collaboration to coordinate the convergence of the two patient administration systems,[1] one developed by ICL Ltd and the other by the North Staffordshire District Health Authority and subsequently adopted by the West Midlands RHA. In mid-1985, the IRC was reconstituted formally under the direction of a management board. Membership included the West Midlands, North West Thames, South East Thames, Northern, and Yorkshire RHAs and ICL Ltd. Oxford RHA, one of the founder members, had left but subsequently rejoined in November 1986.

Wales discussed possible membership of the IRC but never joined. Instead they purchased the source of the IRC product and developed their own version from this.

The remaining RHAs used mainly DEC VAX hardware running a variety of commercial and home-grown products. By the start of the 1990s all the districts in the North Western and South West Thames regions were using the Shared Medical Systems product along with three districts in East Anglia.[2] The other five districts in East Anglia were using the PAS developed by the experimental project at Addenbrooke's Hospital, Cambridge. The Trent RHA had entered into a marketing agreement with AT&T ISTEL to market the Trent PAS system in 1985. By the end of the decade this system was being used extensively throughout the Trent and North East Thames regions as well as in three districts in Wessex, two in Yorkshire and one in Mersey. All districts in the South Western Region were using a system derived from the Trent system and further extensively developed in-house. This system was also used by two districts in Yorkshire and one in West Midlands. Two further minor developments were in place at this time. Three districts in Wessex were using the Abbey system and five districts in Oxford were using OPAS, another in-house system.

This then was the scene at the end of the 1980s. An independent survey carried out in 1989 showed,[3] for the 209 hospitals that responded, a high number of PAS systems and a smaller but significant number of departmental systems such as pharmacy, pathology, radiology and maternity had been implemented. A very small number of order communications systems were operational and there was a considerable lack of integration among the PAS and departmental systems. Indeed, many hospitals had only minimal data communications networks. Most of the system developers had been slow to take up the challenge of order communications systems given that the initial work on PAS, in the West Midlands and at The London Hospital, started in the late 1960s. On reflection it is disappointing that the software platform needed for the system at The London Hospital did not lend itself better to transferability.

THE HISS PROGRAMME

At the start of the 1990s, the NHS was largely divided into two standardized camps: the ICL-based IRC system on the one hand and the slightly more diverse MUMPS-based camp on the other hand. It was into this setting that, in 1988, the Department of Heath set up the HISS project to test the feasibility of defining, procuring and implementing systems that would encompass every operational and information need for major acute hospitals. The project was organized as a formal PROMPT project, with a project board chaired by Michael Fairey, then NHS Director of Information Systems. A central team, led by John Bryant, was established comprising DoH and NHS experts in the fields of procurement, communications, hospital applications, training, and benefits appraisal and realization. The team was to establish and work closely with a programme of local projects to address the perceived questions about integrated systems in hospitals. The following sections describe the nature of the HISS programme, concentrating on the elements that are more relevant to the future development of electronic health records.

What is a HISS?

A Hospital Information Support System, or HISS, was defined as an information and communications technology (ICT) environment that met the real-time operational information needs of health professionals to deliver care to patients, while also providing accurate and timely information for management purposes. The aims of a HISS were:

- to improve the care and quality of services provided to patients;
- to reduce administrative work; and
- to provide better information for resource management, medical audit and medical research purposes.

Thus, it was expected that a complete HISS would include the following components:

- An integrated patient-based ICT environment covering applications of both a hospital-wide (e.g. PAS) and departmental (e.g. pathology) nature.
- A terminal and data communications network allowing access throughout the hospital to the different applications using appropriate networking technology (usually a local area network).
- An order communications system to support the communication of orders from ward users to service departments and the results from service departments to requesting clinicians.

The initial pilot projects

The HISS programme began with three pilot projects: Greenwich District Hospital, Darlington Memorial Hospital and City Hospital, Nottingham. All three were selected within a very short period of time, about two months, because of a perceived need to link to and support the Resource Management Initiative and to make progress as quickly as possible. The consequence of this, along with other political pressures, meant that two of these projects, at Nottingham in particular and to a lesser extent at Darlington, involved significant risk. City Hospital was selected, albeit as a late replacement for another hospital, because it was a very large, complex teaching hospital providing numerous regional specialties. It was, by definition, always going to be a difficult project. Additionally, judged by today's standards, these two hospitals were not ideally prepared to undertake a major computer project.

The first two pilots, Greenwich and Darlington, working independently on producing a comprehensive statement of requirements, both came to the conclusion that their existing systems would have to be completely replaced. The technical difficulty and cost of building on them would have been prohibitive. The Greenwich contract was awarded in 1990 to HBO, a US software company, using their CLINSTAR® and TRENDSTAR® products running on Data General computers. The NHS was both surprised and somewhat alarmed by the estimated £8.5 million cost. (At that point in time HISS was definitely not a commodity product.) However, this represented the capital cost plus the revenue costs over seven years. The Darlington contract was awarded about one year later to another US company, McDonnell Douglas, for their HOMER product at a similar total cost of £7 million. This latter system was already being used at four sites in the UK.

The third pilot project took an entirely different line. City Hospital had already made a much greater investment in computer systems and the associated communications infrastructure. The need to protect this investment was therefore an integral part of their HISS requirements. They chose to express their requirements in terms of generic processes based on the hospital's activities, and used the NHS's Common Basic Specification (CBS) as the model.[4] This complex combination of large teaching hospital, the need to interface with current systems, and the project's status as a pilot for the use of the CBS ensured that the Nottingham project would be extremely difficult. The single-tender contract was eventually awarded to Oracle in February 1992. The overall project cost was estimated to be in excess of £10 million. Some professionals at the time believed that the project was untenable and should not proceed, a view later supported by the Parliamentary Committee on Accounts.

The next steps

This first exploration looked at the implications of starting in a green field site and attempting to achieve a hospital-wide integrated system in one major step – or a 'big bang', as it became known. It was clear from the three pilot projects that many hospitals would find this single-step approach to HISS to be too great an administrative and organizational overhead.

The initial projects required a significant capital outlay at the start and would require anything from three to five years to complete implementation. The programme therefore began to look at more evolutionary approaches to HISS using as many of the existing systems as possible. The next group of projects were selected with regard to their relevance to *Working for Patients*.[5] These projects were much more focused than the three pilots and were selected on the basis that they would provide tangible benefits within one to two years. Specific areas to be addressed included:

- incremental approaches to HISS based on the maximum possible use of existing systems;
- demonstration that identified benefits could be realized in practice; and
- transferability of solutions to other sites starting from the same base.

This second phase of the programme comprised a further seven projects to explore alternative evolutionary approaches to integration of systems and to test improved methods of procurement.

One of the approaches to procurement that was adopted was the establishment of a consortium of three hospitals in the West Midlands. The intention was that the project would demonstrate that a common base system could be implemented across a number of diverse acute hospital environments. The consortium also intended to demonstrate that a phased approach could lead to early benefits. They thought that by merging their resources, particularly specialist development and implementation staff, and with the possibility of attracting discounts for multiple systems, they could generate significant overall savings. The outcome for this consortium, consisting of Walsgrave Hospital, Coventry, the East Birmingham General Hospital, and the South Warwickshire Hospital, was the implementation of the IBM/GTE MedSeries product. The first modules went live in April 1992. Other consortia approaches to procurement were to follow.

It was during this second phase that the HISS project began to focus on the problems concerning the lack of development of order communications systems. A survey commissioned by the HISS Central Team that took place in late 1991 and early 1992 identified order communications systems as a pivotal element of an integrated hospital system and fundamental to meeting a variety of different objectives within NHS hospitals.[6] One of the key questions of that time was whether or not the order communications system needed to be an integral part of the PAS or whether a stand-alone system would function satisfactorily.

Another area of concern was data communications infrastructure. Most hospitals had grown their communications networks in a piecemeal fashion. This had implications for any attempting to implement a hospital-wide integrated information system. The HISS programme initiated a number of projects to explore these questions.

To explore the order communications system questions two projects were initiated. First a study was carried out to explore the possibility of using the IRC PAS as a building block for a HISS.[1] The study concluded that such an approach was feasible but that there were a number of operational and organizational problems. Technical difficulties caused by the lack of a common user interface and the difficulty in linking to other systems, plus the demise of the ICL 2900 Series and the associated problems of migration to the new Series 39 computers, meant that this would not be an easy path. At the same time, changes in the organizational structure of the NHS made it unlikely that collaborations like the IRC would be able to continue.

The second project exploring order communications systems was in the South West Region. The Regional Computer Centre had taken the MUMPS-based Trent PAS, further developed it, and successfully implemented it in each of their 11 districts. The HISS project supported the region in the further development of the PAS system to include a common referrals index, an order communications system and an integrated pathology system. It was felt that the resultant integrated system would meet the requirements of a HISS and be potentially transferable to other sites, particularly those already using the Trent PAS.

The other major project in the second phase was the infrastructure project at the Leicester Royal Infirmary. The procurement and implementation of a GOSIP-conformant local area network at the Infirmary enabled the HISS Project to provide valuable guidelines in this complex but vital technical area.

The third phase

In its third and subsequent phases the HISS programme continued to explore important issues in the area of integrated HISSs, concentrating on consortia and framework approaches to procurement and implementation, the development of information strategy, and the development and publication of a wide range of guidance documents. The extensive library of guides published by the programme covered all aspects of procurement,[7] investment appraisal and benefits realization,[8] organizational development and change management,[9, 10] the use of soft systems methodology,[11] the evaluation of projects,[12] and numerous other topics.

While they were extremely important and valuable contributions to the hospital information systems field, these later phases of the HISS programme are less relevant to the eventual development of electronic health records.

By 1992 the HISS project had identified a framework of levels of integration that described an incremental approach towards a fully integrated HISS:

- **Level 1** – Linkage of patient demographic data between systems: sharing or automatic exchange of patient demographic data across systems.

- **Level 2** – Results enquiry: level 1 plus enquiry access to all relevant information, across different systems, e.g. enabling ward or clinic users to interrogate information about a patient including their test results.

- **Level 3** – Full order communications system: level 2 plus a comprehensive order entry and results reporting system for placing clinical and other orders from ward-based terminals, with automatic transmission of the orders to the relevant departments and results reports available on the wards.

Each of the levels was seen to provide important benefits in its own rights. Higher costs and greater complexity were associated with the higher levels, as were increased benefits. The framework provided hospitals with a set of defined targets as well as a number of options in trying to build towards a HISS in an incremental way. The importance of the framework becomes apparent later.

THE FIRST STEPS TOWARDS THE ELECTRONIC HEALTH RECORD

By the beginning of the 1990s the belief was that health informatics had advanced sufficiently to permit a realistic vision of a new and improved patient record. Those involved thought the technologies were capable of replacing traditional paper-based records with computer-based patient records. It is unfortunate that the credit for the first real step goes to the USA where the Institute of Medicine established the Committee on Improving the Patient Record. The committee's findings and recommendations were reported in 1991:

> *This report advocates the prompt development and implementation of computer-based patient records (CPRs). Put simply, this Institute of Medicine committee believes that CPRs and CPR systems have a unique potential to improve the care of both individual patients and populations and, concurrently, to reduce waste through continuous quality improvement.*
>
> REFERENCE 13

It appeared to take the publication of this US report to stir the DoH into serious consideration of electronic health records.

The NHS took the first tentative steps towards electronic health records in the early 1990s. The long overdue information management and technology (IM&T) strategy was published in 1992.[14] Although the IM&T strategy did not directly propose electronic health records it laid down many of the necessary principles and set out the technology infrastructure that would be needed in the future. The strategy was based on five key principles that originated in the work of Edith Körner:

- Information should be person-based.
- Systems should be integrated.
- Information should be derived from operational systems.
- Information should be secure and confidential.
- Information should be shared across the NHS.

The strategy was pursued through a series of national facilitating projects that included:

- a new-format NHS number;
- a system of NHS-wide networking;
- a thesaurus of coded clinical terms;
- a set of national data messaging standards; and
- a framework for security and confidentiality.

All these elements of the national infrastructure are essential ingredients for an environment within which electronic health records could exist and flourish. However, it was another initiative that first started to address directly issues concerning the implementation of electronic health records.

The US Institute of Medicine report had commented that:

Perhaps the single greatest challenge that has consistently confronted every clinical system developer is to engage clinicians in direct data entry.

REFERENCE 13

By 1992 a number of major computer manufacturers were assessing the need for electronic medical records. In particular image systems that would share the same processor and network infrastructure as the order communications system were being proposed.[15]

Work had begun on clinical support systems and clinical workstations in the early 1980s. This arose from the Programme for Enabling Technology Assessment (PETA) run under the auspices of the DoH's Computer Policy Committee and was subsequently carried on by the Information Management Centre. This work was given greater emphasis by the IM&T strategy with the launch of its Integrated Clinical Workstation (ICWS) project in 1993.

The project was instigated by the NHS Centre for Coding and Classification and aimed to support clinicians at the point of care by ensuring that they could easily access computer technology. The aim of the project was to produce detailed clinical user requirements and develop prototype workstations to be demonstrated at sites in the acute, community and primary care sectors. These were intended to enable clinicians to capture data in a useful and meaningful way that would not hinder their normal working practices. The hope was that this would help the Clinical Terms Project* to gain wider acceptance among the clinicians.

However, with its focus on the broader remit of the adoption of computer technology in the support of clinical care rather than the design aspects of the user interface, the ICWS project missed the target. The Institute of Medicine had already identified the primary barrier to effective use of the technology as the ability to accept and process continuous speech with speaker independence and large vocabularies.[13] Others were also pointing out that physicians' reluctance to use clinical systems was partly attributed to difficulties with the user interface.[16] They indicated that the two major tests for ease of use from the physician perspective were user time to complete a task and memory burden. To achieve acceptance a clinical system would have to pass these tests for both the routine and occasional user, as well as for users whose computer literacy ranged from novice to expert. The demonstrator projects relied on existing graphical user interface technology and never really addressed the fundamental question of clinician usability. In hindsight, it would appear that the project was really more concerned with attempting to gain acceptance of the clinical terms than with providing information technology that would properly support clinicians at the point of care.

---

* The Clinical Terms Project was established at the beginning of the 1990s to undertake redevelopment of the Read Codes to overcome the deficiency of limited levels of detail and to meet the need for more descriptive text.

The Electronic Patient Record programme

Dr Bill Dodd was appointed to direct the Electronic Patient Record (EPR) programme in 1994. This was a direct attempt to move forward from systems that helped manage the logistics of health care delivery to systems that enabled clinicians to provide better care. Building directly on the HISS programme, the EPR programme identified a six-level incremental model for the development of an electronic patient record, later described in *Information for Health*:[17]

- **Level 6 – Advanced multimedia and telematics**

  Level 5 plus telemedicine, other multimedia applications (e.g. picture archiving and communications systems)

- **Level 5 – Specialty specific support**

  Level 4 plus special clinic modules, document imaging

- **Level 4 – Clinical knowledge and decision support**

  Level 3 plus electronic access to knowledge bases, embedded guidelines, rules, electronic alerts, expert systems support

- **Level 3 – Clinical activity support**

  Level 2 plus electronic clinical orders, results reporting, prescribing, multiprofessional care pathways

- **Level 2 – Integrated clinical diagnosis and treatment support**

  Level 1 plus integrated master patient index, departmental systems

- **Level 1 – Clinical administrative data, patient administration and independent departmental systems**

REFERENCE 17

The EPR programme also distinguished, for the first time, between the integrated real-time clinical support system (called the active elements), the passive EPR or actual record, and the data warehouse of data for secondary analysis (previously part of the role of the RMI: see Chapter 6). The programme aimed to influence the development of systems that would be implemented about six years from then. This it hoped to achieve through the efforts of its two core demonstrator sites at Queen's Hospital, Burton upon Trent, and Wirral Hospital, plus a number of EPR associated sites (which were unfunded but agreed to contribute to the overall learning of the programme). The two core demonstrators eventually claimed to have achieved successful hospital-wide implementation of EPR.

While the programme contributed this detailed framework, later to be used for setting strategic targets,[17] it failed to identify and define the contents and structure of the passive record. Nevertheless, the framework and the learning from the programme played a role, if a somewhat confusing one, in the following steps along the path to the current National Programme for IT. The EPR programme was suspended in 1997.

With the publication in 1998 of the DOH *Information for Health* strategy,[17] Government policy, for the first time, specifically identified the electronic patient record as central to future health systems.

> *The NHS in the future must use technology in innovative ways to support this aim. For example:*
>
> *– individualised personal electronic records will be developed to provide NHS professionals with 24-hour secure access to the information important to individual patients' care, when required. This will immeasurably improve emergency care and ensure any professional involved in the care of an individual is up to date with their treatment.*
>
> REFERENCE 17

The strategy went on to say:

> *The arguments for a move towards an electronic record are compelling. Such records are more likely to be legible, accurate, safe, secure, and available when required, and they can be more readily and rapidly retrieved and communicated. They better integrate the latest information about a patient's care, for example from different 'departmental' clinical systems in a hospital. In addition, they can be more readily analysed for audit, research and quality assurance purposes.*
>
> REFERENCE 17

Also, the electronic health record was mooted for the first time as a strategic imperative, using the EPR six-level framework to set national targets for the NHS:

> *By 2005*
>
> *The final phase of implementation will see the completion of the work programme, with comprehensive electronic patient and health records available throughout the NHS to support the delivery of care.*
>
> - *Full implementation at primary care level of the first generation person-based Electronic Health Records.*
> - *All acute hospitals with Level 3 EPRs.*
> - *The electronic transfer of patient records between GPs.*
> - *24-hour emergency care access to patient records (Para 7.11).*
>
> REFERENCE 17

THE ELECTRONIC RECORD DEVELOPMENT AND IMPLEMENTATION PROGRAMME

The final step in the path to the present time, and the precursor to the current Connecting for Health programme, was the Electronic Record Development and Implementation Programme (ERDIP). The newly formed NHS Information Authority was given the task of progressing the outcomes of the EPR programme, in line with the *Information for Health* targets. Additionally, the information strategy had suggested that a number of beacon sites should be set up to explore the issues associated with the development of electronic health records. Finally, there was a clear definition and separation of the concepts:

ELECTRONIC PATIENT RECORD (EPR)

Describes the record of the periodic care provided mainly by one institution. Typically this will relate to the health care provided to a patient by an acute hospital. EPRs may also be held by other health care providers, for example, specialist units or mental health NHS Trusts.

ELECTRONIC HEALTH RECORD (EHR)

Used to describe the concept of a longitudinal record of patient's health and health care from cradle to grave. It combines both the information about patient contacts with primary health care as well as subsets of information associated with the outcomes of periodic care held in the EPRs.

The terminology surrounding electronic records had been problematic for many years, with electronic patient records, computer-based patient records, emergency health records, the electronic patient health record, etc., compounded by the fact that almost all general practices had electronic health records (as defined above) for their patients. Although the separation between what had been the active EPR and the new, longitudinal health record was identified by the ERDIP project, the confusion nevertheless appeared within the programme's projects. The confusion was still apparent in the successive projects: the Integrated Care Records Service (CRS) and the National Care Records Service, and still exists today in terms of what exactly the NHS Care Records Service is intended to be.

Nevertheless, the ERDIP programme made progress. Invitations were made to participate in centrally funded projects in two distinct categories: electronic health record demonstrator projects and EPR focus groups. The key issues addressed by the projects included 24-hour care, patient access, EHRs, level 6 EPR, integrated primary and community care, direct booking, referrals, discharge, pathology messages and technical standards. The ERDIP programme was fully evaluated and the report published in 2003.[18] Although the programme was criticized for lack of clarity of definition and other aspects, it should be recognized that the individual projects were generally successful.

AND THE LESSONS LEARNT?

Two roads diverged in a wood and I –
I took the one less travelled by,
And that has made all the difference.

ROBERT FROST, MOUNTAIN INTERVAL, 1920

It is not the role of a book of reflections to describe the current situation with regard to Connecting for Health – the development and nature of this programme is already well described.[19] What, however, of the electronic health record? Is this concept now properly understood? The current NHS Care Records Service believes so. Others, for example in Canada,[20] are pressing for EHRs albeit for a somewhat different reason: they are concerned with minimizing medical errors. The path towards this holy grail has been fraught with opportunity, both to succeed and to wander off the path.

All of the aforementioned initiatives have been evaluated, in some cases extensively. Each of the Experimental Computer Projects had formal evaluation teams attached in the 1970s (see Chapter 11). At that time there was overemphasis in many projects on simple before and after measures, assuming that these would make possible a scientific comparison and hence judge the relative success of the particular project. Certainly all the projects from the HISS programme to the present have been formally reviewed, some more than others.

The HISS Project was reviewed by the National Audit Office (NAO), which was critical of the speed of selection of the initial pilot projects, the readiness of a number of projects to undertake such complex implementation tasks, and length of time some projects took to achieve implementation and benefits realization.[21] Shortly thereafter HISS was again examined, this time by the Parliamentary Committee on Accounts.[22] The Committee's report essentially agreed with the NAO's findings, adding their astonishment that the NHS Executive had proceeded with the Nottingham project. The Committee also criticized HISS for not fully heeding the lessons of the Wessex RHA's Regional Information Systems Plan, a well-documented failure.[23]

The HISS Project's own review did not appear until 1998.[24] Not unexpectedly, the report's findings concentrated on the comparison of the full-life costs and financial benefits reported by the 18 projects reviewed. Perhaps surprisingly, in view of the extensive criticism aimed at the project, the review estimates showed the estimated costs to be slightly less than the original baseline costs (with the exception of the three pilot projects) and the estimated financial benefits to be almost identical to the baseline estimates (again with the exception of the three pilot projects). The figures for the pilot projects were undoubtedly adversely influenced by the Nottingham project and, to a lesser extent, by the Darlington project. Notwithstanding the various criticisms, the HISS Project is best understood in the context of the current National Programme (NPfIT) as set by Brennan:

In many ways, NPfIT is a direct descendant of the HISS project. Many of the concepts pioneered by HISS are firmly part of NHS-CRS: order communications, integrated applications, a single patient record – all these were tested and proved by HISS. The extraordinary thing about the HISS pilots was really that it took so long before anyone sat back and recognised how successful they had been.

REFERENCE 19

This was, at the time, a very large project having spent approximately £56 million by the time the NAO report was published in 1996. Overall some £48 million were provided in financial support to 16 projects at 25 hospitals. The NHS Executive estimated at the time that these hospitals themselves had spent a further £50 million or more on their projects. The other projects in the path towards the electronic health record were also reviewed, although not to the same extent as the HISS Project. A review of the ICWS project was published at the end of the 1990s. A formal evaluation of the EPR project was commissioned but this was never published in full. Interestingly, the EPR project was the only one to be directly managed by the DoH's Information Policy Unit. Only a limited amount of information was made public. The national evaluation of ERDIP has already been mentioned. Indeed, as late as 2000 ERDIP commissioned a study into approaches to the evaluation of such projects. There is no doubt that, in due course, the same public bodies will wish to examine Connecting for Health, in whose care we now entrust the birth of the electronic health record, some 35 years late.

By comparison with HISS, NPfIT is huge, with estimates of total programme cost ranging from £20 to £40 billion.[19] The health informaticians currently working on this vast programme would do well to take note of some of the latest guidance.[25] Large complex IT projects still fail more often perhaps than they succeed.

REFERENCES

1   Thorp, J. (1990) *The IRC PAS and HISS – the use of the Inter-Regional Collaboration patient administration systems as a building block for hospital information support systems.* HISS Central Team report.

2   Thorp, J. (1991) *MUMPS PAS and HISS – the use of MUMPS-based patient administration systems as a building block for hospital information support systems.* HISS Central Team report.

3   HISS Central Team (1989) *Survey of Acute Trusts.* NHS Management Executive.

4   NHS Information Management Centre (1990) *The CBS: Generic Model Reference Manual.* NHS Information Management Centre.

5   DoH (1989) *Working for Patients* (Cm 555). HMSO, London.

6   Silicon Bridge Research (1992) *Survey of Opportunities for Order Communication Systems.* Silicon Bridge Research, Basingstoke.

7   HISS Central Team (1992) *Procurement Guidelines.* NHS Management Executive.

8   HISS Central Team (1992) *Benefits Management – Guidelines on Investment Appraisal and Benefits Realisation for Hospital Information Support Systems.* NHS Management Executive.

9   HISS Central Team (1995) *Managing Change – Getting the Best from IT.* NHS Management Executive.

10   HISS Central Team (1996) *Making the Most of Information – an Organisational Change Approach.* NHS Management Executive.

11   HISS Central Team (1996) *Guidelines for the Use of Soft Systems Methodology.* NHS Management Executive.

12   HISS Central Team (1996) *PROBE Guidance for NHS Managers.* NHS Management Executive.

13   Dick, R.S. and Steen, E.B. (1991) *The Computer-Based Patient Record: An Essential Technology for Health Care.* National Academy Press, Washington, DC.

14   DoH (1992) *An Information Management and Technology Strategy for the NHS in England.* DoH, London.

15   Moore, C. (1992) *Computerised Medical Records.* IBM internal document. IBM UK.

16   Metzger, J.B. and Drazen, E.L. (1993) Computer-based record systems that meet physician needs. *Healthcare Information Management,* 7(1), 22–31.

17   DoH (1998) *Information for Health: An Information Strategy for the Modern NHS 1998–2005.* DoH, London.

18   PA Consulting (2003) *ERDIP National Core Evaluation Final Report.* NHS Information Authority.

19   Brennan, S. (2005) *The NHS IT Project.* Radcliffe, Oxford.

20   Stuart, N. and Whittick, D. (2004) The pressing case for an EHR – the need to improve patient safety. *Healthcare Information Management & Communications Canada,* vol. XVIII, No. 5, December.

21   Comptroller and Auditor General (1996) *The NHS Executive: The Hospital Information Support Systems Initiative.* National Audit Office.

22   Parliamentary Committee on Accounts (1996) *Seventh Report – The Hospital Information Support Systems Initiative.* The Stationery Office, London.

23   Collins, T. (1996) *Crash.* Simon & Schuster, London.

24   Pareto Consulting and Secta Consulting (1998) *HISS Projects Review – Final Report.* NHS Executive.

25   Royal Academy of Engineering (2004) *The Challenges of Complex IT Projects.* Royal Academy of Engineering.

PART 2

Departmental systems

INTRODUCTION

The title 'departmental systems' indicates a hospital bias, so there may be an overlap of a specialist clinical focus with systems that have a generic approach, such as those covered in the chapters on decision support (Chapter 24), laboratory test ordering and result reporting (Chapter 7) and patient administration (Chapter 4). The specific organizational requirements may also impact on management systems, such as staffing and patient acuity required to calculate the number and expertise of staff needed to balance with the patient's 'conditions'.

Hospital laboratory departments mirrored the 'pure' science laboratories of academic and commercial institutions. The senior scientists who worked there came from a background of research, innovation and improvisation and trained the technicians working with them in such skills. Medical developments also had a tradition of experimentation and it was through collaboration that computer-based developments were introduced. Biochemical and physiological measurement produced a mass of data. The speed of computers in repetitive tasks made them useful for calculations and sequential displays of data. Radiation treatment planning, also, was a natural application for the early computer pioneers as it involved so much computation. The state of this art in 1973 was outlined in a report for the British Institute of Radiology and the DoH by Barry Barber.[1]

At the same time the creation of new forms of images of the interior and functioning of the body required massive processing facilities that were gradually put to work on these problems, leading to the impressive picture archiving and communication systems (PACS) and scanning systems that are available today. Nurse allocation was also an attractive target for operations analysts but, although Alex Barr of the Oxford Hospital Board produced an early program for the Elliott 803, there was a massive programme of data collection and culture change required before it could have been implemented on other than a small scale.

DENISE BARNETT AND BARRY BARBER

REFERENCE

1   Barber, B. (1973) *Computerized Dose Computation.* Report for the British Institute of Radiology & the DoH.

8 Clinical laboratory and other diagnostic support services

ANTHONY WILLS, JOHN NEAL AND ANDREW MORLEY

In this chapter we describe computing developments involving mainstream clinical laboratories, noting some of the benefits and limitations of automation. Some examples are included of services supporting diagnostic imaging[1] and the passing of clinical laboratory reports to primary care general practice systems. We describe the evolving opportunities as networking technology and personal computer (PC) products matured, based on experiences at the Barts and The London NHS Trust (formed through a merger of St Bartholomew's Hospital and The Royal London Hospital). A summary of the main developments in NHS computing in hospitals between 1960 and 2005 is provided towards the end of the chapter. The conclusions offer some observations about hospital computing in the twenty-first century.

REQUESTING DIAGNOSTIC INVESTIGATIONS AND REPORTING THE FINDINGS

Clinicians and nurses make requests for data and specialist opinions from a wide range of treatment plans for their patients. In terms of the number of requests made per day in a teaching hospital, the dominant specialties are typically:

- haematology;
- chemical pathology;
- microbiology;
- cytology and histopathology;
- diagnostic imaging.

An ever changing and increasing variety of investigations is provided within hospital departments and laboratories, usually hosted in one of the above specialties, but also within other units in the NHS, medical colleges and commercial laboratories.

Hospital wards and clinics have, for many years, generated the majority of requests for such investigations; but the workload from primary care trusts continues to increase as more treatments are performed at local surgeries, and as diagnostic imaging services are requested directly by GPs. Effective and efficient treatment for many patients relies on secure and reliable communication links between requesters and providers, ideally including the ability to see how far each investigation has progressed.

The following matters are of crucial importance if such services supporting the front line of health care are to be effective in advising the treatment that will be of greatest benefit to each patient.

- The identity of the patient must be unique and correct at every stage, and be clearly present on the request form; easy to check while the patient is examined or a specimen of body fluid or tissue is obtained; while the investigation is performed; as the report is checked (often requiring reference to previous reports for the same patient); when the requester reads the report via a computer or on paper; and when the report is filed in the patient's permanent medical record.

- The quality of the data and opinions in each report should be to verifiable standards, including accuracy, clarity and completeness.

- Reports should be available for quick review by the patient's doctor as soon as the diagnostic department is prepared to issue them, in case urgent action is indicated.

- The official final copy of the report must usually have been filed in the patient's case notes when needed for the clinician's ward rounds, discharge procedures, outpatient clinics and GP appointments.

A laboratory computer that works without integrated access to the hospital's Patient Administration Service (PAS) will achieve limited success with cumulative reports. If there is no order-entry of requests in the wards, there will also be delays in processing at peak times due to the length of the booking-in procedure.

Ward staff need interactive access to view results as soon as they are ready. If the laboratory computer is unable to be connected to the hospital's communication network, or has inadequate capacity to support access by these users, patient care can suffer. Similarly, staff in outpatient clinics often need to print a copy of reports, because the original report has not been filed in the patient's notes in time for an appointment. When a laboratory computer is able to export all reports to hospital systems, feeder software can automatically update clinical records and management information databases, with considerable benefits to the hospital as a whole.

Until the late 1980s, mainframe styles of computer were the only technological options available for improving these service issues. A large and complex central database supported interactive services in most departments of a hospital.

In the NHS in this period, the variety and complexity of interactive transactions in a teaching hospital dwarfed the technical requirements of banking and other commercial computer applications. However, the available funding for NHS computing facilities was substantially less than the mainframe market was used to. Consequently, most UK and European hospital projects up to the 1990s purchased minicomputers, and those wanting to support more than about 50 concurrent users needed to develop much of their own transaction management software, as well as programming the application services.[2]

In order to reproduce such services at additional hospitals, because so much software was locally developed, success relied on the expertise and continuity of the operations staff and the software. The applications software had to be designed to support the considerable variations in procedures and data formats across hospitals and specialties. The operations department also needed the resources and expertise to install and maintain complex networks and to support the number and variety of users of the services.

The early attempts to procure 'transferable systems' suitable for NHS hospitals and laboratories met with limited success.[3–6]

In the 1990s, as experience in 'client–server' applications matured (involving high-speed networks with PCs as user devices), some hospitals started to find commercially available 'building blocks' of software that enabled bought-in applications to interact dynamically with established multi-user services, exchanging data while ensuring data consistency and integrity at all times. However, in-house expertise was often required to enable such software to interact efficiently and reliably.

COMPUTERS IN CLINICAL LABORATORIES

F.V. Flynn at University College Hospital, London, L.G. Whitby at The Royal Infirmary, Edinburgh, T.P. Whitehead at the Queen Elizabeth Hospital, Birmingham, and I.D. Wootton at the Post-Graduate Medical Centre at the Hammersmith Hospital were key British pioneers of chemical pathology computing from the mid-1960s.

Flynn used an Elliott 803 batch-processing computer to prepare cumulative reports, showing new results for a patient alongside those from any previous investigations.[7] Flynn and Whitby also installed an Elliott 903 minicomputer in each of their laboratories, connected directly to several types of biochemical analyser, to derive the numeric results for each specimen; as the results were printed a copy was automatically punched onto five-hole paper-tape.[8–11] At Flynn's laboratory these paper-tapes were read into the Elliott 803 computer and fresh cumulative reports were printed for delivery to the hospital wards and clinics. For other work by Whitby, see references.[12, 13]

The process-automation capabilities of the minicomputers were successful and versatile at interfacing with several types of analyser, including flame photometers: connecting directly to the colorimeter or flame unit, rather than monitoring a signal that came from the chart recorders used in Technicon and other continuous-flow analysers. However, the limited capacity and power of the minicomputers prevented them from handling database applications for worksheets and cumulative reporting. Elliott

Automation sought to market the 903 system for data acquisition to other laboratories. The computer had a memory of 24 kilobytes of six-bit bytes and no hard disk, plus magnetic tape and paper-tape peripherals.

Whitehead was one of the earliest chemical pathologists to work with Elliott Automation in developing the data acquisition system based on the EA 903.[14–18] He also was one of the first to invest in the automation provided by the main manufacturer of automated analysers, Technicon, integrated within the analysers. This meant laboratory staff could concentrate on developing data-processing facilities for the laboratory rather than the complexities of data acquisition.

Wootton largely side-stepped the online data acquisition developments, concentrating on the growing need for efficient data-processing facilities to ensure the accuracy of automated results and clear reports. He was a major co-developer with Flynn of the Phoenix laboratory system using Computer Technology's (CTL) Modular One computers and the CORAL-66 high-level programming language.[19–21] Flynn subsequently developed a system called Socrates.[22–24]

Early Technicon auto-analysers and other specimen analysers generated analogue signals that were fed to chart recorders. The staff operating the machine needed to:

- detect and measure the height of each peak on each chart;
- derive a numeric value for each parameter being measured;
- associate the values with a patient's specimen; and
- record the results on a worksheet or directly on to the report for sending to the clinician.

By 1971 several NHS laboratories had developed automation for deriving the results from such analogue signals. For example, the chemical pathologists at the Royal Infirmary, Dundee, developed direct links between their analysers and a CTL Modular One computer, using a CAMAC data acquisition front end. Direct asynchronous links from serial ports or teletypes on some of the analysers could also be connected to CAMAC cards. The results were either calculated by software in the minicomputer or assembled from the asynchronous links, and delivered to the reporting programs. Technologists at the Dundee laboratory used the CORAL-66 high-level language for developing some of their applications.

The CAMAC specifications had been defined by the British and CERN nuclear research organizations, and a variety of CAMAC modules were commercially available. Several such modules in a CAMAC 'crate' were able to be operated synchronously, presenting a succession of digitized values to software programs running in the host computer. This enabled signals produced by continuous-flow specimen analysers to be sampled many times per second, and for software programs to derive a value associated with each in a succession of specimens, representing the concentration of chemical substances in these specimens. The supplier of the computer, CTL of Hemel Hempstead, sought to market this set of products, and an early purchaser of one system was The London Hospital, Whitechapel.

A number of hospital laboratories in the UK, Germany, Denmark[25] and elsewhere developed electronic devices to 'pick peaks' from continuous-flow analysers in clinical chemistry and haematology, and calculate chemical concentrations or cell counts. These numeric results were usually presented as a string of characters for passing to a minicomputer that, at intervals, prepared the reports for sending to the requesting doctors.

Over time, the manufacturers of the analysers developed their own electronics for deriving the final results for each specimen. By the late 1970s NHS laboratories were able to install analysers that presented the results for several chemical analyses, or several types of blood cell count, associated with a single specimen as a series of numeric characters. These were suitable for printing directly onto specially pre-printed stationery and for passing electronically to a data-processing computer. Thereafter it was not cost-effective for individual laboratories to develop or buy electronics for deriving results from analogue signals. However, although such analysers increased staff productivity, they introduced the need for additional quality control procedures to ensure accuracy.[26–28]

By 1975 analysers such as the Technicon SMA-12 with a built-in computer became more affordable. Some machines were able to read the identity of a specimen while it was being sampled for analysis, derive numeric values for the concentration of several chemical substances or cell counts in the specimen, and present the specimen identifier (ID) and the string of results to a printer. This greatly reduced the chances of attributing the results to an incorrect specimen and increased the opportunities for monitoring the accuracy of the analyses and automatically adjusting the reported data when appropriate.

During the 1970s and 1980s mainstream clinical chemistry and haematology laboratories became increasingly automated, mainly because of the rising number of requests (typically 400 to 1,500 per day per laboratory). The numeric values forming the reports were easier to derive, print and store electronically than the qualitative information derived by human expertise in specialties such as microbiology, cytology, histology and diagnostic imaging.

Successful reporting services for these latter specialties involved technical staff either recording their findings directly into mainframe-type computer terminals (VDUs), or onto paper documents for transcription by clerical staff into the terminals. J. Davidson and his team at St Thomas' Hospital, London, used optical mark reading worksheets to avoid the manual transcription of the bulk of microbiology and other reports containing coded or multi-choice items.[29]

Computing aids for inputting textual findings and comments were usually confined to selecting standard phrases and statements from a laboratory's local 'phrase book', and very primitive text-editing commands.

In haematology, Coulter® blood cell counters were installed in many larger laboratories for deriving the widely requested blood count reports. If specially prepared multi-part forms were used by clinicians for requesting a blood cell count, and the same forms were referred to by whoever obtained the blood specimen from the patient, the form and specimen could be kept together at each stage of the testing. The form was then inserted into the machine's printer just before presenting the specimen to the analyser's aspiration tube, and the set of results was soon afterwards printed directly onto a pre-formatted area of the actual request form. The form also had space for differential counts of white cells and other manually derived results, plus handwritten comments by the specialist who approved the report for issue. The top copy of the form was then returned to the clinician.

This was a highly effective and efficient operation, complicated only by:

- the need for these special request forms always to be available to clinicians;
- difficulties with providing online links between the Coulter counters and computers for transferring results automatically;
- possibilities of cross-infection through the same paper forms accompanying the specimen and those from other patients during phlebotomy rounds and in the laboratory, and returning to whichever ward the patient was in when the report became available for filing in the case notes.

LIMITATIONS OF DATA PROCESSING

Until the 1990s most clinical laboratory computer systems were able only to handle data generated within one laboratory. They had no information about a request until a handwritten request form plus specimen(s) arrived at the laboratory reception point. There was usually then a critical bottleneck, as the identity and location of the patient had first to be copied into the laboratory computer by the booking-in clerk from the frequently incomplete and poorly legible data on the request form. The appropriate tests then had to be defined, and subsequently a laboratory ID number often had to be allocated to each portion of the specimen to be analysed.

This laboratory ID number was then used in scheduling and testing each specimen, with the results presented to the laboratory database identified solely by this internal laboratory ID number. At intervals during the day, laboratory staff would run batch programs on their computer to assemble the results for each report, approve them for issue along with any clinical comments, and then get the computer to print the reports, which were then mailed to the requesting clinician. Usually the doctor's first sight of the results was when the printed reports had been separated and delivered to the wards.

It was frequently very difficult for the clinicians to ascertain the progress of investigations they had requested, or to find out the contents of the reports before the final printed copies arrived at their destination and been filed in the patient's case notes folder.

To maintain a cumulative history of results for a patient, it is essential that the identity of the patient is clear on every request form, and that the booking-in clerk correctly matches each new request with any previous record for that patient. The hospital case notes number (CNN) ought to be reliable for achieving this linkage, and in many hospitals every request form is supposed to bear a printed patient ID label that clearly includes the patient's name and CNN. In practice, such labels are frequently missing, and handwritten ID details may be poorly legible, incomplete or ambiguous.

When the requests are defined in collaboration with the hospital's Patient Master Index (PMI), and the laboratory reception staff interact with the hospital's PMI when booking-in any handwritten requests, the likelihood of associating each new request with the correct patient's results history is greatly increased. This is only feasible when either the requesting services are part of a hospital-wide computer system that also supports the patient administration functions, or when sophisticated interactive links between the hospital and laboratory computer systems are in place.

Data-processing systems installed in a laboratory often relied on a senior technologist in the laboratory dedicating much of the working day to operating the computer system and dealing with problems and database maintenance.

SERVICES DEVELOPED AT THE JOHN ELLICOTT CENTRE

Context

In 1968, as part of the English DoH's Experimental Real-Time Computer Programme (see Chapter 4), The London Hospital, Whitechapel (which later became part of the Barts and The London NHS Trust) proposed a project to develop a communications network, with computer terminals in all wards and key service departments.[30] After this project was approved, a computer suite with offices for computer systems designers and programmers was built on the hospital site, and the building was named the John Ellicott Centre (JEC).

The first main computer was installed in 1971, and was used for all program development and the subsequent operational services for patient administration. A second computer was installed in 1972, superseding the Elliott 803 batch-processing facilities, and enabling much program development work to take place alongside the operational services on the main system.

In 1973, a CTL Modular One computer with a CAMAC data acquisition unit[31] was installed in the Chemical Pathology laboratory in the hospital, and

149

a Digico minicomputer was used to develop a local network of tailor-made keyboard–printer handsets for Haematology laboratory staff to use when manually counting blood cells through microscopes. These services were implemented only at Whitechapel to start with, and were later extended to the Mile End hospital.

Order entry of laboratory requests for chemistry, haematology and microbiology from ward terminals commenced in 1975, followed by reporting on screen and printed cumulative reports for microbiology in 1976. Operational services for passing laboratory results automatically from analysers in Chemical Pathology to the hospital-wide communications network commenced in 1978. Before 1985, Haematology staff issued cell count results on the manual two-part reports from the Coulter® counting machines, or into terminals on the hospital network, due to difficulties in linking the counting machines and the Digico computer to the main system. Figure 8.1 shows how the services subsequently grew to support a wider range of pathology laboratories, and also laboratories located in other hospitals connected to the JEC.

During the 1990s, the JEC implemented services for online patient administration plus integrated services for pathology at all hospitals in the neighbouring trusts of City & Hackney and Newham. The main computers were installed and operated by JEC staff at Whitechapel, with laboratory computers installed remotely in the laboratories. Service agreements were operated on a commercial basis between these trusts and the JEC.

With the agreement of the Barts and The London NHS Trust, the JEC collaborated with major software houses to enable them to tender for several hospital-wide computer contracts, notably at Greenwich, London, and Clwyd, North Wales, plus several contracts for specialty costing and management information facilities in England. In each case the software house was the bidding contractor. They were to provide implementation and front line user support, with the JEC providing the applications software and second-level support. None of these bids won a contract. Perhaps the software houses wanted a considerable financial safety margin for a first successful bid, and there appeared to be some unease amongst NHS managers that crucial data about their organization might be accessible to staff in another NHS trust, in spite of confidentiality clauses in the contracts.

The development process

In 1971, after an extensive evaluation of tenders,[2, 30] The London Hospital installed a Sperry UNIVAC 418-III real-time computer system, supporting a network of Uniscope-100 VDUs, for developing interactive services in wards and clinical laboratories as well as medical records and administrative areas. Apart from the supplier's operating system, the software environment for applications (transaction processing manager and file handler) was developed in-house, collaboratively with the supplier. This computer initially had just 192 kB of 6-bit memory, two 5 MB fast access rotating drum

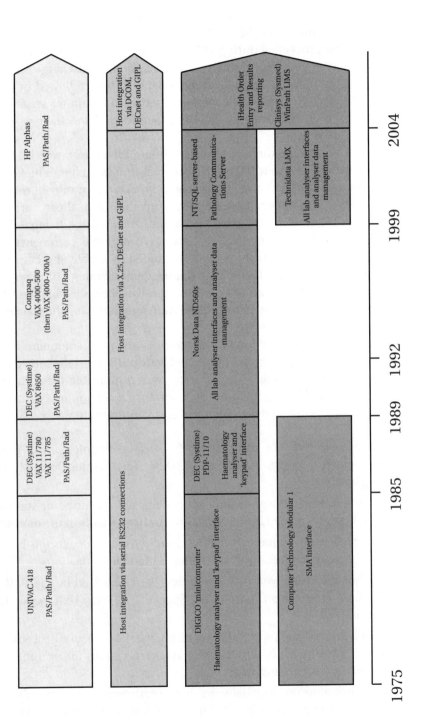

FIGURE 8.1 *A timeline of pathology information systems at Barts & The London NHS Trust since 1975*

stores, and 384 MB of mass storage on a three-ton moving-head pair of drums. A communications cabinet supported clusters of VDUs, connected by special multi-core cables installed around the hospital. The VDUs could display up to 16 lines of 64 characters.

PMI take-on proceeded during 1972, enabling waiting list take-on to be completed also by the end of that year. Admission and discharge services and services in the wards for allocating patients to beds followed in 1973, which was probably the first UK implementation of computer services in wards across a hospital, operated by nurses and clerks at their working locations.

The first implementation of clinical laboratory services was in 1975, when requesting from ward VDUs enabled doctors and nurses to make one or more requests for their patients almost as quickly as writing out manual request forms. The patient was selected from the list of those currently in the ward, followed by the pathology (clinical chemistry, haematology, microbiology, etc.) and then up to four tests; optional comments and specimen collection instructions could be added. A legible and complete order was thus generated, including a unique eight-character ID. A request form was then printed, bearing four self-adhesive labels showing the ID of the request and the patient's name; these were for identifying the specimen containers.[32, 33]

Emergency 'same day' request forms were printed within moments in the centralized laboratory reception office, and were put in an alphabetical rack according to patient name. The requester was responsible for obtaining the specimens and taking them to the laboratory reception office, finding the form in the rack, and delivering form and specimens to the booking-in clerk.

Routine request forms were saved for sorting overnight by a batch program, which then printed in bed number order, request forms in the following batches:

- For each ward, those needing specimens such as urine or sputum, which were obtained from the patients by nurses early in the morning.
- For each ward, requests needing blood specimens collected by the phlebotomy service – most mainstream blood investigations.
- For special units, where all specimens were obtained by the staff in the unit – Intensive Therapy Unit, Special Care Baby Unit, paediatric wards.
- For each laboratory, specimens requiring special collection procedures, which were done by experienced laboratory staff, for example for blood cross-matching and for blood cultures.

These batches of forms were delivered in the early morning to the relevant locations, sorted by patient for efficient collection procedures. As all forms were together in ward and bed number order, staff were able to minimize the number of times any patient was pricked per day for blood specimens.

In all cases, when the forms and labelled specimens were delivered to the laboratory reception office, the receptionist typed the eight-character ID for the request into the VDU and confirmed that all specimens were present, and entries were then automatically made on all relevant laboratory work lists. This fast reception procedure minimized the time-critical bottleneck at each laboratory for the majority of ward requests. Analysis then proceeded as soon as specimens had been prepared and delivered with the work lists to the relevant bench.

Each patient's identity in the PAS was thus securely embedded in each request and carried through into the reports. This enabled a comprehensive cumulative history to be maintained of all reports for each patient across many pathology disciplines. Computer terminals in wards and other areas were able to present to authorized hospital staff access to this cumulative history of validated and approved reports. Additionally, a 'Requests in Progress' service listed all requests made recently for a patient, the date and time each specimen was booked in at the laboratory, when each result first became accessible, and when the report for the hospital case notes had been printed. By selecting a request from this list, any results could be viewed, with those not yet finally released as complete clearly flagged as 'unapproved'.

Doctors and nurses in the wards and clinics were thus able to keep track of what was happening to their requests, and to view results as soon as the laboratories were prepared to release them.

The first laboratory to implement online reporting was microbiology. At that time, in 1976, all results had to be derived by hand, and so there was no immediate alternative to transcribing results from worksheets into VDUs. The data structures needed for culture and sensitivity reports are considerably more complicated than those for other specialties; nonetheless the implementation of cumulative microbiology reports was successful and effective.

For all laboratories the work lists either defined the laboratory ID numbers to be used, or a computer terminal was used by a technician to record them. Where there were automated analysers, the results were passed automatically through online links to the hospital system. Other results were recorded manually on the work lists and copied into a terminal in the laboratory. Validation and approval then took place at a terminal, where there was also access to any cumulative results history for the patient. Overnight, the hospital computer operators printed the cumulative reports for delivery to the wards. Individual reports for outpatients and general practices were also printed centrally and mailed out by the laboratory office staff.

The layout of each type of result was defined by the laboratory doing the tests, but the task of filing reports in the patient's case notes was simplified by the standardization of sheet size and filing method for all pathologies. Cumulative pages replaced any superseded pages inside a folder, which formed the 'investigations' section in the patient's case notes.

In chemical pathology, results from multichannel analysers were transferred to the hospital's main computer via a CAMAC data acquisition unit and a Modular One minicomputer in the laboratory. Should the data acquisition function fail for a period, the analyser's printer was used to punch the results on paper-tape, which was periodically read into the minicomputer; the results were then automatically forwarded to the hospital-wide system. The hospital's UNIVAC 418-III computer was replaced in 1985 by a DEC VAX 11/780 computer linked cooperatively with a VAX 11/785 computer. The main automated analysers were connected individually to the new hospital system and the old laboratory data acquisition unit and computer were discarded. By then the major analysers had the capacity to store a day's worth of results; this enabled automatic catching-up of the transmission of results to the hospital system following all but the most prolonged periods of downtime of the systems or the hospital network (see Figure 8.1).

By 1975, a small Digico minicomputer was installed in the haematology laboratory, and a number of tailor-made handsets were used by technicians at microscopes to carry out differential white cell counts and other investigations. In 1985, this computer was replaced by a DEC PDP-11 minicomputer, which collected results from two Coulter® counters via bought-in interface boxes, and also supported the microscope handsets for deriving and printing differential cell counts and several other types of result.

The laboratory computer forwarded the results within a few seconds to the hospital system, where clinicians could immediately view the provisional results. Once the laboratory staff had completed the requested investigations and checked the whole report for clinical significance on the main hospital system, the provisional 'unapproved' qualification was immediately removed. This service was particularly popular with doctors and nurses in the wards because of its fast turnaround interval between making an urgent request and the results becoming viewable in the wards.

Later, an Abbott ADC 500 machine carried out the differential counts automatically; individual interface conversion units on this and the Coulter® counters were linked directly to the VAX 11/780 + 11/785 hospital system.

Cytology reports usually contain one or more standard phrases, or free text, and these were handled similarly to microbiology on the main computer system. Histopathology reports were usually transcribed from audio dictation tapes into VDUs by specialist secretaries. To provide more effective word-processing facilities to these secretaries a Norsk Data multi-user minicomputer with suitable software was bought and linked to the hospital system (see below). Later developments in the integration of PCs with the hospital system enabled proper word-processing facilities to be used for data entry in many other areas (see Figure 8.1).

In 1989, to provide additional facilities for versatile editing and quality control, as well as local data acquisition, Norsk Data (ND) minicomputers with the Systemator high-level language were installed locally in the histopathology, then in haematology and clinical chemistry laboratories. These were connected to the hospital systems as intelligent clients, using software developed by JEC staff for both the VAX hospital systems and the ND laboratory computers. The X25 protocol was used for messages passed between the VAX and ND computers. Data about each request were sent from the VAX to the laboratory computer as soon as each specimen was booked into the laboratory; approved report data were sent in the opposite direction for viewing and printing by applications running in the hospital computer. There is additional information about these ND computers and their interfaces in the following section titled *Network developments and interconnectivity*.

Similar ND configurations were installed in the laboratories at two other hospitals after they became users of the same hospital-wide patient administration and laboratory services developed and supported by staff at the JEC.

During the 1990s, developments supporting interconnections between computers and greater flexibility in laboratory information management systems (LIMS) began to make it possible to achieve many of the above integrated operations using PAS on the hospital system and a laboratory-based LIMS. This was planned to be implemented in the late 1990s, as part of a large-scale redevelopment of the Whitechapel site under a private finance initiative (PFI) project.

As technology developed and fibre optic cables replaced traditional copper wiring between computers and around the hospitals, networking between computers was able to grow in capacity and speed, and continues to do so.

Network developments and interconnectivity

The UNIVAC 418-III system installed at The London Hospital in 1971 used a proprietary synchronous protocol for connecting VDUs to the computer with expensive multi-core cables. This made it very difficult to connect any equipment other than VDUs, remote job-entry units and serial printers to the network interface units.

The first online link to the 418-III was from a CTL Modular One minicomputer in the chemical pathology laboratory. A program was developed for the Modular One so that it responded like a VDU, formatting results from the analysers and sending them to the UNIVAC system as though they had been typed on a VDU's keyboard. A special application in the UNIVAC system controlled sessions with the laboratory computer and imported the results into the main laboratory database. The UNIVAC protocols included error checks, automatically causing any faulty transmissions to be sent again. This link was used successfully for about eight years (see Figure 8.1).

The DEC PDP-11 minicomputer in haematology initially used a similar principle, as later UNIVAC VDUs could support an auxiliary interface, through which external input and output devices could exchange data with the screen display and mimic control actions such as the 'transmit' keystroke. A card in the PDP-11 computer supported a link to a UNIVAC VDU running nearby and application software running on the PDP-11 formatted a set of results for one specimen from the cell counters, added the 'transmit' command and other control characters, and sent the message to the VDU. A dedicated application in the UNIVAC system maintained the VDU in a state ready to accept a new set of results, and as each one was received without error, an acknowledgement message was sent back to the PDP-11 via the VDU. The next message was then uploaded. This avoided the need for a synchronous interface card in the PDP-11 and a complex VDU emulation program. The laboratory staff could see the messages appear briefly on the VDU screen while the link was operating, and they could use the VDU as an ordinary terminal when the link was not in use.

The DEC VAX computers, bought in 1980 to replace the UNIVAC system, supported asynchronous serial links for input, as well as output to remote printers. This enabled the PDP-11 computer in the haematology laboratory, and the laboratory analysers with a serial feeder port, to be connected directly to the VAX systems over modem links (see Figure 8.1).

The DEC VAX 11/780 computers supplied initially in 1980, with Systel transaction management software, were functionally suitable for the applications developed by the computer staff in the JEC. However, the systems could not sustain the substantial number and variety of concurrent transactions generated by users. A very detailed set of studies and load tests involving a major independent software house (Scicon) was commissioned to recommend how the performance shortfall should be overcome. The eventual solution involved:

- changes to the Systel product;
- certain design standards for JEC applications programmers;
- more powerful VAX computers;
- performance improvements in later versions of the VAX VMS operating system; and
- the development by JEC staff of software that enabled more than one VAX computer to be linked together to share the transaction load.

The initial priority was to implement PAS using the new systems in two neighbouring health authorities. The service to the smaller authority, Newham, was implemented first, in 1981–82. This was satisfactory using a single VAX 11/780 as their transaction load could be sustained with adequate response times at user terminals. The City and Hackney authority implemented PAS on a more powerful VAX 11/785 later, in 1982. Meanwhile, JEC staff developed an inter-processor link (IPL): a piece of software that enabled applications running on one VAX to exchange messages very rapidly, using DECnet™, with applications running on another VAX computer. The network connections used ethernet standards between computers, and later also out to PCs in the hospitals.

The IPL enabled JEC staff to enhance key transactions. For instance, when a doctor in a ward requested a blood test for a patient, even though his/her VDU was connected to the VAX computer running the PAS applications, the request record could be set up and maintained in a different VAX computer to which were connected all the VDUs in the laboratories. Conversely, applications running in the laboratory VAX could interrogate PAS data in the VAX used by the ward users. This was the first use of 'client–server' technology by JEC staff, enabling more than one computer to share data interactively while making sure that the databases in each computer contained synchronized and reliable cross-references to data maintained on other linked computers.

Eventually, after extensive loading tests, a VAX 11/785 linked via DECnet™ and IPL to a VAX 11/780 were shown to be capable of sustaining the full range of applications and transactions per second needed for The London Hospital. During a long weekend in November 1985 the services were closed down on the UNIVAC system. All the application data were converted to the very different data structures used in the VAX VMS systems, and a totally new set of VDUs, printers and other connections was commissioned in all user locations throughout the hospital. Admissions and medical records staff checked samples of data in the new system, and then entered all admissions, transfer and discharge data for the period since closing the old service. Then services to wards and laboratories were opened up in a controlled sequence, to minimize overloading due to peaks of activity during the crucial catching-up period.

The transfer of services from the UNIVAC to the VAX computers was not done in stages, but in a single changeover exercise lasting four days. This was for the following reasons:

- Wards and laboratories had no space for both old and new VDUs in operation for an extended period, and the style of operating services on them was rather different.

- The application database worked as a tightly integrated whole, and to maintain database integrity during parallel running would have required enormous extra development effort.

- The changeover was to be as rapid as possible.

Since then, all major upgrades and additions to the computers and networks have taken place without such a major upheaval, due mainly to the flexibility of modern network technology and the increasing modularity of applications. The future replacement of the PAS services will be a complex exercise, due to the large size of the patient index and episode summaries, which have grown substantially since the initial take-on of data for about one million patients in 1972.

DECnet™ also enabled rapid fallback to alternative computers if any system failed, or needed to be withdrawn for routine maintenance. This further improved the availability of services to the clinical laboratories and other key departments. It also enabled additional computers to be added to the network, so that computer services could be implemented in additional hospitals, and authorized support staff could connect dynamically to relevant systems from their own terminal in response to user calls, and to monitor system activity. Later, some of the computers were moved to an off-site computer room as a disaster protection measure. This ensured that at least key services could be provided to crucial departments in event of a local catastrophe, such as the Baltic Street and Isle of Dogs bomb blasts that took place only a few miles from The London Hospital.

Several clinical laboratories served by the JEC installed Norsk Data (ND) minicomputers in the late 1980s. These computers hosted applications for gathering results from analysers, reviewing and editing blocks of results in spreadsheet-style tables, word-processing for free text reports and approving results for issue. They needed to receive data about specimens booked-in via the hospital system in a secure fashion and to create matching entries in a local database for the laboratory carrying out the tests. Once results had been approved in the laboratory computer, they had to be securely copied to the hospital system where they were matched with their request data and made available for report viewing and printing.

To achieve this client–server level of integration involving computers in user departments, staff in the JEC enhanced the IPL messaging software to support the X25 standards, as well as DECnet™; the new version was called the generalized inter-processor link (GIPL) (see Figure 8.1).

The ND computers also supported the ISO X25 networking standard: designed to provide vendor-independent communication links.

The laboratory booking-in service on the hospital system was enhanced to forward a copy of each request's data via GIPL to the appropriate ND laboratory computer. Approved results were copied to the hospital system over the same interfaces by programs in the ND computers. The application programs using the GIPL in both systems had to be designed to ensure that database integrity was assured across both systems, and that no data were 'signed-off' as reproduced in the 'other' system's database until its receipt had been acknowledged.

The GIPL demonstrated that hardware and software systems of very different architectures could be integrated efficiently, without compromising the completeness and integrity of the data stored and used in each system. A good foundation was thus created for building data exchanges between the hospital databases and additional systems developed by external suppliers.

PCs as HISS terminals and intelligent clients

PCs were introduced progressively across many departments in the hospitals from the early 1980s, primarily for word-processing in secretarial areas. The first relationship to be implemented between PCs and the hospital services on the VAX computers was to enable a PC to be a 'dumb' terminal, able to open a window on its screen that emulated that of a VDU terminal. This enabled consultants to view laboratory results and other services in their offices and other locations already equipped with a suitable PC, just for the cost of installing an asynchronous serial connection to a local cluster controller and a copy of the terminal emulation software. Of course, every additional active 'terminal' increased the loading on the main computer, and periodically additional capacity had to be added to the central systems. The next development was to enable a PC to interact with the main systems as an 'intelligent' client. This required equipping the PC with a DECnet™ ISA card and associated software, and applications that could exchange packets of data with other computers over DECnet™.

The GIPL described above enabled suitable applications running on any networked computer to send and receive packets of data with applications on any other enabled computer.

The first requirement to build a bridge via the GIPL facility from a PC environment came from a Ward Nursing Care Plans application. During the 1980s, McDonnell Douglas Information Systems (MDIS) marketed an impressive application that enabled nurses in hospital wards to create, update and modify nursing care plans for their patients. The London Hospital was keen to implement such services, but wanted the care plan services to use the up-to-date data about the patients that were maintained by the PAS. The requirement was for the PC application to display the list of patients currently in a ward (as maintained by PAS), so that the nurse could just click, with the PC mouse, the name of the patient whose care plan needed attention. Details of the selected patient were then copied from PAS for the next stage of the care planning application running on the PC. The transfer of patients between wards and discharge procedures in the care planning application were also to be integrated with PAS.

MDIS agreed to develop a database object (DBO) in a set of dynamically linked library (.dll) files programmed in the C++ language to enable their applications running under Microsoft® Windows® to communicate over DECnet™ with the JEC's GIPL, and thence to exchange messages with hospital applications running in the VAX VMS/Systel systems.

This interface was highly successful, enabling suitably networked PCs to display lists of 24 or more patients in a ward, using up-to-the-moment data in PAS, responding at least as quickly as if the data were on a local relational database PC file server.

MDIS also agreed to license the use of the DBO by the John Ellicott Centre for additional PC applications. Thus JEC staff became able to develop PC services using remote procedure calls from any desired Microsoft® Windows® applications on authorized PCs to exchange real-time messages, accessing and updating live patient data in the hospital's main computers.

Examples included access to PAS data from within Microsoft® Word for activities such as discharge letter production by medical secretaries and diagnostic imaging reporting.

As the new millennium approached, it became clear that the ND laboratory front end systems were not millennium compliant. They had given about eight years' service and would need to be replaced before the year 2000. Because the major hospital redevelopment PFI project (mentioned earlier in this chapter) was encountering many delays, it soon became clear that an interim replacement for the ND computers was urgently needed. A decision was made to purchase a PC-based Bayer LMX system to replace each ND computer, but the Bayer system required request data and results data to be presented according to the international ASTM laboratory data interchange standards.

Bayer agreed to interface their programs to receive data about each booked-in specimen from the hospital system and to send results back to the main system. This was achieved by frequently exchanging files containing requests and reports in ASTM format. JEC staff built a Microsoft® SQL server-based integration engine called a pathology communications server (PCS), with applications for queuing requests and reports between the Bayer system and the hospital system. PCS communicated with the Bayer system via the ASTM-based interface, while messages to and from the main hospital computer were communicated via the DECnet™-based DBO to GIPL on the VAX VMS/Systel system. The message queues ensured that neither side of the PCS was held up waiting for a response, and data packets were transferred in 'near real time' (see Figure 8.1).

Support for multi-hospital authorities

As outlined above, the combination of DEC's VMS and DECnet™ with the Systel transaction management environment provided excellent options for the JEC to replicate its applications to serve a number of hospitals besides The London Hospital.

A separate Systel environment and VMS database was configured for each NHS Trust. Usually, each trust's system operated on its own VAX (later DEC Alpha) server, although it was possible to share a server during fall-back situations. All the servers were installed in one of the computer suites managed by the JEC on behalf of all members of the consortium. Networked PC servers were usually also installed in one of the protected computer suites, rather than in hospital offices. Laboratory computers were usually installed close to the equipment connected to them, for ease of management and simplicity of communication wiring to their local user terminals.

When there was more than one hospital in a Trust, all the hospitals shared the same database, so that staff could log in at any terminal in the Trust's buildings. Senior staff who practised at more than one hospital or laboratory in that Trust were given access to all the relevant sites under their one password. Hospital staff had no means of accessing or altering data relating to any Trust apart from their own employer. Clinicians and other senior staff practising in more than one Trust had a different password on each system, giving access only to the services each contract entitled them to use.

Each Trust in the consortium maintained a service-level agreement with the JEC, and JEC staff compiled statistics that were reported regularly to each Trust so that compliance with each agreement could be monitored. As mentioned above, support staff could have their VDU or PC configured to switch between more than one VMS/Systel environment, helping them in investigating user queries from more than one Trust.

Automatic updating of clinical department patient records

During the 1980s, a consortium of four NHS Trusts in the North East Thames Region of the NHS went out to tender for an adaptable system for clinical records. This was to enable clinicians in a wide range of specialties to maintain and analyse data about the treatment and progress of their patients. The systems were required to receive data feeds from patient administration and laboratory applications on the hospital systems, automatically maintaining a database of patients treated by each clinical firm, including their laboratory reports. The doctors should then only need to define the clinical data they wished to maintain and ensure they had procedures for entering such data. Discharge letters and other referral letters should be configurable to suit each firm's preferences.

The only remotely viable tender was for PROTON applications running on VAX servers with 'dumb' VDUs, using Read Codes for diagnostic and other clinical data.

The differing brands of PAS used by the trusts in the consortium were already preparing data feeds to a regional standard, and copies of these feeds were processed to supply details of patient movements and discharge coding to that trust's clinical records system.

JEC staff also developed a daily batch feed of laboratory reports. These batch feeds were imported successfully into a collection of specialty databases in PROTON. Several clinical firms agreed to pilot the use of PROTON, including Metabolic Medicine and General Surgery. PROTON included impressive options for displaying numeric laboratory results and other events graphically against an adjustable timeline, which was particularly relevant to the medical firms.

Unfortunately, the following issues combined to cause the project to be terminated:

- The limited options for validating input fields.
- The ways the doctors wanted to work on batches of patients for discharge coding.
- The high computing overheads and poor response times of the interpretive PROTON software when used by more than a handful of concurrent users.

The feed of laboratory results, however, proved to be adaptable, and formed the basis for the service to general practices (see below).

Automated reporting to primary care computers

An increasing number of general practices installed PC-based record systems and transferred their patient records onto their primary care system in the early 1990s. In computerized practices with more than one doctor, there was usually a PC in each consulting room and in the reception area networked to a file server. In London, the most widely used brand of system was from EMIS. Usually, the EMIS file server was connected to a modem and PSTN line, primarily to enable the supplier to provide user support and install upgrades to the software.

EDIFACT standards had been defined internationally for transferring data-sets relating to health care, including reports from clinical laboratories for investigations requested by GPs. Several doctors with systems from EMIS had been asking the clinical laboratories to send reports to them electronically, and JEC staff undertook the development of such transfers using the EDIFACT standards.

There were at the time some PC-based products that could process fixed-format reports from a laboratory computer, and dial up the relevant practice computers to transfer the reports to them. In order to have a linkage that could reliably handle the wide range of result layouts handled by JEC systems, including microbiology, it was decided to purchase a full EDIFACT software package and a small VMS server to run it, and to develop a feeder application to pass laboratory results into this translator. Twice per weekday, batches of reports were transmitted to each practice system using modems with some built-in security checks.

EMIS practices were able to receive these transfers and to update patient records automatically. Until NHS numbers were included regularly in the patient ID on request forms from the practices, and laboratory booking-in staff used them when matching patient ID against the hospital's PMI, a significant proportion of reports needed to be matched manually at the practice before the results could be added to the patient's record.

After a period of successful operation, the EDIFACT server at the JEC was replaced by a commercially maintained server for each laboratory, sending results automatically to relevant GPs.

Implementing services for diagnostic imaging and other specialties

JEC staff developed VDU services to support order entry and reporting of examinations for in-patients, equivalent to the established applications for clinical laboratory investigations.

Most radiologists continued to dictate their reports into audio recorders, and secretaries transcribed the text into VDUs in place of typing paper reports. Basic text manipulation aids and 'boiler plate' phrases and sentences aided this process.

Once PC integration had been developed, as described above, additional services were developed by JEC staff, to print bar-coded labels for films, and to enable word-processing to be primed with a skeleton report including patient and exam details. After completing the typing on a PC, the secretary used a command to upload the provisional report into the hospital system, where it could be checked and approved for issue.

Voice input using Dragon® software was also developed. Some keen radiologists, especially those on emergency duties, used this to dictate their reports directly into the PC, and then check and approve them immediately. The limited power and capacity of PCs available at that time, coupled with the need for the radiologist to have his/her voice pattern on each reporting PC he/she used, meant that most reporting continued to be done via audio tapes.

SUMMARY OF FEATURES EMERGING IN EACH DECADE

1960s

Many hospitals in the USA, that needed to produce itemized bills for patients, installed mainframe computers for batch processing and remote job entry, to assemble priced details of every test and item of care associated with each patient. Laboratories in the UK and the USA started to introduce offline services based on punched cards, sometimes using sorting and tabulating equipment to prepare reports, derive 'normal ranges' for results, develop quality control measures and analyse workload.[34–42] Some UK hospitals acquired small commercial batch computers for finance, payroll and operational research.[30]

From 1967 onwards, several UK clinical laboratories reported the use of computers for deriving results from the analogue electrical signals produced by mechanized specimen analysers, and then preparing printed reports.[9, 31] Some computers assembled results for each patient into 'cumulative' reports, to help the doctors who made the requests track a patient's progress. In 1968, the English DoH invited several hospitals to submit feasibility studies for 'experimental' projects to demonstrate the value of using computers in NHS hospitals in improving patient care.

1970s

In England, the experimental projects succeeded in implementing interactive services, several of which included the work of clinical laboratories.[29, 32, 33, 43, 44] In Northern Europe, several University hospitals developed similar applications, often using 'minicomputers'.[25, 45–49] In the USA, large-scale mainframe-based implementations of hospital information and support systems, including order entry and reporting for clinical laboratory investigations, became quite widely used, with levels of funding substantially greater than those available to non-billing European hospitals.[50]

Microcomputers started to be used in locally developed and commercial signal monitoring boxes, which enabled laboratory computers to import results from auto-analysers and Coulter® counters more easily. However, implementing automated analysers in a laboratory sometimes introduced new problems and overheads, to be balanced against the advantages gained.[28, 51, 52] Automated analysers with built-in computers became more readily available, often making it easier for laboratory and hospital computers to import multi-parameter reports as a string of ASCII characters.[19] Some laboratories producing non-numeric results developed reporting aids using word-processing and text-manipulation facilities for microbiology, serology and histopathology.[53]

After the national moves to 'standard systems' and commercial products in the mid-1970s, from 1984 onwards opportunities for preparing papers and attending conferences petered out. At other hospitals, local development staff were moved to RCCs or they got jobs outside the NHS, leaving many hospitals with legacy systems of limited capability.

1980s

The DoH required RHAs to establish regional computer centres, where 'standard' applications were operated on a bureau basis for all hospitals in each region.[54] Central funding for the experimental programme had tapered off, and hospitals were faced with replacing their computers, some of which had given a decade's service and were becoming difficult to maintain.[55] Hospitals had to finance their own replacement programmes, many moving to one of a few MUMPS-based PAS systems acquired by commercial vendors from the experimental programme. These rarely included modules to help clinical laboratories, leaving most sites reliant on no more than networked access to separate laboratory computers for results queries; provided such systems had the capacity and networking facilities to support such enquiries.[56, 57] A few sites funded their system replacement by undertaking to implement their services at additional hospitals on a bureau basis, with a formal customer service-level agreement. The few authorities who managed to retain their development and support staff, and who could rework their applications and carry forward all the data into their new systems, managed to implement and support their services at additional sites and go on to develop additional services. The Royal London Hospital, Whitechapel, was one such hospital, eventually supporting services on up to 12 sites in the City and East London area. Other hospitals lost many of their computing staff to the regional computer centres and to commercial companies.

Following the success of ARPAnet in academia in the 1970s,[58] an important development was the explosion in the use of the internet in the 1980s and subsequently. Another notable advance in the 1980s was the fully fledged integration of PCs, not only as intelligent terminals to computers in laboratories and hospitals, but also as clients and servers in client–server networked applications. This enabled commercial software that ran in PC and minicomputer environments to exchange data packets with hospital systems at a few English hospitals (e.g. a laboratory PC used to 'book-in' a new request could interrogate the hospital's PMI or list of a ward's current patients, and a laboratory computer could send reports to ward viewing and management information services on hospital systems). Also, the international standards for electronic data interchange (EDI) which evolved during the 1980s, notably EDIFACT, HL7 and ASTM, were implemented in commercial message-handling products and several GP primary care computer systems. Using the new NHS secure national network, automated downloading of laboratory results from hospital laboratories to practices and automated merging of the results into the patient records at the practices began to be implemented. EDIFACT was used for early hospital–GP links, ASTM for clinical laboratory applications, and more recently HL7 has become widely used internationally in health care links.

1990s and the new millennium

By the early 1990s the computer services provided by the John Ellicott Centre to NHS hospital trusts in North East London had grown into a highly functional HISS, with a wide range of services additional to those described above. Other networked systems supported management information, specialty costing, email and intranet services.

Such growth has been possible mainly because:

- The computer hardware from Digital Equipment Corporation, first installed in 1981, was in a family of products that has maintained upwards compatibility as technology developed, through the VAX 4000 range in 1992, and then the DEC Alpha servers, so that individual parts of the systems could be replaced and supplemented in easy stages, and services maintained largely without interruption.

- The VMS operating plus file handling software, the DECnet™ networking product range, and the Systel transaction processing software are fundamentally well designed, and are still supported over 20 years later, so that applications programmed in the early 1980s have continued to perform reliably and responsively over many years.

- Ethernet networking products have been highly effective and constantly enhanced, so that transmission speeds between servers can now be in gigabits per second, and could soon be 100 megabits per second to some user PCs over fibre optic cables.

- Staff employed in the John Ellicott Centre and the Barts and The London NHS Trust have been able and loyal computing professionals, many with long service awards. The Trust's management has understood that the benefits of information technology only start to multiply when most parts of the organization have ready access to all the up-to-date data they need for their work responsibilities, and so the Trust has maintained a viable computing unit with software development as well as operations staff.

- Integration of computing products does not 'happen' – it has to be carefully designed and engineered. Data completeness and consistency across all systems must be maintained, even when errors, user mistakes, equipment and network failures, and other contingencies occur. Interfaces must support 'open' (i.e. non-proprietary) standards wherever possible, so that products from commercial suppliers can be added and integrated, and later replaced, with fairly predictable success.

In the mid-1990s, the Barts and the London NHS Trust recognized the need to plan for the replacement of the core HISS services in a controlled fashion, maintaining integration with the other IT services used by the hospitals. The approach was to 'separate and then replace' modules of the HISS by a mixture of locally developed and commercially available 'best of breed' products, with integration managed by the trust's staff.

Later in the 1990s, the DoH initiated an Electronic Record Development and Implementation Project (ERDIP), and also published a six-level definition of advanced health care information systems titled *Information for Health*. This definition provided a framework around which the trust adapted its own systems modernization programme.

The team designed an integrated clinical workstation as a key component; web-based technology, rather than the DECnet™ proprietary products, became the portal through which all data required by clinicians and other staff were accessed and updated. This also involved a major architectural change; rather than having the integration interfaces mostly in the client workstations, they were implemented in networked servers. The client PCs could then run applications that used only standard internet-style interfaces to access data held on a variety of specialized servers elsewhere in the trust.

This approach has made it possible to replace sections of the monolithic HISS by degrees, while also simplifying the software installed in the ever-growing number of PCs on the Trust's network. The scale, risks and complexity associated with such phased implementations are now far less than with the 'big bang' changeover that was unavoidable in the past.

The first step taken under the new approach was to define and develop in the HISS a library of service components, covering the main actions of viewing and updating selections of HISS data with which modern services could be expected to interact. The integration services were implemented as a set of DCOM-based application program interfaces (APIs), which exposed data in the HISS via a rich set of function calls. Data could be exchanged in a variety of modern formats, including Microsoft® ActiveX® Data Object data-sets and as XML documents (see Figure 8.1).

The first major stage in the Trust's HISS replacement process involved the clinical laboratory services for order entry and results viewing from wards and other user locations. This was to coincide with the relocation of the main laboratories into a newly built pathology centre, and the purchase of a comprehensive laboratory information management system (LIMS).

A contract was signed in 2003 with Sysmed (now Clinisys) for its WinPath system. They were also to include web-based pathology order entry and results viewing services from a New Zealand company, iHealth (now part of iSoft). All these facilities also had to integrate with the HISS for patient demographic and other data, and the viewing services had to integrate transparently with the trust's web-based Electronic Patient Record (EPR) Viewer portal.

The integration of WinPath with HISS was achieved rapidly. There were difficulties in getting the iHealth applications to operate seamlessly as servers to the trust's EPR Viewer application, until the trust's staff provided a module that runs on the iHealth server.

During this period, the NHS Care Records Project was published, somewhat overshadowing *Information for Health*. The Barts and The London Hospital Trust is one of probably a few NHS trusts that are well placed to support any likely nationally agreed patient records and booking services. It remains to be seen if the expensive NHS project to provide countrywide basic PAS services, using products supplied under a PFI contract, will be completed successfully.

REFERENCES

1   Allegaert, P., Kint, E., Willems, J.L., Ponette, E. and Baert, A. (1982) A computer assisted radiology reporting and retrieval system. In O'Moore, R.R., Barber, B., Reichertz, P.L. and Roger, F. (eds) *MIE, Dublin 82: Lecture Notes 16*. Springer Verlag, Berlin. ISBN 0 387112 08 1.

2   Rowson, J.E.M. (1984) The reality of implementing a hospital information system (or: out of the frying pan into the fire?). In Kostrewski, B. (ed.) *Current Perspectives in Health Computing*, pp. 41–8. Cambridge University Press for BCS, Cambridge.

3   Hammersley, P., Roach, M.E., Robertson, V.S. and Campbell, T.H. (1977) Planning for transferability. In Online Conferences Ltd (eds) *MEDCOMP 77*, pp. 259–70. ISBN 0 903796 16 X.

4   Hammond, W.E., Stead, W.W., Walter, E.L., Feagin, S.J. and Brantley, B.A. (1977) Transferability of projects in computer medicine. In Online Conferences Ltd (eds) *MEDCOMP 77*, pp. 271–7. ISBN 0 903796 16 X.

5   Brink, A.M.A., Koedam, J.C. and Slik, W. (1977) Transferring an information system for clinical laboratories. In Online Conferences Ltd (eds) *MEDCOMP 77*, pp. 239–45. ISBN 0 903796 16 X.

6   Mieth, I. and Porth, A.J. (1979) What about 'Turn-key' systems for clinical laboratories? In Barber, B., Grémy, F., Überla, K. and Wagner, G. (eds) *MIE, Berlin 79: Lecture Notes 5*, pp. 394–413. Springer Verlag, Berlin. ISBN 0 387095 49 7.

7   Flynn, F.V. and Vernon, J. (1965) Cumulative reporting of chemical pathology, *J. Clin. Pathol.*, September, 18(5), 678–83.

8   Flynn, F.V., Piper, K.A. and Roberts, P.K. (1966) Equipment for linking the auto-analyser to an offline computer. *J. Clin. Pathol.*, November, 19(6), 633–9.

9   Whitby, L.G. and Simpson, D. (1969) Experience with online computing in clinical chemistry. *J. Clin. Pathol. Suppl. (R. Coll. Pathol.)*, 3,107–26.

10 Whitby, L.G. and Simpson, D. (1973) Routine operation of an Elliott 903 computer in a clinical chemistry laboratory, *J. Clin. Pathol.*, July, 26(7), 480–5.

11 Simpson, D., Sims, G.E., Harrison, M.I. and Whitby, L.G. (1971) Equipment for linking the auto-analyser online to a computer, *J. Clin. Pathol.*, March, 24(2),170–6.

12 Whitby, L.G. (1971). Special requirements for automation and mechanization. *J. Clin. Pathol.*, November, 24(8), 766.

13 Whitby, L.G. (1979) Computer applications in clinical enzymology. *Clin. Biochem.*, December, 12(6),194–6.

14 Whitehead, T.P. (1965) (Title unknown.) In (eds unknown) *Progress in Medical Computing*, pp. 52–6. Elliott Medical Automation, London.

15 Whitehead, T.P., Becker, J.F. and Peters, M. (1968) (Title unknown.) In McLachlan, G. and Shegog, R.A. (eds) *Computers in the Service of Medicine*, vol. 1, pp. 115ff. Oxford University Press for the Nuffield Provincial Hospitals Trust, Oxford.

16 Whitehead, T.P. (1969). The computer in the laboratory. *Practitioner*, September, 203(215), 294–505.

17 Whitehead, T.P. (1969) Symposium on automation and data processing in pathology. Preface. *J. Clin. Pathol. Suppl. (R. Coll. Pathol.)*, 3, xi–xiv.

18 Whitehead, T.P. (1970) Automated clinical chemistry, *Proc. Nutr. Soc.*, May, 29(1),106.

19 Abson, J., Prall, A. and Wootton, I.D. (1977) Data processing in pathology laboratories: the Phoenix system. *Ann. Clin. Biochem.*, November, 14(6), 307–29.

20 Wootton, I.D. (1978) Hospital laboratory computing. *BMJ.*, 1 April, 1(6116), 851.

21 Clarke, A.A., Coleman, S.J., Prall, A. and Wootton, I.D. (1980) Data processing in pathology laboratories: the extension of the Phoenix system into haematology. *Clin. Lab. Haematol.*, 2(1), 63–72.

22 Flynn, F.V. (1969) Problems and benefits of using a computer for laboratory data processing. *J. Clin. Pathol. Suppl. (R. Coll. Pathol.)*, 3, 62–73.

23 Flynn, F.V. (1971) Computer assistance in the chemical pathology laboratory. *Biochem. J.*, January, 121(1), 2P.

24 Flynn, F.V. and Ball, S.G. (1982) Comprehensive computerized data management in a chemical pathology laboratory with SOCRATES. *Med. Inform. (Lond.)*, Oct–Dec, 7(4), 275–305.

25 Jordal, R. (1977) The Laboratory Information System at Copenhagen County Hospitals. In Online Conferences Ltd (eds) *MEDCOMP 77*, pp. 133–40. ISBN 0 903796 16 X.

26 Assicott, M. and Bohuon, C. (1978) Problems with online data acquisition systems in a clinical chemistry laboratory. In Anderson, J. (ed.) *MIE, Cambridge 78: Lecture Notes 1*, (pages unknown). Springer Verlag, Berlin. ISBN 3 540089 16 0.

27 Sandblad, B. (1976). Modelling and simulation for planning and benefit/cost analysis of healthcare service systems. In Laudet, M., Anderson, J. and Begon, E. (eds) *Proceedings of Medical Data Processing*, Toulouse, pp. 117–27. Taylor & Francis, London. ISBN 0 850661 06 4.

28 Cavill, I., Ricketts, C. and Jacobs, A. (1975) *Computers in Haematology*, 'Computers in Medicine' series. Butterworths, London.

29 Davidson, J.M.F., Williams, K.N. and Robertson, V.S. (1979) Table driven and optical mark reading systems for clinical and laboratory applications. In Barber, B., Grémy, F., Überla, K. and Wagner, G. (eds) *MIE, Berlin 79: Lecture Notes 5*, pp. 414–26. Springer Verlag, Berlin. ISBN 0 387095 49 7.

30 Barber, B. and Abbott, W.A. (1972) *Computing and Operational Research at The London Hospital*; 'Computers in Medicine' series. Butterworths, London. ISBN 0 407517 00 6.

31 Griffiths, P.D. and Carter, N.W. (1969) Online acquisition of the output of autoanalyzers. *J. Clin. Pathol.*, 22, 609–16.

32 Wills, A.R. (1977) A computer requesting and reporting system for a variety of clinical laboratories. In Online Conferences Ltd (eds) *MEDCOMP 77*, pp. 151–66. ISBN 0 903796 16 X.

33 Mace, D.R. (1978) A common approach to a variety of clinical laboratories. In Anderson, J. (ed.) *MIE, Cambridge 78: Lecture Notes 1*, pp. 509–19. Springer Verlag, Berlin. ISBN 3 540089 16 0.

34 Neill, D.W. and Doggart, J.R. (1970) The use of an off-line computer in a hospital biochemistry laboratory. In Abrams, M. (ed.) *Medical Computing: Progress and Problems*, pp. 7–23. Chatto & Windus for BCS, London.

35 Whitehead, T.P. (1970) Data processing in clinical biochemistry. In Abrams, M. (ed.) *Medical Computing: Progress and Problems*, pp. 24–8. Chatto & Windus for BCS, London.

36 Flynn, F.V. (1966) Use of a computer by a clinical chemistry service. *Proc. R. Soc Med.*, August, 59(8), 779–82.

37 Whitby, L.G. (1964) Automation in clinical chemistry, with special reference to the autoanalyzer. *BMJ.*, 10 October, 5414, 895–9.

38 Whitby, L.G., Proffitt, J. and McMaster, R.S. (1968) Experience with off-line processing by computer of chemical laboratory data. *Scott. Med. J.*, June, 13(6),181–91.

39 Wootton, I.D. (1969) Computer processing of biochemical information without going on-line. *J. Clin. Pathol. Suppl. (R. Coll. Pathol.)*, 3, 101–6.

40 Wootton, I.D. (1968) Computer science in the biochemistry laboratory. *Br. Med. Bull.*, September, 24(3), 219–23.

41 Flynn, F.V. (1965) Computer-assisted processing of biochemical test data. In Atkins, H.J.B. (ed.) *Progress in Medical Computing*, pp. 46–51. Blackwell, London.

42 Wootton, I.D.P. (1965) Computer-assisted preparation of laboratory reports. In Atkins, H.J.B. (ed.) *Progress in Medical Computing*, pp. 57–60. Blackwell, London.

43 Abbott, W., Barber, B. and Fairey, M. (1974) The London Hospital Computer System – a case study in the installation of a major real-time system. In *Proceedings of conferences held at The London Hospital Medical College*, 27 November 1973 and 24 April 1974.

44 Sargent, S. (1978) Clinical laboratory systems. In Anderson, J. (ed.) *MIE, Cambridge 78: Lecture Notes 1*, pp. 497–507. Springer Verlag, Berlin. ISBN 3 540089 16 0.

45 Heijser, W. (1982) The automation of a number of similar laboratories within a single hospital. In O'Moore, R.R., Barber, B., Reichertz, P.L. and Roger, F. (eds) *MIE, Dublin 82: Lecture Notes 16*. Springer Verlag, Berlin. ISBN 0 387112 08 1.

46 O'Moore, R.R., Clayton Love, W., Ryan, J., Ratcliffe, J., McSweeney, J., Cranney, G. and Field, L. (1982) Experience with a MUMPS based system for clinical chemistry. In O'Moore, R.R., Barber, B., Reichertz, P.L. and Roger, F. (eds) *MIE, Dublin 82: Lecture Notes 16*. Springer Verlag, Berlin. ISBN 0 387112 08 1.

47 Trespe, K.F., Wolters, E. and Tripatzis, I. (1982) Monitoring antibiotic sensitivity and bacterial infections in a hospital. In O'Moore, R.R., Barber, B., Reichertz, P.L. and Roger, F. (eds) *MIE, Dublin 82: Lecture Notes 16*. Springer Verlag, Berlin. ISBN 0 387112 08 1.

48 Loy, V., Gross, U. and Enke, H. (1982) Distributed processing in histopathology. In O'Moore, R.R., Barber, B., Reichertz, P.L. and Roger, F. (eds) *MIE, Dublin 82: Lecture Notes 16*. Springer Verlag, Berlin. ISBN 0 387112 08 1.

49 Brink, A.M.A., Koedam, J.C. and Slik, W. (1977) Transferring an information system for clinical laboratories. In Online Conferences Ltd (eds) *MEDCOMP 77*, pp. 239–44. ISBN 0 903796 16 X.

50 Spraberry, M.N. (1977) A data management system for clinical laboratories. In Online Conferences Ltd (eds) *MEDCOMP 77*, pp. 141–50. ISBN 0 903796 16 X.

51 Berg, K. and Brubakk, A.O. (1977) A study of clinical laboratory computer systems and their manual counterparts. In Online Conferences Ltd (eds) *MEDCOMP 77*, pp. 121–32. ISBN 0 903796 16 X.

52 Baker, R., Upham, J., Schroeder, J. and Stewart, W.B. (1982) On-line quality control in clinical chemistry. In O'Moore, R.R., Barber, B., Reichertz, P.L. and Roger, F. (eds) *MIE, Dublin 82: Lecture Notes 16*, pp. 124–8. Springer Verlag, Berlin. ISBN 0 387112 08 1.

53 Sequeira, P.J.L. (1978) A laboratory data processing system for non-numeric results. In Anderson, J. (ed.) *MIE, Cambridge 78: Lecture Notes 1*, pp. 475–9. Springer Verlag, Berlin. ISBN 3 540089 16 0.

54 Maddison, T.J. (1977) The development of standard National Health Service systems in England. In Online Conferences Ltd (eds) *MEDCOMP 77*, pp. 247–58. ISBN 0 903796 16 X.

55 Wills, A.R. (1982) 'Structured' techniques for design – experiences in their use for an extensive hospital computer system. In O'Moore, R.R., Barber, B., Reichertz, P.L. and Roger, F. (eds) *MIE, Dublin 82: Lecture Notes 16*, pp. 414–19. Springer Verlag, Berlin. ISBN 0 387112 08 1.

56 Drury, P. (1982) The development of a distributed systems approach in a teaching hospital. In O'Moore, R.R., Barber, B., Reichertz, P.L. and Roger, F. (eds) *MIE, Dublin 82: Lecture Notes 16*, pp. 770–5. Springer Verlag, Berlin. ISBN 0 387112 08 1.

57 Harris, N., *et al.* (1982) A coordinated decentralised computer system. In O'Moore, R.R., Barber, B., Reichertz, P.L. and Roger, F. (eds) *MIE, Dublin 82: Lecture Notes 16*, pp. 776–83. Springer Verlag, Berlin. ISBN 0 387112 08 1.

58 Mace, D.R. (1977) Some experiences of a distributed network and their relevance to medical computing. In Online Conferences Ltd (eds) *MEDCOMP 77*, pp. 197–207. ISBN 0 903796 16 X.

9 Radiotherapy planning (1960–1980)

ROY BENTLEY

Radiotherapy is the treatment of disease with ionizing radiation. Because the doctor must have an accurate assessment of the dose delivered both to diseased tissue and to neighbouring normal tissue, a large amount of numerical calculation is necessary. The radiation may be produced by an implanted radioactive source, such as radium. More often a machine is used (e.g. see Figure 9.1), external to the patient, with a radioactive source, normally cobalt-60, which generates a beam of photons. Over the years cobalt machines have been replaced by linear accelerators that can produce photons and electrons, but the numerical problems remain essentially the same. We shall be concerned here mainly with external irradiation, which is the type most frequently used in the treatment of cancer and where many of the early developments in the use of computers began.

Radiotherapy is one of the most numerate branches of medicine and this was one of the first medical applications of computers, along with successful introduction of vaccination and immunization procedures.[1] In radiotherapy it is necessary to calculate the dose at a large number of points to build up a picture throughout the region of interest, so that the dose to a tumour and its surrounding tissue can be assessed.

Most external beam treatments require the addition of three, four and sometimes more individual radiation fields applied from different directions. In the pre-computer age this was done by overlaying charts drawn on transparent sheets in the form of contours showing lines of equal dose, so called isodose curves. For even three fields to be added together, it could be a very tedious procedure. An even more lengthy procedure was the preparation of plans for rotational treatments where the beam source was rotated around the patient. The effect of the moving beam was simulated by adding together 25 or more stationary beams. It was to this problem that computers were first applied, by Wood[2] at Cardiff in the UK and by Tsien[3, 4] in Philadelphia in the USA.

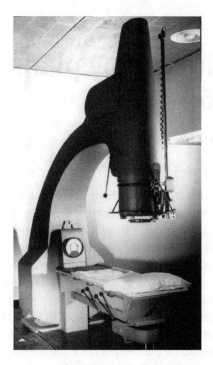

FIGURE 9.1 *Vickers linear accelerator for radiotherapy, installed in the Royal Marsden Hospital in about 1965* (reprinted by kind permission of The Royal Marsden)

THE BEGINNING

Wood described the use of an ICT 555 computer for rotating beam treatments. This machine had to be programmed by plugging wires into a board, rather like an old, manual telephone exchange. Data could only be input from punched cards. There was so little storage in the computer that a deck of punched cards had to be re-entered for each angular position, even though the data remained unchanged. This work was later transferred to a more conventional ICT 1903 computer. Similar programs were developed at a number of other centres in the UK. By 1964 many regional hospital boards had purchased ICT 1900 series machines and various types of computer were available in British universities, all capable of being programmed in Fortran. Well-engineered programs were developed by Orr and his co-workers in Glasgow[5] and these were shared with other centres in a working group known as SCRAP (System of Computing in Radiotherapy and Treatment Planning). Groups at Bristol, Cambridge and UCL, London, contributed to a pool of programs but difficulties in sharing programs between centres with different working practices and with different computers had to be recognized and reconciled.

While these programs were a major advance over the hand-planning methods described above, they all ran in batch mode on mainframes that could be some distance from the radiotherapy centre. This meant that data for input had to be prepared on coding sheets, then transferred to punched cards and then delivered to the computer. After waiting their turn in a queue, sometimes for up to 24 hours, results were delivered back to the radiotherapist. This was long before the advent of the internet, and remote access to a computer over telephone lines was at a very early stage. A further problem was that the results were required in a graphical format; most computers had line printers that could only output fully formed characters on to the paper. Graph plotters were rare and could only be afforded by architects' and engineers' drawing offices, so ingenious ways of using line printers were evolved.

The need for an interactive system

At the Royal Marsden Hospital and other centres, the effort of using the computers was considered worthwhile for rotational treatments, but it was unclear whether much was gained in the daily routine of producing the many fixed-field treatment plans that were needed. However, rotational treatments were going out of favour at this time, because they had been shown to offer little advantage in clinical practice. A different approach to planning fixed-field treatments was needed.

Long before the development of digital computers, a number of workers had been searching for a machine that could produce dose distribution charts. Various designs of analogue computer appeared in the literature, which embodied the concept of rapid calculation and immediate display of results.[6–8] Analogue computers could be extremely, although not infinitely, fast but were generally of limited accuracy. None of these proposals came to fruition before the more accurate digital machine had become readily available at an affordable price.

The minicomputer arrives

In 1965, Cox and his colleagues in St Louis, and Bentley working in the UK, recognized that the solution lay in a minicomputer that could be adapted with suitable devices for immediate input and output. Cox went further and specially designed a programmable digital computer known as the Programmed Console or PC (an unfortunate choice of acronym once the Personal Computer emerged several years later) – details are given by Holmes.[9] Cox believed there was, at that time, no affordable machine on the market suitable for the task. The Programmed Console had its own instruction set with relative addressing, a feature not incorporated into many commercial machines until the PDP-11 was introduced in about 1970. This made it easier to program and possibly to write programs using less memory than a machine with absolute addressing. But this came at the cost of losing all possibility of standardization with the rest of the computer world. This design with its unique instruction set precluded the possibility of any centre obtaining funds to import the Programmed Console into the UK.

Bentley and Milan engineered a similar system in 1971 using a DEC PDP-8,[10] which had been first produced in 1965.[11] With a price tag of around £20,000, this was the first stand-alone, single-task machine that could be afforded (just) by the medical community. The first system developed at the Royal Marsden Hospital, used a PDP-8I which was announced in 1967. Like the Programmed Console, and some competing machines, it had a word length of 12 bits, not in the mainstream of 8, 16 or 32 bits that eventually became universal. Fortuitously, a 12-bit word gave just the required level of accuracy for radiotherapy calculations when it was used in fixed point arithmetic mode. The original system (Figure 9.2), into which the program had to be fitted, had only 8,192 words of memory (equivalent to 12 kB of eight-bit bytes). There were no drums or disks: the only storage on magnetic media was in the form of directly addressable tape (known as DECtape). The only keyboard was the one incorporated into an attached ASR33 teletype machine. Results were displayed on a Tektronix type 611 oscilloscope. This allowed single points or vectors to be plotted with a single computer instruction. Moreover, it could be used in storage or refresh mode or a combination of both in the same image. This was before the now ubiquitous raster-driven visual display terminals were readily available.

The original Royal Marsden Hospital system incorporated many features of the Programmed Console and improved on others. Anatomical data for the patient was input using a rho-theta device. This consisted of an extendable, rotatable arm with two potentiometers so that voltages representing polar coordinates could be fed directly into an analogue-to-digital converter. This was exactly the same as was used by the St Louis

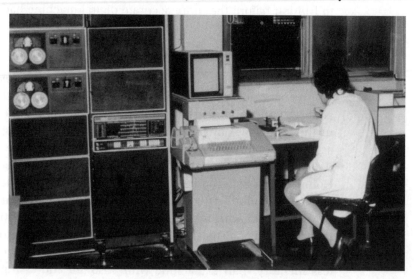

FIGURE 9.2 *Original Royal Marsden Hospital installation (subsequently named RAD-8) From left: magnetic tape (DECtape); main unit, showing switches and blinking lights; keyboard, as part of ASR33 teletype, with four potentiometers and storage oscilloscope above Rho-theta device for graphic input to the right and X–Y plotter at far right* (reprinted with kind permission of The Royal Marsden)

group. Likewise, x- and y-coordinates, and angles of entry of X-ray beams, were input using potentiometers connected to the same analogue-to-digital converter: this was before the invention of the mouse. Dose data for each beam size were stored on magnetic tape. Disks were not available. Programs were written in assembly language for the PDP-8, which had a very simple instruction set, and calculations were carried out in fixed point arithmetic and took about 10 seconds for a typical three-field plan. The results were displayed on the storage oscilloscope. Permanent results were produced on paper with a plotter consisting of a pen driven in x- and y-directions by stepping motors. More details of the programs were given by Milan and Bentley.[12] When DEC started to produce this system commercially, it was named the RAD-8.

> Before the RAD-8 had been developed and even before the Programmed Console at St. Louis, I described, at an HPA meeting, what I thought was needed for treatment planning, which was not so different from what eventually appeared. A radiotherapy physicist was there who said that would be a waste of time and money because his planning technicians could produce plans so quickly, that a computer was unnecessary! He invited me to go and see his staff at work, an offer which I am sorry to say I did not take up.
>
> **ROY BENTLEY**

THE SPREAD TO OTHER CENTRES

Manchester

A number of similar machines were purchased at other hospitals in the UK but the originators of the system encountered similar difficulties to those faced by the Glasgow SCRAP group in extending to other sites that had their own, different, working practices in radiotherapy.

A particular problem was the preparation of dose data for the linear accelerators at another radiotherapy centre. Each individual therapy machine is slightly different, even if two are of the same make and type. In 1972–73, The Christie Hospital, Manchester, was one of the first of a number of UK hospitals to purchase a PDP-8 system based on the Royal Marsden Hospital design. The Manchester group added a wealth of new programs, including those necessary for interstitial radiotherapy. A detailed paper describes the many problems encountered in commissioning this machine.[13] Quoting from a BIR/DoH report,[14] they say: 'a recurring theme was the lack of information about the system'. In their conclusions the authors say that 'implementing the RAD-8 system was not a negligible task'; and comment that this was largely due to the fact that the programs had been written in a very low-level language, coupled with the lack of adequate documentation.

Edinburgh

The next hospital to make a major contribution to the development of the RAD-8 system was the Western General in Edinburgh.[15] The team there realized that if the system was to become truly viable for widespread use the programs would have to be written in a high-level language. By this time an efficient (for that time) Fortran IV had been developed. The slower speed of program execution could be compensated for by adding floating-point hardware to the system. This was also the first interactive radiotherapy system to use a disk for secondary storage, instead of the slow magnetic tape.

This was the beginning of a movement of collaboration between various UK centres with PDP-8 based systems. The Edinburgh group used hardware for input of anatomical details from Glasgow and a Fortran program for generating beam-profiles from Manchester. The conversion to Fortran made the universal adoption of small computer systems viable for radiotherapy. In collaboration with others, new programs for rotational treatments were written in Fortran, as were those for the use of irregular shaped fields (based on the work of Cunningham and colleagues in Toronto[16]) and for interstitial radium needle treatments (based on work in Houston by Stovall and Shalek[17]). A further advantage of the conversion to Fortran was that programs could relatively easily be transferred to new hardware designs as they became available, such as the 16-bit PDP-11, which was purchased by some centres in preference to the 12-bit PDP-8.

CLIENT–SERVER APPROACH

A third approach to the external beam planning problem was reported by Crossland, Haybittle and Jameson jointly working at the Middlesex Hospital, London and at Addenbrooke's Hospital in Cambridge.[18] This was what has now come to be known as the client–server approach. Using a minicomputer as an operator's terminal, it had all the advantages of the Programmed Console and the RAD-8 on the one hand, and the power and programming opportunities of a large mainframe computer on the other. Such a scheme had in fact been proposed as part of the original Programmed Console project in St Louis.

In the Middlesex–Addenbrooke's project, the user was not restricted to any particular mainframe, which could be in a regional hospital, a university or a commercial organization. The only possible drawback was the need for the two computers to communicate using a public network telephone line. It must be remembered that communication between computers, even between two fixed end-points, was in its infancy at this time and, by today's standards, was extremely slow.

For the terminal, a Varian 620/L minicomputer with 8 kB of 16-bit words of memory was chosen. In contrast with the Tektronix storage oscilloscope used in the RAD-8, it had a Hewlett Packard 1310A refresh oscilloscope and an associated Stabletron vector generator. The ability to produce vectors at a single stroke enabled images to be refreshed sufficiently rapidly, and to be stable when viewed on the screen.

Connection to the mainframe computer was via GPO Modems type 1 at each end of the phone line. The GPO acronym reminds us that all telephone communication was at that time a monopoly of the Post Office in the UK, and the modem had to be installed by their engineers. Even then the type 1 modem could only operate at a bidirectional rate of 30 characters per second (300 baud). However, it could be set to operate asymmetrically, 1,200 baud from mainframe to the terminal and 75 baud in the opposite direction. The latter was adequate when the only data to be sent were those generated by keystroke depressions.

It is notable that, even in 1975, the only operator's keyboard was part of an ASR-33 teletype, as in the RAD-8, despite the fact that the earlier Programmed Console had had a keyboard more like those in use today. The only method of storing data external to the machine was by high-speed paper-tape reader and punch. This was needed for keeping details of completed plans; magnetic media such as tapes and disks would have added considerably to the cost.

The initial connection was made to a UNIVAC 1108 in a commercial bureau, but it was later shown that the same Fortran programs could be used on the NHS's ICL and Rank Xerox computers equally viably. An extensive range of programs was written. The external beam therapy suite included the ability to compute photon dose distributions for all megavoltage machines and energies, with isocentric or fixed-source skin distance using plain and wedged fields, with or without tissue compensators. It was also possible to allow for tissue heterogeneities, to calculate the arcs of rotation, and to calculate in the plane of the central axis and in parallel planes.

DATA REQUIREMENTS

Beam data

For computerized radiotherapy calculations, a number of other problems had to be solved. In particular data pertinent to each radiotherapy machine had to be prepared, whether it was cobalt-60 or a linear accelerator. This had hitherto been done by making measurements with an ionization chamber in a tank of water and plotting curves of equal intensity (isodoses) on a sheet of tracing paper. This was wholly unsuitable for computerized methods where arrays of numbers were required for the primary data. Although attempts were made to convert isodose charts into arrays of numbers, they were not very satisfactory and a whole new approach to data preparation became necessary. The Edinburgh group made a major contribution to this with a Fortran program for a computer-controlled dose measuring system,[15] and a similar system was developed for use at the Royal Marsden Hospital.[19]

Patient Data

Before planning for an individual patient could begin, anatomical data were also required for input to the computer. This involved data about the external shape of the patient, the location of the internal organs and most importantly the location, size and shape of the tumour to be treated.

To obtain these external data, a number of ingenious devices were invented. Notable were the rho-theta device from BBN (Bolt, Baraneck and Newman), first used on the Programmed Console at St Louis and later by the RAD-8, and a theta-phi device developed and used in the Edinburgh system described above. The Philips TPS (treatment planning system) used an ingenious ultrasonic device for patient data input.

Data for internal structures were more difficult to obtain, as had been the case for hand-planning before the advent of the computer. This relied on the use of X-ray images superimposed on the drawings of external shapes. Much of this was unsatisfactory, so that when the computerized tomography (CT) scanner became available it hugely advanced the ability to obtain the required anatomical data for individual patients, and offered the much-needed extension into the third dimension.

COMMERCIAL DEVELOPMENTS

The RAD-8 radiotherapy system was developed from the very beginning in cooperation with DEC. It evolved into a commercial product in 1970–72. The company went on to gain great prominence in the computer industry during its formative years, before its eventual demise and absorption by other companies in the 1990s. The BIR/DoH report, the key issues of which were outlined at *MEDINFO 74*,[20] indicated that the RAD-8 was cost-effective where more than 500 plans per year were required. It is believed that more than 100 RAD-8 systems were delivered worldwide: the one at the Royal Marsden remained in use for 15 years.[21]

In the course of time, DEC decided that the development of end-user systems was not within their remit and further developments were taken over by other companies. EMI, who had produced the first CT scanner, completely redesigned the RAD-8 system but retained the basic principles and logic. By now computer technology had greatly advanced and new computers and display devices had become available. Now called EMI-plan, it used a Data General Nova computer.

The EMI-plan used a standard refresh display and, at last, removable disks were used for storage. Programs were, however, written in an unusual language (BCPL, a forerunner of C), and this constrained further development. A very significant development in EMI-plan was the incorporation of CT data with radiotherapy planning data onto the same display. This made it possible for the first time to get accurate information about the anatomy of the patient, both of internal structures and external shape, thereby increasing the accuracy of the planning procedure.

International General Electric took over EMI, renamed EMI-plan as GEplan, and subsequently developed a new system named Target.

> After the RAD-8 came into routine use at the Royal Marsden Hospital, a treatment plan was produced and given to a certain radiotherapy consultant. It was clearly drawn on the *X–Y* graph plotter attached to the computer. On seeing it, he remarked that he was not having plans drawn by computer and threw it in the waste paper basket. He then demanded that it be replanned by the normal hand-drawing procedure. After this the planning staff continued to use the computer, but for this consultant they copied the computer output onto tracing paper before giving it to him.
>
> ROY BENTLEY

Other commercial systems

By 1977 other free-standing, real-time, interactive systems were starting to be produced commercially by other manufacturers, notably by Artronix in the USA (who turned the St Louis Programmed Console into a commercial product), Varian in the USA, Philips in The Netherlands, Siemens in Germany, AEC (Atomic Energy of Canada) and Informatek in France.

The first of these to be imported into the UK was the Philips TPS at Charing Cross Hospital.[22] This was one of the first systems to allow calculations and display of dose distributions in parallel planes.[23]

Although designed around a general purpose computer, the Philips P855, it appeared as a special purpose product and the user had rather less control over its functions than those produced by other manufacturers. A major difference from the PC, the RAD-8 and the Cambridge–Middlesex systems described above, was the use of a generating formula for dose data developed by van de Geijn rather than the extensive storage of tables of pre-measured and pre-stored data.[24] This meant that data for radiotherapy machines were prepared by Philips, rather than by the hospital, thereby relieving local hospital physicists of a great deal of painstaking work. It was possible to store data for as many as 12 different therapy machines on one cassette tape.

ACHIEVEMENTS WITH LIMITED HARDWARE

The systems described above were very slow by the standards of today. The time for a floating-point arithmetic operation (add or multiply) for the systems described would be of the order of 0.1 ms. Hundreds of millions of floating-point instructions per second can be achieved today. Likewise, the memory of these systems was, at most, 256 kB. A desktop computer today often contains 1 GB or more of main memory. Disk capacity was frequently a few tens of megabytes, compared with 200 GB or more now available.

Greatly improved display systems have also been developed using a combination of hardware and ingenious software to allow the rendering of three-dimensional images in real time and instantaneous changing of viewing angle and of other parameters.

It is remarkable that treatment planning systems could be built at all in the 1960s that provided results in an acceptable time, at an acceptable cost. The first RAD-8 cost about £22,000 at 1969 prices. Current machines with vastly enhanced performance probably cost about the same, after allowing for inflation.

SUBSEQUENT DEVELOPMENTS

The foregoing account only goes up to about 1980. Enormous developments have taken place in the past quarter of a century. Treatment planning systems are now universal in the radiotherapy and associated physics departments; many hospitals having several systems. Networking enables the rapid exchange of data between planning computers, CT scanners, radiotherapy simulators and the radiotherapy machines used for clinical treatment. In 1980, planning still remained a two-dimensional procedure; this was one of the principal shortcomings of radiotherapy at that time and was only overcome when CT scanners, magnetic resonance imaging (MRI) scanners and greatly enhanced hardware became available.

There is also a vast literature on improved methods of calculation of the dose delivered at points throughout the individual patient to be treated.[e.g. 25, 26] The ultimate method, now coming into routine use, is the application of Monte Carlo techniques that simulate the interaction of photons or electrons with normal and diseased human tissue and provide more accurate dosimetry data throughout the whole of the anatomy.

REFERENCES

1 Davies, C.K. (1970) Recall procedures – local authority applications. In Abrams, M. (ed.) *Medical Computing: Progress and Problems*, pp. 283–8. Chatto & Windus for BCS, London.

2 Wood, R.G. (1974) *Computers in Radiotherapy – Physical Aspects.* Butterworths, London. p. 76.

3 Tsien, K.C. (1955) The application of automatic computing machines to radiation treatment planning. *British Journal of Radiology,* 28, 432.

4 Tsien, K.C. (1958) A study of basic external radiation treatment techniques with the aid of automatic computing machines. *British Journal of Radiology,* 31, 32.

5 Hope, C.S., Laurie, J., Orr, J.S. and Halnan, K. (1967) The computation of dose distribution in cervix radium treatment. *Phys. Biol. Med.,* 12, 532.

6   Cooper, R.E.M. and Worthley, B.W. (1967) Computer-based external beam radiotherapy planning II: practical applications. *Phys. Med. Biol.*, 12, 241.

7   Skaggs, L.S. and Savic, S. (1963) Use of an analogue computer to calculate treatment dose for multiple fields. *Radiology*, 80, 116.

8   Jones, J.C. and Bentley, R.E. (1966) An analogue computer for the calculation and immediate presentation of dose distributions. *Phys. Med. Biol.*, 11, 156.

9   Holmes, W.F. (1970) External treatment planning with the Programmed Console. *Radiology*, 94 (2), 391.

10 Bentley, R.E. and Milan, J. (1971) An interactive digital computer for radiotherapy treatment planning. *British Journal of Radiology*, 44, 826.

11 Bell, C.G. and McNamara, J.E. (1978) *Computer Engineering*. Digital Press, Elsevier, Oxford. p. 175.

12 Milan, J. and Bentley. R.E. (1974) The storage and manipulation of radiation dose data on a small digital computer. *British Journal of Radiology*, 47, 115.

13 Wilkinson, J.M. and Redmond, W.J. (1975) Experience in implementing the Mark III RAD-8 system for radiotherapy dose computations. *British Journal of Radiology*, 48, 732.

14 Barber, B. (1973) *Computerised Dose Computation*. Report for the British Institute of Radiology and the DoH (available also in microfiche format).

15 Redpath, A.T., Vickery, B.L. and Duncan, W. (1977) A comprehensive radiotherapy planning system implemented in Fortran on a small interactive computer. *British Journal of Radiology*, 50, 51.

16 Cunningham, J.R., Shivastava, P.N. and Wilkinson, J.M. (1972) Program IRREG – calculation of dose from irregularly shaped radiation beams. *Computer Programs in Biomedicine*, 2, 192.

17 Stovall, M. and Shalek, R.J. (1972) A review of computer techniques for dosimetry of interstitial and intracavity radiotherapy. *Computer Programs in Biomedicine*, 2, 125.

18 Crossland, P., Haybittle, J.L. and Jameson, D.G. (1976) An online computer graphics terminal for radiotherapy treatment planning. *British Journal of Radiology*, 49, 868.

19 Milan, J., Dance, D.R., Bentley, R.E. and Lillicrap, S.C. (1976) Online beam data acquisition for a dedicated radiotherapy planning computer. *British Journal of Radiology*, 49, 172.

20 Barber, B. (1975) Radiotherapy computer applications. In Anderson, J. and Forsythe, J.M. (eds) *MEDINFO 74*, Stockholm, pp. 801–5 and pp. 1097–102. North-Holland, Amsterdam. ISBN 0 444107 71 1.

21 Wiltshaw, E. (1998) *A History of the Royal Marsden Hospital*. Altman Publishing, St. Albans. p. 123.

22   Dale, R.G. (1978) Implementation of the Philips treatment planning system for use in radiation therapy. *British Journal of Radiology*, 51, 613.

23   Haarman, J.W., van Leiden, H.F. and Pruis, D. (1975) An interactive three-dimensional treatment planning system. *Medicamundi*, 20, 23–32. Philips, Eindhoven.

24   Van de Geijn, J. (1965) The computation of two- and three-dimensional dose distributions in cobalt-60 therapy. *British Journal of Radiology*, 38, 369.

25   Webb, S. (1993) *The Physics of Three-dimensional Radiotherapy*. IOP Publishing, Bristol.

26   Webb, S. (1997) *The Physics of Conformal Radiotherapy*. IOP Publishing, Bristol.

10 Medical imaging

ANDREW TODD-POKROPEK

While a method for tomographic image reconstruction using analogue and optical techniques is believed to have been proposed in the 1950s, the first uses of digital computers in imaging were in the 1960s. Nuclear medicine was the major application for image processing using computers because of the relative ease in acquiring data. The target device was not the gamma camera, but the rectilinear scanner then still widely used as the workhorse device for imaging. The data were acquired using a counter attached to a paper-tape punch so that the number of radioisotope events detected while the scanner traversed a distance of a few millimetres, could be recorded and input for processing into a computer.

One of the first such installations was at the Middlesex Hospital, London, where the data were processed on the Elliott 803 computer at what was then University College Hospital Medical School. This system, installed in 1962, was in the Medical Automation Unit set up by the DoH and was supported by Elliott Automation under the direction of Professor Payne. The aim was to set up an electronic medical records system for computer-aided diagnosis, patient monitoring and, indeed, image processing. Essentially the data comprised a series of counts, that is how many events had been detected in a certain time over a short distance while the head of the scanner traversed the patient. It was then relatively easy to record that sequence of numbers on a punched paper-tape.

The Elliott 803 computer had a cycle time of about 256 microseconds. To process one of the rectilinear scanner images (typically 60 pixels across by perhaps 100 lines of data) and to perform a simple smoothing operation took about 20 minutes. Input was on paper-tape and the output was in the form of characters printed on paper. While many such scans were processed, the operation was slow and tedious. However, given the very 'noisy' data that were being acquired, even the poor quality output results demonstrated the value of smoothing. The results were first published at the European meeting in Barcelona in 1967. Since only paper-tape input and printer output were required, a terminal connected to a remote computer via a modem could also be used. For this, the GEIS (General Electric Information System) was occasionally used to overprint characters on the teletype printer.

Under the direction of David Keeling, a new idea was explored: that of using more than one isotope and subtracting the results in order to enhance particular structures. The target here was the parathyroid gland and, using the rectilinear scanner, the process was extended to smoothing and image subtraction (still using the same output method). Because the Elliott 803 was very slow, it was decided to use a mainframe to enhance the procedure. Two machines were used, the Atlas computer at the University of London Computer Centre and the IBM 360/65 at UCL, under the direction of Paul Samet.

The process of manipulating the data was even more curious: the paper-tape was first converted to punched cards, which, after a certain amount of manual editing to correct errors on the paper-tape, were then used as the input to the mainframe. A turnaround of about a day could be achieved. The look of horror on Paul Samet's face when he was asked if he could expedite one job since the patient was due for surgery early next day indicated the unusual nature of the use of computers for these applications at that time.

The whole process was time-consuming and cumbersome. A number of minicomputers had started to become available and appeared to be very suitable for connection to the imaging devices, but the main drawback at the time was the lack of a suitable display.

Several groups had used the DEC PDP-8 for such tasks. Several early radiotherapy treatment planning systems had been developed (see Chapter 9). Output from these systems was initially in text, either on printers or on monochrome (green) monitors. However, it was found to be relatively easy to attach and interface (line) plotters so that output in the form of contour maps could be used. These were very appropriate for the radiotherapy applications, but could also be used for imaging.

The Institute of Nuclear Medicine at The Middlesex Hospital Medical School decided to purchase one of the newly available minicomputers in 1969. The one chosen was a Varian 620i, which had the advantage of having a 16-bit memory (as opposed to the 12-bits of the PDP-8) but the system initially delivered had a total of 4 kB of memory and the basic input/output device was a teletype console. It was used for some limited image processing use (smoothing images input from paper-tape) but an expansion was required. As a result of a substantial grant, for that time, the memory was expanded to 8 kB and a hard disk and a magnetic tape drive were added. All the programming was in assembler since no compilers were available. One of the first major software tasks was to write a text editor (in assembler) to accelerate software development. The second was to write a, fairly primitive, operating system so that tasks and images could be stored on the magnetic tape memory and sequences of tasks chained together. Since memory was severely limited, this operating system used the bottom 4 kB of memory while the top 4 kB was reserved for image data. As 4 kB (2k integers) was not enough to store the standard 64 × 64 image from the gamma camera of that time, a virtual memory system (essentially based on the idea used in the Atlas mainframe computer) was developed to page in and out of memory sections of data and program code.

One curiosity of the Varian 620i was the use of a 240-byte memory sector on the hard disk, so that some very odd segmentation of memory was required. This might explain some of the eccentricities of software later developed on commercial systems that resulted from this prototype. An electrostatic printer was also attached.

A group of enthusiastic pioneers met for the first time in 1969 in Brussels at a meeting entitled *Image Processing in Scintigraphy*, funded by Euratom and inititated by François Erbsmann from Brussels. This was the start of the series of meetings, now known as the *Information Processing in Medical Imaging* (IPMI) conferences, which have continued at two-yearly intervals to this day.

CT SCANS

The next major development was the design and implementation of a single photon emission computerized tomography system (in fact still a scanner). The ideas were based on those published by Dave Kuhl in 1968 and a grant of £5,000 was obtained from the Clore Foundation. The prototype system was built by J and P Ltd. It was later commercialized by them and a later system was used by John Mallard (the Aberdeen Scanner). It should be noted that this development was occurring at the same time as work on the prototype CT scanner at EMI Hayes.

Work began to develop a tomographic reconstruction program and, after some false starts, a filtered back-projection system was implemented, based primarily on work coming from Brownell's group at MGH. This allowed 64 × 64 images to be reconstructed on the Varian computer in about 15 minutes. This work was published in 1972 at the same time as the initial results using a prototype CT scanner.

An important development by Nicholas Brown transformed the value of this system by adding a proper colour display. A gamma camera had also been purchased. The gamma camera interface was implemented in CAMAC modules. So, we now had a proper system linking together the acquisition device, the processor with a primitive operating system and a display.

In France, the group headed up by Robert Di-Paola at the Institute Gustave Roussy followed these developments by purchasing a Varian 622i (with an 18-bit word) but linked up with a group at the University of Paris Orsay who had developed a sensible operating system for this computer. This resulted in a spin-off company, Informatek S.A. with Charles Zajde as founding President. Together, with Robert Di-Paola, François Erbsman and myself, we worked for a year in Paris to develop the first more or less complete nuclear medicine processing system with a range of software: from display to image enhancement to quantitation. The system was implemented to process data from a variety of organs: kidneys, livers, brains and eventually hearts.

A new project funded by the DoH started in 1972 at UCH under the direction of John Clifton, head of medical physics. A DEC PDP-11/45 was purchased to support the implementation of four main packages to be run simultaneously in a time-sharing environment. These were: radiotherapy treatment planning; gamma-camera image processing; and blood-gas analysis in neonatal intensive care. The computer was a massive increase in resources in comparison to those previously used: it had 64 kB of memory, magnetic tape and a 2 MB hard disk drive. The gamma-camera interface was purpose built, as were local terminals to be used in the three clinical areas. The software was written in Fortran. It is interesting to note that 64 kB of memory was adequate at that time to support all three applications by using disk-swapping.

During the next decade a number of commercial systems were developed based on the ideas used in the early prototype systems. Examples include: Gamma11 from DEC; Informatek; and here in the UK, Nodecrest, using the Varian computer; ADAC and later on systems closely integrated with the acquisition devices from Siemens, GE, Philips and Toshiba. Many of these commercial developments were spin-offs from the UCH-based groups who had developed the prototype systems.

A prototype CT scanner was developed at EMI by Godfrey Hounsfield. To perform the reconstruction he initially used a mainframe, an ICL 1905, to handle the data. The reconstruction time on this prototype was about five hours. However, the image size was considerably larger than for the nuclear medicine images. The algorithm used was not filtered back-projection (FBP), but an iterative algebraic reconstruction technique (ART) with a relaxation parameter in the modification term. ART had originally been developed by Gabor Hermann in the USA. FBP had been developed by Aaron Klug's group for use in crystallography in the molecular biology laboratory at Cambridge University. There were interesting arguments about the differences between ART and science.

EMI then switched to the use of various models of Nova minicomputers, which eventually developed into the Eclipse-based system in the distributed manufactured system.

At the same time, the 1970s, there was a gradual development of purpose-built boards to do filtering and back-projection in order to speed up the process, as opposed to relying on software algorithms in the minicomputers themselves. The developments in CT were initially the reconstruction methods where, in order to achieve sensible reconstruction times, FBP became the standard tool. Display was also important, initially as output printed on film. The need to be able to connect different CT scanners to different film printers was the original justification for the DICOM standard, then called the ACR/NEMA standard. This initially was a cable plus TCP/IP-like protocols to drive the printer. This hardware approach was barely used, but the idea of a standard for image transfer took off and has now become almost universally used in radiology. Soft copy displays also progressed with various tools for three-dimensional display and various synthetic projection (e.g. Maximum Intensity Projection), surface and volume rendering, and some limited image processing and quantitation tools.

PICTURE ARCHIVING

The development of picture archiving and communications systems (PACS) started much later. Probably the first real attempt in the UK was at St Mary's Hospital in a project initiated by Oscar Craig and Harold Glass in about 1992. Films were digitized on a large drum scanner and reported from hard copy. Tests were made of image quality and network response time. Given the use of a standard 10 MB Ethernet loop, these were rather unsatisfactory and the normal time to image display was 30 seconds. The image quality itself using standard 1 k × 1 k type monitors was likewise poor. Richard Dawood was the radiologist responsible for the testing of this system.

The lessons learned on this system and from experiences in the USA, notably by Huang at UCLA, were used as the basis for the first true PACS implementation in the UK at the Hammersmith Hospital in 1996. This was driven forward by a number of key initiators in PACS, and notably by Harold Glass who was key in obtaining the funding required from the DoH. This system was based on that developed by the US Army and commercialized by Siemens. Dr Nicola Strickland took over the leadership of the project and has been responsible for developing and extending the system over the intervening years. The costs of implementing the hardware needed for such systems have been dramatically reduced, although software costs and in particular maintenance and training have, if anything, increased. But all this has occurred after the initial pioneering phase of computers in imaging and radiology.

PART 3
Clinical specialties

INTRODUCTION

In the twenty-first century, when personal computers are sold through large department stores, and mini-catalogues are tucked inside all kinds of journals and television listing magazines, it may be salutary to hear that in the mid-1970s obtaining one required contacts. For example, hearing of a record shop where enthusiasts in a backroom workshop assembled BBC B/Acorn machines and loaded the Torch operating software to run BASIC. Trying out the machines was a bit difficult as users' computing experience often comprised only of a short computing course or, for example, practice of using mainframes during university summer schools as part of a general Open University degree. Hospital staff with an interest in using computers for their clinical work often had to start with purchasing their own machine for personal use at home.

Small RGB televisions offered the choice of colour for background and text, a big improvement on the single option of green text of the mainframe. The 5¼" floppy disks really were floppy and needed to be handled with care. (Some home computers still relied on cassette tape rather than a disk for data storage, but at least there was no need to worry about 'write-protect' tags. Printing was on daisy-wheel printers, which were fast, but noisy.

The first steps in using computer programs had to be self-taught: learning two or three new key-commands each day and the sophistication of the split-screen editing between two word-processed documents was accepted without question.

In 1985, Torch computers were collaborating with MedStat Ltd, a Nottingham-based company with its origins in the university there, which provided programs to present information in graphical and statistical form for individual patients.[1]

Developing a framework to support care planning for patients in a medical ward with the hope it could later be computerized was a task undertaken at home. It was eventually published as a textbook as the emphasis on Körner data and a centrally funded research programme in an adjacent District made finding the funding an uphill task.[2]

There was a limit on the type and amount of clinical research that could be conducted from home, but knowing how the programs could be used was a great help when attending hospital meetings to discuss the Körner data requirements (see Chapter 4) and the Financial Information Project.[3][*]

Initially, clinical research data collection in the 1980s tended to be manual. The London Hospital's WANG system of word processors was available to prepare reports, but junior staff without personal secretaries or their own computer were reliant on a secretary to type up the material from a Dictaphone or handwritten copy. Physical cutting and pasting of subsequent changes on the draft was still required, but at least these could be made quickly as soon as the busy secretary had a few spare minutes. It was often easier to do it in your own time at home.

Around this time the research committee at North East Thames RHA was receiving increasing numbers of research applications from clinicians that included a microcomputer as part of the list of equipment to be funded. It was difficult to get resources for a computer and hospital computer departments in the Region were not keen to see their proliferation, as they felt they had no control and that there would be unfunded requests for their help. Anecdotes at the time indicated the same problems occurred in other areas of the UK. Much of the initial interest and enthusiasm among the clinical professions was squashed out of existence at this time. If only there had been central funds and courses available to support large numbers of junior and mid-career clinical people, perhaps many of today's difficulties in 'engaging clinicians' would not have occurred.

Enthusiasts among the North East Thames RHA staff for the Apple® operating system encouraged the purchase of an Apple® II home computer to replace the home BBC B. The graphical interface and advances in storage capacity and speed of the operating system made it a much easier machine to use. The hypertext program made linking of data in different files a real possibility. The Apple® II operating system was relatively transparent, so it became easier to understand how the components came together, a facility not offered to those who began their computing experience with the later Microsoft® Windows®. The enthusiasm for Microsoft®'s Windows®, which replaced MS-DOS®, was a bit of a puzzle for those who had been using Apple® systems for some years. The Apple® was described in 1985 as 'one of the most frequently used in medical computing', when it was used for patient monitoring systems.[5] The disks were now small, and with a plastic outer casing were less likely to be damaged in transit. They were 'double density' and held much more data.[6]

---

[*] The NHS Financial Information Project (FIP) set out to provide an information base for monitoring efficiency and effectiveness that was based on the individual patient. The data were used up to health service planning level. The pilot studies in hospital (Coventry HA) and community services (South Birmingham HA), reported in May 1984, showed patient-based costing as a by-product of systems recording patient-related activity (as a proxy for resource use) for management purposes.[4]

For anyone working in a computer-aware teaching hospital, it was no surprise to be invited to the regional office to be introduced to the first Performance Indicators (PIs). It took staff some time to get used to box-plots and the way data were displayed. Hard copy was far less interesting and useful than when it was on the computer where one could 'poke about' in the data. PIs raised questions, rather than provided answers, but they did give a flavour of the effectiveness within a Health District. I remember that it was rather a shock to move to a non-teaching District and find the only machine, with the Torch operating system, capable of displaying the PIs had been 'sold' to the Health Education team for other purposes. A swap was negotiated and it took a while for managers to realize how useful the data might be. Clinicians and administrators quickly caught on to the idea of exploring the PIs for another District before making a job application there.

The civil servants charged with helping to spread the word about PIs were very supportive. The London Boroughs Training Committee asked for two, one-day sessions for community managers to learn about PIs. One of their training rooms was set up with machines carted over from Euston Tower. Just like the hospital staff, once the community folk were able to poke around in other people's data the level of enthusiasm and engagement within the room rose rapidly. All that data definition and supplying of management information suddenly had a personal payoff.

The Joint Academic Network (JANET) made email a reality, at least at work. At last there was a method of communication where the message was delivered quickly but its receipt need not interrupt work. In clinical practice the recipient may be in the midst of patient counselling or coping with a crisis. Email should have been a boon to the clinician. It certainly made multinational communication easier. Working groups within *Medical Informatics Europe* became much easier to sustain, although the academics, civil servants and management consultants had the early advantage of access. Clinical staff in the NHS did not generally have email access except on their personal machines at home.

Interest was beginning to build in defining terms and standardizing systems for European interchange. It was hard to get clinical interest: even the majority of health care academics did not see the potential impact on clinical care if systems were designed by the few without a shared understanding of the meaning of terms.

A project to look at the use of educational materials about HIV/AIDS indicated that nurses did not trust content put out by the DoH,[7] instead they relied on newspapers, radio and television documentaries: publicly available material. Was the DoH support for information technology more of a hindrance than a help? Work for this project led to a review of the readiness of NHS staff for new information technology based on Hall and Hord's work into curriculum change in North American schools.[8] This was a useful model for examining the levels of concern or interest that clinicians felt about different aspects of the introduction of IT to their hospital and its impact, if any, on their personal practice. It suggested a move away from the NHS 'big bang' approach to multilevel analysis and implementation.[9]

From the early work of Professor John Anderson there was a keen interest in the development of departmental and then integrated clinical records. The hopelessly underpowered facilities made available to the King's College Hospital project needed to be fed data from departmental systems before they had any chance of providing the complete medical record so earnestly desired by the pioneers.

It was also necessary for there to be a culture change in respect of typing before clinicians were willing to enter significant amounts of text into their records. The use of a QWERTY keyboard is no longer regarded exclusively as a typist's job, now that so many academic and professional reports are generated by the individuals involved. The issue of input to medical records is now simply a matter of time and logistics rather than culture. Someday, voice input will further improve this situation.

Clinical systems developed from the collection of patient and clinical data for diagnosis, treatment and management, research and planning. As computer use became more prevalent then clinical communications and online interaction of those with a shared interest in a specialty became possible. Medical records stored on computer were seen as providing potential benefits:

- speed of retrieval
- reduction of repetitive clerical work
- identification of high-risk patients, possible drug interactions or sensitivities
- tracking tests and results
- alerts when data fall outside set parameters
- patient recall at set periods
- evaluation of data leading to changes in clinical practice.[10]

The diagnostic services that supported the clinical care of patients began to merge with the minute-by-minute and hour-by-hour monitoring of the condition of the hospital patient. In time, web-enabled technology would allow resiting of the data collection and feedback to community settings, while the high-tech equipment was more centralized.

FUTURE CHALLENGES

As information systems and computer technology become more widespread within NHS hospitals and data sharing with community-based services improves, the challenges in building on the work of the pioneers include:

- adapting laboratory systems to fit with the developing approach of genetically based approaches to individuals, including reaction to medication and nanotechnology;

- adjusting normal values and the related interpretation as clinical testing becomes more representative of the different responses of men, women and children and to age-related physiological changes;

- providing access using internet technologies for patients to their personal clinical records and the related professional support to help them interpret and manage their own care, particularly in chronic disorders;

- providing more accurate statistical evidence of the risks of various components of acute care so that individual informed choices can be made – as in two-person zero-sum games a key issue is the elimination of inferior treatment strategies;

- integrating easy access to evidence-based practice as a resource for professionals to use throughout their working day;

- building routes for individual and local groups of clinical professionals to be actively involved in the development and adaptation of new systems to achieve confidence in computer-based information systems.

DENISE BARNETT AND BARRY BARBER

REFERENCES

1 BJHC (1985) CP85 Stand 34:MedStat/Torch. *BJHC*, 2, 1–32.

2 Barnett, D.E., Knight, G. and Mabbott, A. (1987) *Patients, Problems and Plans: Nursing Care of Patients with Medical Disorders*. Edward Arnold, London. ISBN 0 713144 92 0.

3 Boyde, E. (1987) Financial information project. *BJHC*, 4, 4–34.

4 Fabray, C.E. (1984) Developing patient-based financial information systems. *BJHC*, 1.4, 25–7.

5 BJHC (1985) CP85 Stands 27 & 28:Kone Instruments. *BJHC*, 2, 1–31.

6 Roderick, D. (1988) Disks and drives: a user's guide. *BJHC*, 5, 2, 29–30.

7 Akinsanya, J.A. and Barnett, D. (1992) *An evaluation of educational materials on HIV/AIDS Phase 1*. Anglia Polytechnic, Chelmsford. Unpublished.

8 Hall, G. and Hord, S. (1987) *Change in Schools: Facilitating the Process*. State University of New York Press, Albany NY. ISBN 0 887063 47 0.

9   Barnett, D. (1995) Informing the nursing professions about IT. In Greenes, R.A., Peterson, H.E. and Protti, D. (eds) *MEDINFO 95*, Vancouver, Part 2, pp. 1316–20. AMIA, USA. ISBN 0 969741 41 3.

10 Howard Jenkins (1984) Choosing a clinical department computerised records system. *BJHC*, 1(4), 18–21.

11 Clinical departmental systems

DENISE BARNETT

IN GENERAL

The slow trend towards closing small units and centralizing clinical expertise in larger hospitals encouraged interest in computer-based interpretation of clinical data, such as ECGs where there was no ready access to a cardiologist. Over time this centralizing of expertise paved the way for the adoption of a North American insurance industry product that, once Anglicized, became the support for NHS Direct in England. Occupational health nurses might have a smart electrocardiograph for use in the big London financial institutions where the clinical room had to cope with everything from a secretary's headache to an executive's angina or cardiac arrest.

Multi-centre research requires that data are collected in the same way and that the same clinical terms are used, so interest in definition and standardization developed. Reporting of clinical tests such as electromyography were seen as suitable for a specialized form of word-processing, where the system could store the medical terms such as the muscle names and conduction velocity calculations.[1]

Information-intense specialties perhaps had more need to use technology to handle the volume of physiological data and the clinical response to changes, so intensive therapy and care, and the renal failure treatments such as dialysis and transplant were logical development areas. Where clinical care of the majority of patients followed a predictable path, such as during pregnancy, standard care paths could be computerized.

The development of departmental systems initially relied on individuals who saw the work either as an extension of their interest in clinical computing and looked for aspects of their daily work to use to explore the way information technologies could be used, or those who saw a problem and thought that they might be able to solve it using a computer. Among the former who documented their journey in an amusing manner was David Morrison, Director of Intensive Care at North Manchester General Hospital. He provided a series of articles, from 1984 onwards, in the then quarterly (and free to those in the NHS with an interest) *British Journal of Healthcare Computing*.

In England, by 1984, John Hanby of the DoH was saying:

> *As yet the impact of computing outside traditional administrative domains such as payroll and accounting has been limited. Meanwhile the frustration of those*

197

trying to provide care to patients is growing as they become increasingly aware of what information technology has to offer. This is frequently expressed in the purchase of micros – often not the right solution – in the absence of anything more suitable. One of our Regional Health Authorities has undertaken a survey which has revealed that the total investment in micros in their region is already greater than that in the central mainframe facilities.

REFERENCE 2

The spread of clinicians' own clinical systems was treated as a problem, rather than a strength. Instead of recognizing this crucial resource to improve direct clinical care the NHS strategy in England was to develop proposals to:

- *gear information technology to **management needs** and a drive for improved efficiency and effectiveness;*
- *involve the adoption of common structuring of health information as a major new approach to this end;*
- *shift the emphasis away from centralized bureau facilities handling administrative matters towards meeting **operational and management needs** at the local level;*
- *provide computing facilities directly to users through networks and online working.*

REFERENCE 2

The technical infrastructure was useful, but the civil servants misread the focus of the vast majority of the clinical professionals and support staff: that of caring directly for patients.

Collen in 1983, from a US perspective, defined a clinical departmental system as:

. . . the dedicated use of a computer with associated hardware, software and terminals to collect, store, process, retrieve and communicate relevant information to support primarily one professional specialty group providing direct patient care within a hospital or medical offices.

REFERENCE 3

Common functional requirements were:

- patient identification;
- departmental databases;
- data communication with other clinical support systems, such as laboratories;
- computer-based medical records;
- special data-sets and knowledge bases;
- appropriate terminals to skill levels;
- support such as call/recall;

- fast, reliable 100 per cent access;
- flexible and able to grow;
- affordable cost.

Collen already considered 'in-house' development was a major project that might cost 25–50 per cent more than the original budget.[3]

Professor John Anderson of King's College Hospital, also speaking at *MEDINFO 83* conference, suggested that the growth area in the 1990s would be in departmental information systems to cope with the explosion in specialist knowledge. One driver would be financial stress, which required a department to justify its existence through better management information. He suggested that user-friendly, software tools could be used by clinicians to construct their databases, query them and store them securely. Interest was turning to data dictionaries, definitions and accuracy. The development of local area networks and digital telephone systems made possible the exchange of data with laboratory systems. The introduction of hospital networks to integrate departments with support services including clinical laboratories was underway.

In Anderson's view:

The end-users' view is generally assumed to be that of a passive recipient led by management. In health care institutions, motivation and attitudes are rather different to those of commerce. There is much more emphasis on service to patients, which must have a human interface. Health care professionals are now much more aware of what can be achieved by departmental information systems. Also they are becoming more computer literate, even if they are being educated by children.

REFERENCE 4

It has been difficult to track some of the work of individuals as knowledge of their presence depended on publication of details within the clinical or informatics community. The demise of clinical systems, sometimes linked with the relocation of the interested clinician, rarely appeared in print.

The sheer cost and bulk of the early computers made progress by individuals difficult, unless they were in positions of influence and seniority within the hospital hierarchy. The early machines also depended on machine language to write programs, it was only with the arrival of the Acorn/BBC computers that used BASIC in the late 1970s that clinical staff could really get directly involved. The advent of cheap but powerful personal computers widened the opportunities for staff in other roles to explore their clinical practice, sometimes working at home on personal machines by exploiting the capability of spreadsheets and hypertext.

Software for clinical systems could be based on commercial databases developed for business use but considerable time might be required to adapt it. According to Howard Jenkins:

... some software houses claim to have a fully working system available 'off the shelf' for a particular speciality, when in reality they have no more than a general purpose clinical database system.

REFERENCE 5

The central requirement for standard statistical data that were known by Mrs Edith Körner's name (Körner data) stimulated a lot of departmental work. The pressure to feed the Government department's expectations skewed the priority order and systems that might have delivered direct clinical benefit were passed over in favour of workload indicators. As a result in the 1980s the focus was more on history taking and it was very difficult to get money to develop a home-grown system for wider application. A brake was also applied through the Regional Computer Centres on any new applications that might distract from the Government focus. This was a major lost opportunity as one-off 'demonstration systems' in clinical computing did not progress and enthusiasm faded. Lilford and Chard reviewed the state of play in 1984 and tried to predict the next decade of 'computer revolution'.[6]

Technical factors

Jenkins commented in 1984 that:

Traditionally the policy of many [RHA] computing centres has been strictly mainframe based, and any suggestion of change met with resistance.

REFERENCE 5

- Many of the early computer systems for use in health care were written in the high-level language MUMPS (Massachusetts General Hospital Utility Multi-Programming System), which was developed in the mid-1960s with a standard version agreed in 1972. It was an interpretive language and considered quicker to write than compiled languages such as COBOL or Fortran, but it was also slower in execution and so a problem with programs that needed to perform mathematical calculations.

- The first applications were for laboratory systems, and these remained the most popular applications in 1977, and for medical record systems including data for specialties. It was used for the Health Information Systems (HIS) at Law Hospital, Glasgow, and at UCL, London. By 1978 a MUMPS User Group Applications Library (MUGAL) was being designed to share applications.[7]

- DEC agreed to user's requests to extend its support after problems in using the new faster and more powerful VAX range in 1986.[8]

- By 1990 the Royal Marsden Hospital, in Surrey, was reviewing its 12 years' experience of MUMPS for a large, integrated, hospital information system. There was a tendency to produce cryptic and unmaintainable code and a lack of overall reporting facilities. They used SQL supplemented with Higher Order Software to rewrite their system.[9]

Coding for comparisons

As the central pressure for comparative data about hospital activity increased there was a recognition that greater accuracy was required in coding diagnoses and operative procedures. Hospital Activity Analysis data have been collected since 1969 in England and Wales. As a prelude to contracting within the NHS an assessment of accuracy was undertaken in South Glamorgan Health District in 1990. This found errors in about 20 per cent of the sampled codes, those at the fourth digit level would have led to underfunding. Much of the miscoding was attributed to difficulties for the clinical coders in extracting the relevant data from the case notes.[10] Once managers realized there were financial implications for inaccurate clinical coding there was encouragement to review medical data standards. Clinicians also had an interest in relation to knowledge-based expert systems and multi-centre studies and clinical audits.

In the 1990s, there was interest in how to make data entry by clinical staff so easy it could be part of daily clinical recording. New ways of writing were explored such as using a pen on a graphics tablet.[11] Computer mobility was also important for clinical areas and wireless computers were explored, for example to use during a drug administration round.

External factors

As health managers know, external factors can drive change within hospitals and community settings that may, or may not, be helpful to local developments. The desire of the health departments to control costs and improve productivity was manifest in a number of 'initiatives'. Meanwhile, other Government departments were trying to stimulate the computer industry.

- The Financial Information Project (FIP – see Part 3 Introduction) also had an impact on clinical specialties as efforts were made to collect activity data, particularly with operating theatres and ward nursing systems. A theatre system was tested in five theatres at Cheltenham District General Hospital. The data were collected on paper forms completed by the nurses, and then batch entered, once a week. It was written in Digital Standard MUMPS (DSM). The intention was to link it to the Patient Administration System (PAS) via the patient's hospital number. There were also plans to look at theatre scheduling.[12]

- The first meeting of the FIP Association took place at Guy's Hospital in London in May 1987. It grew out of a project initially funded by the DoH and West Midlands RHA that had started in 1979. When funding ran out the decision was made to continue software research and development through subscriptions from participating regions and districts.[13]

- Clinicians were also caught up in the development of Diagnosis Related Groups (DRGs, see Chapter 6) for comparative costing for the contracting culture that was being introduced. The concept had come from insurance reimbursement schemes for the many private hospitals in the USA. The clinical budgeting data were used by CASPE to explore the differences between hospitals.[14]

- The National Audit Office (NAO) examined the management of operating theatres in 1987 and concluded that efficient use of theatres was the key to reducing the NHS waiting lists and times. In 1991, the NAO reported on progress and the impact of the steering group chaired by Professor Peter Gilroy Bevan. Three existing stand-alone systems were in use: Theatreman, FIP and ORSOS. Two sites had whole-hospital systems with a theatre module: The London Hospital, and the Hammersmith Postgraduate Teaching Hospital. Issues included agreeing coding for operative procedures and ways to capture the data.[15] Only two of the six hospitals visited by the NAO in 1990 had installed and were using computerized management information systems for operating theatres, and one was testing a new system.[16]

- The development of graphics packages allowed clinical systems to become of greater use to clinical staff as trend detection and pattern recognition was easier than viewing tabulated data. These approaches help by predicting changes in the patient's condition from biochemical results such as serum creatinine or alkaline phosphatase.

- The Ministry of Technology had sponsorship responsibilities for a wide range of electronic, mechanical and engineering industries. It included the Computer Advisory Service set up in 1996 to provide broad-brush advice and technical support. One of its roles was to direct the Advanced Computer Technology Project. It also had to foster the development of a British computer industry.[17]

- Interactive terminals for direct use by patients were being explored at the University of Essex in 1969. The system was also intended to meet the needs of clinical teams using a computer as a local controller with a number of terminals.[18]

- The Central Computer and Telecommunications Agency (CCTA) made its impartial technical reports available to the NHS from 1985 onwards, but awareness of this facility did not seem to percolate down to clinical levels.[19]

- In January 1989, a mandatory, systems development method (Structured System Analysis and Design Method – SSADM) was adopted for Government-funded procurements and implementations throughout the EEC.[20]

Funding projects

- The DoH established its computer branch in 1969. The experimental programme for small (mini) computers began in 1968/69. It ended in 1976. Applications included those related to EEG, radiation treatment planning, cardiology, in-patient information systems, nursing and laboratories.[21]

- The Scottish Home and Health Department (SHHD) provided funding from 1979 onwards for computing students to carry out small development projects during their vacations. Many of the projects were related to clinical records.[22]

- The NHS investment in IT for England was under one per cent in 1988 when in business it was between five and ten per cent.[23]

- In England, the DoH encouraged and sometimes funded research and development. The initial policy was to demonstrate the potential and acceptability of computing to health service authorities. The *Strategy for Information Management in the Hospital and Community Health Services* was published in October 1986.[24] Central policy culminated in June 1990 with *Working for Patients – Framework for Information Systems*.[25] By 1996 the research and development work was subsumed under the Integrated Clinical Workstation (ICWS) programme (see Chapter 7). Although a lot of work was done with clinical groups to gather views and ideas the ICWS petered out.

A list of major projects and studies supported by the DoH through the NHS Scientific and Clinical Information Technology Research and Development Programme compiled by Brian Layzell is set out in Table 11.1.[26]

The NHS research and development budget for 1987–88 measured proposals against the following criteria:

- relevance to national priorities
- significance to the hospital and community service in general
- degree of originality
- evidence of managerial ability and user commitment to carry out the work
- need for external funding.

The percentage split between different activities left about 10 per cent (£250,000) uncommitted. The priority areas were:

- management and clinical decision-support systems;
- integration of primary, hospital and community systems;
- screening systems;
- diagnostic applications; and
- portable medical records.[27]

Not much there for small departmental systems.

TABLE 11.1 *Major DoH projects and studies*[26]

Report	Topic	Location	Notes
	Advisory system for breast clinic management	University College Hospital and the Middlesex Hospital, London	1 year
Jul 1986	Help for the disabled	Newcastle uponTyne Polytechnic	Enhanced BARD
1991	Clinical Nursing Information Project	Royal National Throat, Nose and Ear Hospital & Elizabeth Garret Anderson Hospital	MUMPS-based
1985/ 1989	Computer-assisted learning in nurse education	St Thomas' Hospital, London	Acorn, BBC B & Torch systems. Two phases
	Computerization of psychometric tests	Leicester University	Eight tests
Nov 1985	Computer-assisted diagnosis of acute abdominal pain	Leeds University	Eight centre trial
1987/ 1989/ 1991	Computer-readable medical card trials	South Wales GP practices in Exmouth	Smart cards
	Computer recorded information system for psychiatry (CRISP)	St Mary Abbots Hospital, London	
	Dentists/Dental Practice Board electronic links	25 dental practices	Touche Ross evaluation
	Diabetes computer system	St Thomas' Hospital	
	Echocardiograph reporting system	Killingbeck Hospital, Leeds	
	Endoscopy computer system	British Society of Gastroenterology members	Acorn BBC B
1987	Evaluation of the use of an application generator	Hammersmith Hospital, London	Integrated HIS
1988	Expert system for interpreting laboratory data	Royal Free Hospital School of Medicine, London	Fluid-electrolyte & acid–base balance
1989	General Medical Practitioner/ Family Practitioner Committee electronic links	Ten GP practices, two FPCs	Started 1986
1986	Genito-urinary medicine	Middlesex Hospital	MUMPs-based
	Improved user interface for general practitioners	Manchester University	
	Microcomputer analysis of human fetal heart rate	John Radcliffe Hospital, Oxford	Converted to cheaper microcomputers
	Microtext	National Physical Laboratory	Enhanced software for CAL and automated interviewing

Continues

Table 11.1 *Continued*

Report	Topic	Location	Notes
1985	Patient monitoring	Killingbeck Hospital, Leeds	ICU – analogue and computerized
1989	Post-operative management of cardiac surgical patients	Hammersmith Hospital, London	ITU – haemodynamic control
	Radio-immune assay software	MRC Immunoassay Team with Edinburgh University	SHHD involved
1989	Remote data capture in community nursing	Bath Mersey	
1987 1990 1992	Simulation and planning of facial reconstructive surgery and identification of fetal abnormalities	University College Hospital and the Middlesex Hospital, London	Laser & CT scans for orthodontics. Simulation of fetal abnormalities
1988	Waiting list data	West Midlands RHA	1985 start

One central project of relevance to clinical practice was the Electronic Patient Record (EPR, see Chapter 7). This three-year research and development project set out to produce working demonstrator systems in acute hospitals with a view to:

- capturing the interest of the clinical professions;
- convincing managers of the benefits of the EPR systems;
- influencing suppliers to develop the next generation of hospital information technology;
- learning lessons of wider relevance to other person-based systems.

The Project Board commissioned the Systems Failures Group of the Open University to carry out a comparisons study into why other systems in acute hospitals had failed. The broad conclusion was that 'lack of involvement and commitment on the part of key players was the most important feature in the failure of many clinical information systems'. Two major sets of risks had to be balanced:

- Allowing genuine user involvement and risk, so slowing down the design stage and generating conflicts among users, and between users and designers.
- Restricting the involvement in design, and risk producing a system key players would be unwilling to use.[28]

Maturing systems and integration

By the 1980s the clinical laboratories had been transformed from labour-intensive departments to highly automated ones. They could produce vast quantities of accurate and reliable data, relatively quickly and at very little cost per individual test.[29]

Descriptions of integration of systems and data across hospitals began to appear in the journals in the late 1980s. For example, Faukner wrote about the Hammersmith Hospital,[30] and Lilley about The Hospitals for Sick Children, London.[31]

As expert systems and patient data recording systems matured it became possible to link them with clinical laboratory and other diagnostic services (see Chapter 8). The computer systems became more complex and commercial development with transferability to other units and hospitals became an economic necessity to fund the intensive development work of large teams. The NHS Management Centre funded work to develop common basic specifications for systems. One example was the COSMOS project at St Mary's Hospital, London. The clinical process model set out how diagnostic expert systems should fit within the Common Basic Specification to support clinical practice.[32]

Lusted, in 1983, reviewed 25 years of international health care computing and identified that artificial intelligence (AI) was having a similar impact as that of microprocessors.[33] It had become necessary to include a separate stream for AI within the *MEDINFO* conference programme.

Back in 1960 Lusted had listed five principles:

- Adopt computer-aided diagnosis, not computer diagnosis.
- Systems must be acceptable to clinicians.
- Adequate preparation is essential.
- Hardware should be matched to systems and needs – not vice versa.
- Real-life testing is essential.[33]

de Dombal added a further five:

- The computer's role should be closely defined.
- Systems should be transportable.
- Software selection is superfluous.
- The experiment must be adequate to permit valid evaluation.
- The medical content of programs must be impeccable.[34]

McSherry in Belfast designed ALICE (Algorithms through Interaction with Clinical Experts) to allow clinicians to develop simple branching routes to diagnosis. Written in BASIC, it ran on an Apple® II with 48k of memory, a 5¼" disk drive and a video monitor. The clinician with minimal computer skills was to teach ALICE new algorithms through tutorial dialogue using natural language. Algorithms developed by several clinicians could be copied, merged, sorted or classified.[35]

The Directorate of Health Service Information Systems in Scotland was set up in 1987 under the Common Services Agency. Its role was to act as a procurement agent for the Health Boards.[36] In 1990, the National Health Service commissioned them to review and make recommendations on how best to ease the integration of systems. They opted for an open-systems environment for centrally funded developments.[37]

Ownership and control

Involvement in departmental systems has changed over time, from the clinician working with a bioengineer or electrical engineer or physicist to develop programs, to an expanded team including a computer operator to enter the text and other data. Engineers became a hospital's 'IT department' and looked after the mainframe and its location. Large rooms had to be allocated to house the machines, and ventilation to dissipate the build-up of heat was a key requirement. Back-up tapes and archiving were technical tasks, especially when the memory was limited. Then, with the advent of microcomputers, other members of the clinical team were involved in entering or retrieving data, such as midwives and nurses working in intensive care units.

As departmental systems were integrated within larger hospital information systems so trainers became necessary. Commercial systems also required familiarization and training time. Troubleshooting might be by the clinical staff in the department involved in the system's development, the IT department, or over a telephone help-line to the supplier. As storage technologies developed the CD-Rom became a route to upgrades and with it interactive teaching materials and instructions that could be printed by the user. Printed and bound manuals became something that had to be bought separately. There was a steady growth in 'how to' books for the core personal computer systems and applications.

While small minicomputers and personal computers were in the ascendancy there was a need to identify someone to regularly back-up data, to maintain the computer supplies and to 'clean up' the files and to keep virus checking and disinfecting programs up to date. Back-up could be as simple as regularly copying content to a floppy disk. Developed systems that were then integrated came back under the ambit of IT departments. The role, in day-to-day care of the system and data, reverting to specialists within the IT department.

Users of personal computers came together in user groups to share information about programs, for example the Amstrad/Acorn users in 1986.[38]

In the NHS the clinicians only had a short period of autonomy over clinical systems. The centralization of contracts within Regional Computer Centres (RCCs) in the days of mainframes and expensive minicomputers was joined for a few years by the rapid expansion among enthusiasts helped by the advent of the microcomputer. Then the RCCs reasserted an iron grip and control became even more centralized, with the NHS Information Management Group seeking to provide an infrastructure to integrate systems across the NHS. Many clinicians felt squeezed out by the technical focus. The buzz of improving patient care by achieving mastery over the computer, to get the machine to do the lengthy or tedious mathematics and to search, sort and collate, was taken away from the bedside.

In 1991, Lloyd-Williams suggested:

... in fact the stagnation of the market over the last five years has been caused largely by a failure of suppliers to react to a changing pattern of demand. This has in turn led to the phenomenon of 'adversarial procurement' and a further fragmentation as suppliers 'buy' small shares of the market without a pattern .... So the UK has reached a watershed – procurement is strangling what initiatives there are that escape centralist direction.

REFERENCE 39

The Information Authority struggled to retain clinical interest in computing once the focus was on providing information to manage the hospital and, by implication, the clinicians. They became 'human resources', not people with their own clinical interests and enthusiasm for patient care. Research and the development of applications mainly involving content might still be carried out on small, powerful personal computers or even laptop computers by the individual clinician, but success now meant the clinical application was taken away from the individuals or team if it was to have a chance of being spread to other units. It had to be subject to review, testing and the opinion of committees. In part this was a reflection of what also happened within the IT community. The geek in the garage, working on a new application, has given way to the expensive business funded by a venture capitalist.

For a tiny proportion of the health care professions there was a brief period of less than two decades that was computing Camelot. The following chapters in this part of the book seek to capture some of what was achieved, the people involved and the excitement of working together to try to create a better world for patient care.

REFERENCES

1 Product developments (1984) EMG reporting. *BJHC*, 1, 4, 44.

2 Hanby, J.G. (1984) Harnessing technology to health care – the challenge for the future. In Roger, F.H., Willems, J.L., O'Moore, R.R. and Barber, B. (eds) *MIE, Brussels 84*, pp. 616–21. Springer Verlag, Berlin. ISBN 3 540133 74 7.

3 Collen, M.F. (1983) General requirements for clinical departmental systems. In van Bemmel, J.H.V., Ball, M.J. and Wigertz, O. (eds) *MEDINFO 83*, Amsterdam, pp. 736–9. North-Holland, Amsterdam. In two parts.

4 Anderson, J. (1983) Clinical departmental information systems. In van Bemmel, J.H.V., Ball, M.J. and Wigertz, O. (eds) *MEDINFO 83*, Amsterdam, pp. 814–17. North-Holland, Amsterdam. In two parts.

5 Jenkins, H. (1984) Choosing a clinical department computerised records system. *BJHC*, 1.4, 18–21.

6 Lilford, R. and Chard, T. (1984) Clinical computing to the end of the decade. *BJHC*, 1.3, 8–13.

7   Zimmerman, J. and Schutte, F. (1978) Review of the MUMPS language and its applications to medicine. In Anderson, J. (ed.) *MIE, Cambridge 78: Lecture Notes 1*, pp. 303–14. Springer Verlag, Berlin. ISBN 3 540089 16 0.

8   Developments (1986) DEC improves MUMPS support. *BJHC*, 3.2, 6.

9   Milan, J., Munt, C.E. and Dawson, M.W. (1990) A model-based approach to the evolutionary development of a high performance hospital information system. In O'Moore, R.O., Bengtsson, S., Bryant, J. and Bryden, J.S. (eds) *MIE, Glasgow 90: Lecture Notes 40*, pp. 457–61. Springer Verlag, Berlin. ISBN 3 540529 36 5.

10 Wilkinson, E.J. and Harvey, I. (1991) Accuracy of hospital diagnostic and operative procedure coding. In Adlassnig, K.P., Grabner, G., Bengtsson, S. and Hansen, R. (eds) *MIE, Vienna 91: Lecture Notes 45*, pp. 739–43. Springer Verlag, Berlin. ISBN 3 540543 92 9.

11 Cook, C.A., Beresford, R.A., Dodds, S.R. and Thompson, R.L. (1993) Evaluation of a pen-based method of clinical-data capture. In Richards, B. and MacOwan, H. (eds) *Current Perspectives in Healthcare Computing*, pp. 641–9. BJHC for BCS, Weybridge. ISBN 0 948198 14 1.

12 Fabray, C.E. (1984) Developing patient-based financial information systems. *BJHC*, 1.4, 25–7.

13 Boyde, E. (1987) Financial information project. *BJHC*, 4, 4–34.

14 Jenkins, L.M. and Bardsley, M.J. (1985) Using DRGs to separate case mix in the UK from trends in length of stay. In Roger, F.H., Grönoos, P., Tervo-Pellikka, R. and O'Moore, R. (eds) *MIE, Helsinki 85: Lecture Notes 25*, pp. 326–31. Springer Verlag, Berlin. ISBN 3 540156 76 3.

15 NHS Management Executive VFM Unit (1989) *The Management and Utilisation of Operating Departments*. NHS. ISBN 1 851974 99 7.

16 National Audit Office (1991) *Use of National Health Service Operating Theatres in England: A Progress Report*. HMSO. ISBN 0 102306 91 5.

17 Stark, G.M. (1969) Computer advisory service. In Abrams, M. (ed.) *Medical Computing: Progress and Problems*, pp. 200–2. Chatto & Windus for BCS, London. ISBN 0 701115 75 0.

18 Gedye, J.L. (1969) Problems in the design of interactive terminals for direct use by patients. In Abrams, M. (ed.) *Medical Computing: Progress and Problems*, pp. 183–99. Chatto & Windus for BCS, London. ISBN 0 701115 75 0.

19 Developments (1985) CCTA can offer NHS computing advice. *BJHC*, 2, 3–6.

20 Brooks, N. and Reeves, K. (1987) How to use SSADM. *BJHC*, 4.1, 41–3.

21 Abbott, W. (1988) 40 years of IT. *BJHC*, 5.6, 11–12.

22 Williams, S. (ed.) (1986) Scottish Information Technology Users (SHS) News: Project funding. *BJHC*, 3.2, 38.

23 BJHC (1988) NHS 'stone age' IT. *BJHC*, 5, 10, 5.

24 DoH (1986) *Strategy for Information Management in the Hospital and Community Health Services*. HMSO, October.

25 DoH (1990) *Working for Patients – Framework for Information Systems*. HMSO, June.

26 Layzell, B. (1996) The centrally administered research and development programme. In Abbott, W., Bryant, J.R. and Barber, B. (eds) *Information Management in Health Care: Handbook A: Introductory themes*. Health Informatics Specialist Groups, Eastbourne. ISBN 0 901865 85 0.

27 BJHC (1987) £250,000 research money still available. *BJHC*, 4.5, 3.

28 Dodd, W. and Fortune, J. (1995) An electronic patient record project in the United Kingdom: can it succeed? In Greenes, R.A., Peterson, H.E. and Protti, D. (eds) *MEDINFO 95*, Vancouver, Part 1, pp. 301–4. AMIA, USA. ISBN 0 969741 41 3.

29 Pope, C.M., Carson, D.G., Cramp, N. and McIntyre, N. (1983) The role of the clinical laboratory in diagnosis and patient management. In van Bemmel, J.H.V., Ball, M.J. and Wigertz, O. (eds) *MEDINFO 83*, Amsterdam, pp. 991–4. North-Holland, Amsterdam. In two parts.

30 Faukner, J. (1987) Transformation achieved by integration. *BJHC*, 4.6, 19–22.

31 Lilley, C. (1987) HSC pulls the pieces together. *BJHC*, 4.6, 22–4.

32 Cairnes, T., Timini, H., Thick, M. and Gold, G. (1991) A generic model of clinical practice – the COSMOS project. In Adlassnig, K.P., Grabner, G., Bengtsson, S. and Hansen, R. (eds) *MIE, Vienna 91: Lecture Notes 45*, pp. 706–10. Springer Verlag, Berlin. ISBN 3 540543 92 9.

33 Lusted, L.B. (1983) Design for decisions – a 25-year perspective. In van Bemmel, J.H.V., Ball, M.J. and Wigertz, O. (eds) *MEDINFO 83*, Amsterdam, pp. xi–xv. North-Holland, Amsterdam. In two parts.

34 de Dombal, F.T. (1983) Towards a more objective evaluation of computer-aided decision-support systems. In van Bemmel, J.H.V., Ball, M.J. and Wigertz, O. (eds) *MEDINFO 83*, Amsterdam, pp. 436–9. North-Holland, Amsterdam. In two parts.

35 McSherry, D.M.G. (1983) ALICE: a teachable microcomputer system for compiling clinical algorithms. In van Bemmel, J.H.V., Ball, M.J. and Wigertz, O. (eds) *MEDINFO 83*, Amsterdam, pp. 597–9. North-Holland, Amsterdam. In two parts.

36 BJHC (1987) Scotland centralizes healthcare computing. *BJHC*, 4.3, 7.

37 Sayers, R.C. (1992) An open, software-led strategy for the NHS in Scotland. In Richards, B. and MacOwan, H. (eds) *Current Perspectives in Healthcare Computing*, pp. 419–25. BJHC for BCS, Weybridge. ISBN 0 948198 12 5.

38 Nicholls, S. (1986) Amstrad/Acorn users (Letter). *BJHC*, 3.5, 9.

39 Lloyd-Williams, D.R. (1991) The European healthcare informatics market: fact or fiction? What does it mean to the UK? In Richards, B. and MacOwan, H. (eds) *Current Perspectives in Healthcare Computing*, pp. 379–85. BJHC for BCS, Weybridge. ISBN 0 948198 11 7.

12 Clinical systems and audit

DENISE BARNETT

CLINICAL SYSTEMS FOR DIRECT PATIENT CARE

It can be difficult to separate out computer applications that directly contribute to clinical care from those that are discussed in other chapters, such as those that support the management of a hospital or computer-assisted learning for staff and for patients and their family carers (Chapter 23). The development of computer applications for direct clinical care often depended on individuals spending their personal time on developments. The central funding focused on obtaining data to plan or manage the system and finance was a key driver. This and the next chapter do not cover all the clinical specialties where patient data, including clinical histories and laboratory results, were collected, aggregated and used for research, performance measurement and the day-to-day management of services. Nor does it cover the many systems intended to reduce the clerical load. Over time data collection applications based on computers have been used for every body-part.

The focus will be on the early applications that illustrate the main themes and applications that, for one reason or another, had an impact on development. There were many departmental applications for clinic and diagnosis-related registration and administration. Most have been omitted to reduce repetition.

Assessing impact

Some descriptive research projects sought to gain a view of the level of hospital activity and its location, such as the survey by Alan Rector and colleagues for The King's Fund for London.[1] Williams and Web-Peploe suggested in 1981:

> Unless medical computing can be shown to work in a clinical environment there is a danger that computer professionals will increasingly regard it as difficult and unrewarding and that medical staff will continue to look upon all computer systems with suspicion, as time-consuming, expensive and unproductive gimmicks.

REFERENCE 2

The *Hospital Systems* handbook of 1986 provided 14 summaries for managers but only eight on departmental systems, and they all focused on records:

- accident and emergency
- cervical cytology
- transplants
- mental health
- radiation treatment planning
- nursing
- physiotherapy
- occupational therapy.

As a loose-leaf format, pages were replaced with new issues several times a year, so the history of some application areas can only be traced through any pages stored elsewhere after removal. There were 24 updates, the last issue being in December 1995.

Attitudes towards computing among managers and heads of clinical divisions were surveyed in Scotland by Ray Jones in 1989. Half the clinical divisions claimed there was an immediate requirement for at least one clinical system. The obstacles to implementation identified by the clinical divisions were:

- lack of trained staff;
- inexperience of clinical staff in using computers;
- time and money to implement systems;
- lack of suitable systems or at least compatible and integrated systems.[3]

In the late 1980s, the journal focus seemed to be heavily biased towards using computers in managing hospitals, with articles on 'Clinical management systems'. The emphasis was control of money and resources, not on directly improving clinical care, other than through professionals auditing their practice. Ray Rogers stated it clearly in 1989: 'One of the most important uses of information is in helping managers to improve performance.'[4] No wonder it was hard to engage the hearts and minds of clinical professionals who were focused on improving the care they wanted to give to the patients with whom they were in daily contact.

A random sample of hospital consultants within the West Midlands RHA was reported in November 1990. Specialties ranged from A&E to virology. While two thirds (67.5 per cent) used computers at work only a quarter (24.6 per cent) used them directly, the rest delegated their use. Although 31 per cent of the respondents were departmental heads, only 14.4 per cent had departmental systems. Colleagues were the largest single source of information about computers, followed by family/children and medical publications. The authors suggested that:

> ... the way to raise the level of interest in computers is to select interested parties and allow them to gain knowledge and experience which will then be passed on to others.

REFERENCE 5

At the Vienna Congress of MIE in 1991 Stephen Kay, Alan Rector and colleagues from the University of Manchester presented their work on the PEN&PAD prototype patient care workstation and their analysis of the basic requirements of the medical record. This was to support direct patient care: a fundamentally different position to many of the existing records that were derived from the need to aggregate data.[6] They identified, and the team later extended their work on, a model of medical semantics and medical language. (See Chapter 14.)

In 1991, Rector and colleagues suggested:

> ... *to date, the medical informatics community has not been notably successful in producing systems which are widely used in routine clinical practice.*

REFERENCE 7

The Electronic Patient Record Programme (EPR) began in January 1994 and had a three-year research and development programme to help clinicians give better care to patients through the use of electronic records. The EPR programme under Dr Bill Dodds had a seven- to ten-year horizon for the outcome of research to feed into a new generation of clinical systems. (See Chapter 7).

Some of the specialist groups with an interest in information technology produced their own newsletters. The BCS Nursing Specialist Group's journal *Information Technology in Nursing* opened its pages, in 1993, to other health care professionals such as physiotherapists, chiropodists and occupational therapists.

Demonstrating systems

Demonstrating how a computer application might help improve patient care could be a powerful way of spreading news of developments. The annual exhibition associated with the *Healthcare Computing* Conference was one route. Commercial organizations tended to market systems with dummy data at these events. For a brief period there were also demonstrations by clinicians of their own work. Examples included PEN&PAD and expert systems for the treatment of insulin-dependent diabetics.[8, 9]

Conventional clinical teaching uses real-life problems. Simulations could be built involving clinical history, examination, laboratory results and charts. In time, as operating systems and memory developed, video and audio clips could be used. Medical students could assess the data, make a diagnosis and propose the appropriate treatment path. Corbett, from the Department of Surgery, and Taylor, from Computer Science at Liverpool University, used the dialogue programming language, Microtext, on a BBC B microcomputer to create just such a branching program.[10]

An alternative proposal was to take real patient data into the classroom, from the HIS, and for students to plan care based on these data then compare at intervals their proposed plans with the real-life records and outcomes. Patient privacy and consent limited the implementation.

Computers impacting on the professionals

Encouraging busy professionals to learn about new approaches can be supported by activities that provide them with a personal experience and payback. Postgraduate medical education records were one such route. The MEDICS scheme of Viner, Lees and Dick covered one London Region.[11] Bryden and Boyle used the concepts to create a system for Glasgow using dBASE® II. It necessitated building up close contact with the person, in each institution, that recorded each junior doctor's documents. A photocopy was obtained for each six-month tenure for a junior doctor. In 1985, the General Medical Council registration number was not widely used as an identifier, so alternative fields were used to link the records. A publicly identified and personally responsible data holder was used to reassure doctors that their personal data were held safely. This followed a similar system used in Scotland for patients' clinical data. Each doctor received a copy of their own education record.[12]

St Mark's Hospital, London, which specialized in colorectal disease had started to discuss the possibility of computerizing research records in 1978. Williams recalled:

> *Several specialists already had bitter experience of working with* [a] *large mainframe computer department: problems, not of computing, but of human nature – sloth, inefficiency and lack of price amongst other attributes.*

REFERENCE 13

As a result they chose to transfer data-sets to individual, standardized, low-cost microcomputers under the complete control of the relevant researcher. The option was to later network the machines or link to a mainframe. Metasa Ltd took a bottom-up approach, finding out the exact needs of every doctor or researcher. This was slow and expensive, but considered worthwhile.

> *It not only allowed users to have what they wanted but allowed the most enthusiastic departments to trail-blaze whilst others, more reticent about computing, could observe progress and develop their commitment slowly. The interest of less committed departments proved usually to be fuelled initially by their junior members using a microcomputer for word-processing, but then becoming sufficiently intrigued to experiment with database applications.*

REFERENCE 13

CLINICAL AUDIT

The electronic collection of local clinical data within a specialty with the opportunity to sort, select and classify aided the development of clinical audit. The elements include:

- peer review;
- a focus on standards;

- patient-based data;
- systematic studies;
- a commitment to change.[14]

A seminar on implementing clinical audit was held for senior NHS personnel in May 1991. Ian Carruthers, District General Manager in West Dorset, said he viewed the objective of the NHS reforms as a three-dimensional improvement in the quality of service to patients, that is:

- access to services;
- customer relations, that is the treatment of the patient as an individual;
- technical competence.[15]

Obstetrics has a respectable history of audit and confidential review of untoward events related to the reduction of maternal and neonatal mortality. Microcomputers provided a local resource from which data could also be extracted for official reports such as the birth notification and regional returns. One example being the Commodore PET used at St Mary's Hospital, London, where midwives in the labour ward entered their data and clerical staff added discharge summaries for mother and baby. Online vetting was possible, so data could be corrected soon after entry. The accuracy of recording fetal distress was increased by programming the computer to note the presence or absence of three items: fetal acidosis, abnormal cardiotocography, and meconium stained liquor. The Standard Maternity Information Service (SMIS) on the central mainframe was seen as slow to produce reports, the annual one not arriving until mid-year.[16]

Within a few years effective systems were transferred to faster, larger networks of machines, with clinical teams preparing system requirements and seeking tenders from commercial companies.[17]

The St Mary's maternity information system (SMMIS) was in use in all 15 maternity units within North West Thames Region by 1988 and was also used by a further 19 districts outside the region. National and regional users had modified the data items collected, but the fields and definitions were standardized to enable valid comparison. Midwives input the data: automatic data mapping used ICD9 and OPCS4 to code diagnoses. The data fed into a regional congenital malformation register. An integrated neonatal module linked to the obstetric module, and transfer of babies between units could be tracked. A maternity anaesthetic module had been developed based on guidelines of the Obstetric Anaesthetist Association. The paper suggested: 'much of the success of the system has depended on enthusiasm, spare-time work, and "soft" research funding'.[18]

As interest in Archie Cochran's work developed; the type of research method used within studies was seen as important and large randomized controlled trials gained favour. The World Health Organization funded work at the Radcliffe Infirmary, Oxford, to create a register of these for perinatal medicine. The cost of using a Commercial Bureau was prohibitive, so a microcomputer with locally written software offered a cost-effective alternative. It required work to define search terms and add additional classification fields for assessing abstracts drawn from databases such as MEDLINE and MEDLARS. During the project, easy to use information retrieval software for microcomputers became available. The pressure for reliable clinical audit and for access to the work of other clinicians created a need for an evidence-base.[19]

Standard methods for the assessment of clinical conditions, such as perceptual disorders following a stroke, led on to detailed guidance and developing expert systems.[20] (See Chapter 24.)

Microcomputers could be used to assess the reproducibility of measurements in the same patient, both for research and for quality assurance in clinical services. Bolton General Hospital used an incomplete block layout to compare fetal-skull measurements made by ultrasound operators. Each mother was scanned by two of the four operators within the department and the results were compared statistically.[21]

A two-day seminar was held in Oxford in June 1989 to discuss the safety of software controlled medical equipment. There had been a number of deaths from unsafe products. The opening of the European market in 1992 was giving concern that common standards and testing programmes were needed.[22] At the *Medical Informatics Europe* congress in Glasgow in 1990 there were three other papers from the UK on these issues, and six safety first principles set out. Quality assurance was also on the agenda for the suppliers and Geoff Hutt of AT&T Istel described their education and workshop programme.[23]

When looking forward, and back, it may be worthwhile pondering the words of Patricia Tymchyshyn:

- *There is never a perfect time to begin being involved with computers.*
- *It's okay to make mistakes: without them, you have never taken any risks.*
- *You learn by doing, not by waiting for someone to teach you.*
- *The most supportive audience is not found in your home town.*
- *Although some people appear to be experts, you probably know as much as they do.*

REFERENCE 24

An influential report that encouraged the acceptance of clinical audit was the report of a confidential inquiry into perioperative deaths in 1987.[25]

In 1991, the Government's call for increased medical audit, as outlined in the White Paper *Working for Patients*, led to proposals to train junior doctors in computing and audit skills. In a study at the Royal Free Hospital many junior doctors and medical students exhibited obvious anxiety and technophobia when first confronted with a computer. This had an impact on their plan to introduce computer-based audit for vascular surgery.[26]

A review of medical audit systems was reported in 1992. It used eight criteria in harmony with the DoH guidance.[27] An initial list of 59 products was identified. The reviewers also looked at products for quality support and statistics.[28] The product guide to suppliers' systems, published in early 1994, listed 32 systems, 27 of which were aimed at hospitals.

Support services

Support services could have an impact on clinical work such as the ordering of supplies by a ward, one of the earliest uses of computers. Printed sheets were produced and clerical staff in the supplies department entered the data handwritten by the ward clerk or nurses. This was followed by bar-coding and topping-up to a set stock level that could be undertaken by non-clinical staff. The Central Sterile Supplies department were early initiators.

Existing business systems could be adapted for some support services such as pharmacy, supplies and engineering maintenance. The processes were standardized across hospitals so commercial suppliers would have a large enough market for financial viability. By the 1980s computers were part of clinical equipment and the use of microprocessors was expanding.

REFERENCES

1 Rector, A.L. (1978) A survey of developments in medical records and information systems in the United Kingdom. In Anderson, J. (ed.) *MIE, Cambridge 78: Lecture Notes 1*, pp. 91–100. Springer Verlag, Berlin. ISBN 3 540089 16 0.

2 Williams, K.N. and Webb-Peploe, M.M. (1981) The capture and processing of routine clinical data in cardiology. In Grémy, F., Degoulet, P., Barber, B. and Salamon, R. (eds) *MIE, Toulouse 81: Lecture Notes 11*, pp. 936–42. Springer Verlag, Berlin. ISBN 3 540105 68 9.

3 Jones, R. (1989) Clinical computing in Scotland: results from a postal survey. *BJHC&IM*, 6.3, 26–8.

4 Rogers, R. (1989) Indicating comparative performance. *BJHC&IM*, 6.7, 13–14.

5 Young, D., Chapman, T. and Poile, C. (1990) Physician reveal thyself. *BJHC&IM*, 7.9:16, 19–21.

6 Kay, S., Rector, A.L., Nowlan, W.A., Goble, C.A., Horan, B., Howkins, T.J. and Wilson, A. (1991) What should we mean by an 'electronic medical record'? In Adlassnig, K.P., Grabner, G., Bengtsson, S. and Hansen, R. (eds) *MIE, Vienna 91: Lecture Notes 45*, pp. 132–9. Springer Verlag, Berlin. ISBN 3 540543 92 9.

7 Rector, A.L., Horan, B., Nowland, W.A., Goble, C., Kay, S., Fitter, M. and Robinson, D. (1992) A developer's view of formative evaluation: or 'Users are always right . . . except when they're wrong'. In Richards, B. and MacOwan, H. (eds) *Current Perspectives in Healthcare Computing*, pp. 99–106. BJHC for BCS, Weybridge. ISBN 0 948198 12 5.

8 Nowlan, W.A., Kay, S., Rector, A.L., Horan, B. and Wilson, A. (1991) PEN&PAD: a multi-lingual patient care workstation based on a unified representation of the medical record and medical terminology. In Adlassnig, K.P., Grabner, G., Bengtsson, S. and Hansen, R. (eds) *MIE, Vienna 91: Lecture Notes 45*, p. 1043. Springer Verlag, Berlin. ISBN 3 540543 92 9.

9 Lehmann, E.D., Deutsch, T., Carson, E.R. and Sanksen, P.H. (1991) (Title unknown) In Adlassnig, K.P., Grabner, G., Bengtsson, S. and Hansen, R. (eds) *MIE, Vienna 91: Lecture Notes 45*, p. 1045. Springer Verlag, Berlin. ISBN 3 540543 92 9.

10 Corbett, W.A. and Taylor, M.J. (1985) A microcomputer simulated interactive clinical problem. In Roger, F.H., Grönoos, P., Tervo-Pellikka, R. and O'Moore, R. (eds) *MIE, Helsinki 85: Lecture Notes 25*, pp. 113–17. Springer Verlag, Berlin. ISBN 3 540156 76 3.

11 Viner, R.S., Lees, W. and Dick, G.W.A. (1982) A regional medical manpower and training information system. *Community Medicine*, 4, 108, 12.

12 Bryden, J.R. and Boyle, L.B. (1985) A postgraduate medical education database. In Roger, F.H., Grönoos, P., Tervo-Pellikka, R. and O'Moore, R. (eds) *MIE, Helsinki 85: Lecture Notes 25*, pp. 618–23. Springer Verlag, Berlin. ISBN 3 540156 76 3.

13 Williams, C.B. (1986) Computers in clinical research. *BJHC*, 3.5, 25–7.

14 Rigby, M.J., McBride, A.H. and Shiels, C. (1992) The role of computers in medical audit. In Richards, B. and MacOwan, H. (eds) *Current Perspectives in Healthcare Computing*, pp. 398–406. BJHC for BCS, Weybridge. ISBN 0 948198 12 5.

15 Brown, P. and Warren, L. (1991) Implementing clinical audit. *BJHC&IM*, 8.5:14, 17–18.

16 Maresh, M., Steer, P.J., Dawson, A.M. and Beard, R.W. (1982) The logical development of a perinatal database for clinicians. In O'Moore, R.R., Barber, B., Reichertz, P.L. and Roger, F. (eds) *MIE, Dublin 82: Lecture Notes 16*, pp. 156–61. Springer Verlag, Berlin. ISBN 0 387112 08 1.

17 Dawson, A., Maresh, M. and Beard, R.W. (1984) The development of a comprehensive clinical information system for obstetrics. In Roger, F.H., Willems, J.L., O'Moore, R.R. and Barber, B. (eds) *MIE, Brussels 84*, pp. 573–8. Springer Verlag, Berlin. ISBN 3 540133 74 7.

18 Banfield, P.J., O'Hanlon, M., Chapple, J.C., Mugford, M. and Beard, R.W. (1992) How clinicians respond to computer-generated statistics in maternity care. In Richards, B. and MacOwan, H. (eds) *Current Perspectives in Healthcare Computing*, pp. 367–72. BJHC for BCS, Weybridge. ISBN 0 948198 12 5.

19 Mugford, M., Grant, A. and Chalmers, I. (1982) Developing a register of randomised controlled trials in perinatal medicine. In O'Moore, R.R., Barber, B., Reichertz, P.L. and Roger, F. (eds) *MIE, Dublin 82: Lecture Notes 16*, pp. 162–7. Springer Verlag, Berlin. ISBN 0 387112 08 1.

20 McSherry, D.M.G. and Fullerton, K.J. (1984) Knowledge acquisition in the development of an expert system for the management of perceptual disorder in stroke. In Roger, F.H., Willems, J.L., O'Moore, R.R. and Barber, B. (eds) *MIE, Brussels 84*, pp. 371–6. Springer Verlag, Berlin. ISBN 3 540133 74 7.

21 Richards, B., Gowland, M. and Laycock, P. (1985) A computer based quality assurance scheme for assessment of ultrasound measurements. In Roger, F.H., Grönoos, P., Tervo-Pellikka, R. and O'Moore, R. (eds) *MIE, Helsinki 85: Lecture Notes 25*, pp. 692–7. Springer Verlag, Berlin. ISBN 3 540156 76 3.

22 Clark, D.E. (1990) Safety critical systems in medicine. In O'Moore, R.O., Bengtsson, S., Bryant, J. and Bryden, J.S. (eds) *MIE, Glasgow 90: Lecture Notes 40*, pp. 599–602. Springer Verlag, Berlin. ISBN 3 540529 36 5.

23 Hutt, G. (1990) Quality-doing it right. In O'Moore, R.O., Bengtsson, S., Bryant, J. and Bryden, J.S. (eds) *MIE, Glasgow 90: Lecture Notes 40*, pp. 688–92. Springer Verlag, Berlin. ISBN 3 540529 36 5.

24 Tymchyshyn, P. (1984) Computer proliferation: an experience to share. In Roger, F.H., Willems, J.L., O'Moore, R.R. and Barber, B. (eds) *MIE, Brussels 84*, pp. 756–9. Springer Verlag, Berlin. ISBN 3 540133 74 7.

25 Buck, N., Devlin, H.B. and Lunn, J.N. (1987) *The Report of a Confidential Inquiry into Perioperative Deaths*. Nuffield Provincial Hospitals Trust, London.

26 Osborne, M.J., Stansby, G. and Hamilton, G. (1992) Training junior doctors in computing and audit skills. In Richards, B. and MacOwan, H. (eds) *Current Perspectives in Healthcare Computing*, pp. 126–30. BJHC for BCS, Weybridge. ISBN 0 948198 12 5.

27 DoH (1990) *Medical Audit – Guidance for Hospital Clinicians on the Use of Computers*. HMSO, London.

28 Rigby, M.J., McBride, A.H. and Shiels, C. (1992) The role of computers in medical audit. In Richards, B. and MacOwan, H. (eds) *Current Perspectives in Healthcare Computing*, pp. 398–406. BJHC for BCS, Weybridge. ISBN 0 948198 12 5.

13 Specific systems

DENISE BARNETT

MIDWIFERY AND OBSTETRICS

Midwifery

Midwives traditionally looked after women with normal pregnancies and births, with the majority of deliveries being in the mother's home or a small maternity unit run by midwives. As the organization of maternity services moved, with other community services, from the control of the local authority to the health authorities, so there was a review of the number of small maternity units. Over the next decade many were closed. The post-war baby boom was decreasing by the 1960s and by the end of that decade there was a medical 'takeover' of midwives' work with the attitude that, 'a pregnancy and birth is only normal in retrospect'. The proportion of hospital deliveries began to climb, as did the use of episiotomies and Caesarean sections. The policy was that midwives were not allowed to stitch up the episiotomy cut they had made during delivery, so ready access to a doctor required the mother be in an obstetric unit in a hospital. To cope with the increasing workload the length of stay was reduced with domino-schemes introduced to try to maintain continuity with the midwife who gave antenatal care, did the delivery in hospital conditions and within 24 hours then provided post-natal care at home.

Community midwives worked with general practitioners and hospital obstetric departments in cooperative care. This included a 'cooperation card' held by the mother on which clinical details were recorded, so that all parties were aware of the most recent clinical signs and prescriptions. Printouts could be provided to the mother and the GP. The timescale for the introduction of the Körner maternity data-set was seen as interfering with shared care because the systems in general practice were not able to share data.[1] By 1991 Jenkinson was reporting on the use of a credit-card sized, computer readable, patient-held record: the optical memory card. However, this was for data transfer between professionals, so the mother could no longer read her own data, she just became the data transfer mechanism.[2]

In midwifery, the labour record provides a key date and time for the new baby, or babies, and there are statutory records required to notify the birth. As with other areas of health care the Körner data requirements were a big stimulus to reviewing the collection of data. Basing the delivery in hospital made computer-use a cost-effective approach. Data could be incorporated within the developing Hospital Information Systems (HIS) (see Chapter 7) Midwives were involved in the early 1980s in the development of computer-based records for both mother and new baby. Midwifery research was starting to be funded, and the development of care plans and birth plans created with the pregnant woman were being explored.

A midwife from King's College Hospital attended the British Obstetric Computer Society meeting in May 1985 to describe how, with 18 months' data, they were now able to draw up profiles for their local population. Dr Mike Maresh from St Mary's Hospital, London, was able to show differences in the populations of Paddington (St Mary's) and Denmark Hill (King's), such as the proportion of unmarried mothers and its link to the risk of perinatal mortality.[3]

St Thomas' Hospital, London, introduced a midwifery labour-ward audit program in 1990 and reported the first year's data about Caesarean-section rates in spontaneously labouring women, at term, with a singleton pregnancy, and a cephalic presentation. It showed the variables that correlate with risk of a Caesarean were: race, birth weight and cervical dilation. The audit allowed a peer review of existing standards and the clinical management of labour.[4]

In 1996, Helen Betts led the work of the Midwifery Focus Group of the BCS in reviewing 18 maternity information systems. The main benefits identified were related to auditing clinical practice, producing statistics and for research.[5]

King's College Hospital experimented with bar-coded question and answer sheets read with a light pen to create a computer record displayed on the VDU. It was based on the midwife's history taking and clinical examination at booking. The aim was to reduce perinatal mortality and infant handicap. At the time most of the existing obstetric data collection projects in the British Isles were of the retrospective, archival kind, with documents coded by medical staff or clerks and entered into the computer by punched cards or optical mark reader forms. A DEC PDP-11/34 minicomputer was used with Telpens made by SB electronics and a Qume daisy-wheel printer that printed the bar-code labels for the answer sheets and the history for the paper case notes.

By 1982 there were plans to extend the new approach to the labour ward, produce a discharge summary and to include data entry from the ultrasound department and neonatal unit, with a link to data about the baby's subsequent development.[6]

South Warwickshire Health Care Trust set out to use the relational database FoxPro® to record the physical examination of each newborn within 24 hours of birth. This could then be linked to the child health computing system to allow audit and support screening checks at six weeks and nine months.[7] Ultrasound imaging during the pregnancy increased in use and gained popularity with parents who could have an image of the growing fetus and learn the gender if they so wished. The fetal measurements could be used to predict birth weight and identify 'small-for-dates' babies where obstetric assessment was crucial. These data could be attached to the computer-held clinical record.

In early 1985, Amersham Medical Systems announced the release of a system for obstetric ultrasound records that provided automatic plotting of all ultrasound measurements.[8]

Obstetrics

Crawford and Henry, from Dundee, reported in 1974 on their system to record the fetal heart rate and intrauterine pressure from several labour rooms. They used hard lines to transmit the signals some 100 metres to the Elliott 903 computer in the Clinical Chemistry department. There the voltages were fed into an analogue converter.[9]

The Körner data requirements stimulated interest in using computers to record this and other data. As part of implementing Körner the Maternity Project had six pilot sites:

St Mary's Hospital, London (North West Thames RHA)

Bristol Maternity Hospital (South Western RHA)

Princess Mary Maternity Hospital, Newcastle (Northern RHA)

City Hospital, Nottingham (Trent RHA)

Whiston Hospital (Mersey RHA)

Peterborough (East Anglian RHA)

The Computer Policy Committee system backed the development of a perinatal data system developed at St Mary's Hospital Medical School by Dr Mike Maresh. This was adopted by all 20 authorities in North West Thames RHA. It could handle 2,500 patients per year, producing reports for GPs within 72 hours of discharge.[10] Maresh went on to produce an expert system to assess premature labour.[11]

In 1984, Lilford undertook an in-depth evaluation of the benefits of using a computer to take a history at the booking session. He compared his questionnaire at St Bartholomew's Hospital, London, with the data collected at 41 teaching hospitals.[12] The questionnaire covered almost all the historical risk factors recommended by the Working Party of the Royal College of Obstetricians.[13]

The British Obstetric Computer Society was formed in 1981.[14]

Distance monitoring

Telemetry was explored to allow women with antenatal medical problems to remain at home with daily distance-monitoring by the hospital. By 1986 studies included Dalton's work in monitoring the fetal heart, mother's blood pressure in hypertension, or her blood sugar in diabetes mellitus.[15–17]

In 1990, as part of the European Telemedicine project, Sighthill Health Centre, in Glasgow, and the Simpson Memorial Maternity Pavilion, in Edinburgh, were managing the condition of 'at-risk' mothers in their own homes. The midwife could take a portable telemonitor station to the house, attach a belt with a battery of transducers to the woman's abdomen, and view the traces and the interpretation offered by the expert system. If a second opinion was required the hospital alerted the hospital on call.[18]

In 1991, the MUMMIES Project (Maternity Users – Medical, Midwifery, and Management, Information and Expert Systems) was exploring the use of a Common Basic Specification (CBS) for maternity information. The aim was to offer a foundation for developing a consistent set of information systems, by means of a complete set of models for all aspects of NHS data management. This would then support data interchange between systems. The work was supported by the NHS Information Management Centre.[19]

Clinical equipment

In 1986, University College, Cardiff, and the Rosie Maternity Hospital in Cambridge were testing fetal heart signal monitoring, the latter with a Sonicaid D206 fetal heart monitor, for transmitting heart sounds in real time. There was a financial incentive to this work as the initial cost of reusable equipment and the telephone calls were less than the cost of admission and the daily bed-rate.[20]

New screening procedures became available as computers began to make it possible to cope with large numbers of people and tests, for example: radio-immunoassay results from serum and amniotic fluid and alpha feto-protein for screening for neural tube defects.[21]

In the neonatal ICU, by 1983, Bradford had terminals that collected physiological data.[22] Feedback mechanisms then led to automating control.[23]

Labour ward

Sometimes the opportunity to analyse data led to research to collect new forms of data and to create new knowledge that led to improvements in patient care. Continuous fetal heart rate monitoring during labour, using a stainless steel electrode applied to the baby's scalp, was common in hospitals by the 1980s. During labour there are fetal and maternal sources of 'noise' that affect the signal and make it difficult to assess other ECG changes. The Medical Research Council (MRC) helped fund work at Nottingham to develop continuous assessment of the fetal electrocardiogram (FECG) and the intrauterine pressure. Once the new data had been acquired new methods had to be developed for FECG complex recognition, to average it and to evaluate it.[24]

Walker and colleagues in Glasgow developed an expert system on the induction of labour.[25] Richards and Lieberman showed that careful monitoring of uterine pressure, via a catheter-tip pressure transducer, made it possible to reduce the volume of oxytocin.[26] D'Souza and Richards identified that correlating the ECG from the cardiotocograph (CTG) with fetal pH, measured by a scalp electrode, indicated fetal acidosis. They also found babies born to hyponatremic mothers had a higher incidence of jaundice and that Pethidine given to the mother for pain during labour accumulated in the baby.[27, 28] Richards and colleagues also carried out a controlled study to verify the premise that the volume of oxytocin used during labour could be reduced and that this would be beneficial to the baby.[29]

By the 1990s at Derriford Hospital, Plymouth, the analysis of blood drawn at delivery from the umbilical cord was so well automated, with a knowledge-based computer system, that it could be carried out by health care assistants. The same hospital had also developed a decision-support system for intrapartum care with a terminal at each bedside that had a fetal monitor connected to display the cardiotocogram, with textual data, such as the findings at vaginal examination, entered via a touch screen. The midwife could view the information as a CTG trace or a partogram.[5]

Colour doppler scans could be used to show blood flow through fetal organs and the placenta and help detect anatomical abnormalities. In time this would lead to attempts to correct them *in utero*.[30]

Special care baby units

Premature babies and those with known medical problems would be delivered in hospital with a paediatrician and specialist midwife present in the delivery room to receive the baby. Babies with immature lungs require oxygen, but there is a known risk of later development of retrolental fibroplasia, so monitoring blood gas levels is essential. The temperature of the incubator and the baby's body temperature also need to be monitored. As in adult therapy the computer can integrate the many physiological variables that need to be monitored. The calculation of therapeutic doses of electrolyte solutions and drugs can be undertaken, taking account of the baby's age and weight.

Administrative and other systems developed for older children and adults were used in Special Care Baby Units.

In 1988, S.M. Mason from the Medical Physics Department at the Queen's Medical Centre, Nottingham, developed on a BBC B microcomputer an automated routine screening of hearing for neonates in special care, each ear being screened using a machine-scoring algorithm that detects the ABR (auditory brainstem response) waveform.[31]

In 1991, the Paddington Green Children's Unit at St Mary's Hospital, London, compared computer-generated discharge summaries with dictated discharge summaries for babies admitted for intensive and for special care.[32]

INTENSIVE THERAPY

There is a difference between intensive therapy and intensive care. Intensive therapy units (ITUs) have 24/7 medical staffing and a high ratio of nurses to patients. Intensive care units (ICUs) had a number of nurses without the same continuous cover. The terms were often used indiscriminately causing confusion. The patients usually require mechanical lung ventilation and frequent assessment of their physiological and other responses to intravenous fluids and medication. On occasion complications arise rapidly and automated data logging minute-by-minute can provide the only detailed record for later review. Monitors with pre-set limits can sound an alarm when signals from the heart and brain, or pressures within the cardiovascular, pulmonary or nervous systems, fall outside these parameters. As surgery became more complex and developments in immunology supported organ transplants, so the need for intensive therapy beds increased. Head injuries and major trauma from road traffic accidents and from terrorist explosions also had more localized impacts on requirements, such as in Northern Ireland.

Data compression of frequent readings for storage was an issue. In 1972, a neurosurgical ward used four different levels for a pressure signal: one second, 30 seconds, 15 minutes and 24 hours. The one-second data were averaged and then discarded every 15 minutes, with the graphically displayed 30-second averages discarded at two hours. The 15-minute means and their standard deviations were permanently archived.[33]

In 1973, Wythenshawe Hospital in Manchester had installed a mainframe, to run offline data for coronary care and intensive care patients, purchased with a two million pound allocation from the DoH. The software had been developed, by Datasaab, for the Karolinska Hospital, in Sweden. Each bed had an intercommunication set for the nurses to log data instead of using manual charts. According to Dr Clifford Franklin, 10 years later the system needed replacement but the computer industry had not produced an affordable, dedicated system for intensive care, so the hospital was planning to revert to bedside charts.[34]

Dr Alagesin had a program written for the BBC microcomputer that attempted to grade the severity of illness. The scores were arbitrary in 1985.[35] Sometimes the best predictor of the outcome for the patient came from surprising directions. Back in 1969 Joly and Wiel found the 'temperature of the big toe' a good predictor because it reflected the patient's capacity for physiological compensation.[36] In 1980, Clark and colleagues found blood coagulation after head injury was the most useful predictor of death.[37] Comparisons and prediction of the likely outcome of intensive therapy were desired but difficult to achieve until key indicators could be established.

The increasing availability of data made it possible to identify the most useful information, so that the clinical staff were not overwhelmed by the volume.

The Computer Users Group of the Intensive Care Society was founded in 1979, and was open to nurses, physicists, engineers and technicians in addition to doctors.[38] The group set up the Apache Study, to assess the Apache clinical scoring system, coordinated by Dr John Kerr of the John Radcliffe Hospital, Oxford.[39]

Goh and Richards used a computer program in intensive care to check and interpret acid–base and electrolyte values (see also Chapter 24). They had evaluated several existing programs using data-sets from real patients. Their own program drew on the strengths of these. It was written in BASIC.[40]

In October 1983, interest was in the collecting of physiological data and sharing them across a network within the ITU. This would lay a foundation for decision-making systems. The London Hospital had installed the Kontron Patient Management System with a terminal at each bed to collect data from 'intelligent' monitors.[41] Kontron Instruments, based in Watford, was later involved in the INFORM Project to introduce systems to a variety of high dependency units.[42]

Trend recognition was made easier by colour graphics, particularly when any combination of parameters could be displayed. As reliable data accumulated this could provide the foundation for expert systems. In Wakefield, a microprocessor-driven, parameter-dedicated membrane keyboard with a small VDU for data input, simple trends, calculations and storage facilities had been developed. This transmitted data to the ward minicomputer every 15 minutes.[43]

In 1983, at Pinderfields Hospital, Wakefield, Price and Salem set out to provide a practical method of estimating the probability of an individual developing an intracranial haematoma after head injury. They took the 25 routinely collected observation parameters and reduced them to 15, assigning each feature a probability score. The threshold decision level to advise transfer to a neurological unit for a scan and possible evacuation of the clot was set according to the cost of transfer, likely numbers of patients and distance to the unit. A scattergram was produced to allow local decisions.[44]

Optimum treatment was being tempered by the availability of equipment, skills and cost. Knowledge-based systems were needed for interpreting respiratory data, renal function, electrolyte and fluid balance. David Price merged statistical and modelling techniques into a conventional rule-based artificial intelligence shell for use in neurological intensive care. The relationship of intracranial pressure with intracranial volume was used to help select the optimum dose of a drug to control the pressure via a closed-loop system.[45]

Titrating a medication to the physiological response can be difficult where the drug has accumulative effects or a time-lag in its action. Reid and Kenny evaluated a closed-loop computer control of arterial blood pressure after cardiopulmonary bypass with a 'nurse-control' group. They found the computer performed best.[46]

Interest in online monitoring was for respiration, cardiac, ECG, intracranial pressure and arterial pressures.[47] St Thomas' Hospital in London used an investigations trolley, based on a BBC microcomputer.

Once virtually all monitoring equipment had a serial-line interface there was a need to replace the RS232 or RS434 communication protocols so data did not have to be manually entered into the unit information system. By 1990 the Medical Information Bus Standard was being agreed through the Institute of Electrical Engineers. There were problems in the development of biochemical sensors, in catheter tips, for the continuous measurement of blood gases and pH, blood and urinary electrolytes. Some development projects had been abandoned.[48]

As part of the DoH experimental medical computing program, St Thomas' Hospital in London developed an optical mark reader (OMR) form to store and process clinical and haemodynamic data. There was one form for details of cardiac catheterization and two forms for recording the clinical examination and special investigations. These could be used with in-patients or with outpatients. The clinical examination form replaced the need for dictation of the clinical notes. Two lessons were identified: the use of OMR forms meant clinicians did not need to acquire keyboard/typing skills, and the avoidance of parallel running during implementation. Charing Cross Hospital used OMR forms to record pre-, peri- and post-operative anaesthetic histories.[49]

International collaboration

The INFORM project, funded by the European Commission was concerned with the development of advanced information systems in high dependency units. These included intensive care, coronary care, neonatal and burns units, operating theatres and recovery rooms. Five clinical experts from the UK participated in the consortium. The project included creating a conceptual model, defining and identifying the functions required and a survey of existing, patient data management systems.[50]

The data model included three components:

- Clinical patient-related data, with the health care professional caring for the patient incorporating a data element that restricted actions such as prescribing treatment; pre-ICU assessment and a disease severity score such as APACHE II and a disease taxonomic code such as Read or ICD-9.

- Operational ICU management including drug items, consumable items, treatment, monitoring or other equipment including their service plans and maintenance records, and budgets.

- Care evaluation, cost-effectiveness and planning, which included allocation of the patient to a Homogeneous Patient Group (HPG) and statistics relating to all patients with similar diagnoses and disease severity.

The model brought together the disparate aspects of existing systems to provide a specification for databases in the next generation of computer systems for intensive care.[51]

ANAESTHETICS

Anaesthetists work in both intensive therapy units and in operating theatres.

Cardiff began extracting data from anaesthetic records in 1954, storing it on manually punched Hollerith cards, moving on to data entry by operators from a multi-part clinical anaesthetic record.[52]

A computer can display data derived from gas analysis by a mass spectrometer. Closed-circuit re-breathing techniques require close monitoring. ECG, blood pressure and respirations can be displayed and stored. The Ohmeda 'Modulus' had a touch screen and was integral to the anaesthetic machine and provided hard copy. The Kontron PMS 7000 could provide hard copy, or store and download data into other systems.

Reviews of the usability of data capture devices needed to take account of the sometimes cramped working conditions for the anaesthetist in the operating theatre. They included reviews of bar codes, OMRs, light pens, mouse with VDU, a digitizing pad similar to that used in computer-aided design systems, and voice recognition.

In 1985, the Faculty of Anaesthetists Scheme for Telesoftware Teaching started to explore the use of PRESTEL for transmitting teaching programmes. From 1987, this was replaced by a Bulletin Board to download text in ASCII format.[53]

OPERATING THEATRES

Systems for operating theatres were slow to develop in the UK. South East Thames Region attempted to design a theatre information and management system to improve the scheduling of operating time and staff: the 'Lister Project'.[54]

The Financial Information Project (FIP – see Part 3 Introduction) was a driver for work on theatre systems in Walsgrave Hospital, Coventry, but this was modified to provide an operational management system of which cost information was a by-product. Theatre nurses completed paper forms, with the surgeons contributing the coding of the operations based on the Canadian Classification of Surgical Procedures. The screen data-entry form design was slightly different to speed up later data entry by the nursing staff. At Cheltenham the system was run by means of a leased-line on a DEC PDP-11/44 owned by FIP and located in Birmingham.[55]

After FIP came Diagnosis Related Groups (DRGs, see Chapter 6) intended to give greater management control of clinical costs. Patients were to be grouped according to diagnostic categories with homogenous costs based on length of stay and medical activity analysis.[56] Patients were being treated as if they were products of a hospital system, the implications of human variability and multi-diagnosis were to be ignored in the early stages as they added complexity. Yet these were the very aspects that clinical professionals have to take into account during patient care.

SURGERY

In 1969, Aberdeen University was exploring the use of computers for automating the follow-up of patients after treatment for thyrotoxicosis, where up to 40 per cent may become hypothyroid within 10 years of surgery. This included the use of anti-thyroid drugs, autoimmune thyroiditis and lifelong replacement therapy. The patient may attribute the late changes to ageing, and hypothyroidism can lead to changes to the central nervous system, such as depression. General practitioners were sent a form with a patient's details and the eight symptoms and six signs of hypothyroid function. The program to do this follow-up was written in COBOL for an ICL 4/50 computer and required a card reader, line printer, magnetic tapes and replaceable disk backing store.[57]

A collaborative programme involving systems scientists, medical physicists and clinicians was undertaken to help in management of thyroid disease at the Middlesex Hospital. They generated a database from patient data, rather than the literature, that implied there were over 150 different methods for diagnosing thyroid disease.[58]

The Patients' Charter, introduced in 1991, required that information be provided to patients. The right was stated as: 'to be given a clear explanation of any treatment proposed, including risks and any alternatives, before you decide whether you will agree to the treatment'.[59] Computer-based information leaflets (SATIS-FAX leaflets) were produced. One example was the Department of General Surgery at the Friarage Hospital, Northallerton, that produced customizable leaflets for surgical patients, including those for general surgery, urology, ear, nose and throat surgery, orthopaedic surgery, gynaecology and dental surgery. Over 350 operations were covered by the leaflets. The text included a pronunciation guide, and translation into community languages was underway in 1994.[60]

Intravenous feeding

Intravenous feeding can be costly, particularly in prolonged illness with metabolic and nutritional problems. Shearing and colleagues from Newcastle upon Tyne and Lancaster Hospitals reported work on a computer program using the University's IBM 370/168 with the Michigan Terminal System (MTS) that was then transferred to a DEC PDP-8 minicomputer to provide a local service. The program calculated the water, calories, protein, etc., and then, using a list of available fluids, determined a combination of whole bottles to meet the next day's requirements. The work raised issues around data collection times and the relative cost and risks of using cheaper fluids administered by more expensive equipment.[61] The following year Goh and Richards of UMIST reported a computer program to determine the best combination of fluids to prime the extra-corporeal pump for open-heart surgery and for monitoring the patient's acid–base balance and potassium levels during the surgery (see also Chapter 24).[62]

Simulation

Dr George Harvey, Biomedical engineering consultant at Queen's University in Belfast, announced in 1986 a system to allow three-dimensional simulation of hip replacement operations. The system relied on two X-rays of the hip joint taken at right angles to each other. Data processing was based on Computer Vision hardware and CAD3 or CAD4X software. The intent was a pre-operative simulation of the damaged joint and the replacement necessary. It could also be used after surgery to assess bonding of the replacement.[63]

In 1995, the Royal National Orthopaedic Hospital at Stanmore, Middlesex, was using computers to custom-make replacement hip and knee prostheses of metal and plastic. The doctors and technicians used locally written software on an Apple® Mac® to identify the shape of the joints. Prototypes were then made, using a Roland CAMM-3 PPNC engraving and milling machine, to check on the fit before final production. For knee replacements the hospital built a knee simulator to study wear patterns.[64]

ACCIDENT AND EMERGENCY

In the 1960s, a record card was completed for each visit and filed within the Accident and Emergency (A&E) department or, if the patient was admitted, within the main medical record. It could be difficult to track non-accidental injuries, hospital-hoppers, regular drug overdoses and addicts. Staff might recognize a 'regular' but in the middle of the night it could be difficult to obtain the main medical record: supporting the potential for an adverse reaction to the normal emergency treatment in overdoses quickly was impossible, so the patient's life was at risk.

Outputs from a computer-held record, such as the production of a referral letter to the GP within 24 hours, were seen as reducing the number of follow-up patients attending the department. Adding local codes to the ICD primary and secondary codes could provide locally useful information, such as:

- location of accident/injury;
- how the patient was brought in;
- time elapsed;
- for road traffic accidents: whether patient was a pedestrian, cyclist, driver, passenger with or without seat belt.

The collated data could be used to pinpoint places where road accidents frequently occurred.[65]

In the 1970s, Peter Harvey, John Farrer and Jean Roberts at Lancaster Royal Infirmary were using computer data to examine the demands made on services, particularly by people from socially deprived areas. They found problems in using the ICD codes for non-fatal and minor conditions.[66–68]

Bangor General Hospital used an online microcomputer to register and print a casualty record card for patients. In 1982, the Körner Secretariat organized workshops that led to the Computer-based Accident and Emergency Records (CAER) project, chaired by David Wilson of Leeds General Infirmary.[69] Each health district that wished to use the system had to buy its own 16-bit microcomputer. The CAER consortium was established in 1985. It was overseen and financed by a consortium of user health authorities, with the implementation and support coming from the CAER Centre of Development team based at Yorkshire RHA. It used the UNIX® operating system. By October 1986 it was fully operational in five A&E departments.

Trent RHA used the MUMPS operating system on DEC equipment so it could be linked to the main patient administration system (PAS). Three departments were fully operational by May 1986. South Western RHA adopted the same operating system and equipment but only had one fully operational system at that time.

The Derby A&E system was primarily a commercial development with input from two A&E consultants at Derbyshire Royal Infirmary. It was based on Burroughs equipment and operating system.[70]

The Royal Sussex County Hospital, Brighton, wanted a system with more real-time logging of activities than CAER could offer. This allowed staff to monitor the patient's progress through the department. Patients in the waiting area had a television display showing their position in the queue. Based on clustered workstations using micros it allowed small departmental networks to be incorporated for the transparent movement of data without loss of 'local identity'. It was able to link to the central PAS and to other systems. The software was based on the Derbyshire system adapted by Footman-Walker Associates.[71]

In 1992, the National Audit Office concluded that the efficiency of A&E departments would be enhanced by management information systems. Of the six departments visited, only one system could cope with all 15 key items of information and functions. This one had been developed from the consultant's own user specifications.[72]

One of the problems of a stand-alone system was that solutions developed for hospital information systems and PAS were not always readily transferable. One example of this was the difficulty in linking episodes of attendance at the same A&E department. North Staffordshire had found duplicated patient records, so an algorithm was developed to match records using Soundex for surnames to select the sample. It could be run as a background check once a week and for new registrations.[73]

Clinical audit

Computer-aided audit for major trauma was well established by 1992 when Miller reported monthly audits covering a range of cases presented to an A&E department, activity in the observation ward and hospital emergency admissions. Audits of case note samples were undertaken using a pro forma of seven topics, these were scored and entered into a dBASE® III+ program to show the final score achieved by each doctor, for staff development.[74]

The TRISS method of injury scoring combines a trauma score with an anatomical injury severity score. Guy's Hospital, London, used a self-contained computer program written in Borland's Turbo Pascal® for the IBM PC to perform the calculations, feed the result into a database and print charts. The score could also be used in triage.[75]

Poisons information

The National Poisons Information Service, established in 1963 by the DoH, was located at Guy's Hospital. It stored large volumes of product information and data about cases of poisoning. In 1983, it computerized the poisons index listing of symptoms and treatment for over 10,000 products. The data from paper forms, used to record the clinical enquiry (in use for 20 years), were input by a typist.[76]

Dr Alex Proudfoot, of the Royal Infirmary of Edinburgh, directed the development of the Scottish Poisons Information Service. It stored data about 10,000 potentially poisonous substances on a computer for instant interrogation by approved users using Prestel-style sets via a British Telecom public telephone network.[77] By 1985 there was an agreement with Viewtext Ltd of Tunbridge Wells that this could be distributed internationally.[78]

Prestel pages had to be rented and were expensive, so there was a temptation to cram in as much data as possible. Potential users of the system had to purchase their own terminal. By 1984 there were 69 registered for access. The original Incotel system ceased to be marketed, so enquiries from other countries wishing to purchase the database and system had to wait for redevelopment.[79] There were also nationally funded poisons services in Belfast and Cardiff.

Supporting diagnosis

At *MEDINFO 74* de Dombal and colleagues described five years' experience of using simulation techniques in the education of medical students.[80] In 1979, de Dombal and Horrocks of the Leeds General Infirmary reported their work on developing a computer-aided diagnosis system that used Bayesian probabilistic analysis on clinical studies. They used a remote terminal to a time-sharing system operated by the computer centre of the University of Leeds and a back-up system on a desktop computer. The system was used in the emergency ward and led to a drop in admissions, with a significant financial saving.[81]

Surgeons in Glasgow sought to increase diagnostic accuracy in A&E departments for the 10 per cent of patients who presented with abdominal pain. The original system was developed on mainframe computers in Leeds. The team used a DEC PDP-11/10, a Commodore PET and Apple® microcomputers. Data were collected from large samples of patients with a firm diagnosis, such as appendicitis, gallstone pancreatitis, and perforated diverticular disease. The diagnostic system was tested with junior doctors.[82] The interest of the national press was aroused by the thought that computers might be more accurate in diagnosis than junior doctors and the existence of this system was quite influential in raising the profile of clinical computing.

In contrast, in 1970, Evans and colleagues at the West Middlesex Hospital, in Isleworth, had begun exploring computer-based interviewing of patients.[83]

In 1976, the West Middlesex team began to explore the development of questionnaires for patient interviewing to collect medical histories. A four-button box was provided to capture 'yes', 'no', 'don't know' and 'don't understand' responses. The system did not raise clinical interest and seemed to disappear from the computing scene; perhaps the limited patient responses did not fit the reality of the less linear face-to-face interview.[84] However, the team used the computer for interviewing Asian patients in an antenatal clinic and to administer a psychiatric questionnaire. The questions were recorded in Urdu and accompanied by a coloured picture, or a video of the doctor asking the question.[85]

By 1985 de Dombal was able to list decision-support system studies in acute chest pain, jaundice, dyspepsia, inflammatory bowel disease and upper gastro-intestinal bleeding.[86]

RENAL MEDICINE AND TRANSPLANTS

Renal medicine

The early use for computers came in the planning of requirements for haemodialysis and transplant services. Chronic haemodialysis developed in the 1960s and was expanded rapidly in the next decade. Farrow, Fisher and Johnson at The London Hospital used a mathematical model, based on the Markov chain principle, to assess the likely period in dialysis training and then waiting for a transplant, together with survival periods.[87, 88] By 1972 they were reporting that:

> *Many older patients do remarkably well both on haemodialysis and after transplantation. If this is confirmed with larger numbers we shall seriously have to examine the age criteria for accepting patients . . . .*

REFERENCE 89

The opportunity to rigorously examine outcomes through interrogating a mass of data began to have an impact on clinical decisions.

The renal unit at the Princess Royal Hospital in Hull introduced a daily, data-collection system in 1979. It was based on an Apple® II with 6 MB hard disk and an in-house designed and built network of terminals. It allowed 210 active patient records to be accessed from five sites across the hospital complex.[90]

Charing Cross Hospital Medical School used a database to handle clinical renal data on a DEC LSI-11 microprocessor and as early as 1981 they were experimenting with storing compressed digitized photographs.[91]

In early 1984, the British Renal Computing Group reported there were nearly 20 units operating large-scale clinical systems. The Group had been set up in 1982 and was coordinating activity in developing semi-standard programs to handle data, including graphical plotting of biochemical and other data.[92]

Will and Selwood from the British Renal Computing Group reported in 1985 that there was interest in integrating three levels of computer databases: local, national and international.[93] International cooperation grew out of the need to evaluate an experimental procedure: dialysis. There were only small numbers of patients across 43 centres in 1965 when collaboration began, becoming the International Registry of the European Dialysis and Transplant Association (EDTA). By 1985 the Registry held data from over 1,700 centres in 31 countries. Reports on the scientific and administrative data for each centre were useful in negotiating for financial and other resources. The Registry ran on a DEC VAX 7/50 based at St Thomas' Hospital in London and provided the major source of information to the DoH on patient acceptance rates and treatment schedules.

As computing systems become more complex, clinical databases were used for knowledge acquisition, in building expert systems, by generating rules for use within the computer. For example, in Manchester, data from 380 cases collected over six years were used to evaluate machine-learning algorithms: ID3, CN2 and ZEBEL.[94]

As more domain-specific knowledge-based systems became available it was possible to link several of them to clinical laboratory systems. This required new rules to be identified to interpret profiles: renal, bone, liver and cardiac. Clinical computing needed new communication protocols and this supported interpretive reporting of laboratory results.[95]

Renal dialysis for people with end-stage renal failure developed during the 1960s with the creation of purpose-built dialysis units. Patients required treatment three times a week for about six hours. The artificial kidney machines developed with the introduction of disposable membranes and boards. It became possible for patients to set up the equipment and to dialyse themselves, with many going on to do this at home. The training programme initially involved simple aids, such as photographs, films and dummy human limbs with synthetic veins. In 1985, Homer described CELLS – a Computer Enhanced Learning System.[96]

Renal transplants

Festenstein and colleagues at The London Hospital used the Elliott 803 to find a match between a kidney donor and potential recipients. They began collaboration with other centres in 1969.[97] By 1971 this had become a multi-centred collaboration and the first 162 patients were reviewed.[98] In 1974, an international collaboration was reported, between The London Hospital and France.[99]

In 1972, the National Organ Matching and Distribution Service (NOMDS) was set up and combined with the National Tissue Typing Reference Laboratory. It used a DEC PDP-11/34 to maintain the list of patients awaiting kidney transplant with their tissue-matching and other characteristics. When a potential kidney donor was approaching death, a donor and recipient matching run could be requested. As the clinical processes to maintain a healthy kidney during transport improved, so the distance between locations of donor and recipient could be increased. The need to have the patients in adjoining operating theatres could be replaced by theatres two or more hours away by motor cycle courier, taxi or even aeroplane.

The data exchange between national and local systems was initially offline using magnetic tape transfer, and by the 1980s was by online telecommunications on dial-up lines supported by modems, error controllers and file transfer software. The data could be 'out of sync' because of the different frequency of national and international collection. One advantage of having higher-level data has been the possibility of providing data to populate new local computer systems.

The Clinical Data System (CDS) database came out of research at Charing Cross Hospital and provided a basis for local systems that could be run on minicomputers or microcomputers to add the flexibility to reflect the unit's requirements that was not possible with the available mainframes.[93] One strength was that the user could define the information required, another was direct capture of laboratory data. Provided all three levels of system ran in compatible software, time savings could be made in the transfer of data. Thus the data for research could be available to meet management's information needs: a requirement that was reflected in the later NHS Information Authority's national strategy. This possibility then required confidentiality to be addressed.

The UKTS became increasingly involved with heart valve, heart and liver transplants. It also held a national register of platelet and bone marrow donors. In 1984, a new, expanded system was required to meet the needs of the next decade. This was based around two DEC VAX 11/750 machines, running the VMS operating system. Routine enquiries used Datatrieve with statistical runs using a BMDP package. Eight 1,200 bps modems were provided for transplant centres to password-access the system, and two 4,800 bps modems were provided for large file transfer to international data registries. In the event of a queue of users developing the PABX could advise them of their place in the queue.[100]

Potentially toxic drugs may have narrow therapeutic ratios, such as Gentamicin, and 5-Fluoracil, so require careful calculation of the dose for the individual patient. The biochemistry laboratory, the pharmacy and clinicians at the Royal Lancaster Infirmary worked together to develop a computer program to undertake the calculations.[101]

GENERAL MEDICINE

In 1977, the Western Infirmary in Glasgow was using their SWITCH system (System at Western Infirmary for Total computerization of Case Histories) to automatically code data recorded in the weekly peptic ulcer clinics and in wards when patients were admitted for surgery. The system was also used in hypertension and renal outpatient clinics.[102]

Endoscopy

KeyMed was a big mover in the endoscopy field with its flexible fibre-endoscopes. When St Mark's Hospital, London, found customized database developments too expensive, KeyMed provided financial backing, assured that endoscopists in other hospitals were also interested. It produced a database, Patient Endoscopy Data and Record Organizer (PEDRO) based on a Comart 1522 computer with a 20 MB hard disk. This was later upgraded to 40 MB.[103]

In Middlesborough in 1990 a medical audit for quality assurance was being undertaken on data from an Open Access Endoscopy service. General practitioners could refer patients to one of seven medical staff through a simple form. A Compaq 386 was used with Uniplex for the office functions and a relational database: Informix®. Ad hoc enquires of the database could be made using Structured Query Language (SQL).[104]

Computers could be used as research tools to investigate the intestinal tract. Levitt and colleagues in Manchester analysed the effects of variations in gastric and duodenal pH on ulcers in 1986 using electrodes implanted in the patient's stomach and duodenum, and temperature changes within the gut in 1990. For the latter, thermocouples were placed at equidistant intervals within a tube passed down the patient's throat.[105, 106]

Databases tailored to clinical requirements included CRAFT (Clinical Retrieval, Analysis and Follow-up Template) from Ninewells Hospital in Dundee. It could run on an IBM personal computer. It was used in the departments of medicine, surgery, neurology, urology, and obstetrics and gynaecology. Applications included a Crohn's disease register, bladder tumour and follow-up audit, epilepsy study, database for breast disease references, cone biopsy register and for cholelithiasis research.[107]

Intravenous feeding

Prescribing intravenous feeding regimes can be complex. Shearing in Newcastle upon Tyne developed a 'Bottle Selection Programme' in 1976 that was written in Fortran and ran on an IBM 370.[108] This was tried at the Royal Lancaster Infirmary and found to be too large for the hospital's computing facilities. A multidisciplinary team at the hospital developed its own system that included suggestions for various additives at fixed time intervals and reminders to test serum calcium and phosphate levels.[109]

Patients with malnutrition or who require a period of rest for their gut may be treated with total parenteral nutrition (TPN). Biochemical abnormalities can occur quite quickly and the intravenous feeding regime, of around 12 components, needs daily review and adjustment. The West Middlesex Hospital, at Isleworth, worked with the department of Chemical Pathology at Lewisham Hospital, London, to develop a module for prescription and monitoring, and another to store medical and nursing data for audit. Two years of routine use were evaluated. The data enabled the explanation of a number of instances of oedema and dehydration, and their correction.[110]

In 1990, there were four UK expert systems in gastrointestinal diseases among 24 reviewed. These were:

- SUSSEX MEDICINE from Sussex University (abdominal pain) for diagnosis and tutorial;
- GLADYS from South Glasgow General Hospital (Dyspepsia) for diagnosis;
- FIRST-AID from Sheffield University (Gastroenterology) for diagnosis;
- CANSEARCH from Huddersfield Polytechnic (Cancer) for treatment.

At the time the first two were being developed and the second two were in use.[111]

Rheumatoid arthritis

Complex clinical trials to assess the efficacy of different therapies led to the setting up of CELTS (the Cardiff and East Wales Long-Term Study). It used a Database Management System (DBMS) application to follow 600 patients over five years, with as many as 250 items of information collected at each visit.[112]

Cardiology

Electrical signals regulate heartbeats, so changes may indicate pathology and electrical means can be used in treatment. Coronary Care Units were created in the 1960s to provide monitoring for cardiac arrythmias after a heart attack or myocardial infarction. Visual-display heart monitors became more commonplace. Commercial bioengineers worked closely with cardiologists, so these monitors became an early application of computer capability.

The electrocardiograph machine of the 1960s advanced with the advent of microprocessors with functions such as automatic lead checking, baseline correction, noise reduction, hard-copy formatting and online calculation, and display of the heartrate. Twelve-lead ECGs became possible with automatic interpretation, at first on mainframes. Peter McFarlane and colleagues in Glasgow developed a hospital-based system capable of interpreting ECGs from scattered patients. Wythenshawe Hospital used a Marquette mainframe linked by a telephone line to remotely interpret ECGs. Hewlett Packard and Marquette were the early commercial suppliers and over five years incorporated interpretation into portable ECG machines, as did IBM and Siemens.

The bedside monitors of the 1980s provided the capacity to 'freeze' the trace, continuous-loop memory to store the electrical events before any alarm condition, heart-rate display and trends. The individual monitors could be linked to a central console. This required a human presence, but later a display unit showing the traces of several patients could be wall-mounted on a swivel arm, so the nurse could 'keep an eye' on the traces while working elsewhere in the ward.

As many cardiac arrythmias are intermittent, once the technology was capable of being miniaturized it allowed ambulatory monitoring. The body-worn pack allowed all-day monitoring as patients went about their normal daily activities. The 'noise' was rejected automatically by the microprocessor, which then analysed the rhythm and computed the totals for specific cardiac events out of a vast amount of data from recording each heartbeat. This led on to devices that stored only periods of abnormal heart activity.

A continuous ECG while undertaking graded exercise on a treadmill became a standard investigation. Electrical recording from many points over the heart surface became possible with computer processing to produce colour-coded isopotential contour maps.

Early implantable cardiac pacemakers delivered a fixed-rate of stimulation to a single chamber of the heart and needed to be replaced every three years. Digital microprocessor technology allowed for sensing and pacing for both atrial and ventricular electrodes and extended the 'life' of the machine up to 20 years. The device could be reprogrammed using externally transmitted, digitally coded instructions. Anti-tachycardia pacemakers were developed in the 1980s.

Computer-held databases of devices and patients allowed patient recall and device surveillance across Europe. In the 1980s, there was one in Glasgow and another at the National Heart Hospital in London.[113]

Glasgow's cardiology database was held at the Department of Cardiac Surgery at Glasgow Royal Infirmary. Data for the activity audit, which started in 1980, were initially collected, retrospectively, on a three-page questionnaire. Participation in the Cardiac Surgical Research Club Multi-centre Valve Trial required two other forms. In 1985, the system was computerized with data entered prospectively by clinical staff into a database created with Informix® and Uniplex. Summaries included blood usage, specific complications and delayed discharge. Named patients could be selected for discussion at the monthly audit meeting.[114]

The Clinical Information Systems Project (CISP) was set up in 1988 to produce a transferable clinical system. The fieldwork was based in the Cardiology department at Manchester Royal Infirmary. Role Activity Diagrams (RADS) were used to represent the activity within a department, such as an outpatient clinic, so the cooperative process of patient care could be identified and allocated to different roles. The Integrated Process Support Environment initially developed for software production could be applied to ensure the right resources were in place.[115]

EEG recordings

The electrical activity of the brain produces wave patterns on a clinical electroencephalogram (EEG – developed in the 1930s). Experienced clinicians interpreted the complex patterns through visual inspection. Computers provided a route to greater accuracy and consistency in interpretation and at the speed with which this could be provided to guide treatment.

In trying to sort out 'noise' from important events, Hill and Townsend in Edinburgh recorded two-channel electrocortigrams from rats to determine the pattern of epileptic spikes in order to develop a pattern-recognition mechanism. They used equipment supplied by the MRC and a grant from the Scottish Home and Health Department.[116]

MacGillivray and colleagues at the Royal Free Hospital, London, used 16 channels of analogue EEG and other signals, processed the data and compared these against normal reference value using Fourier transform hardware. Clinical data were added at a keyboard by the technician using six, short, data-trees. The computer produced a text description of the record using standard EEG reporting terms.[117]

By 1982 Townsend was reporting a control interface using a Motorola microprocessor to record the EEG and responses evoked by stimuli.[118]

In 1974, Seldrup from the pharmaceutical company Ciba-Geigy, reported on his firm's involvement in a retrospective survey of the treatment of epilepsy in children from Alder Hey Hospital, Liverpool. Initially data were extracted from the records using edge-punched cards but these were insufficient for the volume of data, so four 80-column cards were used. The company's general program for clinical trials was used to analyse over 70,000 items and to identify the 14 medications used alone or in combination.[119]

Electrocardiography

In 1979, the then European Economic Community started a four-year project, involving 18 different institutions, to develop common standards for computer-derived ECG measurements. Experts from North America also collaborated in the work. UK representatives participated at each level: steering committee, referee and working party.[120]

The Cardiology department at Wythenshawe Hospital, together with pathologists and Manchester University, worked to assess the accuracy of a 12 -lead ECG analysis program. By 1980 there were seven systems, from the USA and the UK, using one of three methods of automated analysis:

- a decision tree
- prior probability analysis
- decision rules.

The team evaluated the Bonner program that used decision rules for the Marquette Universal System for Electrocardiography. The pathologists described in detail the cardiac changes found at autopsy. The computer diagnosis was found to compare favourably with the cardiologists' views of the ECG but not with findings at autopsy.[121]

Pulmonary function

The use of computers could sometimes make possible faster testing, with immediate result reporting, because the analysis of the data could be undertaken by the machine. At Bristol Royal Infirmary, Richardson and colleagues used a carbon dioxide re-breathing method to assess cardiac output for a patient using a bicycle ergometer. The computer program was written in MINC BASIC (v1.1) on a DEC MINC-11/03 system. It eliminated the need to manually read the polygraph linked to the mass spectrometer and to transcribe data.[122]

By 1992 the City General Hospital, Stoke-on-Trent, had developed a microcomputer network to support patient care including respiratory function assessment, bronchoscopy, equipment loans and servicing, secretarial work and administration. Direct interfaces collected data from spirometers and blood gas analysers.[123]

North Manchester Hospital participated in the development of a diagramatic representation of the larynx and bronchial tree, with an underlying grid to tag an area to the appropriate anatomical description. Using a mouse to position the cursor, the system automatically displayed the description. The bronchoscopy form system could print a variety of documents including a bronchoscopy report.[124]

Diabetes

In 1979 in Nottingham work started on a clinical information system to ease the problems of a disorganized diabetes clinic with missing case notes and other essential information (such as eye examinations results), patients lost to follow-up, and no feedback on the workload. Ray Jones was asked to evaluate the implementation and published some of the findings in 1990. He found improvements in recording data including examinations, fewer 'lost' patients and an estimate that benefits were at least 15 times greater than the costs. Jones commented on how rarely evaluation studies of clinical systems were reported.[125]

The EURODIABETA project was a collaborative project involving UK hospitals and universities, and commercial companies. It sought to explore the feasibility of a computer-assisted chronic health care environment with diabetes as the focus. Models of activity and data-sets were produced so that information technology could be used to support insulin and diet therapy. The UK worked on the metabolic prototype, short-term insulin adjustment, an insulin pharmacodynamic model for the effect of insulin therapy, and simulations of blood concentrations of glucose, insulin and glucagon.[126]

St Thomas' Hospital, London, undertook medical evaluation of a prototype to provide advice on day-to-day adjustment of carbohydrate intake and insulin. They used a range of clinical scenarios and four clinicians to independently provide suggestions on how blood glucose profiles could be improved. These were compared with the advice generated from the knowledge-based system. On occasion the diabetologists gave quite different advice from each other. The message was that a panel of experts was always needed to agree a standard rather than one expert's views being replicated within the knowledge-based system.[127] The need for collaborative work further reduced the opportunities for clinicians to get involved with clinical computing simply out of local interest.

Registers for monitoring the changes in condition of diabetics could be adapted for shared care between the hospital team and the general practitioner. A pressure-sensitive set of colour-coded documents could be used to provide a copy for the GP, for the case notes and for data to be input to the computer. Middlesborough used this approach to identify patients who lived in areas of social and economic deprivation where there was a greater tendency to ischaemic heart disease.[128]

INFECTION CONTROL

Software to aid monitoring of the incidence and distribution of infection was described in 1988. In microbiology laboratory management it derived antibiotic sensitivity profiles and reported antibiotic usage to help the control of infection.[129]

Useful clinical tools were sometimes developed by one discipline for wider use by other health care professionals. One example was OASIS (Organism Alert and Surveillance Information System) for use by infection control nurses. This computerized microbiology system was implemented in 1990 at Basingstoke District Hospital. It included outbreak control features.[130]

Hospital-acquired infection was at 9.2 per cent in a national prevalence study in 1980.[131]

The pressure to make greater use of beds, the emphasis on reducing length of stay and the increase in drug resistance made control of infection of increasing clinical concern. It was felt that the contracting out of cleaning and domestic services to reduce costs had led to falling standards in cleanliness. By the first decade of the twenty-first century even the Chief Nursing Officer at the DoH was involved in publicity about hand-washing and the prevention of cross-infection.

Glenister and Taylor investigated eight different methods for detecting hospital-acquired infection with a reference method ('gold standard'). The laboratory-based ward-liaison method was judged the most effective and efficient method. A concern was the accuracy of records.[132]

A European system, WHOCARE, produced surgeon-specific infection rates by intervention and contamination class. It gave lists of infected patients and a summary of prophylactic antibiotic usage. The Central Public Health Laboratory in the UK participated in the development. The tool was to help local, multi-centre and multinational studies.[133]

Technology provided a further useful tool for assessing infection: the geographical information system (GIS). *The Chorley Report* provided a review of progress in 1987.[134] A GIS could capture, manipulate, analyse, integrate and display locational or spatial data. Lancaster Health Authority used it to draw attention to the high levels of Campylobacter food poisoning in the district. The address of a case was linked to the postcode and displayed by local authority ward. The postcode was becoming more than a device for directing mail.[135]

AIDS

The first case of Auto Immune Deficiency Syndrome (AIDS) in the UK was reported in 1981. This raised a number of ethical issues for hospital record keeping, research and screening. The variable incubation period made prediction, of the number of likely AIDS cases, difficult. Merseyside set up an Online AIDS Support and Information Service (OASIS) for health professionals to access via videotext technology over a standard telephone line.[136]

In 1990, Ruchill Hospital in Glasgow had developed a clinical information system to improve the quality of case note recording for patients with the Human Immunodeficiency Virus (HIV) and AIDS. One benefit was an improvement in the recording of drug allergies.[137]

Haematology

The Haemophilia Centre at St Thomas' Hospital, London, developed a system to record the anatomical location of each bleed, often into a joint, together with the batch identification for each unit of blood used in treatment with cryoprecipitate to help trace any resulting hepatitis. The patient could also supply episode data during home treatment for later batch input. Problems were found with inconsistent alphanumeric formats for blood source identification. The programs were written in Fortran to ease transferability to other interested hospitals, such as Addenbrooke's Hospital, Cambridge.[138]

In 1988, Richards and colleagues described the before and after situations when manual records in a Manchester haemophilia centre were converted to computer-held ones, using a Commodore 8096 with a Silicon Office database.[139]

PHARMACOLOGY/ADMINISTRATION OF MEDICINES

Important drug sensitivities and drug interactions can lead to unintentional and preventable impacts on patients. Space to record known sensitivities, such as reactions to antibiotics, was provided on the paper drug charts and Kardex systems developed in the 1960s. Potential drug interactions were more difficult to identify and the allocation of named pharmacists to visit wards and inspect the charts was adopted. The London Hospital Formulary was first issued to junior doctors in 1982 and sought to reduce the range used and stocked, and to promote more cost-effective prescribing; the tradition of a local guide reviving that produced between 1882 and 1934. Once computer applications to record the prescribing of medicines were available a logical step was to integrate this with a computer-held patient record.

Henney and colleagues from Dundee University reported their computer-based drug information system at *MEDINFO 74*. A standardized method of recording drug administration to patients enabled the computer to monitor for discrepancies between prescription and administration. A survey of drugs administered in intravenous fluids revealed physico-chemical incompatibilities.[140]

In 1982, the Department of Therapeutics and Clinical Pharmacology of the University of Aberdeen described a system with the aim of improving the transfer of data, between doctors, of clinically important information about high-risk patients. Information for patients from two medical wards was entered from the paper record by secretarial staff to create a discharge summary. The associated approved dictionary of problem titles had 8,000 entries that were matched with ICD-9. The drug listing included: start date, related problem, name, dose, route, frequency and cessation date. Two-line drug information notes, included warnings such as drug–drug and drug–host interactions, were automatically added to selected high-risk drugs. The extension of the system to high-risk outpatients, such as renal, asthma, diabetic and mucoviscidosis clinics, included a hypertension clinic where shared-care with general practitioners in the Grampian Region became feasible using the computer. The facilities of the system were also used to record details of patients willing to participate in undergraduate teaching, to invite them to the ward and to thank them afterwards. The Scottish Health Service Common Services Agency gave financial support to the work.[141]

South Lincolnshire used a 486/33 MHz PC server running DataTree DT Max and networking MUMPS, with a NetBIOS link to a shadower, as well as to PC clients in the wards and departments via the Ethernet network. The server took patient data from the PAS and from the patient history file. There was a link to the pharmacy management system for all the drug information. Security was controlled by an ID code to identify each person and their status: only qualified doctors and midwives could prescribe. The ID acted as an electronic signature, with the system adding it to each transaction. Lists of 'medication due' could be used, with the nurse updating the computer record at the end of the drug round. Other outputs included information for medical audit, Casemix, resource management and costing.[142]

Oncology

The incidence of cancer in the general population is monitored through cancer registries across the UK. Each registry could also collect local data alongside the standard set, for use in local, national or international research. The time lag in collecting and recording data could be up to 18 months, which made it difficult to evaluate new treatments by multi-centre, randomized, controlled trials as quickly as desired. International trials required that standards for classifying patients and for delivering patients were agreed.

University College, London, was involved in the development, over 12 years, of an interactive system that was designed by clinicians for clinicians, initially to study leukaemia and Hodgkin's disease. Its main features included facilities to design datasheets for clinical research and patient management, a user-friendly data-entry system and statistical analysis routines, including actuarial survival. Early access was via a computer bureau, then it was transferred to the Imperial Cancer Research Fund's DEC mainframe computer, accessed by remote terminals. An example study was that of European Bone Marrow Transplant Group into the treatment of lymphomas.[143]

In research it is always difficult to extract data from records that have not been kept specifically for the purpose. Back in 1974, Neal of Brunel University described the problems of analysing data about soft-tissue sarcomas collected over a 21-year period (1948–69) at Mount Vernon Hospital and the Middlesex Hospital. Analysis showed that, in a review of the histological slides, there were 48 per cent of cases where the diagnosis was changed: usually from one type of sarcoma to another. Multiple-choice forms were designed to extract the data from the records and this was transferred to punched cards. A computer graphic output was devised to illustrate the principal events in the patient's history on a timescale.[144]

In 1977, Neal was able to report work using a new language: CODIL. This had been developed for use by medical personnel for whom the computer was just a tool for retrieving information of interest, for example about myocardial infarction, rather than for those involved in the working or programming of the machine.[145]

Microcomputers allowed clinicians to develop their own programs or to use one of the professionally developed systems such as Metabase used in over 70 clinical centres. Expert systems could be developed based on the data and a therapy adviser program was planned as part of the European Strategic Programme for Research and Development in Information Technology (ESPRIT).[143]

The Royal Marsden Hospital, a noted cancer centre, set out to create a system for clinical research with aids to patient care and maintaining medical records as a bonus, rather than the main goal. They found the detail required for research was greater than that needed in day-to-day records so had to employ staff to extract these data from the conventional notes and enter it in the system. It proved difficult to achieve a direct feed of data from the radiology, histopathology, biochemistry, radiotherapy and chemotherapy departments, as the proposed systems were not acceptable to the clinicians. The three main application areas were breast cancer, leukaemias, and testicular tumours.[146]

Databases of clinical outcomes were capable of statistical analysis. For example, Todd and colleagues at Nottingham City Hospital used follow-up data collected over 11 years to study primary, operable, breast cancer. They analysed 30 parameters for each of 1,200 patients to create a prognostic index of likely survival.[147]

Pharmacy

The NHS drug information services of the UK cooperated to produce the Pharmline System of abstracts of medical and pharmaceutical literature. The working party was set up in 1978, and, in 1979, produced a computer-generated index on microfiche. The system was computerized in 1983.[148]

Systems to manage the ward stocks of drugs and medicines were introduced, sometimes to save money. 'Just-in-time' ordering for central stocks was also introduced, so that only two weeks' stock from retailers and four weeks' from manufacturers would be held. St Thomas' Hospital, London, introduced such a system with the intention of saving £350,000 per year. The system had been written to run under PICK for the first time.[149] Other hospital pharmacies made similar changes. The Freeman Hospital, Newcastle upon Tyne, developed Pharmasyst, which included issuing dressings to wards and clinics.[150] Reduced stock levels did lead to the occasional problem when a manufacturer experienced difficulties. There were instances of hospitals within London telephoning other hospitals and community pharmacists in an effort to track down essential medications for patients.

The PICK operating system was in time to proliferate across the NHS as it was adopted by many of the major manufacturers and distributors. Guy's Hospital, London, as a resource management pilot site, introduced the PICK system. (It may be interesting to note it was developed in the 1960s by a Richard Pick.[151]) In 1986, the *Datafile* in *BJHC* listed 15 different systems for pharmacy use that could run on a variety of computers including minis and mainframe, and with languages such as MUMPS and PICK.[152]

The pharmaceutical industry provided information about their products. In 1991, City University in London was involved in a study that compared four databases with an in-house database of a company's products. The external databases were BIOS, EMBASE, MEDLINE and RINGDOC, using clinical papers on Diclofenac. None of the four provided complete retrieval of all the test papers. Once interest in evidence-based medicine developed the reliability of external databases became important.[153]

Hammersmith Hospital developed and implemented a pharmacy computer system over six years. It provided:

- financial reconciliation (expenditure compared with drugs issued);
- workload and performance data;
- ward stock lists;
- updating of the authority's formulary;

- purchasing;
- drug use with trend analysis.

Like most pharmacy computer systems available in 1991, it was not capable of data capture at patient level.[154]

GYNAECOLOGY

Cervical screening systems were set up with the support of the general practitioners and the laboratories. The Family Practitioner Committee provided a computerized registration system. (Information on these developments can be found in Chapter 14).

Machine interpretation of visual information offered the possibility of handling many lengthy and tedious reviews of cells. Abnormal cells were presented to a human operator for verification.

Interpretation could be difficult and over the years there were instances of major recalls of women whose smears had been viewed by individual operators.[155]

Where a smear showed severe dyskaryosis or malignant cells (CIN 2-3) a colposcopy and biopsy were used to assess the cervix. Digital cameras would, in time, make the latter procedure one that could be carried out by a trained nurse, with the images reviewed later by the surgeon or oncologist.

Genetic disorders

The machine assessment of cells could also be used for chromosome analysis, including examining amniotic fluid for chromosome defects such as Down syndrome, and testicular biopsy in cases of infertility.

Northern Ireland had a high rate of certain genetic disorders and congenital malformations including mental handicap, phenylketonuria, homocystinuria, spina bifida and anencephaly. By 1972 computer programs were available to calculate recurrence risks of multifactorial conditions,[156] and genetic disorders,[157] and a geocode index was published in 1976.[158]

In 1977, McDonald and Nevin of Queen's University, Belfast, described a medical genetic patient register. The risk of some genetic disorders had marked geographical gradients. When people came for genetic counselling, data were collected on documents designed in 1975 for punching on cards to be batch run on the University's ICL 1906S computer. The programs were written in Fortran, with SPSS software and other packages used for some of the reports. Record linkage was used for data exchange with the Child Health System established in 1964 and a voluntary Congenital Malformation Notification System.[159]

Richards and Jefferey developed a computer program that enabled paediatricians, geneticists and pathologists to input abnormal characteristics observed in an infant, and to receive a list of the most likely congenital syndromes. This used positive and negative weighting to influence the selectivity of the system.[160]

The Royal Manchester Children's Hospital used a database, on a desktop microcomputer, to hold complete records for children seen in the cystic fibrosis clinic, with the plan that paper records could be eliminated.[161]

HOSPITAL DENTISTRY

There were 17 dental hospitals in the UK in 1986, usually associated with the larger medical hospitals. They were expected to provide Körner data, performance indicators and clinical budgeting information, while reflecting that many of the procedures were carried out by dental students under supervision. Systems could be patient-orientated, dentist-orientated or resource-orientated with a terminal at each dental chair. The DIANE (Dental Institute Area Network) system at The London Hospital provided a mix of the latter two approaches. It also had a link to the Dental Sterile Supply department to monitor the use of expensive instruments linked to patient appointments. On the day of use the sets of instruments were weighed on an electronic balance attached to a computer, and on return they were re-weighed to see if any equipment was missing.

Methods of simplifying data input for the dentist included the use of a touch screen for dental charting. The dentist could use the end of the dental pick used to examine a tooth, to prod the screen and record the findings.[162]

Whipps Cross University Hospital, London, had collected demographic and patient-management data from orthodontic outpatients since 1988. The database application used bar-coded templates for data collection. Adhesive bar-code labels could be printed and stuck to the paper records. Cordless scanner pens with mini-LCD screens (on the side of the pen), and 64 k memory, were used at each workstation so data could be verified before entry.[163]

In orthodontics, computers were being used to support clinical decisions. By the late 1960s microcomputers with an XY-digitizer were used to obtain a geometric appraisal of the relationship of the jaws, and the base of the skull. In the mid-1970s, the tracing could be viewed on the visual display unit and the X-ray was digitized. Attempts were being made to predict facial growth in children with malocclusion. By the 1990s programs allowed the clinician to manipulate the digitized image of the skull to simulate the likely results of surgical correction of severe dentofacial deformity. In 1987, the Consultant Orthodontists Group adopted a single recommended hardware and software package so that digitized cephalometric records could be easily transferred between treatment centres.[164]

Moss and colleagues at The Royal London Hospital combined data from laser and CT scans for a three-dimensional view to assess facial and soft tissues.[165, 166]

Commercial software to capture intra-oral and facial video images of patients was available.

Concern, raised by research findings that a significant proportion of malocclusions were not improved by treatment, had led Bristol consultants to develop an expert system to provide orthodontic advice to general dental practitioners.[164]

General dental practice is covered in Chapter 15.

PHYSIOTHERAPY

Physiotherapy adopted some of the systems developed by other specialties, such as remote cardiac monitoring to assess cardiac efficiency during exercise. By the 1990s testing machines could be found in the private gyms that sprang up in cities and some local council sports facilities. Sports medicine developed as a specialty to help enhance athletic performance.

Goniometry

The Bioengineering Unit at Strathclyde University, Glasgow, developed an electrogoniometry system in 1987 to assess the movement of one or more joints. Repeated measurements over time can be used to assess changes, such as those following joint replacement surgery, or in muscle weakness. A data logger could be attached for a 24-hour record of activity at home or work.[167]

Muscle strength testing

The first isokinetic system was developed by Cybex in the 1970s and used a gear mechanism to alter the resistance offered to a moving limb as the muscle contracted. Later systems used hydraulics and could measure and record the peaks of muscle activity. A computer displayed the data graphically and provide appropriate treatment protocols.[167]

Gait studies

Computer technology helped the development of gait analysis. The Musgrave pressure platform was developed in Ireland, and a colour printer was used to print the scans mapping the different pressure points against time on the sole of the foot. Pressures were recorded in kilograms per square centimetre. The system also provided coloured dynamic footprints in two dimensions. The assessment of gait in care of the elderly on admission was used to maximize mobility with appropriate physiotherapy or orthotic footwear.[168] Walkways were developed to measure temporal and spatial variables of gait, such as stride length, time spent on one leg compared to the other, the distance between the feet and the time taken from 'heel-strike' to 'toe-off'.[167]

Biofeedback

Computer-generated visual and/or auditory feedback from a weight-sensitive platform could show the stroke patient the effect of attempts to restore normal weight distribution when sitting or standing. It could also be used after peripheral nerve damage.[167]

Profoundly handicapped children could learn to impose control over a computer-controlled carriage, on an obstacle course, using a joystick or a mouth switch. Computer games encouraged hand–eye coordination and activity in some muscle groups.[167] Occupational therapy departments also used computers for encouraging independence for tetraplegic patients and life-enrichment for those with severe handicaps.

Information for patients

Centres for displaying aids to daily living were staffed by occupational therapists and advisers. By 1986 there were 17 regional centres. Disabled people could learn about electronic aids such as specialist input devices, including joysticks, voice volume and pitch detectors, foot-operated switches, chin switches, blow and suck devices and the 'twinkle' switch (which used lateral eyeball movement). A person could control, using a single switch, output devices, gain access to British Telecom's Prestel network and to other computers using an electronic mailbox system.[169]

The Special Education Microelectronics Resource Centres were established by the Department of Education. The Communication Aids Centres were funded jointly by the DoH and the Royal Association for Disability and Rehabilitation. There were five in England, the first one was established in Bristol in 1983, and one each in Wales and Northern Ireland. The Department of Trade funded the British Database of Research into Aids for the Disabled (BARD). BARDSOFT provided a database of software applications for the disabled that covered 40 popular computers.[169]

Geoff Busby, a computer professional, Chair of the BCS Disabled Specialist Group and as a person with cerebral palsy, addressed Medical Informatics Europe in 1988 and described how information technology could help disabled people with employment and independence. He made a plea for coordinated effort.[170]

The wheelchair service in Hull computerized procedures for issuing and retrieving wheelchairs, repairs, reviews and information returns.[171]

AUDIOLOGY

The use of a computer to control the stimulus in the assessment of auditory function was being developed for research activities from the early 1980s.[172] Marchbanks and colleagues developed an interest in the tympanic membrane and the reflex response of the tiny bones of the middle ear. The Computerized Audiometric System (CAS) was installed in the Wessex Regional Audiology Centre in the mid-1980s. One strength of a computerized system was that with it only limited training was needed to carry out complex tests.[173]

In time, commercial systems, rather than those developed by NHS clinicians, became the norm for auditory testing.

A BBC B microcomputer was being used in 1986 to develop an audiometer simulator to illustrate various forms of hearing loss as a teaching aid for students.[174]

OCCUPATIONAL THERAPY

The DoH and the Department of Trade supplied 39 free BBC microcomputers to occupational therapy (OT) units in England, Scotland and Wales between March 1983 and 1984. A further three went to OT training schools. The National Physical Laboratory provided some staff training, and also maintained the equipment. A national special interest group was formed in 1984 and produced a quarterly journal and ran an annual conference. Exeter used the Olympus satellite to transmit and distribute television programmes on occupational therapy topics. Teleconferencing with students on clinical placement was also introduced as part of a British Telecom Research Project. The College of Occupational Therapists had a dozen IT-related projects on its Research Register by 1988. However, computers were not then in full and regular use in all OT departments. Training had ceased in 1986.[175]

Carol Roberts, one of the occupational therapists at Odstock Hospital, set up a national special interest working group in 1984, by 1985 there were also regional groups and an annual seminar with 150 participants.

DIETETICS

Dietary analysis is one of the first stages in providing advice to patients on good eating habits and in constructing personalized diets for people with organ damage, such as renal failure or intestinal malabsorption. Graphical displays of the nutrient intake compared with recommended standards could help a patient understand the nature of their current diet. In 1990, the University of Salford was working with a London hospital to test software to construct a palatable diet to meet specified nutrient requirements including patient preferences and the National Advisory Committee for Nutrition Education (NACNE) recommendations.[176]

The 'Microdiet' was intended for use by working dieticians in hospitals of any size, not just those with a large database capacity. Item numbers were assigned to identify the foods and these were already in common use by dieticians. Other uses for the computer were suggested:

- A dietary-product information system.
- Allergic reactions such as in coeliac disease and Tartrazine sensitivity.
- Taking a diet history from a patient.
- Daily menus for patients in metabolic units where only a restricted range of foods may be eaten.[177]

The University of Salford also provided two-day computer appreciation courses for dieticians.[178]

Scotland also had introduced dietary analysis. The system calculated the constituents, energy, vitamins, cholesterol, amino acids and fatty acids. It produced diet totals and daily averages for each patient.[178]

University College London and St Mary's Hospital, London, worked together on the Fluid Electrolyte Balance Estimation (FEBE) project in which a prototype bedside workstation was used to test whether it could improve fluid, electrolyte, acid–base and nutrition therapy in critically ill patients. The project included a high degree of ward staff involvement and iterative testing.[179]

A study of the diet of long-stay psychiatric patients was reported in 1988, stimulated by the Royal College of Nursing's *Care about Food* campaign. Every item of food and drink consumed by a patient during two, 24-hour periods was entered into the COMP-EAT application. Food was weighed before the meal and that left on the plate was weighed after it. Various other factors were assessed, such as dental health, physical activity, height and weight. General nutrient deficiencies were identified, as well as problems for individual patients.[180] Given the continuing concern about nutrition in schools and hospitals it is surprising that this work was not developed into a general application.

OCCUPATIONAL HEALTH

Forth Valley Health Board, in Scotland, installed a system from Abies Informatics to store staff records such as immunizations and for printing invitations to attend appointments. It was based on the Abies GP system (see Chapter 14). It could be used to aggregate data on sickness and accidents and therefore help in identifying risk factors, infection rates and hazards.[181]

PSYCHOLOGY AND MENTAL HEALTH

At the *Spectrum 71* conference, there were a number of issues raised regarding the use of a computer to take a medical history:

- Can the logic of a history taking be easily expressed in the form of a flow diagram that can subsequently be converted into a branching program?
- Can a working program be written for operation on a commercial time-sharing service that does not:
  - take too much time to administer;
  - occupy too much file storage space?
- What is the most suitable method of presenting this information to the patient and what modifications, if any, need to be made to standard computer peripherals?

- What is the patient acceptability of a computer-based history taking system?
- What is the validity of such a method from a medical point of view?[182]

Some of these issues went on to be continuing subjects for academic study of computer-user interaction.

Low-cost microcomputers were used in psychology and psychophysiology departments for diagnostic and therapeutic purposes. The validity of self-reported information from psychiatric patients was investigated at the Maudsley Hospital, London, in 1983 by Carr, Gosh and Ancil. Khan described the work undertaken by the District Psychological Services of the Mid-Staffordshire Health Authority between 1982 and 1983.[183] This linked the patient record with psychometric data, such as questionnaires and part of the clinical interview, and physiological data including from biofeedback techniques. A BBC microcomputer was used with a Torch computer disk unit capable of running CP/M software. The two questionnaires to determine the personality traits (anxiety, depression, sensitivity and assertion) were administered, scored and analysed by the computer.

Mental health

Some clinical systems started with descriptive projects such as that undertaken in the Rugby area in 1974 by Dunn and colleagues. This led on to a simulation to model the flow of information between the patient, GP, consultant psychiatrist and social worker.[184]

Tameside General Hospital explored a psychiatric register based on forms for the previous month's work. This was coded onto magnetic tape and a suite of programs run on the Region's ICL 1904S. Each staff member who submitted forms received three printouts: a list of patients seen with their details, a list of which of 'their' patients had lost contact with the service or missed appointments, and a list of cases referred by other members of the team.[185]

Michael Rigby of Keele University worked on a number of mental health projects, including one by Plymouth Health Authority. He suggested in 1991 that even though mental health had been a stated priority for many years it had failed to attract the degree of attention offered to acute hospital services, and to a lesser degree community services. Protechnic Computers Limited of Cambridge was awarded the contract to develop a patient-based, networked, real-time system (based on the Authority's existing DEC VAX cluster and Ethernet) to record the care plans and delivery process.[186]

In 1991, Alva Grannell described the Speedwell information technology project that was set up to provide employment opportunities. People with mental health problems were taught word-processing, and to use spreadsheets, databases and graphics. The InfoTech Project in Hackney was run on similar lines. At Lancaster Moor Hospital the occupational health department used the Atari ST computer system, with a graphical interface, in a course in computer literacy as therapy.[187]

SUPPORT SERVICES

Catering – menu choices

Northern Ireland had a system giving recipes with ingredients, a menu cycle over six weeks, a menu card reader, and calculation of uptake. It also included stores requisitions, batch-cooking instructions and costings.[188]

The Department of Mathematics at Salford University produced a package to calculate total nutrient intakes and daily averages, with comparisons between the actual and recommended intakes. This could run on a variety of systems including Apple® 11, BBC B, Sirus, DEC Rainbow 100 and Superbrain QD.[189]

Application of technologies, that reduced in price, included automating the collection of data about menu choices through use of optical character readers, thus requiring those in clinical areas to help patients in marking their choices in a way the scanner could read.

In Wales, in 1984, one catering system could produce ward summaries of patient choices, provide meal-cost summaries, and recost all recipes automatically.[190]

Central Sterile Supply Department

A system to provide pack catalogues and issuing information was developed for the Freeman Hospital, Newcastle upon Tyne.[191] The Northern Region had already introduced computerized records for ordering and picking other medical supplies. Other hospitals were trying bar codes on packs, so that ward stocks could be topped up and usage noted by staff, armed with bar-code readers, who visited each ward on a daily or less frequent basis.

Works information

The environment within which clinical care is delivered is important, particularly in health and safety terms for staff and patients as well as in the control of hospital-acquired infection. At ward level in the 1980s, clinical work was temporarily disrupted by the need for works department staff to collect data to feed the NHS Works Information Management System (WIMS). Managers were later presented with the estimates for redecoration, maintenance and investment. Planned closure of wards and clinical areas for redecoration or maintenance could be built into the hospital staffing and cost control measures.

The merger of hospital and community services had led to an unfunded build-up of often very poor quality, former local authority buildings, for clinics and community nurse bases. This had to be remedied and the hospital infrastructure suffered as a result. The hospitals designed in the 1960s had large circulation areas and glass that resulted in high heating costs in winter and uncomfortable heat-gain in summer, particularly for patients. As energy costs rose so there was pressure on clinical staff to save energy, particularly electricity.

WIMS was developed by the DoH. The modules in brief were:

- Assets: to build up an inventory of equipment, plant, work to be done and completed (released March 1981).

- Stores: stock control, and raising and processing purchase orders with cost records (released March 1981).

- Energy: fuels and utility services used, targets set and performance achieved, and including miles per gallon for vehicles (released March 1981).

- Redecoration: surface area of walls, doors, etc. for each ward, room, corridor, etc., to calculate materials and labour costs (released April 1981).

- Budget monitoring (released April 1982).

- Property module: details included types of tenure, rateable values, rights of way (released March 1983).

- Annual maintenance plan for long-term and maintenance work over a forthcoming five-year period (released March 1983).

- Condition appraisal: for a 10-year investment schedule with 'remaining life' and 'cost to upgrade' (released March 1983).

- Electromedical equipment management: including spares costs (released September 1984).

- Vehicle management module: to schedule work on the basis of vehicle mileage (released September 1984).[192]

One of the strengths of the system was that it could run on small, stand-alone microcomputers or major mainframe computers. This type of flexibility was not built in to clinical systems. The first WIMS users' seminar was held in 1984.

The clinical areas provide the workplace for staff and the temporary living quarters for patients. Refurbishment and replacement involved close links between regional estate staff and the individual hospitals. Estate management computer systems in the North Western RHA used: a capital information system to list individual schemes; computer-aided design (CAD) to manipulate graphics for a three-dimensional model of a hospital ward or other area; a system to digitize drawings, and take-off and bill quantity surveyors; and WIMS. Meanwhile, Oxford RHA was investigating the potential of existing CAD systems for estate-control and development plans.[193]

Medical equipment

Centralizing the procurement of electronic and other equipment was seen as an opportunity to reduce costs by using tenders. Greater Glasgow Health Board was able to do this using their Equipment Advisory Service, first set up in 1970, and their Asset Register, in conjunction with the Area Supplies department. The Asset Register had been created locally, using the WIMS module structure, to run on local software to handle the volume of data.[194] In 1987, the Scottish team suggested England should follow their lead.[195]

Newmedics was a system developed for the Freeman Hospital, Newcastle upon Tyne, to hold data on all electronic and engineering equipment. It produced lists as well as maintenance information and service contract renewal details.[196]

Stock-control systems were developed for chiropody, NHS hearing aids (audiology), and medical equipment loans to patients.

Supplies

The National Audit Office noted in 1991 that there had been six major reviews of NHS supplies that included recommendations on the increased use of information technology. Of the 14 regions in England, nine were using the Supply Information Systems and two the newer RESUS, with three choosing alternative systems. The DoH Procurement Directorate had adopted a set of common data requirements in April 1991 and was introducing electronic data interchange (EDI).[197]

Computers provided the ability to centralize the supplies function and, in 1991, a headquarters, with six operating divisions, was set up to take over from the regional and district health authorities. By 1996 this had been reduced to three divisions to supply the whole of England and had already achieved large financial savings and increased value for money.[197]

Ambulance

Patient movement between sites and between hospital and home can have a major impact on hospital bed usage and the efficient running of clinics and day hospitals. In the 1980s, some hospitals set up their own inter-site transport services. By 1987 the NHS Register had two patient-focused applications and four for vehicle management or ambulance scheduling.[198] The ambulance service is covered in Chapter 15.

REFERENCES

1   Fawdrey, R. (1987) Why Körner maternity needs a year's grace. *BJHC*, 4.6, 26.

2   Jenkinson, S. (1991) The optical memory card as an electronic shared-care device for antenatal record keeping. In Richards, B. and MacOwan, H. (eds) *Current Perspectives in Healthcare Computing*, pp. 285–92. BJHC for BCS, Weybridge. ISBN 0 948198 11 7.

3   Richards, B. (1985) British Obstetric Computer Society. *BJHC*, 2.3, 30.

4   Chappatte, O.A., Robson, M.S., Smith, J., Taub, N.A., Maslen, T. and Versi, E. (1992) Midwifery labour-ward audit: who benefits? In Richards, B. and MacOwan, H. (eds) *Current Perspectives in Healthcare Computing*, pp. 373–9. BJHC for BCS, Weybridge. ISBN 0 948198 12 5.

5   Betts, H. (ed.) (1996) *Benefits Realisation Monograph on Maternity Information Systems.* DoH, Information Management Group report IMG A1078.

6   Saunders, M., Campbell, S., White, H. and Coates, P. (1982) A unique real-time method of obstetric data collection. In O'Moore, R.R., Barber, B., Reichertz, P.L. and Roger, F. (eds) *MIE, Dublin 82: Lecture Notes 16,* pp. 143–8. Springer Verlag, Berlin. ISBN 0 387112 08 1.

7   Bannon, M.J. and Ward, M. (1994) A neonatal information system. In Richards, B., de Glanville, H. and MacOwan, H. (eds) *Current Perspectives in Healthcare Computing,* pp. 167–73. BJHC for BCS, Weybridge. ISBN 0 948198 17 6.

8   Product developments (1985) Obstetric ultrasound/records. *BJHC,* 2.1, 55.

9   Crawford, J.W. and Henry, M.J. (1974) Intensive care in the labour room. In Anderson, J. and Forsythe, J.M. (eds) *MEDINFO 74,* Stockholm, pp. 747–53. North-Holland, Amsterdam.

10 Developments (1985) St Mary's perinatal system approved. *BJHC,* 2.1, 5.

11 Maresh, M. (1990) A computerized decision-support system in obstetrics: the management of pre-term labour. *Contemporary Reviews in Obstetrics and Gynaecology.* (Pages unknown.)

12 Lilford, R. (1984) *Microcomputers in obsterics and gynaecology.* PhD thesis.

13 Macnaughton, M.C. (1982) *Working Party on Antenatal and Intrapartum Care.* Royal College of Obstetrics and Gynaecology, London.

14 Lilford, R.J. (1984) The British Obstetric Computer Society. *BJHC,* 1.1, 31.

15 Dalton, K.J. (1986) Telemetry of fetal heart signals from the patient's own home. In Dalton, K.J. and Fawdry, R.D.S. (eds) *The Computer in Obstetrics and Gynaecology.* (Pages unknown.) IRL Press, Oxford.

16 Dalton, K.J. (1986) Computerized home telemetry of maternal blood pressure in pregnancy. In Dalton, K.J. and Fawdry, R.D.S. (eds) *The Computer in Obstetrics and Gynaecology.* (Pages unknown.) IRL Press, Oxford.

17 Dalton, K.J. (1986) Computerized home telemetry of maternal blood glucose levels in diabetic pregnancy. In Dalton, K.J. and Fawdry, R.D.S. (eds) *The Computer in Obstetrics and Gynaecology.* (Pages unknown.) IRL Press, Oxford.

18 Dripps, J., Salvini, S., Williams, H., Fulton, W., Boddy, K. and Venters, G. (1990) Telemedicine in obstetrics. In O'Moore, R.O., Bengtsson, S., Bryant, J. and Bryden, J.S. (eds) *MIE, Glasgow 90: Lecture Notes 40,* pp. 501–5. Springer Verlag, Berlin. ISBN 3 540529 36 5.

19 Lumb, M.R., Currell, R.A., Guist, S.H.F., Dalton, K.J., Fawdry, R.D.S. and Jarman, S.J. (1991) The CBS in maternity care: the MUMMIES project. In

Richards, B. and MacOwan, H. (eds) *Current Perspectives in Healthcare Computing*, pp. 293–9. BJHC for BCS, Weybridge. ISBN 0 948198 11 7.

20 Richards, B. (1986) Obstetrics. In Abbot, W., Barber, B. and Peel, V. (eds) *Information Technology in Healthcare: A Handbook*, B2.7.5. Longman in association with The Institute of Health Services Management, Harlow, Essex. ISBN 0 582061 40 7.

21 Westcott, P. (1984) The register of computer applications. *BJHC*, 1.2, 14.

22 Dugdale, R.E. and Lealman, G.T. (1983) A cotside microcomputer monitoring system for use in the neonatal intensive care unit. *Clin. Phys. Physiol. Meas.*, 4, 4, 372–80.

23 Dugdale, R.E., Cameron, R.G. and Lealman, G.T. (1998) Closed-loop control of partial pressure of arterial oxygen in neonates. *Clin. Phys. Physiol. Meas.*, 9,4, 292–305.

24 Jenkins, H.M.L., Kirk, D.L. and Symonds, E.M. (1982) Continuous intrapartum fetal electrocardiography using a real-time computer. In O'Moore, R.R., Barber, B., Reichertz, P.L. and Roger, F. (eds) *MIE, Dublin 82: Lecture Notes 16*, pp. 136–42. Springer Verlag, Berlin. ISBN 0 387112 08 1.

25 Walker, J.J. (1989) The development of a computerized expert system for the management of pre-labour and induction of labour in a busy maternity hospital. The Second World Symposium: Computers in the care of the mother, fetus and the new born, Kyoto, Japan, 23–6 October. *Journal of Perinatal Medicine*, 17 supplement 1, 103.

26 Richards, B. (1989) Minimizing the volume of oxytocin in labour using a catheter-tip pressure transducer. The Second World Symposium: Computers in the care of the mother, fetus and the new born, Kyoto, Japan, 23–6 October. *Journal of Perinatal Medicine*, 17 supplement 1, 144.

27 D'Souza, S. and Richards, B. (1986) Oxytocinon induction of labour: hyponatraemia and neonatal jaundice. *European Journal of Obstetrics, Gynaecology and Reproductive Biology*, 22, 309–17.

28 D'Souza, S. and Richards, B. (1984) Raised Pethidine concentrations with acidosis in the newborn. *Journal of Obstetrics and Gynaecology*, 5, 98.

29 Richards, B., Cadman, J., Levitt, J. and Lieberman, B. (1991) A computer methodology for reducing the volume of the oxytocin use in labour. In Adlassnig, K.P., Grabner, G., Bengtsson, S. and Hansen, R. (eds) *MIE, Vienna 91: Lecture Notes 45*, pp. 557–61. Springer Verlag, Berlin. ISBN 3 540543 92 9.

30 Richards, B. (1986) Obstetrics. In Abbot, W., Barber, B. and Peel, V. (eds) *Information Technology in Healthcare: A Handbook*, B2.7.5. Longman in association with The Institute of Health Services Management, Harlow, Essex. ISBN 0 582061 40 7.

31 Mason, S.M. (1988) Automated system for screening hearing using the auditory brainstem response. *Brit. J. Audiology*, 22, 3, 211–3.

32 Lissauer, T., Paterson, C.M., Simons, A. and Beard, R.W. (1991) Evaluation of computer generated neonatal discharge summaries. *Arch. Did. Child.*, 66 (4 Spec), 433–6.

33 Price, D. (1986) Intensive therapy units. In Abbot, W., Barber, B. and Peel, V. (eds) *Information Technology in Healthcare: A Handbook*, B.2.7.4–23 issue 12. Longman in association with The Institute of Health Services Management, Harlow, Essex. ISBN 0 582061 40 7.

34 Price, D. (1985) Intensive Care Society Computer Group: meeting at Imperial College, London, 8 November 1984. *BJHC*, 2.1, 52.

35 Alagesin (1985) Intensive Care Society Computer Group: meeting at Imperial College, London, 8 November 1984. *BJHC*, 2.1, 52–3.

36 Joly, H.R. and Weil, M.R. (1969) Temperature of the big toe as an indication of the severity of shock. *Circulation*, 39, 131–4.

37 Clark, J.A., Finelli, R.E. and Netsky, M.G. (1980) Disseminated intravascular coagulation following cranial trauma. *Journal of Neurosurgery*, 52, 266–9.

38 Price, D.J. (1984) Computer Users Group of the Intensive Care Society. *BJHC*, 1.1, 30.

39 Kerr, J. (1986) Intensive Care Society Apache Study. *BJHC*, 3.2, 37.

40 Richards, B. and Goh, A.E.S. (1977) Computer assistance in the treatment of patients with acid–base and electrolyte disturbances. In Shires, D.B. and Wolf, H. (eds) *MEDINFO 77*, Toronto, pp. 407–10. North-Holland, Amsterdam.

41 Major, E. (1984) Bedside computing. *BJHC*, 1.1, 31.

42 Bowes, C., Ambroso, C., Carron, E., Chambrin, G., Cramp, D., Gilhoolly, K., Groth, T., Hunter, J., Kalli, S. and Leaning, M. (1991) INFORM: development of information management and decision-support systems for high dependency environments. *J. Clin. Monit. & Comput.*, 8, 4, 295–301.

43 Price, D. (1984) Bedside computing. *BJHC*, 1.1, 31.

44 Price, D.J. and Salem, F.A. (1983) A pocket microprocessor used for the recognition of patients at risk after head injury. In van Bemmel, J.H.V., Ball, M.J. and Wigertz, O. (eds) *MEDINFO 83*, Amsterdam, pp. 461–3. North-Holland, Amsterdam. In two parts. ISBN 0 444867 48 X.

45 Price, D.J. and Mason, J. (1984) An attempt to automate control of cerebral edema. In Go, K.G. and Baethmann, A. (eds) *Recent Progress in the Study and Therapy of Brain Edema*, pp. 597-607. Plenum, London.

46 Reid, J.A. and Kenny, G.N.C. (1986) Evaluation of closed-loop control of arterial pressure after cardiopulmonary bypass. *British Journal of Anaesthesia*, 57, 247–85.

47 Harrison, M. (1985) Intensive Care Society Computer Group: meeting at Imperial College, London, 8 November 1984. *BJHC*, 2.1, 53.

48 Price, D. (1986) Intensive therapy units. In Abbot, W., Barber, B. and Peel, V. (eds) *Information Technology in Healthcare: A Handbook*, B.2.7.4–23 issue 12. Longman in association with The Institute of Health Services Management, Harlow, Essex. ISBN 0 582061 40 7.

49 Williams, K.N. and Webb-Peploe, M.M. (1981) The capture and processing of routine clinical data in cardiology. In Grémy, F., Degoulet, P., Barber, B. and Salamon, R. (eds) *MIE, Toulouse 81: Lecture Notes 11*, pp. 936–42. Springer Verlag, Berlin. ISBN 3 540105 68 9.

50 Bowes, C.L., Kalli, S., Hunter, J.R.W., Ambroso, C., Leaning, M.L., Carson, E.R., Groth, T., Chambrin, M. and Cramp, D. (1990) INFORM: development of information management and decision-support systems for high dependency environments. In O'Moore, R.O., Bengtsson, S., Bryant, J. and Bryden, J.S. (eds) *MIE, Glasgow 90*: *Lecture Notes 40*, pp. 25–8. Springer Verlag, Berlin. ISBN 3 540529 36 5.

51 Yates, C., Leaning, M., Patterson, D., Ambroso, C. and Kalli, S.T. (1991) Data modelling for intensive care within the INFORM project. In Adlassnig, K.P., Grabner, G., Bengtsson, S. and Hansen, R. (eds) *MIE, Vienna 91: Lecture Notes 45*, pp. 116–20. Springer Verlag, Berlin. ISBN 3 540543 92 9.

52 Mushin, W.W., Rendell-Baker, L., Lewis-Faning, E. and Morgan, J.H. (1954) The Cardiff anaesthetic record system. *British Journal of Anaesthetics*, 26, 298–312.

53 Fisher, M. (1986) Anaesthetics and theatres. In Abbot, W., Barber, B. and Peel, V. (eds) *Information Technology in Healthcare: A Handbook*, B2.7.11–24 issue 5. Longman in association with The Institute of Health Services Management, Harlow, Essex. ISBN 0 582061 40 7.

54 Fisher, M. (1986) Anaesthetics and theatres. In Abbot, W., Barber, B. and Peel, V. (eds) *Information Technology in Healthcare: A Handbook*, B2.7.9–14 issue 5. Longman in association with The Institute of Health Services Management, Harlow, Essex. ISBN 0 582061 40 7.

55 Millar, C. (1985) An operating theatre information system. In Bryant, J. and Kostrewski, B. (eds) *Current Perspectives in Health Computing*, pp. 23–37. BJHC for BCS, Weybridge. ISBN 0 948198 00 1.

56 Sharples, S. (1987) DRGs and computing. *BJHC*, 4.5, 27, 30–1.

57 Weir, R.D., Hedley, A.J., Scott, A.M. and Crooks, J. (1970) Automated follow-up register for the prevention of iatrogenic thyroid disease. In Abrams, M. (ed.) *Medical Computing: Progress and Problems*, pp. 200–2. Chatto & Windus for BCS, London. ISBN 0 701115 75 0.

58 Edwards, P.R., Britton, K.E., Carson, E.R., Ekins, R.P. and Finkelstein, L. (1977) A control system approach to thyroid health care. In Shires, D.B. and Wolf, H. (eds) *MEDINFO 77*, Toronto, pp. 507–11. North-Holland, Amsterdam.

59 DoH (1991) *The Patients' Charter – Raising the Standard.* HMSO, London.

60 Trigwell, P.J. and Edwards, M.H. (1994) SATIS-FAX: the first family of customizable computer-based information leaflets for surgical patients. In Richards, B., de Glanville, H. and MacOwan, H. (eds) *Current Perspectives in Healthcare Computing*, p. 825. BJHC for BCS, Weybridge. ISBN 0 948198 17 6.

61 Shearing, G., Wright, P.D., James, R.M., Roberts, J.M., Cooper, R.I., Bellis, J.D., Harvey, P.W. and Rich, A.J. (1978) A computer program for calculating intravenous feeding regimens and its application in a teaching hospital and a district general hospital. In Anderson, J. (ed.) *MIE, Cambridge 78: Lecture Notes 1*, pp. 631–50. Springer Verlag, Berlin. ISBN 3 540089 16 0.

62 Richards, B., Goh, A. and Doran, B. (1979) The use of computers in open-heart surgery. In Barber, B., Grémy, F., Überla, K. and Wagner, G. (eds) *MIE, Berlin 79: Lecture Notes 5*, pp. 340–44. Springer Verlag, Berlin. ISBN 3 540095 49 7.

63 Developments (1986) Hip operation simulation. *BJHC*, 3.1, 3.

64 Collins, P. (1996) The Royal National Orthopaedic Hospital bone replacement research and development. *BJHC&IM*, 12.7, 38–9.

65 Ahmad, I. (1982) Electronic records in Accident & Emergency. In O'Moore, R.R., Barber, B., Reichertz, P.L. and Roger, F. (eds) *MIE, Dublin 82: Lecture Notes 16*, pp. 450–5. Springer Verlag, Berlin. ISBN 0 387112 08 1.

66 Farrer, J., Harvey, P., Dyer, J. and Roberts, J. (1978) A study of the demands made on Casualty Services by the population of socially deprived areas. In Anderson, J. (ed.) *MIE, Cambridge 78: Lecture Notes 1*, pp. 597–603. Springer Verlag, Berlin. ISBN 3 540089 16 0.

67 Farrer, J.A., Harvey, P.W. and Roberts, J. (1977) Problems arising from the classification in ICD on non-fatal and minor conditions. In Shires, D.B. and Wolf, H. (eds) *MEDINFO 77*, Toronto, pp. 289–92. North-Holland, Amsterdam.

68 Roberts, J., Farrer, J. and Harvey, P. (1977) A progressive study of the Emergency Room demand by the Community. *Medical Informatics*, 2.3, 197–201.

69 Pritty, P. (1984) The Derby accident and emergency system. *BJHC*, 1.3, 32–3.

70 Ousby, J. (1986) Accident and emergency. In Abbot, W., Barber, B. and Peel, V. (eds) *Information Technology in Healthcare: A Handbook*, B.2.7.6–13 issue 3. Longman in association with The Institute of Health Services Management, Harlow, Essex. ISBN 0 582061 40 7.

71 Avila, C.P. (1987) Automating Brighton's A&E. *BJHC*, 4.3, 14–15.

72 National Audit Office. (1992) *NHS Accident & Emergency Departments in England.* HMSO, London. ISBN 0 102158 93 2.

73 Berry, G. (1993) Identification of multiple attenders at an accident and emergency department. In Richards, B. and MacOwan, H. (eds) *Current Perspectives in Healthcare Computing*, pp. 91–100. BJHC for BCS, Weybridge. ISBN 0 948198 14 1.

74 Miller, E.S. (1992) Computer-aided audit in an accident and emergency department. In Richards, B. and MacOwan, H. (eds) *Current Perspectives in Healthcare Computing*, pp. 385–91. BJHC for BCS, Weybridge. ISBN 0 948198 12 5.

75 Montague, A.P. (1992) The use of injury scoring in clinical audit. In Richards, B. and MacOwan, H. (eds) *Current Perspectives in Healthcare Computing*, pp. 392–7. BJHC for BCS, Weybridge. ISBN 0 948198 12 5.

76 Edwards, J.N., Volans, G.N. and Wiseman, H.M. (1984) Poisons information processing: the development of a computer database for case records. In Kostrewski, B. (ed.) *Current Perspectives in Health Computing*, pp. 141–54. Cambridge University Press for BCS, Cambridge. ISBN 0 521267 05 6.

77 Bryden, J. (1984) Medical (Scotland) Specialist Group. *BJHC*, 1.2, 36.

78 Developments (1985) Scottish poisons agreement. *BJHC*, 2.4, 5.

79 Davidson, W.S.M. and Proudfoot, A.T. (1985) The introduction of Viewdata for poisons information. In Bryant, J. and Kostrewski, B. (eds) *Current Perspectives in Health Computing*, pp. 14–22. BJHC for BCS, Weybridge. ISBN 0 948198 00 1.

80 de Dombal, F.T., Horrocks, J.C., Clamp, S. and Storr, J.E. (1974) Simulation techniques and computer-aided teaching of clinical diagnostic process: five years' experience. In Anderson, J. and Forsythe, J.M. (eds) *MEDINFO 74*, Stockholm, pp. 247–52. North-Holland, Amsterdam. ISBN 0 444107 71 1.

81 de Dombal, F.T. and Horrocks, J.C. (1974) Computer-aided diagnosis: conclusions from an overall experience involving 4469 patients. In Anderson, J. and Forsythe, J.M. (eds) *MEDINFO 74*, Stockholm, pp. 581–5. North-Holland, Amsterdam. ISBN 0 444107 71 1.

82 Graham, D.F., Kenny, G. and Wright, R. (1979) Computer diagnosis of acute abdominal pain. In Barber, B., Grémy, F., Überla, K. and Wagner, G. (eds) *MIE, Berlin 79: Lecture Notes 5*, pp. 668–81. Springer Verlag, Berlin. ISBN 3 540095 49 7.

83 Evans, C.R., *et al.* (1970) Online interrogation of hospital patients by a time-sharing terminal with computer/consultant comparison analysis. In *Proceedings of Institute of Electrical Engineers*. IEE, Stevenage.

84 Somerville, S., Evans, C.R., Pobgee, P.J. and Bevan, N.S. (1979) MICKIE – experiences in taking histories from patients using a microprocessor. In Barber, B., Grémy, F., Überla, K. and Wagner, G. (eds) *MIE, Berlin 79: Lecture Notes 5*, pp. 713–22. Springer Verlag, Berlin. ISBN 3 540095 49 7.

85 Evans, C.R., Evans, R.J., Marjot, D.H., Matthews, A.F.B., Somerville, S. and Whitfield, S. (1977) Some experiments in interviewing immigrant patients in their own language using automated presentation of questionnaires. In Shires, D.B. and Wolf, H. (eds) *MEDINFO 77*, Toronto, pp. 225–7. North-Holland, Amsterdam. ISBN 0 444107 71 1.

86 de Dombal, F.T. (1985) Decision support in clinical medicine. In Roger, F.H., Grönoos, P., Tervo-Pellikka, R. and O'Moore, R. (eds) *MIE, Helsinki 85: Lecture Notes 25*, pp. 764–74. Springer Verlag, Berlin. ISBN 3 540156 76 3.

87 Farrow, S.C., Fisher, D.J.H. and Johnson, D.B. (1971) Statistical approach to planning an integrated haemodialysis/transplantation programme. *BMJ*, 2, 671–6.

88 Davies, R., Johnson, D. and Farrow, S. (1975) Planning patient care with a Markov model. *Operational Research Q*, 26.3ii, 599–607.

89 Farrow, S.C., Fisher, D.J.H. and Johnson, D.B. (1972) Dialysis and transplantation: the national picture over the next five years. *BMJ*, 2.3, 686–90.

90 Parsons, R.J. and Kilvington, M. (1985) Renal omission. *BJHC*, 2.3, 11.

91 Gordon, M., de Wardener, H.E., Venn, J.C., Webb, J. and Adams, H. (1981) A graphic microcomputer system for clinical renal data. In Grémy, F., Degoulet, P., Barber, B. and Salamon, R. (eds) *MIE, Toulouse 81: Lecture Notes 11*, pp. 543–9. Springer Verlag, Berlin. ISBN 3 540105 68 9.

92 Will, E.J. (1984) British Renal Computing Group. *BJHC*, 1.1, 33.

93 Selwood, N.H. and Will, E.J. (1985) Computerized information for renal medicine. *BJHC*, 2.1, 11, 13–14, 18.

94 Cameron, C.A., Conroy, G.V. and Kangavari, M.D. (1990) Machine learning techniques for patient and program management in renal replacement/transplantation therapy. In O'Moore, R.O., Bengtsson, S., Bryant, J. and Bryden, J.S. (eds) *MIE, Glasgow 90: Lecture Notes 40*, pp. 286–91. Springer Verlag, Berlin. ISBN 3 540529 36 5.

95 Boran, G., Alexander, D., Grimson, J. and O'Moore, R. (1991) Interpretation of clinical biochemistry test profiles by a method allowing communication between organ-related knowledge bases. In Adlassnig, K.P., Grabner, G., Bengtsson, S. and Hansen, R. (eds) *MIE, Vienna 91: Lecture Notes 45*, pp. 294–8. Springer Verlag, Berlin. ISBN 3 540543 92 9.

96 Homer, G.R. (1985) Computer-aided training in a renal dialysis ward – a unique approach. In Bryant, J. and Kostrewski, B. (eds) *Current Perspectives in Health Computing*, pp. 113–27. BJHC for BCS, Weybridge. ISBN 0 948198 00 1.

97 Festenstein, H., Oliver, R.T., Hyams, A., Moorhead, J.F., Pirrie, A.J., Pegrum, D.G. and Balfour, I.C. (1969) A collaborative scheme for tissue typing and matching in renal transplantation. *The Lancet*, 7617, 389–91.

98   Festenstein, H., Oliver, R.T., Sachs, J.A., Burke, J.M., Adams, E., Divver, W., Hyams, A., Pegrum, G.D., Balfour, I.C. and Moorhead, J.F. (1971) Multicentre collaboration in 162 tissue typed renal transplants. The London and Regional Transplant Group, March 1969–December 1970. *The Lancet*, 7718, 225–8.

99   Dausset, J., Hors, J., Busson, M., Festenstein, H., Oliver, R.T., Paris, A.M. and Sachs, J.A. (1974) Serologically defined HL-A antigens and long-term survival of cadaver kidney transplants. A joint analysis of 918 cases performed by France Transplant and The London Transplant Group. *N. Eng. J. Med.*, 209.18, 979–84.

100 Garwood, D. (1986) The UK Transplant Service. In Abbot, W., Barber, B. and Peel, V. (eds) *Information Technology in Healthcare: A Handbook*, B2.7.12–16 issue 7. Longman in association with The Institute of Health Services Management, Harlow, Essex. ISBN 0 582061 40 7.

101 Daunt, S.M., James, R.M., Roberts, J.M. and Harvey, P.W. (1980) A computerized protocol to assist in the prescription of drugs with a narrow therapeutic ratio. In Lindberg, D.A.B. and Kaihara, S. (eds) *MEDINFO 80*, Tokyo, pp. 52–4. North-Holland, Amsterdam. In two volumes. ISBN 0 444860 29 0.

102 Robertson, J.W.K., Shimmins, J.G., Thomas, A. and Simpson, D.L. (1977) Medical information systems incorporating automatic coding of text. In Shires, D.B. and Wolf, H. (eds) *MEDINFO 77*, Toronto, pp. 271–5. North-Holland, Amsterdam.

103 Williams, C.B. (1986) Computers in clinical research. *BJHC*, 3.5, 25–7.

104 Bell, J., Simpson, D.S., Corbett, W.A., Flavell, J., Simpson, D.S., Ryott, K. and Holland, G. (1990) The planning and development of a clinically based multiuser microcomputer-based system for audit and quality assurance. In O'Moore, R.O., Bengtsson, S., Bryant, J. and Bryden, J.S. (eds) *MIE, Glasgow 90: Lecture Notes 40*, pp. 707–11. Springer Verlag, Berlin. ISBN 3 540529 36 5.

105 Levitt, J.R., Richards, B., Kwiatkowski, R., Salmon, P. and McCloy, R.F. (1986) A computer system to analyse the effects of variations in gastric and duodenal pH on ulcers. In Salamon, R., Blum, B. and Jørgensen, M. (eds) *MEDINFO 86*, Washington, part 1, pp. 517–9. North-Holland, Amsterdam. ISBN 0 444701 08 7.

106 Richards, B., Levitt, J., McCloy, R.F. and Pearson, R.C. (1990) A methodology for use of a computer to assist in the recording and processing of data to temperature changes in the human gut. In O'Moore, R.O., Bengtsson, S., Bryant, J. and Bryden, J.S. (eds) *MIE, Glasgow 90: Lecture Notes 40*, pp. 740–3. Springer Verlag, Berlin. ISBN 3 540529 36 5.

107 Walker, M.A., Bryce, D. and Carter, N.W. (1986) A flexible clinical database. *BJHC*, 3.4, 15–17.

108 Shearing, G. (1976) *NUMAC-RVI infusions program*. Project Note. Newcastle upon Tyne.

109 James, R.N., Roberts, J.M., Cooper, R.I., Park, W.G., Bellis, J.D. and Harvey, P.W. (1980) 'Computafeed' – a method of computer-assisted intravenous feeding. In Lindberg, D.A.B. and Kaihara, S. (eds) *MEDINFO 80*, Tokyo, pp. 55–7. North-Holland, Amsterdam. ISBN 0 444860 29 0.

110 Collinson, P.O., Boran, G.P.R., Gray, T.J., Cock, C. and Harrison, L.A.W. (1990) Design, validation and evaluation of a clinical management and information system for fluid, electrolyte and nutritional therapy combining conventional and novel techniques. In O'Moore, R.O., Bengtsson, S., Bryant, J. and Bryden, J.S. (eds) *MIE, Glasgow 90: Lecture Notes 40*, pp. 275–7. Springer Verlag, Berlin. ISBN 3 540529 36 5.

111 Maceratini, R., Crollari, S. and Rafanelli, M. (1990) Expert systems in gastrointestinal diseases. In O'Moore, R.O., Bengtsson, S., Bryant, J. and Bryden, J.S. (eds) *MIE, Glasgow 90: Lecture Notes 40*, pp. 297–304. Springer Verlag, Berlin. ISBN 3 540529 36 5.

112 Coles, E.C. and Lewis, P.A. (1986) The Cardiff long-term rheumatoid arthritis clinical trial. In Bryant, J., Roberts, J. and Windsor, P. (eds) *Current Perspectives in Health Computing*, pp. 158–64. BJHC for BCS, Weybridge. ISBN 0 948198 01 X.

113 Vincent, R. (1986) Cardiology. In Abbot, W., Barber, B. and Peel, V. (eds) *Information Technology in Healthcare: A Handbook*, B2.7.3:1–12 issue 2. Longman in association with The Institute of Health Services Management, Harlow, Essex. ISBN 0 582061 40 7.

114 Reece, I.J., Spyt, T.J., Wheatley, D.J. and Sharp, W.J. (1986) Cardiology audit. *BJHC*, 3.2, 17–18.

115 Maresh, J. and Wastell, D.G. (1990) Process modelling and the cooperative structure of medical office work. In O'Moore, R.O., Bengtsson, S., Bryant, J. and Bryden, J.S. (eds) *MIE, Glasgow 90: Lecture Notes 40*, pp. 29–33. Springer Verlag, Berlin. ISBN 3 540529 36 5.

116 Hill, A.G. and Townsend, H.R.A. (1974) Determining the patterns of epileptic spikes despite inefficient recognition. In Anderson, J. and Forsythe, J.M. (eds) *MEDINFO 74*, Stockholm, pp. 731–4. North-Holland, Amsterdam. ISBN 0 444107 71 1.

117 MacGillivray, B.B., Wadbrook, D.G., Quilter, P.M. and Douglas, J. (1978) ABSCESS – a system for the analysis by small computer of EEG signals in a clinical setting. In Anderson, J. (ed.) *MIE, Cambridge 78: Lecture Notes 1*, pp. 545–54. Springer Verlag, Berlin. ISBN 3 540089 16 0.

118 Townsend, H.R.A. (1982) An EEG machine with an integral microprocessor. In O'Moore, R.R., Barber, B., Reichertz, P.L. and Roger, F. (eds) *MIE, Dublin 82: Lecture Notes 16*, pp. 245–6. Springer Verlag, Berlin. ISBN 0 387112 08 1.

119 Seldrup, J. (1974) Computer-assisted epilepsy survey. In Anderson, J. and Forsythe, J.M. (eds) *MEDINFO 74*, Stockholm, pp. 879–81. North-Holland, Amsterdam. ISBN 0 444107 71 1.

120 Willems, J.L., *et al.* (1981) Common standards for quantitative electrocardiography: the CSE pilot study. In Grémy, F., Degoulet, P., Barber, B. and Salamon, R. (eds) *MIE, Toulouse 81: Lecture Notes 11*, pp. 319–26. Springer Verlag, Berlin. ISBN 3 540105 68 9.

121 Richards, B., Bray, C.L., Jeffery, C. and Khadri, N. (1980) An assessment of the accuracy of the Bonner/IBM ECG analysis program when compared with autopsy evidence. In Lindberg, D.A.B. and Kaihara, S. (eds) *MEDINFO 80*, Tokyo, pp. 244–8. North-Holland, Amsterdam. In two volumes. ISBN 0 444860 29 0.

122 Richardson, R.B., Sawdon, K.J., Finnegan, O.C. and Lewis, G.T.R. (1982) Computerization of a non-invasive method for the determination of cardiac output. In O'Moore, R.R., Barber, B., Reichertz, P.L. and Roger, F. (eds) *MIE, Dublin 82: Lecture Notes 16*, pp. 102–7. Springer Verlag, Berlin. ISBN 0 387112 08 1.

123 Beech, C.A., Evans, A.H., Hill, V., Mali, N., Bradbury, S.P. and Pantin, C.F.A. (1992) The use of information technology in a department of respiratory medicine. In Richards, B. and MacOwan, H. (eds) *Current Perspectives in Healthcare Computing*, pp. 310–18. BJHC for BCS, Weybridge. ISBN 0 948198 12 5.

124 Vickery, K.J. and Hanley, S.P. (1993) Computer-based bronchoscopy project. In Richards, B. and MacOwan, H. (eds) *Current Perspectives in Healthcare Computing*, pp. 159–68. BJHC for BCS, Weybridge. ISBN 0 948198 14 1.

125 Jones, R. (1990) How do clinical systems measure up? *BJHC&IM*, 7.9, 28, 30.

126 Holland, J. (1990) EURODIABETA: modelling and implementation of information systems for chronic health care – example: diabetes. In O'Moore, R.O., Bengtsson, S., Bryant, J. and Bryden, J.S. (eds) *MIE, Glasgow 90: Lecture Notes 40*, pp. 48–53. Springer Verlag, Berlin. ISBN 3 540529 36 5.

127 Lehamann, E.D., Roudsari, A.V., Deutsch, T., Carson, E.R., Benn, J.J. and Sönksen, P.H. (1991) Clinical assessment of a computer system for insulin dosage adjustment. In Adlassnig, K.P., Grabner, G., Bengtsson, S. and Hansen, R. (eds) *MIE, Vienna 91: Lecture Notes 45*, pp. 376–81. Springer Verlag, Berlin. ISBN 3 540543 92 9.

128 Kelly, W.F., Mahmood, R., Arthur, M., Kelly, M. and Scott, E. (1993) Inner-city deprivation and diabetes control: a computer database. In Richards, B. and MacOwan, H. (eds) *Current Perspectives in Healthcare Computing*, pp. 127–34. BJHC for BCS, Weybridge. ISBN 0 948198 14 1.

129 Clarkson, D. and Thomas, K. (1988) Monitoring infection. *BJHC*, 5.9, 39, 41, 43.

130 O'Brien, G.J. (1991) Organism alert: a computer system for use by the infection control nurse. In Richards, B. and MacOwan, H. (eds) *Current Perspectives in Healthcare Computing*, pp. 249–59. BJHC for BCS, Weybridge. ISBN 0 948198 11 7.

131 Mayon-White, R.T., Ducal, G., Kereselidze, T. and Tikomirov, E. (1988) An international survey of the prevalence of hospital acquired infection. *J. Hospital Infection*, 11, sup. A, 43–8.

132 Glenister, H. and Taylor, L.J. (1992) Assessing information systems for recording hospital-acquired infection. In Richards, B. and MacOwan, H. (eds) *Current Perspectives in Healthcare Computing*, pp. 380–84. BJHC for BCS, Weybridge. ISBN 0 948198 12 5.

133 Johansen, K.S. (1992) WHOCARE: hospital infection surveillance and feedback program. In Richards, B. and MacOwan, H. (eds) *Current Perspectives in Healthcare Computing*, pp. 359–66. BJHC for BCS, Weybridge. ISBN 0 948198 12 5.

134 Lord Chorley (chairman) (1987) *Handling Geographical Information* (The Chorley Report). HMSO, London.

135 Gatrell, A. (1988) Providing healthcare with GIS. *BJHC*, 5.5, 28–9.

136 McWilliams, A. (1988) Medical informatics and AIDS in the UK. In Hansen, R., Solheim, B.G., O'Moore, R.R. and Roger, F.H. (eds) *MIE, Oslo 88: Lecture Notes 35*, pp. 299–302. Springer Verlag, Berlin. ISBN 3 540501 38 X.

137 Christie, P., Heslop, J., Robertson, J., Jones, R. and Gruer, L. (1990) A clinical information system for HIV/AIDS patients at Ruchill Hospital, Glasgow: development and evaluation. In O'Moore, R.O., Bengtsson, S., Bryant, J. and Bryden, J.S. (eds) *MIE, Glasgow 90: Lecture Notes 40*, pp. 567–9. Springer Verlag, Berlin. ISBN 3 540529 36 5.

138 Porter, D.M. and Ingram, G.I.C. (1979) Computer monitoring of haemophiliac bleeds and their treatment. In Barber, B., Grémy, F., Überla, K. and Wagner, G. (eds) *MIE, Berlin 79: Lecture Notes 5*, pp. 427–34. Springer Verlag, Berlin. ISBN 3 540095 49 7.

139 Richards, B., Kiely, S., Hyde, K., Burn, A. and Delamore, I.W. (1988) The computerization of patient records in a haemophilia centre. In Hansen, R., Solheim, B.G., O'Moore, R.R. and Roger, F.H. (eds) *MIE, Oslo 88: Lecture Notes 35*, pp. 377–81. Springer Verlag, Berlin. ISBN 3 540501 38 X.

140 Henney, C.R., Brodlie, P. and Crooks, J. (1974) The administration of drugs in hospital – how a computer can be used to improve the quality of patient care. In Anderson, J. and Forsythe, J.M. (eds) *MEDINFO 74*, Stockholm, pp. 271–6. North-Holland, Amsterdam. ISBN 0 444107 71 1.

141 Petrie, J.C., Robb, O.J., Taylor, D.J., Jeffers, T.A., McLeod, G., Allen, K.F., Dawson, G.M., Murphy, P., Innes, G. and Weir, R.D. (1982) A computer-assisted patient record system. In O'Moore, R.R., Barber, B., Reichertz, P.L. and Roger, F. (eds) *MIE, Dublin 82: Lecture Notes 16*, pp. 41–7. Springer Verlag, Berlin. ISBN 0 387112 08 1.

142 Thompson, B.H. and Chantry-Prioce, A.E. (1993) Interfacing a MUMPS-based drug-prescribing scheme with other health service systems. In Richards, B. and MacOwan, H. (eds) *Current Perspectives in Healthcare Computing*, pp. 434–9. BJHC for BCS, Weybridge. ISBN 0 948198 14 1.

143 Jackson, R.P.P. (1986) Cancer record systems and the evaluation of care. In Abbot, W., Barber, B. and Peel, V. (eds) *Information Technology in Healthcare: A Handbook*, B2.7.10–08 issue 4. Longman in association with The Institute of Health Services Management, Harlow, Essex. ISBN 0 582061 40 7.

144 Neal, L.R. (1974) Computer techniques for retrospective analysis. In Anderson, J. and Forsythe, J.M. (eds) *MEDINFO 74*, Stockholm, pp. 435–8. North-Holland, Amsterdam. ISBN 0 444107 71 1.

145 Neal, L.R. (1977) The computer-handling of medical information for research purposes. In Shires, D.B. and Wolf, H. (eds) *MEDINFO 77*, Toronto, pp. 551–5. North-Holland, Amsterdam.

146 Milan, J. and Bentley, R.E. (1980) A clinical database for research applications using MUMPS. In Lindberg, D.A.B. and Kaihara, S. (eds) *MEDINFO 80*, Tokyo, pp. 480–3. North-Holland, Amsterdam. In two volumes. ISBN 0 444860 29 0.

147 Todd, J.H., Haybittle, J.L., Blamey, R.W. and Pearson, D. (1985) A clinical database for the study of prognostic factors in breast cancer. In Roger, F.H., Grönoos, P., Tervo-Pellikka, R. and O'Moore, R. (eds) *MIE, Helsinki 85: Lecture Notes 25*, pp. 474–8. Springer Verlag, Berlin. ISBN 3 540156 76 3.

148 Rogers, M.L. (1984) The Pharmline drug information system. In Kostrewski, B. (ed.) *Current Perspectives in Health Computing*, pp. 135–9. Cambridge University Press for BCS, Cambridge. ISBN 0 521267 05 6.

149 Developments (1985) Pharmacy computer to save £3,350,000. *BJHC*, 2.4, 7.

150 NHS Register of Computer Applications (1985) Pharmasyst. *BJHC*, 2.4, 32.

151 Clark, R. and Gooch, P. (1988) Take your pick. *BJHC*, 5.4, 33–5.

152 Datafile (1986) *BJHC*, 3.2, 23.

153 Sodha, R.J., Kostrewski, B.J. and Schier, O. (1991) Drug information through online databases: a comparison between external databases and the in-house database of a pharmaceutical company. In Adlassnig, K.P., Grabner, G., Bengtsson, S. and Hansen, R. (eds) *MIE, Vienna 91: Lecture Notes 45*, pp. 908–13. Springer Verlag, Berlin. ISBN 3 540543 92 9.

154 Jacklin, A. and Willson, A. (1991) Sweetening the pill. *BJHC&IM*, 8.10, 23, 25.

155 Rutovitz, D. (1977) Information processing in cytology and pathology: a review. In Shires, D.B. and Wolf, H. (eds) *MEDINFO 77*, Toronto, pp. 569–73. North-Holland, Amsterdam.

156 Smith, C. (1972) Computer program to estimate recurrence risks of multifactorial familial disease. *BMJ*, 1, 495–7.

157 Heuch, I. and Li, F.H.F. (1972) A computer program for the calculation of genotype probabilities using phenotype information. *Clinical Genetics*, 3, 510–14.

158 Department of Manpower Services (Northern Ireland) (1976). *Geocode Index*. DoMS, Northern Ireland.

159 McDonald, J.R. and Nevin, N.C. (1977) Computer-based genetic counselling. In Shires, D.B. and Wolf, H. (eds) *MEDINFO 77*, Toronto, pp. 615–17. North-Holland, Amsterdam.

160 Richards, B. and Jeffery, C. (1980) The computer diagnosis of congenital malformations. In Lindberg, D.A.B. and Kaihara, S. (eds) *MEDINFO 80*, Tokyo, pp. 779–83. North-Holland, Amsterdam. In two volumes. ISBN 0 444860 29 0.

161 Richards, B., Super, M. and Goh, A. (1983) The computerization of the records in a cystic fibrosis clinic. In van Bemmel, J.H.V., Ball, M.J. and Wigertz, O. (eds) *MEDINFO 83*, Amsterdam, pp. 832–5. North-Holland, Amsterdam. In two parts. ISBN 0 444867 49 X.

162 Richards, B. and Khoury, D. (1991) The use of a touchscreen computer for dental charting. In Adlassnig, K.P., Grabner, G., Bengtsson, S. and Hansen, R. (eds) *MIE, Vienna 91*: *Lecture Notes 45*, pp. 145–9. Springer Verlag, Berlin. ISBN 3 540543 92 9.

163 Ash, S.P. and Tan, A.C. (1992) The benefits of using a portable bar-code reader for departmental management and audit. In Richards, B. and MacOwan, H. (eds) *Current Perspectives in Healthcare Computing*, pp. 481–9. BJHC for BCS, Weybridge. ISBN 0 948198 12 5.

164 Stephens, C.D., Williams, J.H.S. and Mackin, N. (1994) Clinical support for clinical decision-making in dentistry. In Richards, B., de Glanville, H. and MacOwan, H. (eds) *Current Perspectives in Healthcare Computing*, pp. 126–33. BJHC for BCS, Weybridge. ISBN 0 948198 17 6.

165 Moss, J.P., Linney, A.D., Grindrod, S.R., Arridge, S.R. and Clifton, J.S. (1987) Three-dimensional visualization of the face and skull using computerized tomography and laser scanning techniques. *European Journal of Orthodontics*, 9, 247–53.

166 Coombes, A.M., Moss, J.P., Linney, A.D., Richard, R. and James, D.R. (1991) A mathematical method for the comparison of three-dimensional changes in the facial surface. *European Journal of Orthodontics*, 13, 95–110.

167 Durward, B. (1986) Physiotherapy. In Abbot, W., Barber, B. and Peel, V. (eds) *Information Technology in Healthcare: A Handbook*, B.2.8.1 issue 9. Longman in association with The Institute of Health Services Management, Harlow, Essex. ISBN 0 582061 40 7.

168 Finlay, O. (1995) The use of computers to assess problems associated with gait in elderly people. *Information Technology in Nursing*, 7.1, 9–12.

169 Gardner, B.P., Jefcoate, R.M., Dyke, R.G. and Krishnan, K.R. (1986) Computers for the disabled. *BHJC*, 3.1. 17–19.

170 Busby, G. (1988) Computer and electronics for the service of the handicapped. In Hansen, R., Solheim, B.G., O'Moore, R.R. and Roger, F.H. (eds) *MIE, Oslo 88: Lecture Notes 35*, pp. 79–83. Springer Verlag, Berlin. ISBN 3 540501 38 X.

171 Ganney, P.S. (1994) Computerization of a wheelchair services centre. In Richards, B., de Glanville, H. and MacOwan, H. (eds) *Current Perspectives in Healthcare Computing*, pp. 192–7. BJHC for BCS, Weybridge. ISBN 0 948198 17 6.

172 Lutman, M.E. (1983) Microcomputer-controlled psychoacoustics in clinical audiology. *Brit. J. Audiology*, 17, 2, 109–14.

173 Marchbanks, R.J. and Martin, A.M. (1985) A computer-based test system for implementing pure-tone audiometry, acoustic immittance and tympanic membrane displacement measurements. *Brit. J. Audiology*, 19, 1, 19–28.

174 NHS Register of Computer Applications (1986) Audiology training. *BJHC*, 3.2, 32.

175 Green, S. (1986) Occupational therapy. In Abbot, W., Barber, B. and Peel, V. (eds) *Information Technology in Healthcare: A Handbook*, B2.8.2 issue 9. Longman in association with The Institute of Health Services Management, Harlow, Essex. ISBN 0 582061 40 7.

176 Bassham, S., Fletcher, L.R. and Soden, P. (1990) Scientific uses of computers in dietetic departments. In O'Moore, R.O., Bengtsson, S., Bryant, J. and Bryden, J.S. (eds) *MIE, Glasgow 90: Lecture Notes 40*, pp. 7–10. Springer Verlag, Berlin. ISBN 3 540529 36 5.

177 Fletcher, L.R. and Bassham, S. (1985) Microcomputers and dietetics. In Bryant, J. and Kostrewski, B. (eds) *Current Perspectives in Health Computing*, pp. 216–23. BJHC for BCS, Weybridge. ISBN 0 948198 00 1.

178 Westcott, P. (1984) The register of NHS computer applications. *BJHC*, 1.2, 10–14.

179 Gregory, R.C., Leaning, M.S., Summerfield, J.A., Browne, J., Wilkinson, D. and McGrath, A. (1991) The FEBE project: designing a clinical information system. In Adlassnig, K.P., Grabner, G., Bengtsson, S. and Hansen, R. (eds) *MIE, Vienna 91: Lecture Notes 45*, pp. 91–5. Springer Verlag, Berlin. ISBN 3 540543 92 9.

180   Brambley, P. and Carlson, E. (1988) PC for diet analysis. *BJHC*, 5.8, 30.

181   Product developments (1984) Occupational health. *BJHC*, 1.4, 42.

182   Evans, C.R. (1971) Psychological assessment of history taking by computer. In Abrams, M.E. (ed.) *Spectrum 71: BCS Conference on Medical Computing*, pp. 9–22. Butterworths for BCS, London.

183   Khan, S. (1985) A computer system for clinical psychology. *BJHC*, 2.1, 35, 37–39.

184   Dunn, T.L., Hughes, N., Luck, M. and Overton, G. (1979) Modelling the mental health referral process. In Barber, B., Grémy, F., Überla, K. and Wagner, G. (eds) *MIE, Berlin 79: Lecture Notes 5*, pp. 9–17. Springer Verlag, Berlin. ISBN 3 540095 49 7.

185   Gleisner, J.W., Marks, J., Tamblin, N. and Binks, K. (1977) A computer-assisted psychiatric care register for service purposes. In Shires, D.B. and Wolf, H. (eds) *MEDINFO 77*, Toronto, pp. 383–7. North-Holland, Amsterdam.

186   Rigby, M. (1991) An electronic patient information system in mental health: an integrated solution for better care and management. In Adlassnig, K.P., Grabner, G., Bengtsson, S. and Hansen, R. (eds) *MIE, Vienna 91: Lecture Notes 45*, pp. 722–6. Springer Verlag, Berlin. ISBN 3 540543 92 9.

187   Grannell, A. (1991) Computers for mental health patients. In Richards, B. and MacOwan, H. (eds) *Current Perspectives in Healthcare Computing*, pp. 17–21. BJHC for BCS, Weybridge. ISBN 0 948198 11 7.

188   NHS Register of Computing Applications (1985) Northern Ireland – RVH metabolic unit/diabetic system. *BJHC*, 2.1, 49.

189   NHS Register of Computing Applications (1984) North Western – Microdiet/dietetics. *BJHC*, 2.1, 49.

190   NHS Register of Computer Applications (1984) Wales – Catering/meals costing. *BJHC*, 1.4, 38.

191   NHS Register of Computing Applications (1985) CSSD. *BJHC*, 2.4, 32.

192   Green, M.F. (1984) The NHS works information management system. *BJHC*, 1.4, 15–17.

193   Benford-Miler, M. (1987) Computerized architecture. *BJHC*, 4.3, 29, 33–4.

194   Shaw, A. and Maxted, K.J. (1992) Cost-effective medical equipment management: the Glasgow experience. In Richards, B. and MacOwan, H. (eds) *Current Perspectives in Healthcare Computing*, pp. 25–31. BJHC for BCS, Weybridge. ISBN 0 948198 12 5.

195   McKie, J., Shaw, A. and Feasey, M. (1987) A medical inventory. *BJHC*, 4.5, 47–8.

196 NHS Register of Computing Applications (1985) Newmedics. *BJHC*, 2.4, 32.

197 National Audit Office. (1996) *National Health Service Supplies in England*. The Stationery Office Limited. ISBN 0 102755 96 5.

198 Datafile 13 (1987) Ancillary services. *BJHC*, 4.4, 25.

PART 4
Primary and community care

INTRODUCTION

Computing in health care outside hospitals and NHS management began in the 1960s and has progressed significantly over the years. Indeed, particularly in general practice, it has led the way in clinical computing. Many of the aspects of care outside hospitals lend themselves to automation and were thus early users of IT systems. Screening, with its call and recall of populations, health promotion and engaging the public all took place in this arena. One of the most successful of the early Experimental Projects (see Chapter 5), the Exeter Community Health Project (see Chapter 4), blazed a trail that the whole of the NHS perhaps should have followed.

Some of the experiences in data collection by community staff for the 1970s should be read by everyone who feels that the collection of data from the 'coal face' is a simple issue in health care. The problems of coordination between community services with their close links to local government and the NHS, which has a very different culture, apply just as much today.

Today we emphasize the role patients can and should play in their own health care. Chapter 16, *Consumer Health*, points out that this view was recognized by the Chinese 2000 years ago. Much of the work outside hospitals has not been published widely. It is the Cinderella of health care computing yet it often contains the details of the glass slipper.

The BCS began to have a major impact in this sector with the formation of the Primary Health Care Specialist Group in 1981. This group became one of the largest of the BCS Specialist Groups with many non-IT staff joining. It was thus a major outlet for the BCS into the health service. It developed links with the other major players in primary care such as the BMA, the Royal College of General Practitioners and the bodies representing secretarial, nursing and administrative staff. It also became a major source of advice to the DoH on IT matters. Through this group many of the technical, implementation and research developments were stimulated. The BCS activities in health also provided the vehicle for all involved in non-hospital computing to publish their ideas and share their experiences. The BCS conferences, newsletters and journals hold a wealth of often unrecognized excellence.

ACKNOWLEDGEMENTS

Many have helped in the preparation of Chapter 14. In particular Hugh Fisher has provided much material and comment. Where appropriate their contributions have been acknowledged in the relevant section. I would particularly like to thank:

    Mr Douglas Ball – previously Director of the PPA

    Mr Tim Benson – Managing Director, Abies Informatics

    Mr Ewan Davis – Chair BCS Primary Health Care Specialist Group

    Mr Hugh Fisher – previously Chief Systems Analyst, Exeter Community Health Services Computer Project, and Manager FPS Computer Unit

    Dr Leo Fogarty – GP

    Mr Ian Herbert – Vice Chair BCS Health Informatics Forum

    Dr Roger Roycroft – GP

    Mr Ian Shepherd – Pharmacist

    Dr David Stables – Medical Director EMIS

    Ms Sheila Teasdale – Director PRIMIS.

GLYN HAYES

14 The history of primary care computing in the UK

GLYN HAYES

This chapter deals with the development of computing in UK general practice (GP), pharmacy and the associated NHS services. It does not include private practice nor the community systems that developed at the same time (see Chapter 15). It covers the period from the late 1960s to 2007. In so doing it highlights some of the jewels in the crown of UK health care computing: the GP computer systems. Developed mainly by the GPs themselves these systems have become internationally recognized leaders in primary care clinical computing. The success of GP computing is in stark contrast to progress in UK secondary care where, despite some encouraging early developments, progress has been slow with clinical computing absent from most UK hospitals other than a handful of honourable exceptions and despite a Government target for 35 per cent level 3 Electronic Patient Record by March 2002.[1]

While the successful implementation of GP computing has been supported by a number of Government and NHS policy initiatives the success in this field has been a result of substantial uncoordinated and piece-meal developments by a number of individuals, and commercial and non-commercial organizations operating in the absence of, or occasionally at odds with, national policy. There is an important lesson to be learnt from this experience for those tasked with delivering the next phase of computerization within the NHS – complex and uncertain problems are often best solved by defining a general direction and allowing a diversity of solutions to develop and naturally converge.

The following story uses, as far as possible, a chronological progression. However, where there are strands that span many timelines, such as coding systems, these are dealt with as separate sections. In many cases the pioneers themselves have described their experiences in their own words. They are acknowledged in the introduction to Part 4.

This period was an exciting time for those involved. There was the fascination of discovering and using new technologies as well as the satisfaction of seeing such systems show they can improve the care of individual patients.

THE DEVELOPMENT OF GENERAL PRACTICE

The inception of the UK National Health Service in 1948 was attended by a degree of political propaganda that still retains the credibility of an approved 'historical' reality. 'I stuffed their mouths with gold', was the comment of Aneurin Bevan when, as Health Secretary, he completed negotiations with the medical professions over their future relationship with their new employer. There was an aura of heroic political achievement in the face of acrimonious opposition. In fact, general practitioners of the day were professionals in an isolated and difficult specialty whose incomes were uncertain and whose status was insecure. It is probably true that most of them favoured the new conditions not only because it made for a fairer delivery of health care, but also because it made their own positions inherently more certain. It is against this background that Bevan's achievement becomes understandable. Far from having pawned the Crown Jewels in order to achieve his socialized medical care, he had, in fact, set the basis for the cheapest and most controllable system of health care in the West.[2]

The elements of general practice structure following the commencement of the NHS included the following:

- Independent contractor status: Each principal in general practice was paid according to a contract with the employer rather than as a direct employee. This meant that they were financially responsible for their own infrastructure as well as for the success of their practices in the face of any competition from other GPs.

- The patient list: In order to have access to health care, each person in the UK needed to register with a general practitioner (and only one) who would then provide both primary care and referral to the more expensive secondary sector.

- Capitation-based payment: The principal basis for a GP's income was the size of their patient list.

- 24-four hour responsibility: Each GP had the responsibility for delivering care to their patients 24 hours a day for 365 days of the year.

- Local management of GP services: Each GP was locally responsible to a centrally funded body, the Executive Council, for their standards of practice and for many of the conditions of their work (such as the ability to open or close premises, to take on new partners and so on).

These structural planks have continued as the basis for UK primary care ever since. They are an essential part of the foundation on which Government control of health care exists. In order for them to work, of course, they are dependent on, firstly, a degree of mutual trust between profession and employer, and secondly, adequate funding in the face of demand. From the very beginning it was apparent that the first of these would be difficult to maintain and the second would never be achieved. Less than 20 years later, the failure properly to fund primary care resulted in the first great crisis in UK general practice (and, almost, in the collapse of the NHS).

The contractual absence of any mechanism for supporting infrastructure development and for encouraging teamwork, and the insistence on capitation-based payments without regard for the requirements of care in a developing world, resulted by the mid-1960s in a set of GPs who were often isolated in small or single-handed practices GPs worked out of poor premises in terms of design and equipment, who had very large list sizes and who had difficulty in achieving any break from the eternal grind of surgery and 24-hour on call. In addition, the income from capitation payments was not increased as it should have been and some practices were near bankruptcy in consequence. In 1966, the British Medical Association, unable to achieve amelioration of conditions through negotiation, organized a concerted response to these difficulties. The majority of UK GPs sent undated resignations from the Health Service to the Minister of State. Rapid and constructive negotiation ensued.

The result was a new contract known as the GPs' Charter. This came into effect in 1967. The basic planks of the contractual relationship were unchanged but the following important modifications were made:

- Support for practice expenses: This covered a number of areas, of which the most important were partial funding of ancillary staff costs, reimbursement of rent and rates and notional rent payments for owned premises. Rent reimbursement was configured in a way that encouraged the building of purpose-built premises.

- Group practice allowances: These encouraged the formation of group practices.

- Item-of-service payments: These covered a small number of activities of public health interest such as contraception, cervical cytology and family planning.

- Postgraduate education payments: for attending a certain number of lectures each year.

- Pay review: A notionally independent body was set up to annually review GPs' income.

These changes, coupled with the burgeoning activity of the Royal College of General Practitioners (RCGP), produced what has been seen as a golden age in UK general practice. The next generation saw nearly all the philosophical and structural changes that form the basis of modern family medicine in Britain. In particular, it was during this time that pioneering work in primary care informatics occurred, which defined most of the elements of current practice computer systems.

Between 1989 and 1991, new reforms of the Health Service were undertaken. The effects of these reforms were far-reaching, lacked political consensus and remain a matter for heated debate. That debate is well documented elsewhere. As far as medical computerization is concerned there were three important parts:

- Computer reimbursement: From 1989, a proportion of general practice IT investment became directly remunerable.

- Contractual change: A new and complex set of changes occurred in the contract between GPs and Government. One thrust of these was to introduce an element of pay that was dependent on reaching targets for population-based care. It was difficult to reach those targets without IT support.

- Purchaser/Provider split: The principal part of the new reforms was the introduction of contracting mechanisms whereby District Health Authorities and some general practices (fundholders) were able to buy hospital care on behalf of patients. The purchasing of secondary care by GP fundholders required the introduction of standardized health care accounting packages within their IT systems. These were fully remunerable. More importantly, the freedom of fundholding GPs to move money within their budgets allowed for greater investment in the wider IT capital of their practices.

These and other changes have made British primary care difficult or impossible to conduct without relatively advanced IT support. They are integral to understanding the relative success of UK GP computing development.

GENERAL PRACTICE COMPUTING

The early steps

It is not surprising that computerization in general practice did not occur in a vacuum. It was built on a number of developments that were necessary precursors. In the 1960s, age sex registers were followed by disease registers.[3–7] By 1985 a survey in the West Midlands found that 52 per cent of practices had registers that they used for screening and surveillance, disease recording and research.[8] As early as 1963 the problems of diagnostic labels in general practice were acknowledged,[7] and this led to the development of the RCGP classification of disease.[9, 10] These developments were crucial in creating the circumstances for computerization in general practice.

One of the first reports of the use of computers in general practice was by Dr Michael Abrams who described a functioning system in 1968.[11, 12] In 1971, Grene described computer-generated patient recalls.[13]

1964 Oxford Community Health Project

The Oxford Community Health Project (OCHP) was established as a collaborative venture between the Oxford RHA (ORHA) and the Unit of Clinical Epidemiology (UCE) at the University of Oxford.[14] Its origins dated back to the time when Dr Donald Acheson, then director of UCE, established the successful Oxford Record Linkage Study, which linked routinely collected abstracts of hospital records, birth records and mortality records, providing a unique (for England and Wales) person-based source of data for epidemiological research.

Dr Acheson saw the potential for extending this system into primary care and in 1966 Dr John Forbes, a GP from Bicester, joined the Unit and developed a prototype punched card of administrative and clinical data from his practice. Other practices joined the project and in 1970 the OCHP was established under Dr Acheson's successor, Dr John Baldwin, with John Perry as its medical director and Dr Alistair Tulloch as a research GP.

The system was based on the ORHA's ICL 1900 mainframe, using a batch system of data entry and used by about 100 GPs. The system was developed over the years and in terms of its concepts, functions and uses, was an important forerunner of the micro systems now in common use. From 1981 the ORHA Regional Computing Unit supported a micro-based equivalent system on an Alpha Micro computer, which was shared by five practices. Twenty-seven practices still used the mainframe system.

To support the data capture Dr John Perry developed the OXMIS codes based on ICD–8 and configured for general practice acceptability.[15, 16]

Dr Perry, a founder member of the Primary Health Care Specialist Group (PHCSG) of the BCS died in 1985.

1970 The Whipton Project

In January 1970, as part of a service to demonstrate the potential of computer systems to prospective customers (in this case the NHS), IBM set up the first real-time GP computer system in the world at Whipton, Exeter. It would be fair to say that no other system provider at that time possessed sufficient expertise and resources to field a system of equivalent utility and reliability in so short a time. The exercise was reported to have cost a quarter of a million pounds. Erhart Lippmann was the IBM Systems Engineer, and John Preece was responsible for the user specification and operation of the system.[17–19]

The IBM Desktop Pilot System (IDPS) was the world's first real-time GP computer system. It is difficult for those who did not witness or read contemporary reports of the IDPS to appreciate its novelty and the considerable impact it had on computer development, because the design principles, which were formulated by it in 1970, have now become established practice and are taken for granted. But it must be remembered that before 1970 the limitations of punched cards and paper-tape as intermediaries between the GP and the computer, with data turnaround times of approximately six weeks, had precluded the development of rapidly interactive desktop GP systems as we now know them. Day-to-day patient management is completely dependent on an immediately updatable patient record, and if the latter is to be computerized, it requires the provision of VDUs, keyboards and real-time computer access.

In a single-handed Whipton practice, doctor and receptionist were provided with VDUs, keyboards, and a printer that communicated through telephone lines with an IBM mainframe computer in Croydon. During surgery hours computer records were used as the source of data for patient management, augmented only by newly arrived paper hospital reports and pathology test results. Despite the distance travelled by the data, fast reactivity was achieved and the system proved remarkably user-friendly and reliable – more so than some of the systems in use today.

The IDPS established for the first time that:

- electronic records are much quicker and easier to access, both in terms of pulling and refiling, and in data appraisal and updating on display;

- electronic patient record files became the hub of the GP's computer system;

- patient record data should be distributed as discrete subsets, whose design is dictated by the needs of rapid appraisal by the doctor when the patient attends for a subsequent consultation. The subsets used in the IDPS were priority details and summary history, continuing medication, drug idiosyncrasies, immunizations and consultation log (the latter a chronological log in free text of all attendances).

The IDPS was the first system to use:

- GP summary records and GP problem-orientated records in electronic format;

- cut and paste in GP records;

- morbidity codes to standardize terminology in working GP electronic records;

- an audit of workload from a GP computer system;

- tetanus recall from a GP computer;

- electronic back-up in a GP system.[20, 21]

Despite IBM's world lead in computer expertise in 1970, the contract for further development of GP systems within the NHS was given to a British company, hence the use of ICL technology in the later Ottery St Mary project. This should not be allowed to obscure IBM's justifiable claim to have mounted the world's first real-time GP computer system with all the highly innovative activities which that involved. Through the IDPS, IBM was also instrumental in giving the UK the lead in GP computer development, which it has retained ever since.

The Whipton experiment showed that a real-time system could be used at the time of consultation. It could be argued that the actual design contributed in only a small way to the design of future systems as the system only held 1,200 male patients.

1972 Exeter Community Health Project

The Exeter Project had its origins in the 1960s when a survey conducted on disease in Exeter by Exeter University resulted in a feasibility study, in 1967, into data and data collection. This reported in 1969 and a bid was submitted to the DoH for possible funding. The proposed system was to cover two hospitals and two health centres. In 1969 the Regional Hospital Board took responsibility for the staff involved and a systems analyst was recruited. Jack Sparrow became project director and Hugh Fisher its technical director. After another report in 1970, the project was fully integrated into the DoH Experimental Programme (see Chapter 5). Funding for buildings and computers was made available and an Operational Requirement written. Systems analysts were recruited. Programmers were recruited in 1972.

The project was unique in attempting to amalgamate all the information related to a patient's health, wherever generated, into a single record. This was based on the philosophy that the patient is at the centre of any well-directed system of medical care, and that information about a patient should be readily available, where and when needed, by any provider of medical services.[22]

The system was developed involving two health centres (Ottery St Mary, Devon, a rural practice, and the urban practice, Mount Pleasant, Exeter) and two hospitals (Royal Devon and Exeter (Wonford) and Princess Elizabeth Orthopaedic Hospital (Exeter)). Computer terminals were installed in all four centres by 1975 to be operated by the existing NHS personnel. The design work was to achieve a fully integrated approach. It started with the creation of identification information at the patient's first point of contact with the NHS locally and was maintained by the addition of new information at all subsequent contacts at primary or secondary level of care. Each section of the organization maintained its own records but linked them on the computer files with other sections. Appropriate access to the information was provided when necessary in the interests of improved patient care. [23]

The hospitals had a full registration system installed that maintained up-to-date patient identity information and links with other parts of the hospital. Labels were produced to use on the front of case notes, appointment forms, etc. An immediate improvement in patient flow was realized directly as a result of the new system whereby all sections of the Medical Records departments had simultaneous access to all patient identity information.

A computerized waiting list system was introduced for the Ear, Nose and Throat (ENT) specialty at the Royal Devon and Exeter. This had the unexpected and unexplained effect of halving the waiting list. The DoH would not fund an investigation as to the cause of this effect.

Also introduced at the Princess Elizabeth Orthopaedic Hospital was a Nursing system. This created and maintained a complete and up-to-date record of all nursing care given to the patient. The record was created from standard phrases agreed with the local nursing management. It was noted that the system provided a clear record of care and would form the basis for reporting and evaluating care.[24]

The intention of a comprehensive patient information system thus came close to realization (on an ICL 1904A) but further implementation was hindered by the apparent cost of this course of action, uncertainty as to the technical future of computing as the age of the micro dawned, plus 'political' difficulties over the integration of primary and secondary care. Thus in 1982 the Exeter centre ceased to have any further responsibility for hospital systems. In 1983 the centre assumed the title of the Exeter FPS (Family Practitioner Service) Computer Unit with Hugh Fisher as the manager.[25] The Exeter Unit continued to have responsibility for its GP computer system until this was hived off as a commercial concern in 1986.

The continuation of the story of the Exeter Unit is later in this chapter.

The Ottery St Mary Practice and the computer: the GP perspective

In the middle of the 1960s a survey of the prevalence and distribution of disease in Exeter was undertaken, by a group of doctors, by the painstaking method of the local doctors listing on paper the illnesses that they were seeing and treating. As you can imagine this painstaking task took a great deal of time to do and to interpret. Dr N.G. Pearson, Director of the Institute of Biometry at Exeter University, decided that time and effort would be saved by recording illness types and numbers on a computer, so that analysis would be easier.

In the 1960s, computers were large and slow. They stored data with difficulty and were unusual as well as being regarded with suspicion. In 1968, Dr Pearson chose the Ottery St Mary Practice for his proposition because it was the first Health Centre in Devon and had a track record for innovation. He proposed that they kept a second record of each consultation on a suitable form, which secretarial effort would put in a 'batch' on the computer that would then analyse it.

They realized that in a busy practice the likelihood of maintaining two written records at each consultation would be impractical and said that they would only help if they had a screen on the desk into which the doctors would type the records (so-called 'real time'). The practice took great care that confidentiality was complete with passwords and other safeguards. This revolutionary decision to 'computerize' was the subject of much speculation and suspicion from fellow medics, but is now regarded as the norm.

The Government of the day financed the scheme and, in the 1970s, the practice started designing the record, meeting every Monday evening with the Exeter Project team of computer analysts.

By 1974 the record was designed and the practice started summarizing 11,000 paper records to put into the electronic record. This was a considerable task and involved hours of work each evening for a year. In 1975, they went 'live' as one of the first fully computerized GP record-keeping systems in the world. They then produced the first printed prescriptions in the world (to the considerable relief of pharmacists who could at last read them with ease).[26, 27]

The project won the British Computer Society Award for 'computing most beneficial to mankind'. The team's ideas have now become the standard for GP record keeping in this country, being reflected in all the other systems in use and the Exeter System (as it is called) is continually developed and improved.

Thus Ottery St Mary developed and used what was a very unusual and innovative medical record-keeping system a long time before the rest of the world caught up.

The Exeter Primary Care System has been evaluated as follows:

- *The structure of the computerized patient record leads to improved quality of the record.*

- *It gives a more continuous picture of a patient's history.*

- *It is more legible and easier to search.*

- *The time to find a piece of information is almost halved.*

- *Almost half (45.8 per cent) of manual records cause difficulties in assimilating and correlating facts. These problems disappeared with computerization.*

- *Referral and other practice communications should benefit from computerized information.*

- *Computer records are easy to analyse.*

- *Management of recall schemes becomes realistic.*

- *A fully computerized record system reduces receptionist workload by two hours per week per 1,000 patients.*

- *Job satisfaction in the practice improves.*

- *Better control over drug prescribing. A saving of at least 2.7 per cent of drug bill.*

- *The cost, based on a mainframe system with remote terminals, was £1.44 per patient per year. (1980 levels).*

REFERENCE 28

1981 Viewdata and Prestel

Viewdata is now an almost forgotten information technology but it was one of the first to find widespread application in general practice. It used the still familiar Teletext display format which provides 24 lines of 40 character text in six colours with limited block graphics. However, unlike Teletext which is limited to a few hundred pages, which it transmits serially using unused sections of the broadcast television signal, Viewdata could support unlimited numbers of pages which it delivered on demand via a modem and the user's phone line. Viewdata also had the major advantage of being a two-way medium allowing basic interactive services and simple email.

Finding information on Viewdata systems required that the user knew the page number of a suitable starting point from where you could follow its hierarchical tree structure using numeric menus. Each screen of information could point to up to 10 other pages (choices 0–9) and each page could have up to 26 screens accessible sequentially (labelled A–Z).

Communications with the Viewdata system used a 1200/75 full duplex modem providing 120 characters per second (cps) to the user with just 7.5 cps back from the user. This allowed a typical screen to draw in between five and eight seconds with the return channel just about able to match the typing speed of a user. The modems used had no error correction and while they worked well much of the time, a noisy line added spurious characters, requiring the user to manually refresh the page from time to time.

Early Viewdata terminals were adapted televisions and most services required only the input from a numeric keypad on the television remote. However, full keyboards were available and these enabled the user to enter text, which allowed interactive services that required more than numeric entry and a basic email service.

The emergence of Viewdata in the late 1970s was only slightly ahead of the appearance of the first PCs. Increasingly people started to access Viewdata services on these new machines and even to download software using a crude but effective telesoftware protocol that enabled the transmission of binary data and provided error correction. The BBC Micro, launched in 1983, provided a particular boost to Viewdata as it provided hardware support for Viewdata graphics and was available with an optional Viewdata adaptor (modem) at a price well below that of Viewdata capable televisions.

There were a number of private Viewdata services but by far the largest and most significant was the public Prestel service offered by the forerunner of British Telecom; General Post Offices Telephones (the GPO). Anyone with a few thousand pounds to spend could become an information provider (IP) and would be allocated their own three-digit root page. These IPs could in turn make space available to others and so it was possible to get a presence on Prestel for a few hundred pounds.

The GPO charged IPs for each page used, on top of a basic subscription, and also gained revenue from users through a charge for time online (5p per minute during the working day) as well as picking up additional phone call revenue. IPs were able to apply a charge per page view (with a cut to the GPO). While page charges could be up to 99 pence, most pages were free and charges, where levied, were usually in the 0.1–5p range. Prestel also supported closed user groups, which allowed private areas to be established on the service, either to restrict access to sensitive information or to support a subscription-based revenue model.

Pharmaceutical companies were among many that experimented with the new medium and it was their funding that placed many Viewdata terminals in health care. Those involved included Glaxo, who gave terminals to Postgraduate Medical Centres. E.R. Squibb provided many thousands of terminals to GPs (initially television adaptors but later BBC Micros) in support of the post-marketing surveillance of their products.

Meditel, which was later to spawn the major GP system supplier AAH Meditel, was the first independent medical information provider on Prestel. It hosted pages for the BMA General Medical Services Committee, the RCGP and the BCS Primary Health Care Specialist Group, as well as a number of pharmaceutical companies and other medical organisations. Baric, a part of the then leading British IT company ICL, were also significant players in health care with a private Viewdata system that offered better integration with software running on mainframe systems. This was used by many pharmaceutical companies to provide online data collection for a number of market and clinical research projects.

Prestel is no more and while Viewdata services survived until the late 1990s, notably in the travel sector, the technology and the business models it tried to promote have now been replaced by the internet which provides bandwidth and capabilities not dreamt of by the early Viewdata pioneer.

Despite its disappearance, Viewdata in general, and Prestel in particular, was one of the first examples of a publicly accessible and widespread online system which gave many their first experience of IT, online services and electronic mail. It stimulated thinking which contributed to the development of the World Wide Web. In UK primary care, as well as generally raising IT awareness, Viewdata resulted in the large scale donation of BBC Micros to GP practices. This provided a platform for a number of the early GP practice systems and played a significant part in increasing the penetration of such systems from under one per cent to about five per cent by the mid-1980s.

EWAN DAVIS

THE FIRST COMMERCIAL GP SYSTEMS

The advent of the microprocessor allowed the creation of the first personal microcomputers and a new generation of low-cost minicomputers. These systems were affordable by individual enthusiasts and created the prospect of a viable commercial market. During this period many hobbyists developed systems, which some started to market, typically as a sideline to their day job. Also, some of those who had implemented experimental systems before 1980 took the opportunity to interface them to the new low-cost hardware and offer them to the market.

These systems fell into two broad classes: single-user systems typically running on now famous earlier US micros (Apple® II, Commodore PET and Tandy TRS 80) and more advanced systems running on a new generation of microprocessor-based minicomputers, now long forgotten.

Systems in the first of these classes were seriously constrained by the capability of the hardware, operating systems and development tools. (By the end of this period 8 kB of RAM and a 360 kB floppy disk was a high-specification system. Such systems were firmly single user.) The functionality of these systems was limited, typically handling basic patient age sex registers, call and recall, and repeat prescribing in the back office.

Systems in the second of these classes, while still constrained by the hardware, offered greater power and the major advantage of being able to support multi-user operation via dumb serial terminals. They could also support more mature development tools originally developed for minicomputers, in many cases providing a platform to which early GP applications, developed on traditional minicomputers, could be connected, making such systems affordable to the broader market.

Although LAN-based systems and PC workstations would eventually replace them, single processor multi-user systems with dumb terminals would continue to dominate the market until the mid-1990s. AAH Meditel System 5, VAMP Medical, and EMIS LV, running on Xenix/UNIX®, BOS and MUMPS, respectively, were all being examples of such systems, with the first versions of VAMP IGP appearing in 1985, Abies version 3 in 1981 and EMIS M in 1988.

LAN-based systems were complex and very expensive in the early 1980s and the only early GP systems that took this route to providing multi-user operation were the Update PCS system and the RCGP system at Clifford Kay's practice in Manchester.

By the end of the 1980s there were upward of 40 systems available, either commercially or as shareware, and the first 'professional' suppliers with full-time staff and customers in double figures had appeared: Abies, BMDS, Update, VAMP.

This phase ended with the 'Micros for GPs' scheme.

1982 The micros for General Practitioners scheme

In 1982, following a demonstration to Kenneth Baker of the Exeter System at the *IT82* conference, the Department of Trade and Industry, with the involvement of the RCGP, announced the 'Micros for General Practitioners' scheme to provide partially funded systems to GPs. At that time it was estimated that about 50 practices were using computers. One thousand and fifteen practices applied for places on the scheme, and over 140 systems were installed. This represented a quantum leap in GP computerization.[29]

Before this the number of suppliers involved in GP computing was small and the systems were expensive. The effect of the scheme was to mean that all those suppliers not involved, with the exception of Abies, left the market. The two suppliers in the scheme, CAP and British Medical Data Systems (BMDS) developed systems that did not find great favour with the GPs. A lot of lessons were learnt from the project,[29] but neither the suppliers nor the systems themselves survived the scheme. CAP left the medical market altogether and BMDS concentrated on the hospital sector, eventually being taken over by the US company SMS, which survived in that sector until the end of the century.

PHARMACEUTICAL COMPANY INVOLVEMENT

By the mid-1980s pharmaceutical manufactures were taking a greater interest in GP computing. Interestingly, few recognized that GP systems would alter GP prescribing habits, the interest being focused on marketing opportunities to be gained by tapping into GPs interest in the new technology and the opportunity to gather clinical and market research data.

Many pharmaceutical companies supported meetings and publications in the area and there were a number of small donations of cash or equipment at a local level to support the computerization of individual practices. Indeed ICI Pharmaceutical provided facilities and funding for the first PHCSG conference. However, there were also some more significant strategic engagements.

Capoten PMS

E.R. Squibb (now BMS) carried out a large-scale post-marketing surveillance (PMS) study of their new anti-hypertensive Capoten®. This study used a private Viewdata system, provided by Baric, to collect data from GPs about their patients taking Capoten®. Initially Squibb issued participating GPs with a Viewdata adaptor that plugged into a television and telephone line, but they subsequently replaced this with the Viewdata-capable BBC B micro. It is understood that in total Squibb supplied 3,000 BBC micros, giving many GP practices their first experience of computing and providing a platform for which a number of suppliers made a basic GP system available.

Ciba Geigy

Ciba Geigy sponsored a national series of educational meetings during 1984 delivered by Meditel. They followed this initiative by providing a subsidized GP system based on the GPASS product, developed by Dr David Ferguson, supplied through their advertising agency Lavey, Wolfe and Smith (LWS). Ciba Geigy's actual subsidy was modest but they were able to gain economies of scale that allowed complete systems to be sold at significantly less than a GP practice would have been able to independently buy the bare hardware. Over 800 practices had taken the Ciba Geigy system by the time Ciba Geigy withdrew. At this point LWS also withdrew selling the GP system so the firm's management created Advanced Medical Computing (AMC). AMC went on to develop a multi-user system, AMC 2000, and grow its customer base to 800 practices before it was acquired by AAH Meditel in 1992.

Abies

I joined the Charing Cross Hospital in January 1974 as the leader of the computer evaluation team. In that role I met the team responsible for the Exeter (Ottery St Mary) GP computing project. I visited Dr Larry Weed's PROMIS problem-oriented primary care system in Vermont in 1977, and evaluated Dr Christopher Evans' MICKIE (Medical Interviewing Computer System), which was developed at the National Physical Laboratory (NPL). MICKIE was probably the first medical application of a microcomputer.

Dr Chris Evans died at the end of 1979, and I began negotiations with the NPL to obtain a licence to market and develop MICKIE. I demonstrated MICKIE at the RCGP's GP-INFO-80 conference in March 1980: a seminal event in the history of GP computing. At that conference, I resolved to establish Abies and to develop a microcomputer-based GP system. The software was written by Joy Healy, a programmer at Charing Cross Hospital, and ran on SWTP 6809 microcomputers.

The original Abies system allocated 128 bytes per patient to cover registration data, three recall dates and 96 on/off flags. It also supported a variable number of repeat prescription records. The system ran on floppy disks.

By 1982, Abies had sold about a dozen systems (customers included Dr James Read, later of Read Code fame, and Dr David Markwell, who later became their system designer) when the Government announced the 'Micros for GPs' scheme. Rather than give up, Abies obtained a grant from the DTI and raised money in the city in 1983 to develop a new generation of consulting room systems for GPs. The core idea was to keep all of the clinical data in a single fixed length record structure using a common coding scheme.

During 1984–86, I collaborated with, David Markwell and James Read to develop what became the Read Codes (see Medical Termiology).[30] The Read Codes were eventually sold in 1989 to James Read, who sold them to the DoH in 1990. This work has now been incorporated into SNOMED CT® (see Chapter 22).

In 1987, AAH Meditel acquired exclusive rights to use the Abies GP software (then at Version 4.1.8) and Abies agreed to develop a new problem-oriented system which linked medication and diagnostic data. This version (System 5) was written by David Markwell.

Abies went on to develop a range of clinical systems for use across 20 hospital specialties, using MICKIE as an intelligent data-entry system. I began to explore the requirements of electronic shared care and standards for clinical messaging.[31] I developed these ideas as convenor of CEN TC251 WG3 (Messaging and Communications Standards) and the core concepts were taken over to HL7 and incorporated in HL7 Version 3.

TIM BENSON

The free schemes

Noting the pharmaceutical industries' interest in GP computing and the potential of GP systems as tools in clinical and market research, at least four organizations independently developed the 'free scheme' concept. Only two of these schemes came to market (both in the same week in May 1987): AAH Meditel and VAMP (see below).

They used the same broad concept: GPs leased or rented a comprehensive multi-user GP system and associated training and support, and received a payment broadly equal to the cost of the system if they agreed to provide 'anonymized' data and participate in research projects.

The response to these schemes was overwhelming (AAH Meditel obtained an 80 per cent positive response to its first mail shot). During the period 1987–90, each scheme attracted about 1,000 participating practices.

While the two schemes were broadly similar, there were a number of differences. Participants in the VAMP scheme entered into leasing agreements with an independent finance company. Participants in the AAH Meditel scheme entered into a rental agreement contractually linked to the data provision agreement. This difference was to be critical when the schemes eventually folded. VAMP relied on the extraction of data to floppy disks that were then collected by the company. AAH Meditel used EDI via the Istel Fast-Track network. By the time the schemes folded, AAH Meditel had 800 practices connected to the network, which was also used for early experiments with pathology laboratory messaging and to collect data for the RCGPs' weekly epidemiological monitoring service. It was to be nearly 10 years later before the NHS matched this level of connectivity in GP–HA links and 12 years before NHSnet matched it.

The free schemes had a dramatic effect on the level of GP computerization. Firstly, they computerized between 1,500 and 2,500 new practices. (The figures are unclear because the proportion of practices that were already computerized before switching to the free schemes is lost in the mists of time – probably about 20 per cent). Secondly, and perhaps surprisingly, the schemes provided a boost to many other suppliers as the schemes changed the question in GPs minds from 'Shall we computerize?' to 'How should we computerize?'. Some practices chose to go elsewhere. This included those who had ethical objections to the free schemes but who wanted a sophisticated consulting room system.

While the GP response more than met the schemes' requirements, take-up of the data services by the pharmaceutical industry was slow. The schemes were also dogged by the DoH's unhappiness over issues of data ownership and confidentiality (which were not shared by participants and their professional bodies). The last straw for the schemes came with the introduction of direct reimbursement for GP systems, by the DoH, which was structured to deduct data payments from any external reimbursement paid, significantly reducing the financial incentive for participating in the free schemes.

VAMP were the first to pull the plug on the free scheme in January 1991, offering their participants a stark choice, with just one week to decide. Either 90 per cent of them would need to agree to waive their rights under the existing data provision agreement or the company would close. In either case GPs would be left with an ongoing lease commitment. Only by waiving their rights was there any real prospect of further support for their computer system that many had now discovered was bringing benefits to their practice and patients. The company worked closely with the user group and managed to hammer out a deal that allowed the data provision scheme to continue, under a profit-sharing arrangement, and gave user-group representatives seats on the company's board. Miraculously this was all done in a week and allowed VAMP to secure the reduced funding it needed to continue.

VAMP continued to pursue its data business under the new agreement. However, pharmaceutical revenues remained disappointing and the data collection scheme was gifted over to the DoH when the company was bought by Reuters in 1993. It continues as the General Practice Research Database, operated as a stand-alone division of the Government's Medicines Control Agency.

AAH Meditel followed soon after. The financial strength of AAH and the different contractual arrangements meant that the closure of the AAH Meditel scheme was a less frantic affair. Participants were given three months' notice at the end of which they could return their system with no ongoing commitment or keep the system at their own expense. However, like VAMP users, most AAH Meditel customers had come to value their systems and, despite having an easy way out, 85 per cent chose to pick up the ongoing funding for their systems, with a remarkable number taking up the company's offer of an 'at cost' upgrade to their file server.

AAH Meditel's contract meant that they were unable to continue to have any involvement in ongoing data collection. However, two independent schemes emerged, which continued to offer AAH Meditel customers the opportunity to contribute to the creation of an epidemiological database. The first of these was operated by IMS (the leading information provider to the pharmaceutical industry) and the other by the Doctors' Independent Network (DIN – a 'not for profit' organization formed for the purpose) working with another pharmaceutical information provider, Compufile. Both these schemes continue today, providing information based on 'anonymized' data collected from Torex systems for a range of commercial and academic research purposes.

AAH Meditel

Meditel became involved in a number of early projects in primary care informatics and created AAH Meditel as a joint venture with AAH in 1987. (AAH were the UK's largest pharmaceutical distributor.)

Following the closure of the 'free schemes', most of the GPs kept their systems and AAH Meditel, now a wholly-owned subsidiary of AAH, grew with continuing sales of System 5 and the acquisition of AMC, to become the largest supplier of systems for a brief period before finally falling back to being the third largest behind EMIS and VAMP. AAH Meditel was the first supplier to spot the likely merger of GP and community systems and acquired the leading community supplier, Peak Systems, in 1994.

AAH Meditel survived the takeover of AAH plc by GEHE AG (Europe's largest pharmaceutical distributor) in 1996 who continued to support the development of System 6000. However, after three years, and as a result of poor performance and a change in corporate strategy, GEHE decided to dispose of AAH Meditel, which was sold to Torex in July 1999. System 5 was well ahead of its time and remains in use in many practices today. System 6000 lives on as the basis of Torex Synergy.

EWAN DAVIS

VAMP Health – Reuters Health Information – In Practice Systems

VAMP Health was originally formed in 1985 by Peter Williams, Ian Collins and Dr Alan Dean and operated out of Ian's spare bedroom. The original Integrated General Practice (IGP) software had been developed by Alan and Jan Boda with the intention of collecting 'anonymized' patient data from practices for pharmaco-epidemiological research. IGP, later known as VAMP Medical survived the 'free schemes' and, in 1993, VAMP began the development of an entirely new Windows® GP system, Vision.

In late 1993 the company was sold to Reuters and later became known as Reuters Health Information. Vision was launched in 1995 and quickly became established as a leading edge GP system. When Reuters purchased the company, the VAMP Research Database was taken over by the Government and continues to this day as the General Practice Research Database, now hosted by the Medicines Control Agency.

After five years ownership, Reuters decided to sell the company on 31 December 1998 to Cegedim SA, a French company quoted on the Paris Bourse (stock exchange) and involved in health care software and services throughout Europe. At that point the UK company was renamed 'In Practice Systems'. Vision has continued to grow its user base and is now the most widely used Windows®-based software in General Practice in the UK.

MIKE ROBINSON

The Microtest Practice Manager system

In 1979, Greg White and Geoff Rhodes were computer science lecturers at Mid-Cornwall College of Further Education when a GP, Dr Andrew Crawshaw of Mevagissey, attended an evening class to learn how to write computer programs. During the course the students were asked for an example program to be written as a class project – Andrew suggested a repeat prescribing program.

Several months later Andrew reappeared at the college with a system he had written for himself based on the example produced on the course. He convinced them that it would be a useful program for all GPs to have so they rewrote a new version from scratch, the first version of what was soon to become 'Practice Manager'.

Greg and Geoff ran their own company, G+G Software, and worked with a local company, Microtest Ltd, who provided the hardware and increasingly the support services to practices. This early version of Practice Manager sold to many practices throughout the country, with the software being especially well received in the South West of England. The last two working BBC versions were only finally prised away from practices some seven years ago with these being replaced with the latest Practice Manager Windows® version.

In 1982, the first MS-DOS® version of Practice Manager became available and, in 1985, Microtest installed the first multi-user UNIX® versions using, what for time were large amounts of memory (1 MB), supporting excitingly large numbers of users (20). Working increasingly closer together, in 1989, G+G Software merged with Microtest, enabling the company to combine the hardware, software and support activities under one roof. In the 23 years since the GP system work started they have increased the number of employees from two to 62 and support an area from the Isles of Scilly to Cumbria and from Pembroke to Suffolk with regional staff based in different parts of the country.

The current version of their GP system was one of the few systems to be accredited to RFA99 v1.2, and, apart from providing a comprehensive core clinical system, it also provided integrated document scanning, workflow, patient paging systems, unattended patient arrival systems, mobile systems on PDAs and notebooks and much more.

The new Windows® GUI system, Practice Manager Evolution, was released at the beginning of 2005 and computerizes virtually every aspect of life in today's busy practice. Apart from being able to support single or multiple practices in health centres, it is also able to run remotely from a Data Centre.

Unusually for GP system suppliers, Microtest provides every aspect of service to practices themselves. They employ their own software designers and developers, software testers, installation and support engineers, network engineers and trainers and even build and maintain all their own hardware.

They are still based in Bodmin and continue to support practices throughout England.

ROGER ROYCROFT

Update PCS

Work started on what became Update PCS in December 1977 at Exeter University under Prof John Ashford. The initial design was by Dr John Preece, (see the Whipton Project earlier in this chapter) and it appeared on BBC TV's *Tomorrow's World* in 1978. The microprocessor used was an Intel 8080, using either Intel assembler or PL/M. In early 1978, sponsorship was obtained from Philips (the electrical consumer goods company), and by the end of 1978 the system was redesigned, basing it on a formal data dictionary. This lasted until 1980, when Philips Medical Systems realized that they hadn't got the hardware needed to make it viable in the GP market, and were hard-pressed to fund further development of their tomography kit. It went on the shelf until, with the agreement of Philips, it was picked by Dr Abraham Marcus of Update in early 1981. I, having been involved from the beginning, was asked to lead the software development, and we rewrote the application in UCSD Pascal for the first IBM PCs, while still based alongside the Operational Research department at Exeter University. They installed the first system in Dr Stan Shepherd's practice at Whiteladies Health Centre, Bristol, in December 1981. I went on to become the medical director of Update Computing. Dr Marcus decided that he could get his software developed cheaper by external companies, so the Exeter 'factory' closed in March 1986. The system continued to develop and become one of the top sellers of the 1990s. It was eventually bought out by Torex Health.

IAN HERBERT

It was Update that developed the concept of opportunistic screening described later in this chapter.

MicroDoc

Microdoc was another early success. Developed by Gary Everson, in Sheffield, he developed a comprehensive clinical and practice management system that was in wide use by many practices.

Its main claim to fame was that it was the computer system in use by the serial killer Dr Harold Shipman. One of the major pieces of evidence against Shipman was the evidence from the Microdoc audit trail that he had been deliberately falsifying the patients' records to try to cover his tracks.

Microdoc was eventually taken over by Torex Health.

Exeter Systems

Exeter Systems originated in 1975 having developed from the system in use by the Exeter Community Health Project. It was, therefore, a comprehensive system that gained much acclaim from its users. Twelve original practices formed the Exeter Systems User Group Ltd (ESUG) and, since 1986, have been responsible for direction and development of Exeter Systems. The system has continued to be developed and is still in active use.

The Exeter GP system, still written in MUMPS (later renamed M) was included as an option in the 'Micros for GPs' scheme. Unfortunately a decision was made, not by the Exeter team, to implement it on a Comart Micro. This machine was never powerful enough. However, the users liked the applications so much that they campaigned to have the system installed on hardware that was much more appropriate. At the same time the DoH was trying to divest itself of any responsibility for application development. The users took the option of setting up a company with their own support and sales staff. A licence agreement with the DoH was set up so that the ownership of all existing application software was vested in the DoH, however, all new applications became the property of the ESUG.

The ESUG then set up an agreement with a company that had MUMPS-based finance systems for GPs.

The ESUG always suffered from the fact that all its directors were GPs first and company directors second and it became obvious that an alliance with a larger company would be an advantage. Accordingly, in 2002, a merger was arranged with Protechnic of Cambridge.

Recent developments in NHS computing have led many to conclude that size is important and what was Protechnic Exeter has now been subsumed by Ascribe plc, an international organization. Throughout these changes the system has continued to develop.

M-TEC

M-TEC was a company run by Gerry Perry and Doug Mounter (both ex-Royal Air Force). They originally developed interpreters for BBC operating systems which allowed an early GP system, Hungerford Medical Systems, to run on IBM machines. Through this M-Tec took over the development of Hungerford systems, which continued to use BBC BASIC very successfully for many years. They were taken over by Torex Health in the 1990s.

Roneborough Computing

Roneborough was one of the many projects that came and went with hardly a trace except for the knowledge gained by those involved. It was started when a small system supplier, run by Ian Elliot, persuaded his GP, Glyn Hayes, to help him develop a system for GPs. It was based on a DEC PDP-11/23, a minicomputer that was housed in the practice. There were dumb terminals in all the consulting rooms, reception and the nurses' room. It was a full consulting room system with appointment systems, registration facilities, full clinical records and prescribing. One of its claims was that it was the first system in the world to print acute prescriptions in the consulting room. (The existing systems either did not cater for acute scripts or printed them outside the consulting room on a system printer.)

Many of the features now regarded as normal were there:

- The clinical record used the RCGP Codes and free text.
- Users could enter as much free text as they needed.
- The doctor could see the state of their progress through a surgery from the appointments module.
- Users could select the next patient from the appointments module.
- The actual codes were hidden from the user who chose from a list of rubrics.

The system developed well but the climate introduced by the 'Micros for GPs' scheme meant that there was no longer a commercial market for such large-scale systems and the business was wound up.

EMIS

EMIS began 20 years ago in a rural dispensing practice in Egton, near Whitby in North Yorkshire, where Dr Peter Sowerby and myself wrote the software.

Initially, the system was written for two GPs looking after a few farmers on the North York Moors. However, it soon became apparent that some other doctors found it useful too, and Peter Sowerby set up the company, which started trading in 1988.

It became strangely popular in deprived inner city boroughs such as London's East End, and over time the number of practices choosing the system grew until it became the most commonly used system in the UK.

More recently EMIS has redeveloped the system for other primary care professions, and it is designed to manage primary care records for populations in towns and medium size cities.

EMIS never bought other IT companies or systems because it never had the money and thought the clinicians wouldn't take kindly to being told what to use.

DAVID STABLES

Growth: 1992–present

Rapid growth, investment from the pharmaceutical industry and fund holding created an environment where a number of GP system suppliers were able to flourish. However, as the market started to reach saturation in the 1990s it became increasing difficult for smaller suppliers to survive and a process of consolidation began in the market with the number of suppliers beginning to fall for the first time.

There are difficulties in getting an estimate of the number of systems and suppliers at this peak. The 1993 DoH survey lists 103 systems and suggests that in addition an excess of 200 practices were using systems they had written themselves,[32] giving a total of more than 300 different software packages in use. However, nearly half of the suppliers listed in the DoH survey were only mentioned by one respondent. The survey achieved a 60 per cent response rate and only 39 suppliers got more than five mentions. It seems likely that many of the suppliers listed in the 1993 survey represent GP hobby activity rather than organizations with any serious marketing intent and it is the author's view that at the peak the number of organizations with serious plans to distribute GP systems was about 50. This estimate is supported by the list of GP suppliers maintained by the DoH and published by the General Medical Services Committee (GMSC) in February 1992.[33]

The launch of Microsoft® Windows® version 3 in 1990 and its rapid adoption as the de facto standard desktop GUI highlighted the fact that all of the leading GP systems were running on ten-year-old technology and that, while performing well, they were looking jaded and out of date. While much was done to make the legacy systems integrate more easily with Windows® and its standard desktop applications (dumb terminals were replaced by Windows® workstations and a range of Windows® frontends from simple terminal emulators to full-blown Windows® clients), it was clear that substantial new investment would be required to develop a new generation of GP systems. Such investment was beyond the reach of many smaller suppliers and this combined with the growing requirements of the Requirements for Acreditation (RFA – see later in this chapter) beginning to bite by the mid-1990s to increase pressure towards consolidation.

The change of Government in 1997 and the end of the GP fundholding scheme removed an important source of revenue for GP suppliers and led to an acceleration of the consolidation process. By July 2002 there were nine suppliers with 14 RFA99 accredited systems and only three suppliers having significant market share.

EWAN DAVIS

The growth in GP systems is shown in Figure 14.1.

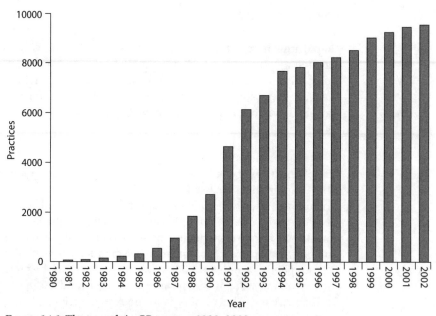

FIGURE 14.1 *The growth in GP systems 1980–2002*

THE SCOTTISH EXPERIENCE

The view from GPASS

The General Practice Administration System for Scotland (GPASS) was accepted by the Scottish Office in 1984. There had been a divergence of view, between England and Wales on the one hand and Scotland on the other, as to how best to introduce computer systems to GPs. The English view was informed by the market-oriented philosophy of the Government of the day. Scotland authorized a standard system available without cost to Scottish practitioners. It remains to be seen which path will be the more successful.[34]

The origins of GPASS date back to 1984, when Dr David Ferguson, a Glasgow general practitioner, developed a repeat prescribing program for his own use. The system was offered to and taken up by an increasing number of general practices with a growing requirement for support. Dr Ferguson then made an offer of GPASS to all UK Health Departments. In Scotland this offer was accepted, to follow on from the Scottish 'Computers for GPs' scheme, and a temporary project, GPASS, was established.

The initial project proved successful and, in 1986, it was established in the West Coast Computer Consortium based in Argyll and Clyde Heath Board. By 1989, 200 practices were using a single-user version of GPASS. At this time the Scottish Office undertook its first review of GPASS, and a management 'buy out' was proposed. After consideration this was rejected and instead GPASS became the direct responsibility of the Scottish Office Health Department.

By 1994 the system was in use by 800 practices in its multi-user form. Despite a programme of ongoing developments, there was a view, particularly from GP stakeholders, that GPASS was under-funded and lagged behind other systems. At that time, key omissions were: lack of a GUI; lack of support for fundholding; and lack of an effective appointments system. An independent SWOT review of GPASS was commissioned by the Scottish Office. It found that GPASS was a cost-effective asset that should be retained and built on. Two major actions were undertaken: firstly, to instigate a change programme to refocus the organization, and secondly, to migrate GPASS to a new technical platform.

The development and implementation of the 'New GPASS' product raised greater awareness of its functions and capabilities within the NHS in Scotland. Developments started to extend GPASS use to practice nurses, and investigation into the potential use within the wider community (health visitors, community nurses, home helps, etc.) is also ongoing. Other primary care clinicians (dentists, pharmacists and chiropodists) saw the product and were attracted by its features.

The secondary care sector (hospital clinicians, A&E, outpatient clinics) also expressed keen interest, and a study of the potential use of 'New GPASS' in these areas was commissioned by the Management Executive. It was hoped that all of these initiatives underlined the potential of 'New GPASS' to facilitate end-to-end 'joined up working' when dealing with patient records and patient care.

GPASS was unique in several respects. The GPASS organization maintains very close links with their active and vocal GPASS User Group via a network of local area coordinators. This was to try to ensure that GPASS practices had a significant role in ensuring the development of the system matches the requirements of general practice in Scotland.

Their help desk and second line support services provided locally based on-site support and through their team of Field Support Officers who had a wealth of general practice experience. GPASS software and support were provided free of charge to practices in Scotland, offering a cost-effective solution to both practices and Primary Care Trusts. Their training team provided a portfolio of formal training courses to meet the training needs of GPASS customers. Courses were delivered by professional trainers at Hillington, at training venues within each of the Health Boards areas, or, if preferred, on site.

Their large market share meant that GPASS provided a common data-set across 80 per cent of patients in Scotland enabling the effective collection and analysis of national morbidity and prescribing data.

GPASS promoted national standards such as Read Codes and the Primary Care Information Unit (PCIU) Drug Dictionary and helped the collection and analysis of national data by the Electronic Questionnaire. New features provided decision support and eased the use of national guidelines to encourage best practice and provide support for clinical governance.

The GPASS software ran on industry-standard systems and the hardware, operating systems and communications standards supported by GPASS were all based on open systems standards. It was hoped that this would result in long-term stability, choice for the practice, as well as clear and flexible upgrade paths.

Data extraction from GPASS

The GPASS Data Evaluation Project (GDEP) was initiated in 1988.[35] Its purpose was to extract information from GPASS practices and to provide national and individual practice feedback on clinical management of patients and to aid planning for population care. Initially, data were collected from GPASS practices via an electronic questionnaire. Later in 1994, additional software was developed to extract registration data, encounter data, referral data and drug, and Read Code data expressed as a series of Date, Age, Sex and Postcode (DASP) for each drug or Read Code found. DASP data were used to collect anonymous morbidity and drug data at patient and postcode level for a subset of participating practices across Scotland. DASP data are anonymous and patients can only be re-identified via their GPASS system identifier in the computer of the originating practice. The original GDEP project has now evolved to deliver data for the Scottish Programme for Improving Clinical Effectiveness in Primary Care (SPICE-PC) under the auspices of the Primary Care Clinical Informatics Unit (PCCIU). Data-sets similar to DASP are extracted for SPICE-PC, but the postcode has been restricted to postcode sector to reduce the possibility of inadvertent patient identification.

Has GPASS worked?

For any review of GPASS it is important to appreciate the history and landscape in which GPASS is located. It is essential to understand, at least to date, that:

- migration and rollout have been a significant drain on resources and development;

- given that demand for development has always outstripped capacity to deliver, and given the conflict between stakeholders' priorities, the needs of users have been placed alongside those of national initiatives. This is in stark contrast to commercial systems, where the reverse applies;

- where clear focus and objectives have been set, they have been successfully delivered. For example, the only area of enhanced functionality targeted for 'New GPASS' was the appointments system.

There is both the requirement and the opportunity to provide the appropriate framework for GPASS to operate in for the future, thereby ensuring that the system firstly meets the needs of patients for primary care, clinical staff and then other stakeholders.

Perhaps the surprising thing, considering its support from the Health Executive and its free nature, is that GPASS is only used by 80 per cent of Scottish GPs. The other 20 per cent purchase commercial systems with their own money. Recently the Scottish Executive Health Department has changed its rules to allow all Scottish GPs a free choice of computer systems. As a result a large proportion of Scottish GPs have opted to install other systems. Perhaps central control and administrative supervision by central government do not produce systems of the same quality as those developed according to market forces?

THE DEVELOPMENT OF USES OF COMPUTERS IN GENERAL PRACTICE

Age sex registers

Before computers most GPs had no way of knowing what were the issues and problems facing their patients in total. The more academic practices used card indexes with various manual methods of extracting the information on them. However, the computer made such facilities available to all. The list of patients registered with the practice was entered manually in the early systems but a download of the practice list soon became available from the Family Practitioner Committees. Suddenly a practice knew the precise age, sex, etc. breakdown of their list. It was also easy to add a limited amount of clinical data to these computerized registers to allow for simple screening and audit of their patients.

Prescribing

Prescribing was soon recognized as the most significant advantage of computerizing a practice.[36] The drugs being prescribed could be chosen from a list generated from a standardized database rather than written in longhand. The computer could check that each drug was suitable for the patient by means of cross-checking for interaction with other drugs the patient may be taking, avoiding known allergies and contraindications if the patient had a disease that suggested that drug was inadvisable. For the doctor the production of a printed prescription was a major advantage.

Standards for prescribing by computer were published in 1985.[37]

One of the most time-saving effects of computerized prescribing was the impact on repeat prescribing. About 60 per cent of prescriptions issued in general practice are repeat scripts when the patient does not see the doctor but just requests a repeat of their long-term medication. This was a labour intensive task with both administrative staff and GPs having to spend time handling and checking such items. The computer allowed for automatic checks on the acceptability of issuing a repeat script, including that it was authorized to be issued, that it was not too soon since the last script and that the patient was not due to see the doctor personally.[38]

A further advantage of the computer system is that the patient can be given a copy of the prescription, which can then be used to request the next prescription.

Call and Recall

The advent of computerized age sex registers and some clinical data allowed practices to have much more effective screening programmes. Instead of paper records and relying on patients to remember when they needed care the computer could control the whole process. It could determine the cohort of patients who needed a particular intervention, produce the letters and labels to send the reminders and then keep a check that such actions actually took place.[39] Computers revolutionized the ability of the practice to manage preventative medicine for their patient lists. The most important and effective programs developed initially were cervical smear management and childhood vaccinations.

Recall systems for influenza and pneumonia injections help provide better care for the elderly and in rural locations recall systems for tetanus inoculation are also important. Patients who do not visit their GPs frequently can be gratified by being 'called' out of the blue.

Clinical notes

The early systems, based on microcomputers, had little room for storage of patient records and so concentrated on limited functionality. Indeed the early versions of the Read Codes were designed to store clinical data without taking up much disk space. Thus the early experience of full clinical notes came using the mainframe and mini-based systems. However, as disk space became more available it was apparent that in general practice the more data stored on the computer the more useful the information derived from it became. Gradually the structure and format of such electronic records was developed, so by the mid-1980s systems that could store a whole patient record became more widely available. This had many advantages. The ability to display patient information in many ways, the prompting and alerting based on this more extensive data, and the legibility and availability of patient records, all helped to stimulate GPs to record more and more. As time went by an increasing number of GPs gave up using paper notes.

GETTING RID OF PAPER CLINICAL NOTES

The GPs' *Terms of Service* (i.e. the details of the contract held by GPs with the NHS) stated that GPs would keep records on forms provided by the Secretary of State. This meant that in effect all GP records were supposed to be kept on the 'Lloyd George' record sheets or A4 sheets provided by the NHS. Thus any GP keeping the records on computer, unless they also kept a paper copy, was in breach of this contract. Many GPs ignored this as they felt the benefits of electronic records outweighed any contractual rules. Indeed the DoH was careful not to make a public issue of the problem as they also felt that Electronic Patient Records (EPRs) were of benefit. There were many attempts to change the *Terms of Service* but these always fell foul of political manoeuvring. It was not until 2000 that the regulations were changed and GPs could keep records electronically without the need for a paper copy.[40] The Government removed the legal requirement for paper-based records allowing General Practices to legitimately become paperless,[41] subject to their meeting conditions to be established by their Local Health Authorities. Part of the changes included the publication by the RCGP of the Good Practice Guidelines,[42] which outlined best practice in keeping electronic records.

In the 1990s, the DoH stopped commissioning its regular surveys of computer use in general practice, so changes in levels of computerization are speculative. In 1998, Richards *et al.* showed that a cohort of Scottish GPs were using their computers in about half their consultations, mainly for prescribing.[43] However, by 2004 it could be stated with confidence that almost every practice had a functioning clinical computer system doing, at least, repeat prescribing.

However, the most remarkable change has been in the move to 'paper light' operations. A 'paper light' practice makes all its clinical entries on the computer alone. Practices needed the reassurance that their computer records alone would be acceptable for medico-legal purposes and confidence in the robust and continuing operation of their hardware. Many then chose to take advantage of the full capacity of their software systems.

A 'paper light' practice scans all paper correspondence into the record; it has a link to the local pathology laboratory to take lab results; it has a computerized appointments system; and it regularly monitors its activities and performance.[44] Clearly this requires the computer to be a central part of every consultation, including in the patient's home.

In a research project undertaken in 2002, a team in Nottingham recruited practices entering all their clinical records in the computer record, and compared their records with those still using paper records or a mixed system.[45] They had no problem identifying 'paper light' practices for the study and in fact found that their problem was that many paper practices were converting to 'paper light' as the study occurred.

It is conjectured that about half of all English general practices are now 'paper light'.

Computer use in the medical consultation

The use of computers during the consultation was described by Bolden in 1981.[46] Hayes described the impact a computer terminal can have on the delicate interrelationships of the clinician–patient consultation.[47] He went on to examine the best ways to minimize potential damage to the consultation by adopting new skills and by appropriate positioning of the terminal.[48] One of the most important aspects of this impact on the consultation is the on-screen sharing of the medical record by both the doctor and patient. Scott and Purves examined this relationship of the patient, clinician and computer, and coined the term 'triadic' relationship.[49]

There have been several studies looking at the effects of the computer on the consultation, from a variety of angles.[50] They concluded that the GP spent more time looking at the computer screen than at paper notes, thus increasing the length of the consultation. Other research found that the doctor spent less time directly interacting with the patient as a result of looking at the screen.

It was in the 1980s that the rising tide of computerization was spawning research into its effects within the consultation. Initial enthusiasm for computer use was tempered by concerns. A RCGP Working Party said in 1980:

> We have one important reservation about this development. We do not know whether direct input to the computer during the consultation will have an effect on doctor–patient communications.

REFERENCE 51

A detailed study by Robinson and Booth described the techniques that could be used to develop good consulting techniques with a computer present.[50]

Word-processing

Hospital referral letters, letters to patients and the daily business of general practice had meant a lot of typewriting had always taken place. The advent of word processors took away this need but it was the linking of word-processing with the clinical record that had the major impact. Letters to patients could be mail-merged from the computer system. Also, in the more advanced systems, referral letters could extract salient parts of the clinical record to add to the information being sent to hospitals.

Appointments

The use of loose-leaf books full of appointments was a common sight in general practice. The first computerized appointment systems came in the late 1970s. The advantages soon established included:

- Appointments could be booked wherever there was a computer terminal (whereas a book could only be used by one person at a time).
- The appointment system was easier to amend and could be configured to suit different doctor's preferences.
- The system could note when the patient arrived, when they were seen and when they left. This allowed the practice to audit their appointments better and also allowed individual doctors to see in real time how they were progressing through their list of patients.
- The clinical screens could be used to select a patient record from the appointment list without the doctor having to key-in the patient's details.

Initially the systems printed out lists from the appointment systems, which the doctors could use in their consulting rooms. As consulting room use of computers became universal the lists were displayed on screen.

Opportunistic screening

Although the concept of screening of defined cohorts by call and recall was well understood and accepted the computer brought another method to general practice. It was Stan Shepherd from Nailsea who thought up the concept of 'opportunistic screening' using his Update system. He persuaded the system developers to add new functionality that, when producing the appointment lists, would append to each patient appointment any outstanding action according to criteria defined by the practice. Thus if a child was behind with its vaccinations, or a patient over 55 had not had a recent blood pressure test, that omission would be noted on the patient list for the doctor to consider during the consultation even though the patient had attended for some unrelated problem.

This idea soon spread and vastly improved the ability of the practice to provide good preventative care. Once computerized appointments were handled on screen rather than as printed lists, the EPR itself developed the ability to prompt for outstanding actions or issues.

Medical audit

Sheldon described five steps to medical audit in 1982.[52] The implications for computer support took some time to be defined but it was realized early on that to be effective it must be a by-product of routine data capture.[53] It was also realized that it needed to be cyclical and feedback to the user was vital – the audit cycle was clearly described in many publications.

One of the early pioneers, Bob Johnson, a GP from Oldham, started using computers for audit in 1967. Originally using punched cards he transferred to an early micro. One of his main points was that computers are good at handling numbers but very poor at handling free text. He devised his own coding system, which allowed him to collect signs and symptoms gleaned from his patients many years before electronic medical records, as we know them today, came into being. He included drug therapy in the array of data that he collected.[54] He analysed his data and reached some conclusions, far-reaching in their impact. He quoted one elaborate manually based audit system that demonstrated that the commonest clinical condition was the common cold. He was able to show from real data that in fact the commonest condition was pain, of which back pain was by far the most frequent.[55] His data collection scheme was never taken up more widely and like so much early work has disappeared.

Patient interviewing

Geoffrey Dove played a key role in the early development of GP computing. In the mid-1970s he met up with Dr Chris Evans of the National Physical Laboratory (NPL), Teddington. Chris Evans was a pioneer of human–computer interaction, and this led to Geoffrey becoming the first GP, anywhere in the world, to use a computer to question patients. He was astonished that it had a genuinely therapeutic effect.[56] This work paralleled that of Warner Slack in the USA.

Chris Evans also wrote *The Mighty Micro* and *The Making of the Micro*,[57, 58] and developed the original MICKIE (Medical Interviewing Computer), which had its own specialized text language and was probably the first ever medical application of a microcomputer.[59] Chris died in late 1979 and shortly afterwards Geoffrey called together a meeting of various people who had been involved with Chris to discuss how to take forward the work without him.

Discussions were started with the people from the NPL to take on the commercial development of MICKIE, which led directly to the establishment of Abies by Tim Benson in March 1980 (see earlier in this chapter).

Messaging

In 1987, as part of the Oxford GP links project, about 30 practices connected to the John Radcliffe Hospital for the purpose of receiving pathology reports using the ASTM 1238 standard message. The first experiments successfully demonstrated the concept with links between the John Radcliffe Hospital in

Oxford and the Wolfson Laboratory in Birmingham and local GP practices in the two cities. By the end of 2002 most GP practices were able to receive pathology results messages conforming to an agreed NHS standard. Reaching this point had taken nearly 15 years. The story as to why it took so long is an interesting one from which much has been and still can be learnt.

The technology was not the problem. While communications technology was much less capable than that available today, networks in the late 1980s using 300 baud modems and proprietary store and forward systems were capable of meeting the underlying requirement. AAH Meditel was already using such technology for data collection, and EDI links between community pharmacies and their wholesalers were already the norm.

Standard formats available at the time lacked the flexibility of today's XML, but fixed format and character delimited standards, allowing text files to be used to carry structured data between systems, were well established.

The problems were in three areas:

- Agreeing the data items that should be included in messages and the granularity at which items should be recorded. There was a long and amazingly acrimonious debate between those who preferred the flexibility of breaking down terms into indivisible atoms and those who preferred the relative simplicity of compound terms (both groups had a point and both found compromise difficult).

- Agreeing how data items should be represented (i.e. coding and terminology). There was a little less acrimony here but a dearth of suitable coding schemes.

- Integrating electronic messages into existing workflows and processes (both manual and already computerized).

In 1993, GP/FHSA (Family Health Service Authority) links were implemented. This was the commencement of linkage of general practices to their local authorities for the purpose of registration/deregistration of patients using UN/EDIFACT messages. This enabled practices to electronically inform the FHSA of patients' movements or changes, and vice versa. The result was the first true population index for the NHS, kept up to date by the practices, which needed to do so to ensure they were being paid for the patients they looked after.

This was later to expand to claims for items of service. This enabled a practice, when they had performed a procedure which warranted payment, to send the information, directly from the EPR, to the FHSA who would then pay.

The standards needed for both the above were developed by collaboration between the DoH, the profession and the suppliers. They were then enforced by the Requirements for Accreditation (RFA – see Standards for GP Systems).

In 1995, the UK National Standard Clinical EDIFACT Messages were published. These covered Referral/Discharge and Pathology/Radiology

Request and Reporting. First implementations of these occurred later that year but there were problems about terminology and granularity of the data.

In 1997, the General Practice Provider Links (GPPL) project was instigated by the NHS Executive as part of its response to *Patients, Not Paper*, the first of two Cabinet Office efficiency scrutinies. It used the Trial NHS Standard messages, which had been agreed by the clinical professions and which enabled laboratories to communicate results of investigations to GP surgeries. A 'bounded list' of coded terms, which identify pathology investigations, was selected from the Read Codes and piloted in the Avon region. This resolved the terminology and granularity disagreements.[60]

STANDARDS FOR GP SYSTEMS

Analysis of GP systems

In 1989, the DoH decided it needed to establish the system requirements of a GP system. It set up a project, the GP Minimum System Specification Project, to evaluate this. Although it was never acknowledged that the purpose was to link system specification with reimbursement this was felt to be behind the aims of the project. A huge amount of work was done, much of it lost, but it moved far beyond a minimum specification. It was driven by management consultants, who developed a full-blown, but incomplete, GP system data model. When the budget holders realized the project had gone out of control it was abandoned. The data model was published for members of the PHCSG as the *Analysis of Activity in General Practice*. It did not provide what was needed to determine specifications that could be linked with reimbursement so the Requirements for Accreditation process was commissioned instead.

The Requirements for Accreditation

The Requirements for Accreditation (RFA) was a set of standards published by the DoH, which any GP system had to meet if it were to satisfy the requirements agreed by the profession and the DoH. The NHS (England) Primary Care Computer Systems RFA was put in place in 1990. The initial specification was written by David Markwell and Glyn Hayes. After consultation it was published by the DoH in 1992 (but is now unavailable).

The first systems to undergo the accreditation process were arranged for 1993. It was initially voluntary, but subsequently enforced as only accredited systems qualified for GPs to receive remuneration for the systems. There were several iterations as the requirements developed. It was at version 4(1997) when the RFA development stopped, awaiting national standards for the National Programme for IT in the NHS.

The RFA initially concentrated on core content. It was subsequently extended to include:

- messaging standards for GP to Health Authority and GP to hospital communications;
- data standards;
- prescribing criteria to manage according to prescribing regulations.

The RFA can be considered in several parts:

- The core data-sets for patient, staff and organization identification.
- The clinical functionality – this is in great detail with an emphasis on standards for functionality to ensure patient safety.
- Messaging standards.
- GP–Health Authority Links for patient registration and GP payments.
- GP to laboratory – this describes the message content and format (currently using UN/EDIFACT), the terminology to be used (a bounded list of Read Codes to ensure semantic consistency between GPs and the laboratories) and the functionality at the GP end to provide workflow support and appropriate handling of incoming messages.

Impact

In 1990, there were said to be 90 suppliers of GP systems. In 2003 there were three major suppliers and two or three minor suppliers left in the market. The systems have advanced clinical functionality driven by the RFA.

Funding for GP computer systems

GP practices are independent contractors to the NHS and as such are responsible for procuring the necessary facilities (buildings, staff, equipment and computer systems) that they require to deliver the service they have contracted to provide. Payments to GPs include an average amount to cover some expenses paid to all GPs (indirect reimbursement) and specific payments for other items paid to individual GPs in relation to actual cost incurred (direct reimbursement). Before 1 April 1990 computing costs were not directly reimbursed so practices purchasing computer systems did so out of practice profits, seeing only a small percentage of the cost actually incurred returned to them as part of the indirect expenses element. This arrangement, combined with the fact that 70 per cent of staff costs (and 100 per cent of employers' Earnings Related National Insurance Contributions) were directly reimbursed, substantially undermined any financial case for computerization based on savings in staff time and acted as a significant disincentive to computerization.

However, in 1990, it became increasingly clear to the DoH that future plans for primary care would require widespread computerization and new rules were introduced (backdated to 1 April 1989) providing 50 per cent direct reimbursement for approved computing costs. With the introduction of a new GP contract, also in 1990, it was widely recognized that a computer system would be an essential tool for a practice which wished to maximize income under the new contract (both to help them hit the targets set for incentive payments and to gather the evidence to prove they had done so). These changes provided new incentives for GPs to computerize and allowed the momentum created by the free schemes to continue and suppliers to prosper following the collapse of the free scheme, in part brought about by the way the new reimbursement arrangements took account of payments for data.

The introduction of direct reimbursement also saw the creation of the GP Systems RFA. In theory reimbursement would only be available for systems that met the RFA in practice, however, it was to be over five years before the regulatory link between reimbursement and the RFA was established.

Funding for systems was piecemeal as funds transferred to Health Authorities for reimbursing GPs were often used for other purposes. Under their *Terms of Service* the GPs were reimbursed a proportion of the costs. By 2003 this had risen to 50 per cent of the costs if the Health Authority could find the money.

INFLUENCES ON THE UPTAKE OF SYSTEMS

The 1990 contract

In the last decade of the twentieth century, GP computerization was driven by the need for practices to achieve the target payments in the 1990 *New Contract for General Practice*, to take part in clinical audit,[61] and to comply with clinical governance.[62] Practice staff became skilled in using the practices' computers and general practitioners valued the facilities for clinical care and medico-legal protection.

All major computer suppliers developed refinements, on a concept known as templates, to meet the requirements of this contract. These templates allowed the entry of a diagnostic or procedure Read Code to be linked to data-entry screens that prompted for relevant data items. A review of a person with diabetes would prompt, for example, for blood pressure, weight, smoking, HbA1c and microalbuminuria among other entries. This simple innovation enabled practices to standardize around local protocols based on national guidelines, improving data completeness and quality.

Limited list prescribing

The introduction of restrictions on what medication a UK GP could prescribe, the *1984 Limited List Prescribing*, caused confusion amongst doctors. Many of the products no longer available were standard preparations in use for many years. One of the benefits of prescribing from a database was that the database could have such items flagged as non-prescribable thus saving the doctor time and confusion.

Items of service

The 1967 GP Charter introduced additional payments for specific public health-related services such as contraception, cervical cytology and family planning. This led to a requirement for better recording of population-based health promotional activity for GPs. Computers enabled practices to record this information more easily.

Reimbursement

From 1989 a proportion of General Practice Information Technology investment became directly remunerable. The computer reimbursement scheme currently offers partial recompense for the cost of computerization conditional on the system implemented being RFA-compliant to the decreed level.

Health promotion/Financial incentives

Contractual changes between GPs and the Government led to an element of pay being dependent on GPs reaching set targets for population-based care in the late 1980s and early 1990s. It was difficult to demonstrate reaching these targets without IT support.

Fundholding 1990

General health service management changes introduced a purchaser/ provider split. April 1990 saw the commencement of GP Fundholding. This invention by the Conservative Government allowed fundholding GPs to purchase care on behalf of their own patients. This resulted in the requirement for participating practices to introduce health care accounting packages within their systems. The cost of purchasing these systems was fully remunerable. A change of Government saw the abolition of Fundholding.

DATA STANDARDS, QUALITY AND EXTRACTION

PRIMIS

Lincolnshire Medical Audit Advisory Group (MAAG) was one of the first to run comparative audit with graphical feedback. A team was formed with Sheila Teasdale (then a practice manager), Dr Mike Bainbridge (a GP who was working in Derby on sharing data between practices and the local hospital diabetic clinic using Sophie protocols), Dr John Williams and Cheryl Cowley, to look at comparative data across the 23 practices with which they were working.[63] They devised Clinical Audit days at PHCSG conferences in 1992 and 1993, which started the work of formalizing the kind of process a practice needed to go through in order to optimize chronic disease management. This was published as *Dynamic Data* in 1994.[64]

By this time, the educational side of information management was high on the list of interests, both through the MAAG experiences and the *Dynamic Data* work. A small study was carried out for the then NHS Training Division on how GPs and their teams liked to learn new skills, and it found that they didn't like videos, books, lectures or even CD-Roms, but they really liked experiential learning, preferably in their own practice with their own team and their own tools. Out of this came a contract to devise a primary care informatics curriculum – now known as JIGSAW[65, 66] – which in 1995 brought together a team that included Mike Pringle, Mike Bainbridge, John Williams, Shelia Teasdale, and a new addition, Dr Pete Horsfield, noted for his expertise with primary care clinical systems and Read Codes. (This was about the time when the expression 'information proficiency' came into currency.) This curriculum specification was extensively used in 1998 to create the curriculum outlined in *Learning to Manage Health Information*,[67] designed to incorporate health informatics concepts into medical and nursing under- and postgraduate education. It was an examination of the role of education and training in improving data quality in primary care clinical computer systems.[68]

Along a parallel track, starting in 1992, the tool that became known as MIQUEST was being developed.[69] This was designed to provide comparable data from disparate systems (and there were about 90 types of GP computer systems at the time). David Markwell and Nick Booth were a couple of the prime movers for this work, ably helped by Andrew Perry.

Almost all of the above-mentioned then worked together in 1996 on what has become known as JIGSAW 2: an options appraisal for the electronic transfer of clinical data between GP clinical systems.[70] This work is still being used by the GP-to-GP record transfer working group.

One of the justifications for the development of MIQUEST was the possibility of extracting data from GP systems regularly and routinely; a programme of work was being considered in 1993–94 which eventually resulted in the Collection of Health Data from General Practice (CHDGP).[71] In early 1997, Mike Pringle, Mike Bainbridge, Pete Horsfield and Sheila Teasdale won the contract to carry out a pilot study designed to devise and test educational methodologies for improving data quality and information management, to test and improve MIQUEST software, and to devise and test information feedback mechanisms. The team was expanded to include two trainers and an information analyst. This work ran for a total of three years: the original two-year pilot and a third year while the business case for the national roll-out was developed.

Thus PRIMIS (Primary Care Information Services) was born. The impact of PRIMIS is now widespread, with over 400 information facilitators working in 90 per cent of Primary Care Trusts (PCTs) in England. But the beginnings of the successful partnership between the NHS Information Authority (IA) and the University of Nottingham in early 1997 were relatively small.

In April 2000, the NHS IA contracted the University of Nottingham to deliver PRIMIS nationally, and learning consultants, information analysts and support staff were recruited to join the original CHDGP team. Primary care organizations in England were encouraged to employ information facilitators who, with support and training from PRIMIS, would help GP practices to improve patient care through the effective use of their clinical computer systems.

In November 2000, PRIMIS released the first coronary heart disease query set for its Comparative Analysis Service (CAS). Designed with the aid of the PRIMIS Clinical Advisory Group to support practice and PCT review of National Service Frameworks, the PRIMIS CAS provides practices with detailed information about their achievements, as well as comparisons with practices in their own and other PCTs. By the time of the first CAS, the PRIMIS training team had held over 100 training sessions, and by March 2001 PRIMIS had achieved its targets for the first year, having trained more than 90 facilitators and signed up over 60 primary care organizations as active PRIMIS schemes.

The third annual PRIMIS Conference, in 2002, was attended by almost 400 delegates. This confirmed the growing influence of PRIMIS in the field of primary care informatics. In order to cater for the growing number of interested stakeholders, PRIMIS launched a new website in December 2002 with the help of the NHS IA web design team. The site soon became an increasingly important source of information for PRIMIS facilitators as numbers increased, with 350 having been trained by the end of the third year. This was also the year in which PRIMIS launched *Rush for Practices*, a set of feedback tools to enable practices to view and analyse their own data in support of better patient care.

The PRIMIS service continued to develop with the release of the CAS diabetes query set in June 2003. As well as CAS, the Information Team manned a busy helpdesk for facilitators and analysed data for feedback to practices. In August 2003, the Technical Support staff at PRIMIS launched a new web-based discussion board, designed to help facilitators and others find answers and solutions to their queries and problems. The PRIMIS Learning Consultants developed new training modules on topics of importance to practices, such as going paperless, information governance, patient safety and the quality and outcomes framework of the new General Medical Service contract. Response to these new modules was described as 'overwhelming'.

In September 2004, PRIMIS released CHART, a new upgraded version of *Rush for Practices*. Later in the year, a longstanding ambition to have PRIMIS training accredited was fulfilled when the first PRIMIS facilitator gained the Professional Certificate in IM&T(Health). At the end of 2004, it was announced that the PRIMIS service would move to the National Programme for IT in the NHS (NPfIT) when the NHS IA closed in March 2005. PRIMIS were now helping practices use their high-quality data not only for the new GMS contract, but also for other national audits, such as the NCASP Diabetes audit, the national flu campaign, and the smoking and obesity reports to support PCT Local Delivery Plans.

PRIMIS has played a key role over the last five years in helping practices to improve data quality and make sound decisions about patient care. The statistics are impressive: since April 2000, over 5,000 training sessions have been held; almost 600 facilitators have been trained; approximately 8,000 helpdesk calls have been answered; more than 30 million patient records have been analysed from over 4,500 practices; and the PRIMIS website had 10,000 visits during December 2004.

The role of PRIMIS is expected to continue to develop in line with new programmes from NPfIT and an increasing level of work to support national audits. However, the emphasis will remain on individual facilitation and support for members of primary care teams that has been the hallmark of PRIMIS's success.

SHEILA TEASDALE AND WENDY SUTTON-PRYCE

HEALTH INFORMATICS PROGRAMME FOR CORONARY HEART DISEASE

This pilot initiative used an approach that focused on improving the whole business of primary care, its processes and its people. The Health Informatics Programme for Coronary Heart Disease (HIP for CHD),[72] addressed the two faces of clinical governance but had a prime focus on the development of learning organizations. The project has developed a methodology and an associated set of tools that it has tested and evaluated at a small number of pilot sites. The work of HIP for CHD was focused on coronary heart disease but the methodology is equally applicable to other clinical areas. In particular, HIP for CHD provided an approach that allows the diverse strands of all of the National Service Frameworks to be handled in a 'joined-up' way in primary care.

RCGP WEEKLY RETURNS SERVICE

The Royal College of General Practitioners Weekly Returns Service (RCGP WRS) is a network of sentinel general practices providing weekly data on illnesses diagnosed in general practice across England and Wales.[73] It was established in 1963 using paper records. As more GPs became computerized standard extraction routines were implemented for the systems of the contributing GPs.

The WRS contributes to the surveillance of infectious disease, most notably influenza.

The network comprises 92 practices, with a registered population of approximately 680,000, well distributed throughout England and Wales. These practices monitor the morbidity presented at every consultation, distinguishing between new episodes of illness and ongoing consultations.

DOCTORS' INDEPENDENT NETWORK

The Doctors' Independent Network (DIN) is a registered charity, which collects morbidity data from 100 representative practices selected from 250 Torex Meditel System 5 donor practices nationwide.

Complete individual patient data-sets are collected (four-byte Read Codes without free text additions). Additionally, practice characteristics are kept on the DIN membership database. A total of 1.7 million patient years of continuous patient data is kept. This database ('Deep Thought') can be accessed via a dial-up telephone connection. The data are extracted daily from the GP systems and transmitted to DIN overnight. The extraction is invisible to the practice.

Patient data-sets are identified by:

- the donor practice's own internal patient number;
- patient's year of birth; and
- patient's sex.

Historically, patient data-sets have been sociologically classified and roughly located using a stripped postal code. However, more recently an ACORN CACI social status descriptor code is allocated before the data are extracted from the practice system. This enables adequate de-identification and greatly improves the potential for understanding the patient's social status and its relationship with their morbidity.

DIN states that they collect the data with the express intention of using it for the public good. The DIN network is funded by their providing huge amounts of real-life and (almost) real-time data for epidemiological and clinical research. Their stated aim is to be able to undertake automated clinical audit via the internet. Historically, DIN has been partly funded by St George's Department of General Practice via the South Thames NHS Exec R&D budget. However, to date, DIN has failed to attract the attention of GP researchers in any significant way. This is ascribed to changes in post, illness and other such factors.

GENERAL PRACTICE RESEARCH DATABASE

The General Practice Research Database (GPRD) was originally set up by the VAMP Software Company in May 1987. The GPRD is currently owned by the DoH and is administered by the UK Medicines Control Agency (MCA). The DoH collects doctor-diagnosed illness, prescription and outpatient referral information from participating GP practices. These practices follow agreed guidelines for the recording of clinical data. These data are routinely checked for accuracy and validity. There are currently over 400 practices across the UK submitting data to the GPRD. Consequently the GPRD has over four million subjects and over 25 million person-years of data.

USER INVOLVEMENT

One of the most successful features of GP computing has been the heavy involvement of the users, GPs, practice staff, health authority staff and nurses in its design, development and strategy. The national arena has been dominated by the Primary Health Care Specialist Group of the BCS (see Chapter 1).

Supplier-specific user groups have been fundamental to the development of individual systems, and electronic user groups, particularly GP-UK, have allowed much wider participation than would otherwise be the case.

Supplier user groups

From the early beginnings of commercial systems the suppliers accepted the need to involve GPs in the development of their systems. At first a few interested GPs worked directly with the suppliers. As numbers of users grew they formed themselves into dedicated user groups. Most of the user groups have been independent of the supplier but the degree to which the suppliers get involved has varied.

Such user groups aim to represent the users to the supplier, putting pressure on system development, service standards and other issues. There have been cases where the supplier ignored the user group, usually to the supplier's detriment. Equally there have been occasions where the user groups have successfully defended the supplier against officials from the DoH who did not like the commercial basis on which suppliers had to operate.

These user groups have also been pivotal in educating about computing, running conferences and workshops to help their members learn to use their systems better, and promoting understanding of wider computing and health service issues.

The other role of user groups, only beginning to have an effect since 2000, is in national negotiations and discussion about NHS IT.

Electronic user groups

GP-UK, the first internet listserver for UK general practice, was inaugurated in October 1994 by members of the PHCSG. It continues to thrive.

Membership of the list is growing steadily and it now has well over 200 subscribers. As well as strong local support it has attracted members from the USA, Canada, Australia, New Zealand, South Africa, Hong Kong, the Netherlands, Malaysia, Iceland, Norway and Ireland. It has attracted a high calibre of membership.

The GPUK.net website (www.gpuk.net) includes details of how to make best use of the mailbase server and includes hyperlinks to various PHCSG pages.

Discussion topics on the list have included NHS-wide networking, coding and security issues, evidence-based medicine and the future of electronic publishing. GP-UK is an academic list and the quality of articles posted has been consistently high. A number of clinical issues have also been raised.

There has been some cross fertilisation with its American counterpart, Fam-Med (http://fpen.org/fam-med), and the list is now accessible via hyperlinks from several other websites.

It first published news of its activities in *GP Magazine* in 1996. Following a letter to the *British Medical Journal*,[74] a further news article appeared in *Pulse* magazine. It continues to maintain links with the popular GP press to try and encourage as broad a membership as possible.

It is anticipated that, in the not too distant future, GP-UK will need to divide into a number of daughter lists reflecting the major interests expressed in the discussion threads to date. Members will then be able to choose which discussions they wish to participate in or subscribe to the 'super list' and read the whole lot.

The *British Medical Journal* itself (in part) is now published on the web, raising the possibility that discussion lists, such as GP-UK, may become associated with it. The list owners can be contacted by email at gp-uk-request@ mailbase.ac.uk.

IAN TRIMBLE

THE ROYAL COLLEGE OF GENERAL PRACTITIONERS AND THE BRITISH MEDICAL ASSOCIATION

The Royal College of General Practitioners (RCGP), the learned society of GPs, and the British Medical Association (BMA), the trade union side, have not had a huge impact on the development of GP computing. However, they have made a few fundamental steps in moving the whole process forwards.

1980 The RCGP report

A RCGP working party published *Computers in Primary Care* in 1980.[51] As a review of the function and potential of clinical computers in general practice, this report was the first articulation of the link between information systems and quality of care:

The development of general practice computer systems and the parallel development of clinical standards to which the whole profession is already committed are closely interrelated.

REFERENCE 51

317

ICI Computer Fellow

In 1983, the RCGP established a position of Computer Fellow, sponsored by ICI. This part-time post was filled for a period of two years by Norman Stoddart, a Commodore PET user GP from Nottingham. He did much to educate GPs on the merits of using computer systems in their practice.

Adoption of the Read Codes

There was little agreement over which coding systems should be in use until the Joint Computing Committee of the BMA and RCGP produced a report analysing the options. It was this report that caused the DoH to purchase the Read Codes.

Data protection

The Data Protection Act 1984 caused confusion and anxiety amongst GPs. They needed advice on how to register under the Act and they worried about the Subject Access provisions. The BMA produce an excellent set of advice that made managing compliance with the Act much easier.[75]

The Joint Computer Group also examined the issue of confidentiality and security of the increasing use of general practice data for non-clinical purposes. They produced a paper with basic principles that still apply today.[76]

Computer reimbursement

For many years GPs wanted to have direct reimbursement of computer systems. At first this was roundly rejected by the DoH because GPs as independent contractors had to buy their own systems. Initially full reimbursement was also rejected by many GPs who were afraid that this might cause the Inland Revenue to redefine their status as 'employed'. Eventually the need for data from systems became stronger and the BMA negotiated the first steps towards financial support for computer purchase.

MEDICAL TERMINOLOGY

It was recognized by John Perry in the Oxford Linkage Project that it was necessary to limit the way data were recorded if it could be used for later analysis. The record needed to be coded from a list, which had standardized meanings that the computer could recognize. This resulted in the OXMIS codes, which were designed specifically for data collection. The RCGP also wanted to code information collected manually and so developed the RCGP codes. Although developed for manual data collection they later developed into a version for computer use. In the early 1980s, the users of the Abies system decided they also needed to be able to code data items for later use. At first they thought they would only need about 20 codes for this purpose. They set up a small group, which was led by a Loughborough GP, James Read, and soon discovered that they needed many more codes. The codes they devised were not only for the purpose of later analysis, that is reporting, but also so that the system itself could use standardized data to provide warnings and other functions to aid practice management. The Read Codes were the end result.

An international activity under the World GP body (WONCA) wanted more than a simple coding system. They wanted a full classification system that could be linked to the standard International Classification of Disease codes (ICD). Thus arose the International Classification of Primary Care (ICPC).

Why do we need a thesaurus of codes?

Competent computer programs can search for anything that has been typed into it, but typing mistakes, different words for the same thing and other human foibles mean that a system is required to minimize these errors. Such a system is the coding of terms. Searching a database (of patients' illnesses or medication) using a particular code is much quicker than looking in the written notes. Ideally, codes should be added to freely typed (free-text) entries but many clinical computer systems require the selection of a code that has an accompanying string of text known as the 'rubric' before they will allow any note to be made. Usually the rubric will be the phrase you would use yourself, in which case it doesn't take much longer.

OXMIS

Developed under the Oxford Community Health project, the OXMIS codes were originally devised for use by GPs and were based on ICD-8 codes and OPCS operation codes. They were used by the early versions of VAMP systems but fell out of use when Read Codes became the NHS standard.

RCGP

The RCGP defined its first set of codes for use on paper in the early 1960s. It was for GPs to add the relevant code to their manual age sex register for research and audit. Based on ICD-8, the codes were a true classification and did not easily meet the needs of clinicians wishing to record their consultations. In the early 1980s, it became clear that the future lay in the development of a coding system for use by computers. This required a more comprehensive, understandable yet still consistent set. This more extensive set was still under development when the DoH decided to standardize on the Read Codes. The RCGP codes, although still used in academic practices for some time, gradually ceased to be used.

Read Codes

Read Codes are a coded thesaurus of clinical terms, which enable clinicians to make effective use of computer systems. The codes ease the access of information within patient records to enable reporting, auditing, research, automation of repetitive tasks, electronic communication and decision support.

The Read Clinical Classification was developed by Dr James Read, in 1986, for use in General Practice. Initially designed for the Abies system it was thus in use in the Meditel System 5.[77]

In 1988 the Joint Computer Policy Group of the BMA and the RCGP carried out a review of all the coding systems in current use. The result of this review was a recommendation to the DoH that the Read Codes be adopted as the national standard coding system for GP systems in England and Wales.[78] The DoH purchased a licence for the codes in the same year for £1.25 million. A commercial company, Computer Assisted Medicine, was set up by James Read to develop the codes and handle licences for use other than in English and Welsh GP systems. Subsequently the NHS Centre for Coding and Classification was established by the DoH with James Read as its director.

The initial version of the Read Codes (about 30,000 terms) used four-character (byte) codes and a hierarchy that was based on ICD for disease codes and the British National Formulary structures for drug codes. However, it did not follow these structures exactly.

The problem with using just four characters was that the limit this imposed on the number of concepts that could be contained was too small for all of the medical terms needed. Also the overall structure of the terms meant that the rubrics (text descriptions) of the terms were limited. As a result, in 1991, Read Code version 2 was developed using five characters (bytes) and had the ability to contain much longer rubrics. Its structure of the hierarchy was much cleaner. This version of the codes (about 100,000 terms) was adopted by some GP systems and some hospital systems. The new five-character version mapped more precisely to Read 1, ICD-9, OPCS4, CPT-4 and BNF (British National Formulary).

This new set of codes was still too limited for all of medicine. Also the simple hierarchical structure could not cope with some of the complexities of medical data. For example a code for the condition, pneumococcal pneumonia needed to be in two places, in the chapter on lung disease and in the chapter on bacterial diseases. In 1994, it was therefore decided to develop a set of codes based on the theory of an 'acyclic directed graph' or a multiaxial structure, which would allow codes to be seen as children of more than one parent. It used core terms and qualifiers that allowed for significant granularity but also combinations that were meaningful to clinicians. The Clinical Terms Project, was set up in which hundreds of workers from all the fields of medicine described the terms they needed to be able to record. The result was Read version 3 or the Read Clinical Terms, at a cost of £3.7 million. This was a huge effort that produced a very comprehensive set of terms and codes in a very complex structure. It was used by a limited number of both GP and hospital systems but did not gain general usage.

In 1999 it was decided to amalgamate Read Codes with the American classification system SNOMED® RT, and this process has yet to be completed. Almost all UK General Practice computerized systems use Read Codes for diagnosis and symptoms recording. However, there is less commonality where pharmaceutical coding is concerned.[79]

International Classification of Primary Care

The first edition of the International Classification of Primary Care (ICPC), published in 1987, was accompanied by an electronic text file.[80] It also included the mapping or conversion structures from ICPC to the ICHPPC-2 coding system, the RCGP codes and ICD-9. The limitations of the mapping were stated explicitly: only the conversion between ICPC and ICHPPC-2 could be relied on as complete. The relationship with ICD-9, and consequently with the RCGP codes and ICD-9-CM, was incomplete and only worked from ICPC towards ICD-9 and its related systems. Many gaps existed in the other direction, for example for episode-oriented epidemiological purposes, morbidity studies and patient documentation.[81]

The ICPC–ICD relationship underwent a substantial change when, in 1993, Oxford University Press published the *International Classification of Primary Care in the European Community*,[82] with a reliable and complete electronic conversion in both directions between ICD-10 and ICPC (the ICD chapters on external causes were not included). Through a careful use of the signs, the editors made sure that the mapping contained no errors other than those based on arbitrary interpretations by the individuals responsible for comparing the two systems.

The most important advantages of this mapping were that it provided access to ICD-10 as a nomenclature for ICPC, and allowed both the multi-language layer of ICPC and the national translations of the book to relate to the translations of ICD-10.

ICPC is the ordering principle of the domain of international family practice, providing logically structured classes for this domain's common symptoms, complaints, diagnoses/health problems and interventions. It lacks, however, the specificity needed for documentation at the level of an individual patient in EPRs. ICD-10, as a nomenclature with approximately 12,000 labels that cover medicine at large, obviously provides far more detail. Family physicians who use this mapping can be certain that the diagnostic label chosen from ICD-10 is included in the clinical content of the related ICPC class. In fact, ICD-10 thus serves as a terminology for ICPC because of the careful mapping of the clinical content rather than only the coding structure of both systems.

ICPC contains many 'rag-bag' rubrics, in most of which many ICD-10 classes are included: these contain specific disorders as well as ICD-10 'rag bags'. Obviously, in an electronic patient record, a patient should not be labelled with a 'rag-bag' diagnosis, but with a specific diagnosis attached to an ICPC code to structure the database, and an ICD-10 label for additional specification.

Of course, even ICD-10 labels often will not suffice for the documentation of the full medical history of an individual patient, and free text will be necessary.

Code conversions

In 1995, an early attempt was made to convert Read Code version 1 (four-byte set) patient databases to Read Code version 2 (five-byte set) patient databases. It soon became apparent that a straightforward conversion using a mapping table had significant problems.[83] There was not a one-to-one match between codes in the two sets. Even worse there were children codes in the four-byte set that if converted as children of their parents produced errors in the database. Thus XXX became YYY. One of the reasons for these problems was an assumption that the code itself was sufficient to pass on the meaning as recorded by the doctor. As the designers of systems and codes had not always kept the code hierarchies and the associated meaning together, a code in one system may have had a different meaning when viewed in another code set. The only answer, and one that is now recognized as essential, is to transfer both the code and the rubric that described its meaning when doing any code conversions. It stresses the need to always store the actual rubric added by the user as well as the code. Relational mechanisms tend to make designers assume that the relationship will always remain the same and that it is only necessary to store the code.

ACADEMIC PROJECTS

PEN&PAD

There have been relatively few health informatics projects that have researched user interaction and user interfaces of clinical information systems. A notable exception was the successful PEN&PAD projects of the late 1980s and mid-1990s.[84] These projects were initially conceived of as a set of experiments in usability, focused on practitioners entering notes (i.e. PEN) and accessing data (i.e. PAD) by means of a prototype clinical workstation.[85] The close relationship between data entry and data display led to the projects being marketed and regarded by external observers as a single entity; in reality PEN&PAD is the overall name given to a set of projects that had very distinct aims and objectives, but which were conducted and led by the same research group.

The first funded project, in 1988, concentrated on consulting room systems in Primary Care and the role of the GPs in directly entering data. Historically its funding from the DoH began in response to the awareness that both the 'Micros for GPs' scheme and the pharmaceutical investment in data collection systems for GPs were faltering. At that time, the perceived bottleneck to practitioners engaging fully with computers was related to the issues of 'data entry'. British Computer Society support, Research Council funding, and later European project and DTI funding enabled the data-entry focus of PEN&PAD to be expanded to incorporate presentation, and later to change emphasis altogether to models of the medical record and clinical terminology.

PEN&PAD researchers had long held the belief that direct-structured data entry by clinicians would have to become the norm within clinical practice; to them this approach seemed to be the only feasible way of capturing clinical information. It still does. At that time, however, major obstacles to success were the current limitations of the technology and design knowledge. Consequently the PEN&PAD researchers chose to conduct laboratory-based usability experiments using the latest technology available (i.e. object-oriented programming for rapid prototyping, windows-oriented desktops, and large-screen workstations with mouse and keyboard input), predicting that this would be commonplace within five to ten years. The techniques and technology used were criticized by some observers as being too 'blue-sky'. This was a legitimate but ultimately unrealistic reaction, reflecting more on the immediacy of the problem than on the nature of the proposed solutions. Pragmatically, the choice of techniques and technology was governed a) by the nature of the experiments to be conducted, and b) by the capabilities of the research setting; these required and satisfied, for example, large-screen demonstrators and rapid prototyping.

User-centred design and formative evaluation

User-centred system design was the philosophy used within PEN&PAD. In addition, an original method of formative evaluation was developed within the projects that involved iterative cycles of requirement and design workshops.

The funds for the GP project supported an inner circle of three doctors, an external evaluation team, and the logistics of reaching larger audiences (i.e. an outer circle) to review the emerging research results. The requirement and design workshops permitted the inner and outer circles to be critical of the latest interface presented to them and they provided refined and additional requirements for the next round of development. Typical tasks involved clinicians interacting with a mouse (a novelty then) to enter clinically sensible descriptions, with minimal keyboard usage, into the EPR. Tasks were scripted as realistic scenarios that might commonly arise in clinical practice.

Relatively early, the research team became aware that the structure of the clinical record and the nature of clinical terminology (topics that are very closely associated with 'look and feel') affect how users performed tasks associated with data entry and display. This was an important finding, and partly explained why the emphasis of PEN&PAD was to migrate towards record structures and clinical terminology.

The research study was in many ways seminal in that it brought together the intellectual outlook of the programmers with the intuitive skills of the diagnosticians in the team. In doing so it highlighted, probably for the first time, the rather obvious fact that patients complain of symptoms (e.g. 'a pain in the side of the chest') while doctors were taught and continue to think in terms of clinical systems and pursue their questioning to elucidate the truth before deciding on the correct label to apply to the consultation. One outcome from this was the development of the GUI whereby a simple mouse click on the appropriate area of an outline manikin provided a list of possible causes of the symptom described which could lead the questioning if desired. As a corollary the key questions that need to be asked but could easily be overlooked in a busy surgery were particularly highlighted.

From an ergonomic standpoint the lack of typing skills amongst doctors was confirmed and the use of a mouse and menus (structured to the complaint presented) helped greatly to reduce the barrier that computer systems present between patient and advisor. This was very clearly demonstrated when the system, even though accompanied by the investigating team, descended on an unscripted real-life surgery and was awarded what in theatrical circles would be described as 'a rave review'.

To gild the technical elegance of the system the programming team produced grammatically correct sentences from the usual disjointed observations that are recorded such that: 'Mr X had had a burning pain in the left side of his chest for some six months which is aggravated by meals and by bending down'. They also showed that, by a simple mouse click, the output could easily be translated into other European languages.

There was no doubt that PEN&PAD was a great step forward in the development of medical computing but unfortunately the Government of the day looked on this breakthrough in the same way as other Governments before them had looked on radar, the jet engine and the hovercraft, and promptly withdrew their funding. There being no free market in health care provision in the UK and none of the incentive that private business provides, the profession could not see PEN&PAD as being cost-effective and so it was allowed to die. Perhaps its principles may be revived at some point in the future?

ROGER ROYCROFT

PRODIGY

PRODIGY was an initiative by the English DoH to encourage GPs towards better and more effective prescribing.[86] It was set up as a collaborative venture between the DoH, the University of Newcastle, the profession and the system suppliers. It was this broad collaboration that is worth examining. The first version of PRODIGY released was a straightforward decision-support system driven by guidelines from the university. It was rolled out to all of the suppliers' sites and subsequently included in the RFA.

The DoH (and Treasury) needed to control drug costs, and the profession's desire for safer prescribing meant that they set up a project with all the stakeholders intimately involved. The university designed the specification and the medical knowledge to drive the modules, the profession agreed they were appropriate and the suppliers then were supported in development. The project ultimately failed due to the difficulties of improvising guidelines for the wide range of conditions intended and the desire to extend the functionality into untried technology. However, the first version of PRODIGY still stands as a way for the health service to work with all the stakeholders. The suppliers received financial support in the development as long as it was in the research phase. Once it got into the RFA this support ceased. During the research phase the suppliers built several prototypes and they were also fundamental in getting feedback for the user community.

Exeter smart card project

The Exeter Care Card Project was a DoH sponsored pilot trial to investigate the potential for a patient-retained medical smart card within the NHS. The card contained the following:

- Registration details
- Ethnic group
- Religion
- Language spoken
- Health Care Professional Identifiers
- Medication/Prescriber details
- Allergies/Sensitivities
- Symptoms
- Examination findings
- Diagnostic findings
- Laboratory data
- Employment details
- Administration details
- Emergency data.

It involved 9,000 patients in Exmouth and the equipment to read and write to the card was installed at the group practices, Exmouth Hospital, a dental practice and every pharmacist in Exmouth.[87, 88]

Analysis of the project showed no drop in consultation rates but the number of new visits and telephone consultations dropped significantly. Hospital referrals were not affected but there was a drop in the number of investigations performed. The number and cost of prescriptions fell at a time when the average cost was rising. It was felt that the technology had proved itself in a testing environment.[89, 90]

It was felt at the time that this was a significant breakthrough but no attempt was made by the DoH to take this technology forward.[91] It was rumoured that the actual cost of a widespread implementation of smart cards was not affordable.

OTHER PRIMARY CARE SYSTEMS

Computing in the pharmacy

In order to understand the reasons for the development of applications to support the practice of pharmacy one should be aware of the scope of activities undertaken by the average community pharmacy. In addition to providing the NHS dispensing service which itself makes significant demands of information from the contractor, the pharmacy provides a retail environment for the sale of medicines, chemists' sundries and increasingly cosmetics and photographic items, all of which require systems for their procurement, stock control and efficient sale.

It was the efficient procurement of medicines for dispensing which promoted the use of computers for online stock ordering in the late 1970s. Pharmacies typically receive two or three deliveries each day of dispensing items in order to quickly fulfil the needs of patients. It is not possible to keep the enormous range of products available from stock at all times and so any reduction in the lead time for deliveries had significant impact on the efficiency of the service. Wholesalers employed teams of telephone operators who telephoned pharmacies every few hours to collect orders. As well as being inefficient for the wholesaler, this took the pharmacy staff away from their dispensing activity and demanded that they prepare an order while the operator waited on the end of the telephone. One of the first computers used widely for this task was the Grundy NewBrain introduced in 1980.

Orders were placed into the computer as they arose and the wholesaler was able to retrieve the orders from the computer at their leisure, once connected by modem link. One of the changes necessary to implement such a system was the need for coded stock identifiers, since the full text of each item would have been unmanageable. These codes were first developed for the verbal telephone ordering systems and were imprinted on the dispensary staff's brain: such codes as ALD22 for 'Aldomet Tablets 250 mg' are still remembered by many. However, each supplier had different code sets. Within a few years the need for a universal identifier was required since most pharmacies receive their stock from a number of different sources. The seven-digit PIP code was introduced by the *Chemist & Druggist* publication to fulfil this requirement and is still the most widely used identifier in community pharmacy.[92]

Few had employed the computer for anything more demanding than ordering until the Royal Pharmaceutical Society introduced a policy of requiring machine printed labels for all dispensed medicines in the early 1980s.[93] A number of systems were developed, some little more than typewriters. In fact, a number of pharmacies used special typewriters for several years. Quite why this happened is somewhat of a mystery since these typewriters with their special small font and special character spacing were very expensive. Perhaps it was because of the lack of confidence in the emerging computer systems? However, the benefit of being able to store label and patient details for an increasing number of patients who required repeating medication was soon realized. This required some mechanism of storage and retrieval of patient medication records. A large number of suppliers started to develop and sell systems to meet this need. At one point in the early 1980s there were over 50 suppliers in this marketplace.

The early systems were based around the early computers of the day such as the BBC Micro, the Tandy TRS80, the Apple® II and the Commodore PET.

The necessary printing was accomplished using adapted dot matrix printers. Adaptation was necessary to accommodate the rolls of self-adhesive labels. Latterly the use of inkjet and, rarely, laser printers has overtaken the use of dot matrix devices; however, these have brought their own issues. One returning issue is the lifespan of readability of inkjet printed labels. Fading and difficulty of reading has been the subject of professional guidance and is still causing some disquiet. As more printers are developed for the smaller specialist markets such as pharmacy labelling, the situation is much improved with the latest ink sublimation type seeming not to suffer from the fading problems of the inkjet variety.

There was little standardization of the stored data-sets, or indeed the method of operation of the different systems, and each was differentiated in the marketplace by an increasing number of 'useful' facilities. Most offered some commercial benefits in being able to speed up the operation of dispensing or make easier stock control. A few concentrated on the benefit to patients such as the availability of personally tailored patient information leaflets. One system which provided patient leaflets in a customer-focused, easy to read format received criticism from industry licensees, since the information contained in the leaflets did not necessarily totally agree with that provided by the manufacturer. The difficulty of providing useful and accurate patient information still remains a challenge even though all patient packs are now provided with a patient leaflet containing information which forms part of the licence of the medicine.

Another system offered to increase the NHS income of those community pharmacists who used it by efficiently managing the information provided for the reimbursement of dispensed medicines to the Prescription Pricing Authority (PPA). Following a challenge from other system suppliers, the Advertising Standards Association ruled that the advertised claims made for the system were unfounded. That said, the PPA instigated a special section to deal only with prescriptions submitted by users of the system.

Initially, there was no recognition of the benefits of computerization of the pharmacy by the DoH other than some support for the machine printing of labels. Eventually, however, there was a small yearly payment made as part of the professional allowance paid to a pharmacy contractor for maintaining patient mediation records for 100 'at risk' patients. No specific criteria were given for such patients and why the arbitrary number of 100 was chosen remains a mystery.

With the opportunity to manage the dispensing of nursing and residential homes came the requirement to keep accessible records of the patients in order that the dispensing operation could be as efficient as possible. The use of reminder aids (Monitored Dosage Systems or MDS) to patients such as calendar blisters, or other calendar-based reminder packaging, required more stringent record keeping and more advanced printing routines to help the production of the special packaging. More and more features were incorporated into these systems as each supplier tried to maintain their market share.

By the end of the 1990s the number of active developers of pharmacy systems had reduced to less than ten. This had mainly come about by a process of amalgamation and takeover, with customers being 'sold' on to the larger suppliers. This migration took its toll on several of the larger players whose insistence on an upgrade to their preferred ongoing system resulted in numbers of customers being lost to other players.[94]

As the market was now mature and few opportunities existed for new business, several of the system developers focused on related business developments and started to redevelop and enhance their electronic point of sale (EPOS) systems in order to secure their position in the market. By the middle of the decade, most of the system suppliers had either developed their own, or were partnering with, an EPOS system. Few opportunities were exploited at the time, to link the two systems, however. One developer did link the pharmacy patient medication records to the point of sale system enabling prompts such as when Paracetamol was requested over the counter by a patient already taking a similar medicine by way of prescription. While this was welcomed by the profession, few other links have been made by developers between the 'clinical' and 'commercial' activities of the community pharmacy systems.[95]

Most of the developers had received pressure to develop graphically based systems also incorporating the ability to undertake several concurrent tasks. Almost universally the systems adopted Microsoft® Windows®.[96] Use of touchscreens was tried to overcome the difficulty of using mouse-like input devices in the hostile (dusty and maybe wet) environment of a dispensary, but these brought their own problems and were initially expensive. It is interesting to note that in the USA, where similar pressures were evident, UNIX® was the preferred underlying operating system.[97]

Apart from the pressure for a GUI, several other professional system requirements were emerging, such as: the ability to take part in piloting electronic transmission of prescriptions;[98] workflow management; more sophisticated financial management (for NHS claims); ability to share patient records between pharmacies (of the same owner). As before, most of these initiatives were approached in a piecemeal fashion – each developer taking their view of the initiative and developing essentially parochial solutions for their user base. There was little or no standardization of approach – either from a system specification or technical perspective. The focus on delivering standardized data-sets and functional specifications in pharmacy systems late in the 1990s was slow to gain support, and work to deliver international standards in these areas of system development has only recently started.[99]

Alongside the development of the systems themselves, others were developing decision-support tools, information resources and audit tools to be used in conjunction with the pharmacy systems. Use of these resources prompted some standardization of approach by the developers, but there still remained much customization of the resources and add-ins to suit the various systems. The high cost of this customization has pushed developers to adopt standard interfaces in order to make use of more standard (and cheaper) offerings. One of the challenges that remains is how to interact effectively with the increasingly important audit mechanisms being employed by the NHS. As new services are delivered by pharmacies they will be required to provide audit information in a form which is useful to, and can be easily assimilated by, the service commissioner.[100] As the development of the data-sets has, to date, been largely piecemeal and led by market forces, it is difficult to forecast how the necessary quality and completeness of this information can be assured. Many of the developers of supporting drug information have taken the portable route: developing resources for use on Personal Digital Assistants (PDAs) or small 'pocket PCs'. The full integration of these standard reference resources with the clinical processes has been slow to develop.

Quite a different approach has been taken within the acute sector.[101] Here, the developments of pharmacy systems have continued to concentrate on stock control, acquisition and financial control. As hospital-wide information systems continue to progress toward integration of information centred around the patient, those parts of the pharmacy system holding patient related information have begun to be integrated with the hospital-wide record of care. The need for facilities such as patient specific labelling is much reduced in hospitals and the continuation of the ward-based medicines stock systems results in specific labelling only being required for outpatient and named patient supplies, where the medicine is not normally available for administration on the ward. System developers in the acute sector have been much more active in adopting new technology in order to integrate the functions of the clinical pharmacy service with the wider hospital patient record, by use of wireless and portable computing devices. Here also the number of developers has rationalized over the past 20 years with only a handful of suppliers covering the whole NHS. The acute sector has been swift to experiment with, and embrace, many new technologies. The focus on hospital pharmacy systems on the accurate and timely supply of medicines to patients has prompted the widening use of robotics for helping the warehousing and selection processes. Significant patient safety benefits together with staff productivity gains are claimed by the use of closed-loop systems, incorporating both medicine and patient identity coding and by the use of automated robotic picking systems.[102]

With the advent of required standardization across the whole health care technology domain to promote safe sharing of records, it will be interesting to see how each of the systems retains its perceived market differentiators while complying with the need to take part in wider dialogue with other national organizations, which will be using many standard components within their systems.

IAN SHEPHERD

Prescription Pricing Authority

The Prescription Pricing Authority (PPA) was established as a Special Health Authority within the NHS under the NHS Act of 1977.

The PPA's main functions in 2005 were:

- checking, pricing and payments to contractors for the supply of drugs and appliances prescribed under the NHS;
- production of the Drug Tariff (this contains the reimbursement and remuneration rules for prescribable items);
- production of information for GPs, PCGs/PCTs, HAs, the DoH and other NHS stakeholders about prescribing volumes, trends and costs;
- to administer the NHS Low Income Scheme, the purchase of prepayment certificates, and the issue of medical and maternity exemption certificates;
- to deter, detect and seek redress for prescription charge evasion by patients: to promote compliance by contractors with their Terms of Service;
- to provide European Health Insurance Cards to UK residences.

The processes which were totally manual remained undisturbed until 1978.

The birth of prescribing information on a national scale was born out of a change in the school leaving age. The PPA in the 1970s used to recruit mainly young females straight from school. The Government decided to raise the school leaving age and that for a period dried up the PPA workforce market. The PPA got behind with its processing and Mr Tricker from the Oxford School of Studies was asked to review the organization and come up with some recommendations. Out of which was born the team which would computerize the PPA and provide the first national electronic prescribing analysis system of its kind in the world. The PPA had been manually producing prescribing information which involved sorting through 10,000 contractor accounts, containing about 1000 forms per month, looking for a particular doctor's prescriptions and tabulating the results. It used to take about nine months to provide the details for a limited number of prescribers.

The Tricker report (1976) concentrated on the premise that if every doctor received prescribing information then they would voluntarily manage their own prescribing. The report also suggested that if they did not do it voluntarily then having the right level of prescribing information would enable the appropriate management controls to be put in place.

A team of IT experts from a range of disciplines, led by senior civil servant Eddie Arthurs, was brought together to determine the feasibility of computerizing the process of capturing information from prescription forms to pay dispensing contractors. The ability to store and analyse vast amounts of data was never in dispute. The team were also asked to investigate whether they could demonstrate that providing prescribing information would lead to better prescribing and reduce inappropriate prescribing. The complication, apart from the quality of the doctors' handwriting, was how to capture the information from prescription forms. Today this is less of an issue since most prescriptions are now computer-produced (although it would help if GPs replaced their printer ribbons/ink cartridges more often).

Capturing information from a prescription form is complex. Additionally, staff have to input what a dispenser should be paid, according to the drug tariff rules, based on what is both prescribed and dispensed. The feasibility study to computerize the PPA started in 1978 and was complete by 1980. The computerization of the data capture and payments to contractors then followed on and was completed in 1984. This involved developing the systems, retraining 1400 staff, and installing 36 data-capture systems supporting 1300 terminals: all without any disruption to the service. During that period the mantle passed to myself to lead the IT development and computerization programme. The programme was completed within budget and with a lower running cost than the manual system.

In 1978, the PPA was processing 240 million prescriptions. By 2005, the PPA processed over 700 million prescriptions with even less people. Those IT systems have not stood still and, by a combination of technology and new business processes, individual member of staff output has risen from 1200 prescriptions per day to over 4800 prescriptions per day. A quiet IT success story within the NHS.

In 1986, the first national computerized prescribing reports (PD1, PD2 and PD8) were born: replicating the existing manual service, but faster and covering all GPs. While in presentational terms it was not a radical step, it proved that the functions of the PPA had been successfully computerized and set the stage for the development of PACT (Prescribing Analysis and CosT).

The DoH created a user group with representatives from the General Medical Services of the BMA, DoH, PPA, RCGP, The Pharmaceutical Society of Great Britain and the Society of Family Practitioner Service (FPS) Administrators, to help develop the business requirements for PACT: the basis for all prescribing type information provided by the PPA. There were a number of key players, led by civil servant Geoffrey Podger. Jane Richards on behalf of GMSC, as both a PPA Authority member and a representative of the GMSC, gave tremendous support in developing the prescribing information systems.

PACT was born in 1986 using the British National Formulary as the basis for analysing all prescribing. There were three types of prescribing reports available:

- Level one: A compact four-page document providing an overview of a GP's prescribing with comparative information to enable the prescriber to compare themselves against their colleagues.

- Level two: These reports were sent automatically to any practice that exceeded predefined thresholds.

- Level three: A full catalogue of prescribing that introduced the concept of 'prescriber units' (PUs) to counteract the impact of different patient list profiles (initially the impact of the over 65s) that could be considered when comparing prescribing by GPs. These reports were sent to every GP every quarter and were available via the Family Health Service Authority (FHSA) and nationally. In the first year of issue PACT was credited with reducing the drugs bill by £80 million. The ability to provide prescribing, dispensing and financial information electronically was the springboard for a totally different approach to managing and improving prescribing and the treatment of patients.

The PACT system then provided the basis of the Indicative Prescribing Scheme. In 1991, a new system was developed to provide monthly reports to GPs, FHSAs, RHAs and the DoH to monitor prescribing costs, either against Indicative Prescribing Amounts or Fund Holders' budgets. To achieve the timetable for the provision of the new service required re-engineering all the major PPA business processes, new methods of working and new IT systems, reducing the time it took to provide the necessary information from six weeks to four weeks. The PPA had to achieve all these changes within a year.

At the same time a system called Prescription Costs Analysis was implemented that provided information on all drugs dispensed in Primary Care on behalf of the NHS for the whole of England, Scotland and Wales. It provided information to the DoH on the volume and cost of drugs enabling:

- an analysis of drug prices;
- national forecasting;
- monitoring new products;
- provision of data for monitoring adverse drug reactions for the Committee on Safety of Medicines;
- an analysis of claims made according to patient exemption categories;
- flexibility for ad hoc analysis;
- the potential for research;
- the identification of possible fraud indications or irregularities.

The system took 70 man-years of development over a two-year period. It now covers only drugs dispensed by Primary Care on behalf of the NHS in England for the past three years (1.8 billion items).

Once the information on prescribing and dispensing became available the philosophy with regard to prescribing changed, from self-policing to organizational policing, to enable consistency and reduce waste. With this change of emphasis it became obvious that there was too much data and not enough information. The move towards prescribing advisors brought a new set of business requirements, which needed addressing. The next stage of development was about access to meaningful information but in a flexible manner.

The philosophy of the NHS towards prescribing had changed. Having created a huge prescribing and drug database with two billion items (currently three billion items) which provided regular reports to prescribers and their respective management organizations, it became apparent that the PPA needed to develop tools which made it easier to manipulate data. To resolve this problem the PPA developed its first online system called PACTLINE. This was based on an application developed by Dr Archer for his local FHSA. The application went through a number of iterations to meet changing requirements. It was a thick client type system, in that each user who wished to access the PPA database had to have the application loaded on their PC with all the associated set-up problems that came with different systems. The next generation was a thin client system called ePACT, which in turn has now spawned a GP version called Electronic Practice Financial Information on Prescribing (ePFIP).

Moves towards greater financial management of prescribing have led to a range of financial and budget type reports being produced monthly to enable GPs, practices, PCTs, Strategic Health Authorities and the DoH to manage the drugs budget element of the overall Primary Care spend.

The PPA has been heavily involved in Electronic Transmission of Prescriptions, and in setting standards and message definitions. While the various projects have come close to implementation the business case has never been fully justified.

The PPA is responsible for the development of the NHS Dictionary of Medicines and Devices which is setting the standard for a single drug descriptor throughout the whole of the NHS. It has been adopted as the national standard for *Connecting for Health* and if properly implemented will save lives. Patients should no longer die because of misunderstandings caused by ambiguous drug descriptions.

DOUGLAS BALL

FAMILY PRACTITIONER SERVICES COMPUTING

The first computerized Family Practitioner Committee (FPC) system was commissioned by the DoH from the Dental Estimates Board. The brief was technically limited, that is it was to be a batch system and the existing FPC manual system with two card indexes was to be replicated. The main index (the nominal index) had a card for each patient filed by patient name. The data were repeated in a separate index where patient cards were filed in groups according to the GP with which they were registered. The system was implemented in West Sussex FPC where it was shown not to be cost-effective.

The DoH then decided to commission a real-time system from Trent RHA. This system was implemented in about half a dozen FPCs. Those implemented in the South Western RHA got their support from Exeter. When the Exeter FPC Unit was established, it took over responsibility for developing this system for general NHS use in England and Wales. Funding came from the DoH. At the time there were about 10 FPCs who had taken or were preparing to take the Trent system. It ran on DEC PDP-11s and used

Digital Standard MUMPS-11. Exeter considerably rewrote the Trent system and the basic system contained:

- general user information;
- patient information;
- lists and indexes;
- tables and references;
- analysis and print routines;
- passwords.

Driven by unique keyboard keys it displayed screens full of data rather than just individual data items.

This basic system became accepted as the standard very quickly and within two years the number of FPCs using it had risen to 50.

The vision behind the registration system was far-reaching. It was recognized even then that confidentiality issues had to be considered with the development of what would become a national community index. It was proposed that links to GP systems would update both systems appropriately. This became the GP–Reg Links process, which has worked well ever since. It was even proposed that for those GPs who did not have their own computer system they would be able to remotely access the FPC system to view their patient lists. Downloads from the FPC system were written to seed the initial database of a GP practice system.

CERVICAL CYTOLOGY CALL AND RECALL

Trent RHA incorporated into their FPC system a very inflexible recall-only system for cervical cancer screening. When responsibility for FPC systems was transferred to Exeter, responsibility for the recall system was transferred as well. DHAs and GPs complained about the inflexibility of the system. In addition they were adamant that a call module was also needed. The DoH, for reasons that were never explained to the Exeter team, insisted that the system should only deal with recall. This seemed to be so contrary to the users requirements that the manager, at Exeter Family Practitioner Services Computer Unit (FPSCU), obtained funding from the Imperial Cancer Research Fund (now Cancer Research UK) to develop the call module.

This was a flexible system allowing women to be called or recalled according to age bands and type of result (see Chapter 15). Although the call and recall was run on the FPC computers, the cervical screening process itself was the responsibility of the District Health Authorities and mostly carried out by GPs. The advantage of running the system on the FPC systems was that a woman's cervical cytology history was available no matter where her test had been done and also if she moved practice. Very shortly after this the call module became Government policy.

Contractor payments

Payments for pharmacies, based on printed schedules from the PPA significantly reduced paper work. So also did the dentist payments based on data from the Dental Estimates Board.

In time, systems evolved to cope with the GP payments. This required the linkage with GP systems to collect Item of Service activity. This, plus the practice lists on the FPC computer, automated the process very effectively.

NHS CENTRAL REGISTRY

The NHS Central Registry (NHSCR) began life as the National Register,[103] which was used to issue identity cards and ration books and to help with the call-up for the Armed Forces during the Second World War. When the NHS was set up in 1948, the National Registration numbers were used to ensure that each patient's record had a unique identification. After rationing ended, the register was retained as the Central Register for the NHS for England and Wales. This ensured that FHSAs maintained up-to-date lists of patients resident in their areas. Subsequently the NHSCR, at Southport, provided a comprehensive system to help with NHS patient administration in England and Wales. Hence, the NHSCR's main purpose is to help the Health Authorities (formerly FHSAs) in the transfer of primary care medical records within the general practitioner network. As part of this administrative function, it is routinely updated with details of patient's births, deaths, name changes, and movements between health authorities and related events.

These data are exploited in various ways:

1948–1990

The NHSCR was formed in 1948 to record all residents of England and Wales in registration books containing a single line entry for each individual. These records were continually updated with all births up to 1990. Each book corresponded to a particular birth registration district and time period and was labelled with an alphanumeric code, which until 1996 formed the first part of an individual's National Health Service number. From 1971 to 1990 cancer registrations notified to NHSCR were recorded on the relevant registry entry. Death registrations were similarly linked to the relevant entry.

Before 1990 records were kept in transcript books structured by National Health Service number (allocated at birth registration) with a unique entry for each person. Each transcript book holds registrations for one registration area for a specific time period.

1991– 2005

The NHSCR database now contains over 60 million records from all health authority databases.

Since 1 January 1991, a new computerized NHS Central Register was compiled by aggregating the computer records of all FHSAs. It includes a record for all individuals in England and Wales alive on 1 January 1991, together with immigrants who have registered with a NHS GP, and all births since 1991. As far as possible all duplicate registrations were removed, and mistakes corrected.

The new NHS computerized register, the Central Health Register Inquiry System (CHRIS), records the births and deaths registered, and contains flags relating to both cancer registration and those which indicate membership of any existing medical research study.[104] Entries on the NHSCR include NHS number, name, sex, date of birth and current Health Authority (and date of acceptance) of the patient.

Updating an entry: from Health Authorities

The entries on the NHSCR are updated on receipt of information from HAs. If the updating involves a change of HA (because the patient's new GP falls within a different HA) a 'migration' record is created. In addition to patient moves, migration records are created when patients remove themselves from NHS doctor's lists to enter the armed forces, and then again when they return to civilian NHS doctors. Finally, migration records are created for those emigrating and later returning, and for new immigrants from outside the UK. However, some patient moves do not count as migrations. For example, moves to long-stay psychiatric hospitals and imprisonment. In addition, internal migration estimates do not include the movements of armed forces.

Updating an entry: from the Registrar of Births, Marriages and Deaths

Registration offices of births, deaths and marriages now enter registration details directly into a computer database, and the data are transferred to the NHSCR. For example during the period November 2000 to October 2001, 530,000 death notifications were received in this way. Additionally, the NHSCR receives death notifications from the HAs directly, when the GP deducts their patient from their list giving the reason as 'death'.

What's been delivered?

In 2001, it was estimated that 70 per cent of practices had been computerized for over 10 years. Indeed, at that time, over 95 per cent of practices had an EPR capable of paperless operation. Estimates varied, but some 30 per cent to 50 per cent of practices had given up using paper records. It was known that most clinicians were using computer systems at the point of care, although the degree to which they used them was variable. It was also known that 90 per cent of prescriptions were produce electronically (data from the PPA) and so had been through the automatic checking for contraindications, interactions and sensitivities, which all the current systems contained. By that time all computerized practices connected to NHSnet and so had access to the internet. The features in common use included:

- call and recall programmes;
- repeat prescribing;

- electronic appointment systems;
- opportunistic screening;
- online clinical protocols;
- clinical governance support;
- pathology links;
- GP–HA links;
- ubiquitous clinical coding;
- decision support.

Practices had become completely dependent on current IT for much of their basic administration and indeed the way they provided services to their patients. The value of GP data has been demonstrated via the many data extraction initiatives, although the importance of training users in the recording of quality data had also been demonstrated.

UNFULFILLED PROMISES

Despite the significant successes since GP computing started there are still some areas of importance that had not been achieved and other, once promising areas, which had not turned out to be as useful as had been expected.

GP to GP record transfer

Perhaps the major missing area is the ability of one EPR to be transferred electronically to another practice when a patient moves. Attempts to define this process and enable such transfer have been considered several times.

The TextBase project was a laboratory experiment to assess the feasibility of a common exchange format for sending a transcription of the contents of the EPR between different general practices, when patients move from one practice to another in the NHS in England.[105] The project was managed using a partnership arrangement between the four EPR systems vendors, who agreed to collaborate, and the project team. It lasted one year and consisted of an iterative design process followed by creation of message generation and reading modules, within the collaborating EPR systems, according to a software requirement specification created by the project team. The resulting paper described the creation of a common record display format, the implementation of transfer using a floppy disk in the laboratory, and considered the further barriers before a national implementation might be achieved.

This project did not result in a long-term solution but did outline many of the issues that stood in the way. Some of the problems related to the different ways in which individual suppliers had built their systems. The manner in which the relationship between data items was recorded was often very different and incompatible. Yet without these relationships the meaning (context) of the information was lost and this was not acceptable. Another problem with automating record transfer was that one practice may not be willing to accept the automatic inclusion of data coming from another practice into their own records. Doctors want to assess information before it can be included as they may have different standards and acceptance criteria. The major problem, however, was logistical. For an automated transfer of records to take place there needs to be a management mechanism, so when a patient moves the systems can determine who was the previous GP, where they work and how to access the previous record. This is a huge task in the diverse and complex nature of the NHS. It is hoped all these problems can be solved in the *National Programme for IT* in the NHS (NPfIT).

Electronic transfer of prescriptions

The generation of a prescription by the computer is well developed but results in the printing of a paper script. This script is then taken by the patient to a pharmacy, where the pharmacist enters the same data into a computer. The form is then passed to the PPA where it is again entered into a computer database. This is clumsy, time-consuming, expensive and does not help in the management of the whole process. Several projects have attempted to solve this. The major one being Pharmed.[106] However, none of these has been accepted by the DoH and manual re-entry of data is still the norm. It is hoped this issue will be resolved by NPfIT.

Hospital to GP messaging

Messaging between practices and health authorities has been working well for many years. So well in fact that people tend to forget about its success. Messaging between practices and pathology laboratories is also working well. However, there is little or no messaging between the hospitals and the practice. This is a problem as vast numbers of messages are passed between the two elements of the NHS every day. Patient referrals, discharge notifications, discharge letters and letters between hospital staff and GPs produce vast amounts of paper and result in a slow and unwieldy system. The main hindrance to automating this process has been the lack of electronic facilities at the hospital end. Most hospital records are in paper form and even when letters are generated by word processor they are not suitable for automated transfer. The electronic National Data Spine of NPfIT will help to solve this problem.

Expert systems

Since the early years of computers there has been a dream that expert system technology will significantly improve patient care (see Chapter 24). The early work of de Dombal, showed that it could help in training doctors but had less impact on experienced staff. Much work in the USA on Iliad, QMR and

DxPlain encouraged the belief that machines could make diagnoses that were better than mere human doctors. However, none of these sorts of systems have made significant inroads into actual patient care. Even the PRODIGY protocol driven system has not been used widely, even though all accredited GP systems have it built in. The main reasons seem to be that doctors do not make diagnoses by a logical algorithmic process. They find that being forced to think in a machine-driven way damages their care processes. The other major factor in primary care is that it requires knowledge of every aspect of medicine, psychology and sociology, and this wide range is not easy to incorporate into an expert system. Decision support by means of improving the display of patient data, alerts, prompts and watchdogs is a more effective way to help clinicians provide better patient care.[107]

Another review stated that the place for expert systems themselves should be limited to:

- *laboratory systems: clinical laboratories have proven to be a fertile domain for the use of expert systems;*

- *drug advisory systems: there is a clear opportunity to design expert systems that will assist clinicians with the prescription of medications and selection of the most cost-effective treatments;*

- *signal interpretation: the development of interpretive alarms for real-time clinical signals in areas like the intensive care unit will offer some assistance with the task of clinical vigilance;*

- *quality assurance: there will be a need to check that all different types of knowledge-based systems in clinical settings remain up to date;*

- *education: the need continually to educate both patients and clinical professionals offers significant opportunities for automated assistance. Indeed, education for most clinicians is an ongoing process. Expert systems should be built with this type of support in mind.*

REFERENCE **108**

Patient interviewing

The initial success of patient interviewing and patient education by computer applications has not resulted in widespread use. The main problem is that most general practices do not have the room to allow patients to sit with a computer in private: a necessary prerequisite of such tools. They are also quite staff-resource intensive as patients need someone to guide them through the process.

Voice recognition

When early voice recognition software was developed it was hoped that it would help those doctors who found using a keyboard difficult. It was assumed that a doctor in the consultation would find it easier to talk to their computer than to type with keys. The main problem has been that voice recognition software itself has taken a long time to be accurate enough for real-time use. The other issue is that the doctor talking to a computer during the consultation is quite distracting and disturbing for the patient. Voice recognition has found uses in dictating letters and reports.[101]

Significant under exploitation of existing systems

Despite the advanced functionality contained within UK GP systems it is known that many of these functions are not widely used. Indeed it is common for users to request new functionality that already exists on their systems. The main reason seems to be lack of education and training. GPs and their staff are fully committed to patient care. They do not have the time or the inclination to spend significant amounts of time learning about something that is a tool they have to use. It is also very difficult in a small practice to provide the back-up needed when a member of staff is released for training. As a result they learn the minimum to get by and do not realize what else is available. It is also a problem that there has never been enough funding for training. The suppliers would usually provide the basic training as part of the purchase price of a system. However, they charge for extra training and the money was always in short supply. It is a common problem that a member of staff may learn only the bare essentials. When a new member of staff arrives they are trained by the existing members and thus only learn the basics, often not even the most efficient way to use the system. Therefore poor usage, training and attitude tend to be passed on year on year.

LESSONS LEARNT

Government involvement

The impetus for the development of GP computing, particularly in the period up to 1990, came initially from far-thinking enthusiasts working in general practice and from some academics. The commercial sector recognized the opportunities soon after, with serious commercial players entering the market after 1980, leading to an investment in excess of £40 million (mainly by pharmaceutical interests) by 1990. While the DoH took an interest in the development of GP computing and funded some modest research activity it provide no significant targeted financial support (even though the indirect expense payment to GPs meant that the Government was passively picking up 100 per cent of the cost) until 1990 (the support was backdated to 1April 1989).

The only material Government involvement in this first 10 years was the 'Micros for GPs' scheme run, not by the DoH, but by the DTI whose interest lay with promoting the UK computer industry, particularly hardware manufacturers, and not with the promotion of better health care. While usefully raising the profile of GP computing, the 'Micros for

GPs' scheme was launched with little consultation and two suppliers were chosen without any form of open competition. While the systems supplied through the scheme were certainly not the best available at the time, the Government subsidy and endorsement meant that these systems soaked up available demand in the market, seriously undermining more promising systems – many fatally.

The 'Micros for GPs' scheme was to be the first of a number of Government interventions and not the only one that would have a negative effect on development. However, the scheme did result in the formation of the GP Computer Suppliers Association (GPCSA) and the opening of a positive dialogue between suppliers and the DoH. This grew in to a three-way dialogue between DoH, suppliers and the professions, which has been successful in promoting cooperation between suppliers and has helped ensure better alignment of policy with both professional and commercial interests.

The next major intervention by the DoH was the introduction of the GP systems' reimbursement scheme. While the professional bodies were unhappy that scheme did not provide 100 per cent reimbursement, this first substantial funding from the DoH targeted at GP computing was welcomed and provided a boost to the market. Substantial new money as ever attracted new suppliers, some of whom were more interested in short-term profits than a long-term engagement. As a result, the early 1990s saw a number of cheap, poor-quality systems introduced, particularly where systems were installed by health authorities to meet central targets rather than in response to the needs of GP practices. This problem was exacerbated by the fact that the intended link between reimbursement and system accreditation was not in place, as intended, and did not in fact appear for some years.

As indicated above, the original intention of the DoH was that reimbursement should only be available for accredited systems. However, two issues delayed this process and a requirement that reimbursement should be made only for accredited systems did not emerge until 1997. The first of these issues was the failure of the DoH to deliver the necessary GP minimum system specification (GPMSS). This project was originally intended to provide a minimum specification of functionality that would become mandatory for all GP systems that would attract reimbursement. However, the project scope drifted and it only actually produced a GP systems' data model. While this model was of some limited interest it certainly did not provide a minimum system specification. Rumours at the time indicated that certain senior officials were very unhappy about this, however, the public position was that the outputs delivered were what had always been intended and that the only problem was a misleading project name. To clear up any misunderstanding the project was renamed an 'Analysis of Activity in General Practice' (AAGP) and a new project was started, 'The GP Systems Requirement for Accreditation', which would eventually deliver a GP minimum system specification.

The delays in the availability of a GP minimum system specification, however, meant that many GPs went ahead and bought systems that did not comply with the RFA. This lead to the second problem: the DoH could not mandate the link between RFA and reimbursement without being willing to fund the cost of upgrading or replacing non-compliant systems. It was not until later when extra funding was found that a link was established with the RFA, which by this time was at version 4.

The leading GP suppliers and the GPCSA had supported the RFA but were unhappy with the delays in establishing the link with reimbursement. Supplier support had been secured on the understanding that only those suppliers who invested in meeting the DoH requirement would have access to reimbursement. These suppliers were unhappy at being undercut by suppliers of poor quality systems who had not invested in RFA compliance. This issue seriously undermined the RFA and explains the drop in the number of compliant systems between RFA and RFA2, at a time when the number of systems in the market was actually rising.

A good example of the 'management' of a complex adaptive system

The NHS does not function as a tightly managed system. It is not a 'command and control' organization. It is more of a 'complex adaptive system'. This is particularly true of general practice where the GPs are independent contractors working under a contract, not directly controlled. It is well understood that a complex adaptive system should be managed by carrots not sticks. The carrots that drove the computerization of general practice were the need for GPs to collect data for their target payments and their interest in anything that improves their care of patients.

Engagement of clinicians and other front line staff is critical

The GP systems have been extremely successful and widely adopted because they provide the functionality needed by the end-users: GPs and their staff. This happened because the GPs themselves did the work to explore what could be done and often invented features themselves. Even when commercial suppliers began to dominate the market the designers were often GPs and there was a very strong involvement from the users in what the suppliers developed.

REFERENCES

1   DoH (1998) *Information For Health: An Information Strategy for the Modern NHS 1998–2005*. DoH.

2   Fogarty, L. (1997) Primary care informatics development – one view through the miasma. *Journal of Informatics in Primary Care*, January, 2–11.

3   Pinsent, R. (1968) The evolving age sex register. *Journal of the Royal College of General Practitioners*, 16(127), 34.

4 Burdon, J. (1961) Display by spectrum. *Journal of the Royal College of General Practitioners*, 4, 106.

5 Eimerl, T. (1960) Organized curiosity. *Journal of the Royal College of General Practitioners*, 3, 52.

6 Morgan, R.J. (1965) The use of an age sex register as a practice index. *Journal of the Royal College of General Practitioners*, 9, 64.

7 Watford, P. (1963) The practice index. *Journal of the Royal College of General Practitioners*, 6, 32.

8 Cooper, R.F. (1985) Do most practices have an age sex register? Results of West Midlands age sex register study. *BMJ*, 291 [16 November], 1391–3.

9 Royal College of General Practitioners (1963) Classification of disease. *Journal of the Royal College of General Practitioners*, 6, 24.

10 Royal College of General Practitioners (1984) *Classification of diseases, problems and procedures 1984*. Occasional Paper 26. RCGP, London.

11 Abrams, M. (1968) A computer general practice and health information system. *Journal of the Royal College of General Practitioners*, 16, 27.

12 Abrams, M. (1972) Computer terminals in a health centre. *Health Trends*, 4, 18–20.

13 Grene, J. (1971) Automated recall in general practice. *Journal of the Royal College of General Practitioners*, 21, 352–5.

14 Coulter, A. (1988) The Oxford Community Health Project. *Newsletter of the PHCSG*, 7, 4.

15 Perry, J. (1972) Medical information systems in general practice: a community health project. *Proceedings of the Royal Society of Medicine*, 65, 241–2.

16 Perry, J. (1974) Uses of the Oxford medical information system in the support of primary medical care. *Colloques I.R.I.A.*, 1, 367–8.

17 Gillings, D.B. and Preece J.F. (1971) An analysis of the size and content of medical records used during an online record maintenance and retrieval system in general practice. *Int. J. Biomed. Comp.*, 2, 151–65.

18 Lippmann, E.O. and Preece, J.F. (1971) A pilot online data system for general practitioners. *Computers and Biomed. Res.*, 4, 390–406.

19 Preece, J.F., Gillings, D.B., Lippmann, E.O. and Pearson, N.G. (1970) An online record maintenance and retrieval system in general practice. *Int. J. Biomed. Comp.*, 1(4), 329–37.

20 Preece, J.F. and Lippmann, E.O. (1971) Record design for the computer file in general practice. *Practitioner*, August supplement, 3–12.

21 Preece, J.F. (1972) The computer file in general practice. *Update*, July, 155–66.

22 Clarke, D. (1982) The evolution and features of a MUMPs-based primary care system. *Meth. Inform.*, 7, 2, 127–47.

23 Sparrow, J. (1976) The Exeter Community Health Services Project. *Medical Record*, 17,4, November, 205–8.

24 Procter, P., Jarvis, J. and Head, A. (1982) The computerized nursing record – an effective means of communication. In O'Moore, R.R., Barber, B., Reichertz, P.L. and Roger, F. (eds) *MIE, Dublin 82: Lecture Notes 16*, pp. 309–16. Springer Verlag, Berlin. ISBN 0 387112 08 1.

25 Fisher, H. (1984) A computer system for the family practitioner service. *The British Journal of Healthcare Computing*, 1, 4.10, 13.

26 Bradshaw-Smith, J.H. (1976) A computer record-keeping system for general practice. *BMJ*, 1, 1395.

27 Bradshaw-Smith, J.H. (1982) The role of the computer in general practice. *Practitioner*, 226, 1211–3.

28 Kumple, Z. (1981) *Evaluation of Primary Care Systems: SWRHA Exeter Community Health Services Project*. Exeter Project internal document.

29 Project Evaluation Group (1985) General practice computing: evaluation of the 'Micros for GPs Scheme'. HMSO, London.

30 Read, J. and Benson, T. (1986) Comprehensive coding. *BJHC*, May.

31 Benson, T. (2002) Why GPs use computers and hospital doctors do not. *BMJ*, Nov., 325(7372), 1086–9.

32 DoH (1993) *Computerisation in GP Practices*. 1993 survey. DoH.

33 General Medical Services Committee (GMSC) (1992) *Developing a Practice Computer Record System*. British Medical Association, London.

34 Webref (accessed August 2005) www.gpass.co.uk/

35 GPASS Review Group Primary Care IM&T in Scotland (2002) *Promoting Progress, Securing Success Report of the GPASS Review Group, June 2002*. NHS Scotland, Information and Statistics Division, Edinburgh.

36 Hayes, G.M. (1985) Prescribing systems. In Sheldon, M.G. and Stoddart, N. (eds) *Trends in General Practice Computing*, pp. 18–27. RCGP Books, London.

37 GMSC/JCGP Joint Computer Policy Group (1985) Standards for computer issued prescriptions. *BMJ*, 290, 1252–3 (drafted by Hayes, G.M.)

38 Malcolm, A. (1985) Repeat prescribing. In Sheldon, M.G. and Stoddart, N. (eds) *Trends in General Practice Computing*. RCGP Books, London.

39 Pringle, M. (1985) Preventative Medicine. In Sheldon, M.G. and Stoddart, N. (eds) *Trends in General Practice Computing*. RCGP Books, London.

40 NHS Executive (2000) *Electronic Patient Medical Records in Primary Care. Changes to the GP Terms of Service*. National Health Service (General Medical Services) Amendment (No. 4) Regulations 2000 – Statutory Instrument No. 2383. NHS Executive, Leeds.

41 Shaw, N.T. (2001) *Going Paperless: A Guide to Computerization in Primary Care*. Radcliffe Medical Press, Oxford.

42 Webref (accessed 2005) Good practice guidelines on completion of GP reports from the Joint GP IT Committee. www.bma.org.uk/ap.nsf/Content/GoodPracGPreports0804

43 Richards, H., *et al.* (Date unknown) Computer use by general practitioners in Scotland. *British Journal of General Practice*, 48, 1473–6.

44 Purves, I. (1996) The paperless general practice. *BMJ*, 312, 1112–13.

45 Hippisley-Cox, J., *et al.* (2003) The electronic patient record in primary care: regression or progression? A cross-sectional study. *BMJ*, 326, 1439–43.

46 Bolden, K. (1981) Computers in the consulting room. *Update*, 1672–33.

47 Hayes, G.M. (1993) Use of the computer in the consultation. *Update*, 44, 4, 465–8.

48 Hayes, G.M. (1993) Computers in the consultation: the UK experience. In Safran, C. (ed.) *Proceedings of the 17th Annual Symposium on Computer Applications in Medical Care*, pp. 103–6. McGraw-Hill, New York.

49 Scott, D. and Purves, I. (1996) Triadic relationship between doctor, computer and patient. *Interacting Comput.*, 8, 347–63.

50 Webref (accessed March 2007) www.gp-training.net/training/iicr/iicr.htm

51 The Computer Working Party of the RCGP (1980) *Computers in primary care*. Occasional paper 13. RCGP, London.

52 Sheldon, M.G. (1982) *Medical Audit in General Practice*. RCGP Occasional paper 20.

53 Hayes, G.M. and Robinson, N. (eds) (1991) International Primary Care Computing. *IMIA 1991*, pp. 63–5. Elsevier Science Publishers.

54 Johnson, R.A. (1974) Detailed computer analysis of drug reactions during three years of general practice. *Symposium on Medical Data Processing*, March. Taylor & Francis.

55 Johnson, R.A. (1974) A method of evaluating treatment in General Practice. *J. R. Coll. of Gen. Pract.*, 24, 832–6.

56 Dove, G.A., Clarke, J.H., Constantinidou, M., Royappa, B.A., Evans, C.R., Milne, J., Goss, C., Gordon, M. and de Wardener, H.E. (1977) The therapeutic effect of taking a patient's history by computer. *J. R. Coll. Gen. Pract.*, August, 27, 181, 477–81.

57 Evans, C. (1980) *The Mighty Micro* (new ed.). Coronet Books. ISBN 0 340259 75 2.

58 Evans, C. (1981) *The Making of the Micro: A History of the Computer*. Van Nostrand. ISBN 0 442222 40 8.

59 Somerville, S., Stewart, J.S. and Raine, M.M. (1977) MICKIE–experiences of taking histories from patients using a microprocessor. In Grémy, F., Degoulet, P., Barber, B. and Salamon, R. (eds) *MIE, Toulouse 81: Lecture Notes 11*, pp. 713–22. Springer Verlag, Berlin. ISBN 3 540105 68 9.

60 Webref (accessed 2005) Robinson, D., Holland, M. and Pill, S. Supporting a set of Read-Coded terms for use in pathology messaging. www.primis. nottingham.ac.uk/informatics/feb2000/feb4.htm

61 Pringle, M. and Harriss, C. (1992) *Current Capability of General Practice Computer Software to Perform Medical Audit.* RCGP Books, London.

62 NHS Executive (1998) *A First Class Service: Quality and the New NHS.* Stationery Office, London.

63 Cowley, C., Bainbridge, M.A., Johnson, A.B. and Peacock, I. (1993) The Derbyshire initiative. *Diab. Nutr. Meta.,* 6, 343–4.

64 Teasdale, S., Williams, J., Bainbridge, M. and Cowley, C. (1994) *Dynamic Data.* Primary Health Care Specialist Group of the BCS, Worcester.

65 Teasdale, S.J., Bainbridge, M.A., Horsfield, P., Simpson, L., Teasdale, J.K. and Williams, J.G. (1997) JIGSAW information management and technology programme: foundations and quality of care programmes: curriculum specification. *Institute of Health and Care Development,* Bristol.

66 Webref (accessed Dec 2006) Development of the JIGSAW curriculum specification. www.primis.nhs.uk/pages/jigsaw.asp

67 Department of Health (1998) *Learning to Manage Health Information.* Department of Health, Leeds.

68 Teasdale, S. and Bainbridge, M. (1997) Improving information management in family practice: testing an adult learning model. In Safran, C. (ed.) *Proceedings of the AMIA Annual Fall Symposium 1997,* 19, 687–92.

69 Webref (accessed Dec 2006) MIQUEST www.nhsia.nhs.uk/miquest

70 JIGSAW team (1996) *Options Appraisal of Means for the Electronic Transfer of Computerized Medical Records Between General Practices.* NHS Executive Information Management Group report.

71 Webref (accessed Dec 2006) Collection of health data from General Practice. www.nottingham.ac.uk/chdgp/

72 Cowley, C., Daws, L. and Ellis, B. (2003) Health informatics and modernization: bridging the gap. *Informatics in Primary Care,* 11, 4, 207–14(8).

73 McCormack. A., Fleming, D. and Charlton, J. (1995) *Morbidity Statistics from General Practice 4th National Study 1991–1992.* UK Government Statistical Service, London.

74 Trimble, I. (1995) Letter. *BMJ,* 311, 512–13.

75 BMA General Medical Services Committee (GMSC) (1984) *1984 Data Protection Act – A Code of Practice For General Medical Practitioners.* British Medical Association, London.

76 Joint Computer Group of the GMSC and RCGP (1988) Guidelines for the extraction and use of data from general practitioner computer systems

by organizations external to the practice. Appendix III in *Committee on Standards of Data Extraction from General Practice Guidelines*. Joint Computer Group of the GMSC and RCGP.

77 Booth, N. (1994) What are the Read Codes? *Health Libr. Rev.*, 11, 3, 177–82.

78 Technical Working Party of the RCGP/GMSC Joint Computing Group (1988) *The Classification of General Practice Data*. RCGP, London.

79 de Lusignan, S. (2005) Codes, classifications, terminologies and nomenclatures: definition, development and application in practice. *Informatics in Primary Care*, 13, 1, 65–70(6).

80 Lamberts, H. and Wood, M. (eds) (1993) *International Classification of Primary Care*. Oxford University Press, Oxford.

81 Lamberts, H. and Hofmans-Okkes, I.M. (1996) Episode of care: a core concept in family practice. *J. Fam. Pract.*, 42, 161–7.

82 Lamberts, H., Wood, M. and Hofmans-Okkes, I. (1993) *The International Classification of Primary Care in the European Community with a Multi-language Layer*. Oxford University Press, Oxford.

83 Hawking, M. (1995) Code conversions, data stability and the future – an agenda for discussion. *Informatics in Primary Care*, June, 3–5. Webref (accessed Dec 2006) www.primis.nottingham.ac.uk/informatics/june95/june3.htm

84 Horan, B., Nowlan, A., Wilson, A., Sneath, L., Rector, A.L., Goble, C.A., Howkins, T.J. and Kay, S. (1990) Supporting a human impossible task; the clinical human–computer environment. *Interact-90*, 247–52.

85 Rector, A.L., Kay, S. and Howkins, T.J. (1988) A human–computer interface strategy for doctors. In Richards, B. (ed.) *Current Perspectives in Health Computing*, pp. 213–23. BJHC for BCS, Weybridge.

86 Webref (accessed Dec 2006) www.prodigy.nhs.uk/indexMain.asp

87 Hopkins, R.J. (1989) The Exeter Care Card Project – one year on. In (eds unknown) *Proceedings of the PHCSG of the BCS*, pp. 71–4.

88 Hopkins, R.J. (1989) The Exeter Care Card – a patient held medical smart card network. In (eds unknown) *Proceedings of Smart Card '89*.

89 Hopkins, R.J. (Date unknown) Assessment of user needs and impact of patient data cards in healthcare systems. In (eds unknown) *Proceedings of AIM Euroforum*.

90 Hopkins, R.J. (1991) The Exeter Care Card. In (eds unknown) *Proceedings of the 3rd Global Conference on Patient Cards*, pp. 327–32.

91 Hopkins, R.J. (1991) Shared information – a foundation for shared care. In (eds unknown) *Proceedings of the 3rd Global Conference on Patient Cards*, pp. 423–6.

92 Webref (accessed Dec 2006) www.dotpharmacy.co.uk/pip_change.html

93   Webref (accessed June 2007) www.rpsgb.org

94   News report (2002) System allows repeats via the internet. *The Pharmaceutical Journal*, 268, 7184, 161–7.

95   Shepherd, I. (1998) *A Review of the Use of IT in Pharmacy*. MEL Research & Aston University, January.

96   Webref (accessed Dec 2006) www.ndchealth.com/products_services/ pharmacy/002.htm

97   Webref (accessed Dec 2006) www.dh.gov.uk/PolicyAndGuidance/Med icinesPharmacyAndIndustry/Prescriptions/ElectronicTransmissionOf Prescriptions/fs/en

98   Webref (accessed Dec 2006) www.iso.org/iso/en/stdsdevelopment/tc/ tclist/TechnicalCommitteeDetailPage.TechnicalCommitteeDetail?CO MMID=4720

99   Webref (accessed Dec 2006) www.psnc.org.uk/index.php?type=page& pid=67&k=11

100   Meeting report (2003) Pharmacy automation. *The Pharmaceutical Journal*, 271, 7272, 590–1.

101   Hayes, G. and Bainbridge, M. (1994) Has multimedia a place in the primary care electronic medical record? *In Proc. Ann. Conf. PHCSG of BCS*, Cambridge, pp. 136–9.

102   Webref (accessed Dec 2006) http://homepages.newnet.co.uk/dance/ webpjd/intro/nhscrweb.htm

103   Hedley, A.J. and McMaster, W. (1988) Use of the National Health Service Central Register for medical research purposes. *Health Bulletin*, 46, 63–8.

104   Boden, P., Stillwell, J. and Rees, P.H. (1992) How good are the NHSCR data? In Stillwell, J., Rees, P.H. and Boden, P. (eds) *Migration Processes and Patterns, Volume 2: Population Redistribution in the United Kingdom*, pp. 13–27. Belhaven Press.

105   Booth, N., Jain, N.L. and Sugden, B. (1999) The TextBase Project – implementation of a base-level message supporting electronic patient record transfer in English general practice. *Proc. AMIA Symp.*, 691–5.

106   Daly, M. (2000) Where to start with electronic prescribing – learning from the difficulties experienced by others. *The British Journal of Healthcare Computing and Information Management*, May, 17.4, 40–1.

107   Young, A.J. and Beswick, K.B. (1995) Decision support in the United Kingdom for general practice: past, present, and future. In Greenes, R.A., Peterson, H.E. and Protti, D. (eds) *MEDINFO 95*, Vancouver, Part 2, pp. 1025–90. AMIA, USA. ISBN 0 969741 41 3.

108   Webref (accessed Dec 2006) Metaxiotis, K.S., Samouilidis, J.E. and Psarras, J.E. Expert systems in medicine: academic illusion or real power? www.primis.nottingham.ac.uk/informatics/feb2000/feb3.htm

15 Community and public health

DENISE BARNETT

Reading the newspapers of the mid-1800s, when town councils were created, illustrates the direct funding and involvement that our great, great grandparents offered to the developing health services. The creation and improvement of roads and drains, burial boards and fund-raising events to build a hospital for the town were recorded in detail. Atlee and Robson, writing in 1925, said:

> The development of our local government system has been very largely due to discoveries in the field of preventative medicine and their application during the nineteenth century. The district council was originally a purely sanitary and health authority.
>
> REFERENCE 1

The development of information technology for community-based services was influenced by reorganizations of local government and the NHS and the related changes in funding sources. The community-based clinical professions had an even harder struggle to develop clinical systems to meet direct patient needs than did their hospital colleagues.

Community health services in the 1960s were provided by qualified staff who had already gained several years of hospital experience. They were expected to work autonomously and, in some rural areas, might provide triple duties: nurse, midwife and health visitor. Local authorities, in building new estates might include a 'nurses' house' with an annex providing a clinical room. Such professionals were never 'off duty' to their neighbours.

Medical Officers of Health (MOH) provided a wide range of advice to local government officers and councillors. They sometimes had additional roles such as Port Medical Officer. The MOH was relocated to the NHS in 1974, perhaps on the assumption that as new antibiotics and other medications began to reduce the incidence of some infectious diseases they could refocus on epidemiology and health service planning. The District Medical Officer might have one or more community medicine specialists working with them on health promotion and the prevention of disease. Some struggled and Donald Acheson's report, in 1988, reflected disappointment with the role of the community physician, recruitment difficulties and problems with outbreaks of communicable diseases, such as the outbreak of food poisoning at the Stanley Royd Hospital, Wakefield, in 1986.[2]

Environmental health remained with local government.

When the health services were reorganized in the 1970s there was a growing interest from managers, often from hospital backgrounds, in directly managing the work of community staff. To do this information was required: data were collected on paper forms completed by the practitioner at the time care was provided, or that was the fond assumption of the managers who then planned services according to the aggregated data. Clinical staff sometimes did not set much store on data collection for others and there is no doubt that some filled in the forms retrospectively, making up data they could not remember.

Local councillors serving on health authorities in the 1980s came from a different culture to those with a health service background, who formed the majority in the health authorities. Some councillors tried to use the authority meetings for point scoring and politicking, particularly when local elections were in the offing and the press were present. This could be very irritating when proposed policies involved patient care issues, and delaying decisions until the next meeting meant patients might wait a lot longer for important services to start. Computer-based services were common within the local authorities and health authority information was sparse in comparison: this may have caused frustration to the councillors in turn. Yet in their list of priorities computer developments were low: there were few public votes in it.

MANAGING THE COMMUNITY HEALTH SERVICE

Looking back over the past few decades of computing, the structure of the nations' health service and the associated power to make autonomous decisions has changed. Before the National Health Service Act 1946, hospitals in the UK were run by local authorities and voluntary bodies. They could raise income by levying rates, making charges and appealing for funds. From the inception of the NHS the hospital funding came via central government and the Minister for Health was in charge, with the hospital management committees and boards acting as his agents. The counties and county boroughs had health responsibilities before 1948, after it the local authorities were responsible for:

- ambulances;
- domiciliary midwifery;
- home nursing;
- health visiting;
- the welfare of mothers and young children;
- vaccination and immunization;
- the care and after-care of the mentally ill and mentally defective;
- the home help service.

Over the next few years there were proposals, reports (including a review of the NHS by the Guillebaud Committee), legislation and a flow of circulars setting out new arrangements for hospital services. By 1974, in anticipation of the recommendations of a Royal Commission to examine local government in England outside London, a green paper proposed an integrated health service. That change seemed to unblock the log jam and the frequency of restructuring seemed to increase. Managers felt they were subject to political control and an increasing stream of 'initiatives' and targets set from above. To demur might lead to a loss of career at the next reorganization, when incumbents often had to reapply for their jobs. The titles of posts and the organizations, including the acronyms, changed with each reorganization. (It may be helpful to remember that the central administration had responsibility for the standard of all health care provision, including that provided by independent practitioners and independent hospitals, although the focus of the politicians was mainly on the NHS.)

In the 1960s, as health care computing began to develop, the large teaching hospitals had a direct relationship with the DoH through their hospital management committees. Community and public health were part of local government and were far more influenced by local politics. Local authorities were very active computer users for the many office and business-type services, but it was less easy to access funds to develop large health projects.

The tripartite system of the NHS involved:

- regional hospital boards and hospital management committees;
- executive councils for the family practitioner services;
- local health authorities for community personal health services including public health, home nursing, health visiting and domiciliary midwifery.

The changes brought about by the National Health Service Reorganisation (England) Act 1973 and the National Health Service Reorganisation (Scotland) Act 1972 came into effect on 1 April 1974. Local government in England was reorganized on the same date, although in London the Greater London Council and 16 district councils had been reformed in 1965. The intention was for local government counties to correspond geographically with county and district councils.

The new health structure in England was:

- regional health authorities (RHAs) (14) each with 14 to 18 members and having one or more university medical schools within its boundaries;
- area health authorities (AHAs) (90) with boundaries to match the non-metropolitan counties and metropolitan districts of local government;
- health districts each with a population of 150,000 to 300,000;
- family practitioner committees; and
- community health councils for public participation.

The RHAs were responsible for the allocation of resources to the AHAs and for monitoring their activities. RHAs also provided computer services, and were responsible for the planning, design and construction of large health service buildings, and for coordinating the ambulance service and the blood transfusion service. The RHAs were accountable to the Department of Health and Social Services (now the DoH).

In Scotland, the structure was simpler with health boards, health districts and family practitioner councils, as well as local health councils for public participation. A new body, the Common Services Agency (CSA), provided some of the functions that in England were provided by the RHAs including computer services, the planning, design and commissioning of health service buildings, the ambulance service, and the blood transfusion service. Health boards had direct access to the Scottish Home and Health Department.

In Wales, the structure changed a year later, on 1 April 1974, to provide area health authorities and health districts. There were eight AHAs with responsibility for planning and managing the NHS under the control of the Welsh Office. Five of the AHAs also had health districts: two had four districts, three had two districts. Each area had an area ambulance officer. The Welsh Health and Technical Services Organisation (WHTSO) was set up at the same time to provide similar services to Scotland's CSA, including computing. The Blood Transfusion Service for Wales was provided by the South Glamorgan AHA.

Proposals for the next restructuring in England were set out in a consultative document, *Patients First*, issued in December 1979.[3] The hard-bitten health service manager may reflect on the scene in the television programme *Yes Minister* when Sir Humphrey suggested it was politic to put the difficult bit in the title and then not mention it again in the report. The intent of *Patients First* was to reduce the size and direct involvement of the DoH, abandon the area tier and create district health authorities (DHAs). As a result, coterminosity with local authorities was lost. Ambulance services were managed by one DHA on behalf of the others in a multi-district area. These changes came in to effect on 1 April 1982.

At regional level some of the changes set out in *Patients First* were quietly abandoned following new appointments of Secretary of State and Health Minister. Each year the regional chair and team of officers was to meet the Minister, review the previous year and plan in the light of official policies and agree objectives for the coming year. Computing developments, or lack of them, within the NHS in England have thus been very strongly influenced by Government.

The introduction of general management within Scotland was slower and not achieved until mid-1987. In Northern Ireland reorganization of the services occurred in similar ways to those in England and Wales, with the major difference being that, since 1948, the four boards, which reflect local government arrangements, have been responsible for both health and social services.

In 1983, the Secretary of State invited a small group of businessmen, under Roy Griffiths of the supermarket chain Sainsbury's, to advise on the management of the NHS in England. This led to the introduction of general management. In 1988, the Government published the white paper *Promoting Better Health*, which set out changes in primary health care.[4] That same year Sir Roy Griffiths, as he was by then, published *Care in the Community: Agenda for Action*.[5] This focused on the confusion in the provision of community care for the elderly, the mentally ill and others where both health services and local authority social services were involved.[6]

FAMILY PRACTITIONER COMMITTEES

In 1984, management consultants Arthur Andersen studied family practitioner services for the DoH. Their report included a strategy where the costs for new computer systems could be met by administrative savings by reducing the 9,450 staff by a third. A central Family Practitioner Committee (FPC) computer system capable of local adaptation was proposed, with a communications network to link medical and dental practices.[7]

The Körner Steering Group reports stimulated an interest in providing community management data. The fifth report recommended that from April 1988 information should be collected on: services provided to the community; and patient care in the community. The services provided to the community included immunization, health surveillance and early detection of disease, health promotion and education, and professional advice and support.[8]

The Scottish Health Service reviewed information technology in its 15 health boards in 1986 with the introduction of general management to the Scottish Health Service. As part of its CSA, it set up a Directorate of Health Service Information Systems (DHSIS). A data administration unit was also set up within the Information and Statistics Division (ISD). Together DHSIS and ISD were responsible for the development of information systems and IT strategies. Commercial suppliers were seen as important contributors to the development process. A procurement exercise was instituted within Tayside and Forth Valley Health Boards to establish a communications network to act as a model for the other boards.[9]

Immunization and vaccination

The local authority health services managed child health services. The call and recall facilities offered by computers were used for immunization and vaccination records. West Sussex introduced this in 1964. No manual records were kept by the county or by the GPs where the child could be taken to receive the treatment. The immunity indices increased and by 1967 were the highest for any local health authority in England.[10]

A copy of the full file was printed out periodically for administrative purposes and an enquiry program was used to extract individual records from the tape file and print them out as a special computer run.[10]

Cervical cytology

West Sussex also introduced the call and recall facility for cervical cytology in July 1967. The county MOH was convinced the service would be most successful if all eligible women were given direct invitations to have the screening tests, rather than depending solely on the response to a publicity campaign. The initial data, punched on cards, were drawn from the electoral register and women under the age of 35 were asked to consent to the tests in due course when they became eligible. Women aged between 35 and 70 who consented to the test were sent an invitation with a timed appointment at a clinic or GP of their choice, and if they had not responded within a month were sent a second one. If this was ignored their names were added to a health visitor's follow-up list for a personal visit. Appointment lists were drawn up and balanced to ensure the capacity of the clinics, GPs and pathology laboratories were not exceeded. A multi-copy laboratory request form was also prepared.

Smear results were sent back to the County Health Department to be entered on the computer. Negative results were sent to the patient and to her GP. Where results were positive, the notification was done by the medical staff of the health department.

The West Sussex County Council developed and introduced a multi-access terminal enquiry service. This allowed the county MOH to type in a query and print out the results from the immunization and cervical cytology systems. It started with 15 telecommunication lines to the computer, with security measures so that only authorized users had access to the specific file data from their terminal. Each record had a unique identifier and in 1970 there were plans for a cross-indexed master file with Soundex code order.[10, 11]

By late 1984 many computerized cervical cytology recall systems were in use, such as the Harrogate HA system, which ran on an Equinox microcomputer, using a hard disk and a daisy-wheel printer.[12]

The Exeter Family Practitioner Service Computer Unit was funded by the DoH and was created from the original Exeter Project in 1983. By 1985 the Exeter team had rewritten and enhanced the program and extended the system to include cervical cytology. It provided screens for:

- recording a cytology test result;
- results of previous tests and test-due dates;
- amendment of previous test records;
- relevant information and test-due dates for new patients;
- amendment of test-due dates, including suspension.

The data belonged to the HA and the FPC acted as its agent.

It was anticipated that a standard system would be in all FPCs in England within three years.[13] The DoH decided to invest about £6 million to achieve implementation in all FPCs in England and Wales by April 1988.[14] (For more information about the Exeter Project see Chapter 14.)

In April 1985, political pressure emerged after the death of a cervical cancer victim who had not been informed of positive smear results. Oxford RHA linked positive screening to the electoral register. The new system was installed by the region as the Reading FPC was not yet computerized and could not link to the Exeter cytology module. At the time only a third of FPCs had computers.[15] It was not until late 1987 that all 98 FPCs had computers, even then there were some 19 registers still to 'go live'.[16]

In late 1987, the DoH purchased the software developed by Oxford RHA that allowed a direct link between the cytology laboratory records and the family practitioner service computer system. This helped identify women who had not been screened.[17] For more about the GP side of cervical screening see Chapter 14.

The problems with tracking women for cervical screening and the interpretation of abnormality in the smears continue to require local service reviews and recalls for groups of women. In 1998, the National Audit Office reported on the performance of the screening programme in England.[18]

CHILD HEALTH

The school medical service began in 1907 after a public outcry at the number of men found unfit for the army in the Boer War. Its aim was to detect deviations from normal development patterns. Health visitors were involved alongside clinical medical officers.

In 1967, Gloucestershire began creating computer records for every child born within the county. Every birth had to be notified within 36 hours by a person present at the event, usually the midwife. Details of children moving into the county with their families were taken from a health visitor's paper record from the area from which the family moved. Updating was undertaken for some children from the notification of discharge forms passed from the midwife to the health visitor, and for premature babies, from the health visitor's notification 28 days after birth. A separate booklet for vaccination and immunization included a number of 'certificates', designed as computer input documents, sent on completion of a course, to the health authority. In 1971–72, as children born in 1967 entered school a computerized school health record was created. This could be completed during a medical examination in a clinic, using 'bar-marking'. Routine audiometry could also be added to the record. One of the outputs could be a child's history for a paediatric consultation. Another could be the monitoring of children with potential or identified handicaps.[19]

Islington HA had a batch-processing child health system that used the services of the local authority, but by 1981 it was difficult to enhance and expensive to maintain. In 1982, the North East Thames Regional Medical Officer's discussion group on child health recommended asking Islington to pilot a new system on a microcomputer-base to replace the National Child Health system. In the end, a DEC PDP-11/44 was selected with Rapid

Programming Language (RPL) as the development software. Initially work was carried out at the Regional Computer Centre while a room was prepared in Islington and a telephone line installed for remote support. The system went live on 1 January 1984 with data converted and transferred from the local authority system.

The Islington immunization recall system moved away from a schedule of immunization frequencies, based on age and intervals, to a 'see again' date based on the first immunization. This helped with part-doses and unusual times, such as unscheduled visits. Appointments were made based on a computer-held 'diary'.[20]

Gloucester DHA developed a DEC-based front end to allow interfacing between the ICL mainframe running its local copy of the national child health standard system and its District DEC hardware.[21]

The child health modules, which were community based, could then be linked to hospital records for a child where appropriate. The early Gloucestershire work had provided a framework for the later sequential use of the modules within the national system. These could be implemented, just in time for use, with the first cohort of babies being added to the register module, then receiving their initial immunizations coordinated by the next module. The pre-school immunization module could be implemented, tested and ready as the cohort grew old enough to start school. This could then be followed by the school health module.

In January 1987, Bloomsbury HA announced it was installing the first commercial MUMPS-based child health system which had been written by Abbey Business Consultants for British Telecom (BT). This was BT's entry to the health care software market.[22] At this time 122 districts in England and Wales were using the child register and immunization modules of the national child health computer system. A rapid increase in the use of the pre-school module (24 districts) and the school health module (11 districts) was forecast for the next 12 months.[23]

COMMUNITY NURSING, HEALTH VISITING AND MIDWIFERY

Leicestershire had used a comprehensive patient record storage and statistical information package since 1973. It produced a daily visit schedule for nurses. Workload information was also collected and manipulated with standard general-purpose packages such as SuperCalc® II for spreadsheets and dBASE® II for database management. WordStar was used for word-processing. Silicon Office was also frequently used. Six of the 10 Wessex districts collected statistical nursing workload information on microcomputers linked to the regional mainframe via the telephone network.[12]

Bromsgrove and Redditch HA developed COMCARE as part of the Management Information Pilot Project (MIPP) that began in 1983. This used input forms (an initial staff registration form, a patient registration form completed at the first contact, and a daily activity sheet, which used treatment/activity codes) on which each member of staff on duty recorded their work location and time taken, along with each patient's details. Travelling time was also recorded, and there was space for other activities such as case conferences and meetings. Forms were submitted weekly and input by a central office, where there was also some data-quality assessment. Most reports were generated monthly, but a weekly analysis of patient contacts could be generated.

One big drawback in collecting data through field staff was the lost patient-contact time. East Dyfed HA undertook a trial of hand-held data recorders to assess whether this would speed up the process.[24] The Psion Organiser was considered to have worked and be acceptable to community staff, so a second phase was funded for East Dyfed to develop additional features.[25] Community staff liked having their own caseload readily to hand, having other facilities (such as a watch, calculator and diary) on the machine, and being saved the chore of completing travel claims.[26]

Joyce Wiseman of Mersey RHA estimated 30 per cent of a community nurse's time was spent on recording, and this could be equated to approximately £3,000 a year. She suggested the hand-held Husky at £750 was a significant saving. The Psion Organiser was even better at £150.[27] A national COMCARE user group met at intervals in 1988–89 to discuss enhancements to the next version. (Chapter 20 also covers many of the community nursing computer developments.)

Staff within Islington HA were keen to improve their ability to respond to changes in patients' needs and preferences. They defined the need for a relational database, to set up a client-based community and paramedical information system (CPIS), and a large group were actively involved in the data structure and requirements well beyond the Körner data-set. The system was developed jointly by the HA staff and Oracle to provide a 'skills transfer'. Outputs included examination of how an individual patient used the entire range of services available from the HA. Terminals were located in each health centre so practitioners could enter and access their data.[28]

Taking stock of progress by 1985

The Körner Steering Group report stimulated an interest in providing community management data. Some existing systems, such as those in Hillingdon and North Birmingham, were able to provide compatible data.

In 1985, Paul Catchpole, from Darlington Health Authority, reported the results of a survey to create a National Register of Community Systems. He undertook this as part of a PhD. He achieved a 95 per cent response rate from a postal questionnaire sent to chief nursing officers. The batch-operated National Standard Child Health System was in use in over 60 per cent of the authorities. It had five modules:

- a child register
- vaccination and immunization
- pre-school health

- school health
- statistics.[12]

At this time some authorities were already using microcomputers to capture their local data and then transmit it to the regional mainframe for processing.

Only four authorities reported stock control systems for the loan of nursing aids (these included the Green Walk Aids package in Islington that ran on a Minstrel microcomputer). The South Birmingham HA had a loan system, for nursing equipment and domiciliary incontinence supplies, integrated with their Financial Information Project (FIP) home nursing records. This was written in MUMPS and therefore could only run on compatible hardware such as DEC computers.

Many authorities maintained a mental handicap register, including Croydon, which ran the MENINDEX system. Bromsgrove and Redditch HA were involved in the DoH-funded MIPP, which had a specific community module covering all levels of staff, designed to produce the percentage of time spent on various patient groups as well as the Körner minimum data-sets. This ran on ICL DRS 20 computers.

Catchpole reported 'one-off' applications that included:

- a centralized chiropody service in Shropshire;
- a dental workload system in Clwyd;
- a survey analysis system in Central Nottinghamshire;
- a 'children at risk' recall system in Portsmouth and South-East Hampshire.[12]

By 1987 the community systems datafile listed 14 systems and included management systems: for chiropody, clinics and community nursing. There were also a case list register for district nurses, a system to record child injuries and one for diabetic patient care.[29]

GENERAL DENTAL PRACTICE

Most dentists, like general medical practitioners, ran their own businesses. When mainframes were the only available computers, developments were limited to academics who had access to these machines. As microcomputers became affordable, the *Symposium on the Application of Computers to Dentistry* was held in 1972. The Dental Estimates Board had installed a mainframe to handle NHS payments, one practice had installed an advance system and a dental hospital was conducting feasibility studies in patient administration. In the 1980s, cheaper microcomputers brought with them many systems designed for use in general dental practice. A patient's record was a series of treatment plans from a limited range of items of care. NHS patients could receive treatment at a reduced rate with the dentist being reimbursed the remainder of the set charge. In 1985, the FP17 examination form was redesigned so that it could be completed on a computer printer. In

1986, there were proposals for the electronic transmission of dental claims and for standards for the patient record, for example a tooth-numbering system.[30]

Dental practices as businesses

As businesses, a single-handed or group dental practice might be expected to be interested in separate or integrated systems for:

- practice administration (a patient register, financial accounting and stock-control);
- appointments;
- patient records.

However, take-up of computers within dental practices was slow. The British Dental Association commissioned a study of use, by Scicon Consultancy International in 1981, which emphasized education and step-by-step implementation.[31]

The 1984 report from Arthur Andersen suggested three developments:

- Electronic transmission of digital images of radiographs.
- The use of holograms to transmit and store dental models.
- 'Expert' computer systems to replace dental advisers.[32, 33]

In 1991, a survey of the use of computers in general dental practices found that 22 per cent of practices (about 2,000) had a computer and that 27 per cent of the dentists had a home computer.[34] The first three computer-assisted learning (CAL) programs developed were:

- an orthodontic assessment and treatment planning program (by Prof C.D. Stephens and Mr P. Rigg of Bristol University);
- a biopsy and surgical endodontics program (by Prof A. McGregor, Dr A. Long and Ms P. Mercer of Leeds University);
- a pulpotomy for deciduous molar teeth program (by Mr A. Gould, Ms P. Smith and Dr M. Cox of King's College, London University).

As a result of the evaluation of these programs the DoH provided funds for a further 55 standardized computer workstations to be placed in Dental Postgraduate Centres in all the English regions. Plans were also made to provide more CAL programs and to issue them on CD-Rom.[35]

On 1 January 1994, there were still only 1,566 practices, representing 3,931 dentists (out of a total of 15,680 with NHS contracts) transmitting treatment data to the Dental Practice Board. A.M.J. Lynn considered the possible reasons for this slow take-up and suggested:

- limited time when inundated with other changes;
- low-cost systems purchased with the DoH grant of £900 were sometimes of poor quality with limited support;
- too many systems (160 in 1991) with 26 companies certificated to transmit treatment claims;

- slow data-entry, especially via a keyboard, and infection control for input devices;
- guidance awaited on clinical records kept only on computers;
- slow development of audit in dentistry;
- limited use of email by the Dental Practice Board, few systems with online stock ordering facilities;
- few CAL packages;
- few systems to help in practical dentistry.[36]

Individual dental practitioners did use their computers to develop new tools.

Kaye used patient's post-adolescent photographs, scanned them to create a digital image then manipulated this followed by a colour print, to help improve the standard of denture construction.[37]

INTEGRATING INFORMATION

Rowntree, from the DoH, identified that, although consideration had been given in the 1970s to linking the records of individuals, there had not been much progress. The Oxford Record Linkage system had not been followed widely in England. The population register concept had taken root in the child health field but not elsewhere. It was admitted that data quality was a considerable problem for central policy making.[38]

In April 1985, the FPCs became autonomous bodies in England. There were tensions between them and the District Health Authorities over the 'ownership' of information, with a fear of duplication within the NHS. In Scotland there was only one shared database to provide a Community Health Index (CHI).[39] This was designed to allow a health board or consortium to follow every person from birth to death, including identifying 'at risk' groups. Tayside Health Board developed the forerunner: a master patient index. It had been identified in the late 1960s but there was then insufficient computer power to run such a big system. The CHI at the East Central Computer Consortium in Dundee, Scotland, went live in June 1987. It covered the health records for the whole of a geographically defined population. Links to hospital PAS were introduced.[40]

The East Dyfed HA work on an integrated community health system identified the need for an index of all clients who were or could be receiving health care services. The Scottish CHI system was adapted to fit the different organizational structure and meet the needs of the family practitioner services, hospital and community health services, becoming the District Health Index (DHI). Dyfed's FPC provided tapes for a batch interface to load and update the system. To maintain up-to-date data there was an interactive application data interchange (ADI) with operational systems, such as to COMCARE, to the standard child health system, to a social services client information system for mentally handicapped clients and to the ABIES general practitioner system. By 1988 there were plans to link to PAS, the

MERIT radiology system, the CAER system in A&E as well as to pathology, nurse information and clinical management systems.[41]

Resource management

Farrington suggested that the community was treated as an enormous hospital ward in calculating staffing requirements, yet different areas had population changes that had a big impact on resource needs. He cited the St Paul's Cray estate in Bromley HA, which had been built in the late 1940s, where, by the 1980s the children had left home and ageing parents were requiring increasing amounts of district nursing care. Bromley had rejected COMCARE and opted to pilot the FIP community system for resource management. The district was divided into three geographical areas with sufficient terminals for community staff to key in their own data. This provided reports on contacts by geographical location to reflect the different requirements.[42] Other tools were required to help the community medical teams identify medical need within population groups. These included the development of geographic information systems (GIS). *The Chorley Report* (see Chapter 13) strongly recommended the adoption of the postcode as a location identifier.[43]

In 1977, the Labour Government set up a working group on inequalities in health, chaired by Douglas Black. The report bearing his name was published in 1980, but by then the Conservatives were in power and very few copies were released. However, in 1982, Townsend and Davidson published it as part of a Penguin book, *Inequalities in Health*, and there was then a lot of media interest in the scandal of trying to hide the impact of poverty on health.[44] *The Black Report*, published in 1980 highlighted the effects on public health of income, poor housing, educational disadvantage and occupation. Direct impact on many of these aspects had been lost at the 1974 reorganization of health and local government.

The Black Report proved influential among public health specialists and thus to the geographical aspects and eventual use of GIS.

Joint funding and planning had proved difficult in many localities. By 1988 nearly half the metropolitan authorities in England and Wales had established health committees to promote public health.

Health education

Health education officers helped in trying to achieve the World Health Organizations target of *Health for All by the Year 2000*. They often provided an information service for other staff, which included videos, leaflets and teaching aids, as well as coordinating health campaigns. As a result they were involved in the development of computer-assisted learning (CAL) for health promotion purposes.

Ambulance service

Emergency ambulances in the 1960s were staffed by qualified ambulance personnel, trained at regional centres. There was a growing pressure to provide enhanced training. Experienced officers began to spend time in A&E departments to observe intubation, infusion and defibrillation. In time this developed in to the paramedical cadre, with extended training to deal with cardiac and trauma cases as well as multiple injuries. In 1986, this was reflected in a tiered salary structure according to the kind of patients to be transported and the skills that might be required: accident and emergency; general outpatients; or day patients.

The equipment in the vehicles could also be related to the possible patient requirements. Accident and emergency vehicles could include ECG machines and defibrillators, and more recently tiny video cameras worn on a headband that are able to transmit images of the accident victim for remote diagnosis and advice from a senior doctor.

South West Thames RHA was responsible on behalf of the four London Regions for the ambulance service in London. It tried to introduce a major computer-based change in the service with the London Ambulance Service Command and Control system. This ran on two Eclipse 280 and two 120 computers. The system was claimed to be able to validate an address in the Greater London Council area using a database of 200,000 streets, places, names and public houses. The intention was that on receipt of a emergency telephone call the system could propose the best ambulance source from any one of 200 located across 75 depots. If the central control operator decided to use the selected vehicle the system then automatically sent the relevant details to that depot. Development cost £4 million, which included £1.065 million on computer hardware and software.[45]

Problems with effective ambulance despatch leading to serious delays and patient harm caused concern in the media and ambulance personnel were unhappy about relocation and merging of services, not just in the London area. The National Audit Office published a review of patient transport services in 1990.[46]

In 1985, Surrey Ambulance Service used geographic information systems for automatic vehicle location using DATATRAK and mobile data communication. These reduced the time it took new ambulance despatch staff to develop local knowledge to make the best use of the available vehicles.[47]

Mobile data communication also allows the receiving A&E hospital staff to view an ECG and other data before the arrival of the patient, so that appropriate equipment and medication can be ready for immediate use.

By 1986 the *BJHC Datafile* listed three ambulance-control and five scheduling systems; with another for fleet management included stock control. There was one system to manage car loans and another for the Crown fleet management, plus one for vehicle maintenance and a more inclusive WIMS vehicle management system. A comprehensive system was in use in Cambridgeshire and Norfolk. There was also one for ambulance bonus payments and another for statistics.[48]

Blood Transfusion Service

DONALD (DONor Appointment and Linkage of Data) or DRACULA (DonoR Appointment and Call-Up Linkage Activities), were, respectively, the official and unofficial titles given, in 1969, to J. Emlyn Jones of the Manchester Regional Hospital Board (which administered the local Blood Transfusion Service). As Regional Treasurer, he had been involved in the computerization of their donor panel. The output included a session slip for each donor expected to attend, panel status tables, alphabetic lists of donors, laboratory information and statistics.[49]

In 1984, North East Thames RHA inaugurated a computerized system (MITHRAS) at their regional centre in Brentwood, Essex, starting with donor call-up and registration. The system helped with session management, donor laboratory input, fractionation procedures, blood issue procedures and stock control. The intention was to extend it to tissue typing and plasmapheresis.[50]

Automated blood grouping machines were used in 1988 within the Regional Blood Transfusion Services in England, as were bar-coded labels and readers for sample identification. Laboratory test results and blood donor data were captured on computer. Having acted as a loose confederation of 14 centres, a national management structure was agreed, with operational responsibility vested in the NHS Management Board. In Scotland the five Regional Transfusion Centres had formed a National Service, while in Northern Ireland there was one centre.

THE FUTURE

The exchange of data between the different components of the NHS in England, Wales and Scotland to provide a computer-based health record has been slower to develop than was envisaged in the national computer strategies. Negative publicity around major computer system failures in the NHS and in Government departments has not helped. Information technologies including secure virtual private networks have enormous potential to reduce duplication of patient data and to provide ready access to authorized clinical staff. The closer liaison between those working in local government and in the health service over such aspects as child protection is again on the agenda in the 2000s. The history of these two major public services might provide encouragement to further develop informatics-based routes to enhance cooperation.

REFERENCES

1   Atlee, C.R. and Robson, W.A. (1925) *The Town Councillor*. (Out of print.)

2   Acheson, D. (1988) *Public Health in England*. HMSO.

3   HMSO (1979) *Patients First*. HMSO.

4   HMSO (1988) *Promoting Better Health*. HMSO.

5   Griffiths, R. (1988) *Care in the Community: Agenda for Action*. HMSO.

6   NHS (1989) *NHS Handbook* (4th ed.) The Macmillan Press Ltd for the National Association of Health Authorities in England and Wales, London. pp. 3–9. ISBN 0 333484 44 4.

7   DHSS Leaflets Unit (1985) *Report of a Study of Family Practitioner Services and Administration and the Use of Computers*. Arthur Andersen for the DHSS.

8   NHS (1984) *A Report by the Steering Group on Health Services Information on the Collection and Use of Information About Services for and in the Community in the National Health Service*. HMSO, London.

9   Knox, C. (1988) New-look IT strategy for Scotland. *BJHC*, 5.9, 13, 16–17.

10 Davies, C.K. (1970) Recall procedures – local authority applications. In Abrams, M. (ed.) *Medical Computing: Progress and Problems*, pp. 283–8. Chatto & Windus for BCS, London.

11 Galloway, T.McL. (1963) (Title unknown.) *The Medical Officer*, 109, 232.

12 Catchpole, C.P. (1985) Community computing. *BJHC*, 2.2, 14–15.

13 Fisher, R.H. (1984) A computer system for the family practitioner service. *BJHC*, 1.4, 10–12.

14 Venning, R.W.D. (1987) Cytology call and recall (letter). *BHJC*, 4.2, 13.

15 Developments (1985) Cytology links to electoral roll. *BJHC*, 2.2, 5.

16 Developments (1987) All FPS computers installed. *BJHC*. 4.6, 5.

17 Developments (1987) Cytology link goes national. *BJHC*, 4.6, 10.

18 National Audit Office (1998) *The Performance of the NHS Cervical Screening Programme in England*. The Stationery Office. ISBN 0 102945 98 5.

19 Livesey, F. (1977) Getting child health records 'off the mark'. In Shires, D.B. and Wolf, H. (eds) *MEDINFO 77*, Toronto, pp. 539–43. North-Holland, Amsterdam.

20 O'Brien, S. (1985) An interactive child health computer system for Islington. *BJHC*, 1.3, 19, 21.

21 BJHC (1986) Child health system may go DEC. *BJHC*, 3.5, 3.

22 BJHC (1987) Bloomsbury HA pilots MUMPS child health system. *BJHC*, 4.1, 6, 16.

23 Starkey, A. (1987) Welsh update. *BJHC*, 4.1, 13.

24 Clarke, D.A.J. (1986) Community Körner data. *BJHC*, 3.6, 37, 39–40.

25 Green, M. (1986) The Psion Organizer for community data collection. *BJHC*, 3.6, 36.

26 Goldberg, B. and Savill, A. (1987) COMCARE on hand-held. BJHC, 4.3, 23–5.

27 BJHC (1988) Community computing saves time and money, conference told. *BJHC*, 5.4, 9.

28 Watson, L., Richardson, V. and Golding, C. (1988) Community care. *BJHC*, 5.10, 31–2.

29 Datafile 12 (1987) Community. *BJHC*, 4.3, 27.

30 DHSS (1986) *Computers in General Practice: A Pilot Trial of the Electronic Transmission of Dental Claims – System Specification*. Department of Health and Social Security, London.

31 Palmer, P. (1981) *Computing in General Dental Practice: A Report for the British Dental Association*. Scicon Consultancy International, London.

32 Arthur Andersen & Co. (1984) *A Study of Family Practitioner Services Administration and the Use of Computers*. Arthur Andersen & Co. ISBN 0 946539 04 9.

33 Smales, F.C. (1986) Computers in dentistry. In Abbot, W., Barber, B. and Peel, V. (eds) *Information Technology in Healthcare: A Handbook*, B2.7.11–24 issue 5. Longman in association with The Institute of Health Services Management, Harlow, Essex. ISBN 0 582061 40 7.

34 MORI (Date unknown.) *Report of a Survey of the Use of Computers in Dental Practices in 1991*. Commissioned by the Department of Health and carried out by MORI's Health Research Unit.

35 Eaton, K.A. (1994) CAL for general dental practitioners: an evaluation of the current scene. In Richards, B., de Glanville, H. and MacOwan, H. (eds) *Current Perspectives in Healthcare Computing*, pp. 97–102. BJHC for BCS, Weybridge. ISBN 0 948198 17 6.

36 Lynn, A.M.J. (1994) Applications of computers in dentistry: where are we at? In Richards, B., de Glanville, H. and MacOwan, H. (eds) *Current Perspectives in Healthcare Computing*, pp. 107–12. BJHC for BCS, Weybridge. ISBN 0 948198 17 6.

37 Kaye, M. (1994) Computer facial imaging: dental usage and other possibilities. In Richards, B., de Glanville, H. and MacOwan, H. (eds) *Current Perspectives in Healthcare Computing*, (pages unknown). BJHC for BCS, Weybridge. ISBN 0 948198 17 6.

38 Rowntree, J.A. (1983) Using health activity statistics in national health care policy making: some problems of concept, quality and relevance. In van Bemmel, J.H.V., Ball, M.J. and Wigertz, O. (eds) *MEDINFO 83*, Amsterdam, pp. 288–91. North-Holland, Amsterdam.

39 Penfold, G. (1987) Learning by example. *BJHC*, 4.5, 25–6.

40 Hunter, J. (1987) CHI goes live. *BJHC*, 4.5, 23–4.

41 Savill, A. and Goldberg, B. (1988) Integrated chips. *BJHC*, 5.9, 37.

42 Farrington, K.J. (1987) Community management budgeting. *BJHC*, 4.2, 23–5.

43 Gatrell, A. (1988) Providing healthcare with GIS. *BJHC*, 5.5, 28–9.

44 Whitehead, M., Townsend, P., Davidson, D. and Davidson, N. (1992) *Inequalities in Health: The Black Report and the Social Divide.* Penguin, London. ISBN 0 140172 65 3.

45 BJHC (1985) NHS register of computing applications: South West Thames: LAS Command and Control. *BJHC*, 2.1, 49.

46 National Audit Office (1990) *National Health Service Patient Transport Services.* The Stationery Office. ISBN 0 102565 90 2.

47 Foy, M. (1993) The application of GIS and associated technology in the Surrey ambulance service. In Richards, B. and MacOwan, H. (eds) *Current Perspectives in Healthcare Computing*, pp. 245–50. BJHC for BCS, Weybridge. ISBN 0 948198 14 1.

48 Datafile (1986) Transport. *BJHC*, 3.5, 32.

49 Jones, J.E. (1970) The use of computers in the Blood Transfusion Service. In Abrams, M. (ed.) *Medical Computing: Progress and Problems*, pp. 298–307. Chatto & Windus for BCS, London.

50 NHS Register of Computer Applications (1984) North East Thames – MITHRAS/Blood Transfusion. *BJHC*, 1.4, 38.

16 Consumer health

RAY JONES, BOB GANN, MARK DUMAN AND JEANNETTE MURPHY

INTRODUCTION

In 2007, patients may browse the World Wide Web for information about their condition, contact other sufferers by email, send an email to their doctor, complete computer interviews giving a preliminary diagnosis and advice before the consultation, or use interactive television in the home or touch-screen kiosks in the supermarket to get health information. Patients may be given an 'information prescription' by their GP to go to the library to research their condition. The UK Government uses all forms of media to promote a healthier diet.[1, 2] Patients may log on to their medical record via the internet and add details from home blood glucose monitors. This chapter seeks to describe how different strands of work have come together, influenced by economics, policy, developments in technology, and the media, over the last 40 years to form this field of consumer heath informatics.

We have identified six different 'starting points' for the practice and research of consumer health informatics:

- **Provision of consumer health information.** This is often associated with libraries and information providers such as self-help groups. These services tended to be developed by people other than the clinical professions.

- **Public health and health promotion.** These tended to be both regional and national initiatives between NHS and academia to promote healthy lifestyles for whole populations. They were driven by multidisciplinary groups in public health and can be differentiated from the first group by thinking of this as 'information pushed' whereas consumer health information was designed to meet the needs of consumers ('information pull').

- **Chronic disease management.** Within this category we might identify three separate developments: (a) patient-held records, (b) registers and clinical information systems, and (c) computer-based patient education. Although there was some overlap with the public health agenda, the chronic disease agenda has been mainly driven by hospital clinicians working as part of multidisciplinary teams. Computer-assisted learning is described in detail in Chapter 23 but its influence on consumer health informatics will be described in this chapter.

- **Computer–patient interviewing**. This use of technology was again driven by clinicians: often surgeons or those dealing with acute care, who were seeking to make the collection of signs and symptoms from patients (and subsequent diagnosis) more effective.

- **Clinician–patient communication and decision-making**. This approach grew from the study of how clinicians (mostly doctors) and patients communicate and arrive at a decision (e.g. whether to operate, treat medically, or leave). Like the other strands it involved many disciplines but the clinical drivers tended to be in general practice and concerns centred on the psychology and sociology of communication, power sharing, and doctor and patient values. Compared with the chronic disease management strand, which was concerned with information sharing and disease control, the communication strand tended to focus on particular events and decisions (such as treatment decisions).

- **NHS computing**. This was the more managerially, medical records driven (but nevertheless multidisciplinary) strand, which aimed to find ways of using computers in patient administration and to develop clinical records. This aspect of health informatics is described in Chapters 4 and 7, but two aspects that contributed to consumer health informatics are described here. Firstly, those concerned with data capture and the quality of the information within these systems turned to the patient to provide data in more systematic ways (computer–patient interviewing) or to check their own data held on the computer. Secondly and more recently, legislation has required systems designers to find ways of giving patients access to computer-held information.

As with all the health informatics developments described in this book, consumer health informatics activities have to be seen against economic, technological, societal and NHS policy changes. We will briefly review the influences of:

- economics and policy;
- technology;
- mass media.

At the end of the chapter we will consider how these developments have come together to form consumer health informatics and why we are likely not to hear this term for much longer.

PROVISION OF CONSUMER HEALTH INFORMATION

Health information has been available for the (Chinese) public for two millennia. Between 100 BC and 200 AD, a book entitled *Huang Di Nei Jing* ('The Yellow Emperors' Inner Classic') described the laws of Yin-Yang, the Five phases, the nature of Qi (or vital energy), applying these fundamental principles to the prevention and alleviation of illness. Since then thousands of books have contributed to the development of Chinese medicine.[3] Looking to Europe, over 15 treatises on the theory and practice of medicine have been attributed to the Rabbi Moshe ben Maimon (born 1135, Cordoba, Spain). Among these are works on poisons and their antidotes that were used throughout the Middle Ages, a discourse on asthma, and a list of hygienic regulations that would lead to a healthful life. His rabbinical works also contain many references to medicine and its practice and numerous rules for healthful physical well-being including *Sefer HaMadda* ('The Book of Knowledge').[4, 5]

In the UK, there has been a flourishing trade in lay health care guides since the Middle Ages.[6] The *Regimen Sanitas Salernitanum* was a thirteenth-century collection of Latin verses giving practical advice on a health lifestyle.[7] In the seventeenth and eighteenth centuries there were a number of 'best-selling' family health guides with titles such as *Everyman His Own Doctor* and *The Compleat Family Physician*. Despite the availability of written advice, physicians have not always encouraged patient education and autonomy. Hippocrates, for example, while recognizing that 'The patient must cooperate in fighting the disease', advised physicians to 'perform duties calmly and adroitly, concealing most things from the patient'.[8] Professional acceptance of the need for information for patients improved with the awareness of tuberculosis as a leading cause of death in the late 1800s.[9] The chronic nature of the illness and the need for patients to take medication over long periods made some doctors realize that patient education was the key to controlling tuberculosis. The level of activity and support for health and patient education has waxed and waned with the strength of public health. Between 1930 and 1960 it was relatively weak as advances in Western medicine enhanced the belief of a medical cure for everything. In the 1960s, radical movements questioning society (including deference to the professions) and new studies linking smoking and lung cancer brought a greater understanding of the role of lifestyle and health.[10] In addition, studies showed that the vast proportion of morbidity, related to both common illness and chronic illness, was not treated by medical practitioners and that they only dealt with the 'tip of iceberg'.[11] These pointed to the need for ordinary people to have better access to health information. Furthermore, governments recognized that the costs of health care were rising inexorably and that they needed to find ways to contain costs. Getting patients or citizens to take more responsibility for their own health was one strategy.

In the USA, in the early 1960s, voluntary agencies and the US public health service funded successful patient and family health education projects in heart disease, stroke, cancer and renal dialysis.[12] In 1969, the Kaiser Permanente Medical Center in California was the first hospital to provide a service to the public.[13] By the mid-1970s public health information in the USA was a major issue and legislation and funding resulted in a number of library-developed health information services (e.g. Boston).[14] The late 1970s saw UK libraries start to develop community information services, including health information. Two pioneers were the Health Information Service developed at the Lister Hospital, Stevenage, and *Help for Health* in Southampton. The Stevenage service was developed initially by Sally Knight from an index to health information in the general press and popular medical books. In the 1980s, patient libraries and outreach services were developed.[15] *Help for Health* was established following a research project directed by Roy Tabor, which showed the wealth of support available for patients and the lack of an effective mechanism for communicating it.[16] Although run from a hospital the project received a large number (the majority) of its enquiries from the public and health professionals in the community. Run by Bob Gann, it developed the largest collection of health literature in the UK and a database of self-help groups (*Helpbox*), initially on disk and later online. From the late 1980s onwards the number of consumer health information centres increased rapidly, providing advice, booklets, leaflets and often using *Helpbox* as the source of contact details. An early example was the Sunderland Health Information Centre.[17]

A variety of other media and methods have been used in both the 'pull' and the 'push' of consumer health information. Magazines and newspapers have, of course, been a source of health information for decades. In 1979, problem page columnist Claire Rayner estimated that she personally received 40,000 letters in that one year, most of them on health problems.[18] Health promotion units at both local and national level used television, radio, billboards, leaflets, health fairs and other methods to try to get across their healthy lifestyle messages (see Public Health and Health Promotion).[19] Numerous self-help groups for particular conditions developed telephone helplines, leaflets and booklets. For example, by the early 1990s, BACUP (now called cancerBACUP), established in 1985,[20] was receiving 22,000 enquiries a year; by 2005, combined annual enquiries to the helpline, local centres, and by email, reached 64,500. Its website, developed in 1997, now has an average of 250,000 unique visitors a month.[21] Other telephone helplines, such as *Healthline* (from the College of Health) and *Healthcall*, both offering advice on a wide range of health conditions, were set up in the 1980s.[22] Users called *Healthline* and an operator chose and played an appropriate pre-recorded tape. With *Healthcall* users directly dialled a number specific to a topic and a digitized recording was automatically played.

One 'crossover' from another strand was the development of public access, general health information, touch-screen kiosks. *Healthpoint* was a touch-screen kiosk first developed in an academic department of public health at Glasgow University in 1988 and evaluated in a 1990–91 study. Ray Jones had been trying since 1984 to get funding to develop and evaluate a more disease specific 'patient workstation' giving patients access to their medical record and related information. This would have been the natural development of information for patients with chronic disease (see Chronic Disease Management). Frustration at the lack of success and scepticism from funders about the ability of patients to use computers necessitated a change of approach. Through student projects and eventual Scottish Office funding,[23] it was demonstrated that members of the public of all ages and social class would use touch-screen kiosks to access health information in public places (e.g. supermarkets, post offices, pubs).[24] *Healthpoint* was developed and installed at about 50 sites (bought mainly on health promotion or new 'NHS Hospital Trust' budgets), before being abandoned in 1999 with the launch of NHS Direct and commercially backed kiosks (such as *In Touch with Health*).[25]

This 20-year evolution of consumer health information in the UK reached its most comprehensive development in the late 1990s with the establishment of NHS Direct. NHS Direct now provides 24-hour access to nurse advice and health information using a range of channels including telephone call centres, a website (NHS Direct Online),[26] a digital television service (NHS Direct Interactive) and a printed self-help guide. The first pilot call centres were set up in 1998, with the website launched soon afterwards in 1999 and the digital television service going live on digital satellite in December 2004. The scale of the service is impressive: in a month there are over 600,000 calls to the NHS Direct telephone service and over one million visits to the website, while the self-help guide has been delivered to 18 million homes.[27]

PUBLIC HEALTH AND HEALTH PROMOTION

During the 1970s the contribution of modern medicine in improving the health of people was called into question. For example, it was shown that mortality rates in England and Wales declined long before the advent of modern medical care during the twentieth century and the negative consequences of some aspects of modern medical care were exposed.[28, 29] The health promotion movement had its origins in a Canadian government report in 1974.[30] This was the first recognition by a national government of the limitations of modern medicine. The report emphasised the role of health promotion as a key strategy for improving health, looking beyond biomedical antecedents to the psychological, social and economic dimensions of health and illness. The Ottawa Charter for Health Promotion followed in 1986 and defined health promotion as the process of enabling people to increase control over, and to improve, their health through building healthy public policy, creating supportive environments, strengthening community action, developing personal skills and reorienting health services.[31]

Health promotion in the UK tended to be through regional and national initiatives between NHS and academia to promote healthy lifestyles for whole populations. These were driven by multidisciplinary groups in public health and can be differentiated from the first strand of consumer health informatics in that it was 'information pushed' whereas consumer health information was meeting the needs of the consumer ('information pull').

By 1992 there was a considerable range of health promotion software suitable for use in schools, mainly on BBC computers, but also increasingly on IBM PCs.[32] Topics included AIDS, alcohol, diet, food hygiene, sources of infection, smoking and others.

The move from stand-alone computers to software on the internet from the mid-1990s onwards has fostered the development of consumer information (above) rather than health promotion. The main difficulty of computer-based health promotion remains engaging people who are not necessarily seeking health information (i.e. 'push' rather than 'pull').

The touch-screen kiosk, *Healthpoint*, was developed from a public health perspective, and in the mid-1990s was bought by a number of health promotion departments. Sited in shopping centres, pubs, sports centres and elsewhere, it managed to catch people's attention and its most frequently chosen topics were public health topics such as drinking, smoking, sexual health, exercise and diet. Nevertheless, television, radio and the press remain better media for health promotion 'push'. Health authorities have been using television adverts to campaign on health issues since the drinking and driving campaign adverts in 1964 and now regularly use it for campaigns on sensible drinking, protection against skin cancer, sexual health, etc.[33] The mass media is a major influence on knowledge, beliefs and attitudes to health (see Health as a Mass Media Topic).

CHRONIC DISEASE MANAGEMENT

The changes in Western medicine from the middle of the twentieth century, moving from a focus on infectious and acute disease to the maintenance of people with chronic disease, is well known. The management of chronic disease has been one of the drivers of consumer health informatics. Within this strand, we might identify a number of separate developments: (a) registers and clinical information systems, (b) patient-held records and (c) computer-based patient education. Although there was some overlap with the health promotion strand, the chronic disease agenda was mainly driven by hospital clinicians operating as part of multidisciplinary teams.

Registers and clinical information systems

In the 1960s, clinicians dealing with chronic diseases such as diabetes, renal disease and thyroid disease saw the importance of a systematic approach to their care. They began to explore the potential for the use of computers in the follow-up of patients, and the involvement of the patient in that process. The 'thyroid follow-up system', developed in Aberdeen by Tony Hedley *et al.*, was one of the earliest examples of a register.[34] A register for newly diagnosed diabetic children was another early one.[35] A number of centres developed 'registers' for diabetes or thyroid disease with increasing amounts of clinical data. For example, in the late 1970s, three diabetes centres, King's College Hospital and St Thomas' Hospital,[36, 37] in London, and the Queen's Medical Centre in Nottingham,[38] began to develop registers and clinical information systems. Similarly, renal units made widespread use of a clinical information system and were among the first widespread users of departmental systems (see Chapter 13).[39]

Patient-held records

Others involved in chronic disease (also in general practice) had recognized the potential of patient-held records and had started to use cooperation cards or patient-held records by which the patient may record home measurements.[40–42] The Nottingham diabetes system brought these ideas together with a clinical system that produced a clinical summary for the hospital notes, GP notes and for the patient.[43] This raised the question of whether or not patients should have access to their medical record (which had been previously debated in a non-computer-based context).[44–47] In the late 1970s and early 1980s most hospital case notes in Britain still had 'NOT TO BE HANDLED BY THE PATIENT' in large letters on the front. With the introduction of computers and clinical information systems to issue patient-held records it was possible to study when and why doctors withheld information.[48]

Computer-based patient education

Computer-assisted learning (CAL) for medical, nursing and other students and staff has been in development since 1966 (see Chapter 23). Patient education was recognized from the 1960s or before as a very important part of chronic disease management and as computers became easier to use and applications of CAL more widespread[49] various groups found ways of using them for computer-based patient education. For example, in the mid-1980s, Peter Wise *et al.* developed a computer-printed paper-based feedback system for informing diabetes patients and Homer developed a simulation of a dialysis unit to train renal patients.[50, 51]

COMPUTER–PATIENT INTERVIEWING

Computer–patient interviewing started with a different motivation: to use the computer to take a patient history that was consistent, accurate, and complete and where the computer might suggest a likely diagnosis. Work on this started in the early 1960s with Warner Slack and others in the USA,[52, 53] and by Chris Evans (and the Mickie program),[54, 55] Wilfred Card, Robin Knill-Jones and others in Glasgow (the Glasgow Dyspepsia System – GlaDys).[56, 57] Although this work never seemed to gain momentum beyond specific projects, in 1994, Bill Dodd and Peter Drury, from the NHS Management Executive showed great interest in getting patients to interact with computers as a way of entering information into computerized clinical records. A review of the literature was able to cite 150 papers demonstrating the potential of direct patient input,[58] but on contacting authors of systems still in use (for a workshop), it became evident that the main problems were in implementation and maintenance of such systems. In order to get critical mass, it would be necessary to build larger systems, addressing a range of conditions, which could be merged with other record and patient access systems.

It was clear that the use of computers by patients to give information in this way would only succeed if they enabled patients to also get information. Some experimental work was carried out in the 1990s to explore different types of interfaces.[59]

Computer–patient interviewing has now started, at last, to go 'mainstream', at least in the USA where a number of systems such as Instant Medical History are being implemented at numerous sites.[60, 61]

CLINICIAN–PATIENT COMMUNICATION AND DECISION-MAKING

This approach grew from studies of how clinicians (mostly doctors) and patients communicate. Philip Ley was one of the earliest pioneers of this work.[62, 63] Medical power has waxed and waned.[64] Things have certainly changed since the late 1800s when Wendell-Holmes announced to a class of medical students:

> *Your patient has no more right to all the truth you know than he has to all the medicine in your saddle bags . . . he should only get so much as is good for him.*

Studies of shared agreement between patient and doctor can be traced from the work of Balint in the 1950s and 1960s,[65] through Starfield in 1981,[66] to Charles and Gafni in 1997,[67] and to Marinker and Britten.[68, 69]

Compared with the chronic disease management strand, which was concerned with information sharing and disease control, this strand tended to focus on particular events and decisions (e.g. whether to operate, treat medically, or leave). Like all the strands that have fed into the development of contemporary consumer health informatics, it involved many disciplines but the clinical drivers tended to be in general practice and concerns centred around the psychology and sociology of communication, power sharing, and doctor and patient values.

Research in the USA in the 1970s, carried out by the epidemiologist John Wennberg and colleagues,[70] demonstrated that clinical outcomes varied across the country. In subsequent initiatives these clinicians aimed to involve patients in decisions about evidence-based treatment options and developed interactive video discs to help this. In 1992, when the concept of evidence-based medicine was gaining ground in the UK, the King's Fund agreed to test one of these. In 1995, the Fund established a development initiative, called Promoting Patient Choice, to develop a series of information materials for patients based on the principle of shared clinical decision-making. With additional funding from the Gatsby Charitable Foundation, preliminary work established a number of criteria for good quality, evidence-based, patient information. Seven projects were selected for development based on these criteria:

- Anxiety in Asian women – audiotapes, a directory of services and booklets in four Asian languages for Asian women (Redbridge and Waltham Forest Health Authority).

- Bedwetting (nocturnal enuresis) – multimedia package for children (Nottingham University/Nottingham City Hospital).

- Colorectal cancer – multimedia package (Hull University/Castle Hill Hospital).

- Post-operative pain relief – booklet (Queen's Medical Centre, Nottingham).

- Ulcerative colitis – personal organizer (Manchester University/Hope Hospital).

- Urge incontinence – multimedia package (Bristol Urological Institute).

- Hormone Replacement Therapy (HRT) – trial in general practice in Oxford of the interactive video on HRT produced by the Foundation for Informed Medical Decision Making.

In parallel with this work, colleagues at the King's Fund conducted a study comparing patients' and clinicians' views on existing patient information showing that quality was generally poor.[71] This influenced thinking at the Department of Health (DoH), which, in 1995, set up a Centre for Health Information Quality,[72] whose aim was to determine, and share, standards for quality in health information. Independently, the King's Fund ran conferences and published specialist reports, with the aim of disseminating as widely as possible, the concept of patient choice based on good quality evidence.[73]

In 1999, the DoH funded a programme of 12 research projects aimed at exploring the complexities of patient–professional interaction around treatment decisions and the challenges faced by organizations with an interest in and commitment to involving patients and the public in health care decision-making.[74] DoH policy and initiatives in patient involvement and patient partnership since have been both at the individual (treatment) level as well as at public (policy) level. The latter is outside the scope of consumer health informatics.

In the 1990s, the clinical professions started giving higher priority to patient participation and the relevance of good-quality patient information. Work and publications on concordance were developed through a number of routes including academic and professional bodies.[75] Initiatives included the Doctor Patient Partnership (now known as Developing Patient Partnership), established by the British Medical Association in 1992, and

Patient Liaison Groups, established by a number of the Royal Colleges. In 1999, the *British Medical Journal* (*BMJ*) dedicated its September issue to the theme of embracing patient partnerships.[76] (Four years later the *BMJ* published a 'patient issue' and an issue on concordance.[77, 78])

Towards the end of the 1990s this strand of development started to merge with others such as Consumer Health Information. For some years Bob Gann had a column in the *BMJ* on 'what your patient was reading' and there was growing recognition of the need to integrate developments in patient education and consumer health information into clinical practice.[79] In 1998, a multi-centre study was set up to identify the learning needs of clinicians and the organizational changes needed if patients' information needs were to be fulfilled. Funded by the Enabling People Programme, the Universities of Glasgow, Birmingham and Nottingham, together with Great Ormond Street Hospital and the King's Fund, interviewed some 100 clinicians and identified learning objectives and the need for considerable organizational change.[80] These results were shared at a conference in 1999 called *Help! Does my patient know more than me?*[81]

NHS COMPUTING

This was the more managerial, medical records driven (but nevertheless multidisciplinary) strand, which sought to find ways of using computers in patient administration and in developing electronic clinical records. Drivers from within health informatics came from those concerned with data capture and the data quality of the information within these systems. Drivers from without came from those engaged in chronic disease management who were struggling to merge disease specific and hospital or primary care systems. These projects are described in more detail in Part 1 but it is worth noting here some developments and how these merged with consumer health informatics.

Although the USA was further advanced in the development of departmental clinical systems,[82] there had been pioneering efforts in the UK, such as the hypertension system in Glasgow.[83] However, it was not until the technological advance from punched cards to direct data entry in the late 1970s that clinical systems began to be more widely used. Chronic disease systems started to develop as 'departmental systems' and in the early 1980s, in various hospitals, these began to be linked with the hospital patient administration system. For example, the Nottingham diabetes system, which had been a 'stand-alone' clinical system, was merged with the new hospital administration system in 1982, demonstrating a lack of consistency between the two sources. Patient involvement (the norm in diabetes care, in the form of a computer-produced patient-completed questionnaire) was introduced at a hospital level to improve the quality of information on the hospital patient master index.[84] Others had noted the potential and importance of patients being able to check their own computer-held information.[e.g. 85]

Throughout the last 30 years, data capture has remained one of the major problems of NHS computing and this was one of the motivators for NHS interest in computer–patient interviewing. Those responsible for the Information Strategy (Bill Dodd in Scotland and Peter Drury in England) commissioned a review and workshop.

Another driving force in the last 20 years has been the Data Protection Act 1984 introducing Subject Access Rights.[86] Long-term champions of patient-held records such as Brian Fisher (see also Patient-held records) together with the push from legislation and other strands helped to bring to fruition recent initiatives such as the *Copying Letters to Patients* guidelines.[87, 88]

OTHER DRIVERS ON CONSUMER HEALTH INFORMATICS

The six strands described above represent to some degree the 'membership' of consumer health informatics. There were clinicians, academics and other health professionals who identified with work in one or more of these strands who would see themselves as having contributed to the development of consumer health informatics. There were, and continue to be, other 'drivers'.

Economics and policy

Because of the way health care is funded in the UK, Government policy has had a big impact on NHS computing. Especially in recent years, priorities tend to be set centrally and these central priorities have favoured developments in consumer health informatics. This shift in policy can be seen in the increasing emphasis on the role of the patient in NHS policy documents from Körner through to *Information for Health*.[89, 90] Although the Körner Committee recognized the role of the patient in checking the quality of hospital data, the emphasis was on managerial information.[91] Priorities then shifted to the needs of clinicians and not until 1998 do the information needs of patients, carers and citizens become central.

For those working in the field it may be satisfying to think that this has all been a result of politicians noting the results and potential of clinical and academic development and research. (A quarter of a century for research and innovation to become embedded in policy is not unusual.) Realistically, however, the long-term economic situation of a health service with infinite demand and an ageing population clearly is a major driver towards a greater role for the patient or consumer.

Technology

The developments in technology from mainframe to microcomputer and the internet, and the merging of telecommunications, television, computing and other technologies are well known. These developments have clearly been essential to the development of consumer health informatics. If the famous prediction by Professor D. Hartree ('We have a computer here in Cambridge, there is one in Manchester and another at the National Physical Laboratory. I suppose there ought to be one in Scotland, but that's all' said in 1951)[92] had come true, patients would never have had ready access first to microcomputers, then to the internet, and more recently to mobile phones and interactive television (and who knows what devices in the future?) for health information. It is clear that the technology drives this area of work. Mobile phones become ubiquitous so researchers start to see what health information-need they can meet using mobile phones.

HEALTH AS A MASS MEDIA TOPIC

In 2001, a survey of NHS Direct kiosks revealed that 1666 users, in the preceding two weeks, had obtained health information from:

- health professionals, family and friends (33 per cent);
- printed information (15 per cent);
- television (6 per cent);
- computers (4 per cent);
- the telephone (2 per cent).[93]

However, this is probably an underestimate of the influence of television because of its impact within fiction and 'entertainment'. One episode of *Coronation Street*, in which a character died from cervical cancer was attributed with a rise of 21 per cent in the uptake for cervical screening.[94]

There is a substantial literature that looks critically at the role of mass media in conveying images/information of health, illness and medicine. Anna Karpf reviewed a number of different themes and found that, between the 1920s and the 1940s, the media were promoting fitness and healthy eating.[95] Early radio talks about health in 1927 attracted an audience of 10 million and generated 20,000 letters in three months. In the early days of broadcasting, the BBC was preoccupied with health rather than medicine because, Karpf suggests, of the economic conditions of the time. Austerity gave rise to 'a punitive individualism where each citizen had a moral duty to make themselves fit'. In the early days of the NHS, there was very little public debate about deficiencies in the health care system but, by the late 1970s, consumer critiques of health had become commonplace.

In 1976, the BBC reviewed its medical programmes and their effect on lay audiences.[96] At that time the main television 'scientific' health output came from the *Horizon* series, a quarter of which was on health and included, in 1975, asbestosis, drug abuse, lead in water, heart transplants, cannabis and medical negligence. Other 'science' outlets for health included *Tomorrow's World, Tonight, Panorama, Nationwide*, and special series such as *All in the Mind*.

Television and other media aim to entertain while being responsible in their output, but throughout television's history there has been some friction between television and the medical profession. In 1975, a BMJ editorial said:

Many television documentaries ... have done a first class job on health education, but there has been a trend inside journalism for self-appointed and self-taught experts to investigate a subject and reach an opinion . . . and then to promote that view with all the skills of modern propaganda techniques.

REFERENCE 97

Similar articles/editorials have appeared over the last three decades.[98]

It may seem suspect to suggest that media representations of health (be they through the news and current affairs or through popular entertainment such as 'soaps') are part of consumer health informatics, but the merging of technologies to provide digital television, internet television, programme websites, etc. means that we need to understand more about how health is portrayed in the media. Although there has been understandable concern about the inaccuracy of some health websites,[99] the impact of such information on population knowledge and behaviour is likely to be small compared with unbalanced reporting or inaccurate representation of medical or health information within popular television fiction.

Further study of the relationship between television and other mass media and the internet and other computer-based means, as sources of consumer health information, are likely to become more important as the technologies continue to merge.

THE DEVELOPMENT OF CONSUMER HEALTH INFORMATICS

During the late 1980s and 1990s there was increasing crossover between the hitherto separate strands, and a coming together of people from the different 'camps'. For example, Ray Jones and others in Glasgow shifted from focusing on clinical records for health professionals to exploring patient online access to their medical records.[100, 101] The aim of this work was to develop a 'patient workstation',[102, 103] which would link their medical record to health education and other appropriate materials tailored to their needs. They drew on a variety of streams: work on patient-held records, clinical record systems, patient education and computer patient interviewing, and brought in newer techniques from computer science.[104]

Work originating from consumer health information and the 'library'-base merged with the shared decision-making strand and together these have formed the basis of the development of much of DoH policy and initiatives such as NHS Direct.

Various forums have been set up for researchers and practitioners in consumer health informatics, such as the consumer-health-informatics discussion list,[105] set up after the 1997 King's Fund conference, *Promoting*

Patient Choice Together. These have helped to bring together the different strands of consumer health informatics over the last 10 years.

Consumer health informatics has now 'come of age', both academically and in terms of practical outcomes. The NHS has NHS Direct as one of its major 'flagships' providing the largest telemedicine service in the world. On the academic side, consumer health informatics interventions are being evaluated with randomized trials with the same rigour as drug trials.[e.g. 106] Consumer Health Informatics was a topic of a paper in the *British Medical Journal* in 2000 in which Eysenbach defined it as:

> . . . *the branch of medical informatics that analyses consumers' needs for information; studies and implements methods of making information accessible to consumers; and models and integrates consumers' preferences into medical information systems.*

REFERENCE 107

The terminology, however, is changing and the phrase consumer health informatics itself may be consigned to history. Some people, such as the NHS Service Delivery Organization programme, now tend to use the term 'ehealth' to refer to most of what was considered health informatics, while others equate ehealth with consumer health informatics.[108]

In 1895, the *British Medical Journal* commented:

> *In everything else machinery is used, why not in medicine? So the public drops its penny and receives its prescription. It is not difficult to foresee the time when the prescribing chemist will be a thing of the past. There will, in fact, be no need for him; his shop will be crowded with penny-in-the-slot machines, by the use of which the patient's weight will be taken, his eyesight tested, his urine examined, his vital capacity ascertained, his muscle power measured, his knee-jerks recorded, his pulse trace taken, and ultimately his prescription written plainly by a typewriter, so that it can be made up by an assistant at a mere living wage, while the chemist, free from all responsibility, will take the fee and flourish exceedingly.*

REFERENCE 109

A hundred years later the *BMJ* returned to the theme with Richard Smith describing the change from industrial to information age medicine.[110] The most likely reason for 'consumer health informatics' to be abandoned as a concept is that in the future it could become the biggest and dominant part of health informatics. Patients using computers in all aspects of health care will probably become so commonplace that there will no longer be a need for a special name.

REFERENCES

1   Webref (accessed May 2007) Self care. www.dh.gov.uk/PolicyAndGuidance/OrganisationPolicy/SelfCare/fs/en

2   Webref (accessed May 2007) Better information, better choices, better health: putting information at the centre of health. www.dh.gov.uk/PublicationsAndStatistics/Publications/PublicationsPolicyAndGuidance/PublicationsPolicyAndGuidanceArticle/fs/en?CONTENT_ID=4098576&chk=hXI1Uf

3   Webref (accessed May 2007) www.senhealth.com

4   Webref (accessed May 2007) www.jewishhealing.com

5   Webref (accessed March 2005) www.rambam.co.il

6   Gann, R. (1987) The people their own physicians: 2000 years of patient information. *Health Libraries Review*, 4(3), 151–5.

7   Lafaille, R. and Hiemstra, H. (1990) The regimen of Salerno: a contemporary analysis of a medieval lifestyle programme. *Health Promotion International*, 5(1), 57–74.

8   Hippocrates (1967) *On Decorum and the Physician*. Harvard University Press, Cambridge MA.

9   Bartlett, E. (1986) Historical glimpses of patient education in the United States. *Patient Education and Counseling*, 8, 135–9.

10 Doll, R. and Hill, A.D. (1950) Smoking and carcinoma of the lung. *BMJ*, 221(ii), 739–48.

11 Last, J.M. (1963) The 'iceberg': completing the clinical picture in general practice. *The Lancet*, 2, 28–31.

12 Fiori, F. (1974) Health education in a hospital setting. *Health Education Monographs*, 2, 11–29.

13 Collen, F.B. and Soghikian, K. (1974) A health education library for patients. *Health Service Reports*, 89, 236–43.

14 Gartenfield, E. (1978) The Community Health Information Network (CHIN): a model for hospital and public library cooperation. *Library Journal*, 103(7), 1911–14.

15 Varnavides, C.K., Zermansky, A.G. and Pace, C. (1984) Health library for patients in general practice. *BMJ*, 288, 535–7.

16 Gann, R. (1981) *Help For Health: The Needs of Practitioners for Information about Organizations in Support of Health Care*. British Library R&D Report 5613. British Library, London.

17 Childs, S.M. (1990) *Health Information in the High Street. A DIY Manual for Setting Up and Running a Health Information Centre. Based on Information Provided by Susan J. Milner from her Experience in Setting Up and Running the Sunderland Health Information Centre*. Health Information Series No. 4. (Unknown source.)

18 Rayner, C. (1979) Reality and expectation of the British NHS consumer. *Journal of Advanced Nursing*, 4, 69–77.

19 Gann, R. (1986) Information services and health promotion: what libraries can do. *Health Education Journal*, 45(2), 112–15.

20 Clement-Jones, V. (1985) Cancer and beyond: the formation of BACUP. *BMJ*, 291, 1021–3.

21 Leibowitz, A. (2005) *CancerBACUP*. Personal communication, June.

22 Jones, R.B., McGhee, S., Hedley, A.J. and Murray, K.J. (1988) Patient access to information. In Richards, B. (ed.) *Current Perspectives in Health Computing*, pp. 206–12. BJHC for BCS, Weybridge.

23 Jones, R.B. and McLachlan, K. (1990) Helathpoint: a public access health information system. In Richards, B. (ed.) *Current Perspectives in Health Computing*, pp. 65–9. BJHC for BCS, Weybridge.

24 Jones, R.B., Navin, L.M. and Murray, K.J. (1993) Use of a community-based touch-screen public-access health information system. *Health Bulletin*, 51, 34–42.

25 Webref (accessed May 2007) In Touch With Health. www.intouchwith-health.co.uk/default.htm

26 Webref (accessed May 2007) NHS Direct. www.nhsdirect.nhs.uk

27 Gann, R. (2004) Enabling patient access and expertise. In (eds unknown) *Thought Leaders: Essays From Health Innovators*. Cisco Systems.

28 Mckeown, T. (1976) *The Role of Medicine: Dream, Mirage or Nemesis*. The Nuffield Provincial Trust Hospital, London.

29 Illich, I. (1976) *Medical Nemesis: The Expropriation of Health*. Pantheon Books, New York.

30 Lalonde, M. (1974) *A New Perspective on the Health of Canadians*. Health and Welfare, Ottawa.

31 World Health Organization (1986). *Ottawa Charter for Health Promotion*. WHO, Geneva.

32 Leonard, J. (1992) *Health-related Computer Software*. Health Education Authority, London

33 Berridge, V. and Loughlin, K. (2005) Smoking and the new health education in Britain 1950s–1970s. *Am. J. Public Health*, 95(6), 956–64.

34 Hedley, A.J., Scott A.M., Deans Weir, R. and Crooks, J. (1970) Computer-assisted follow-up registry for the North East of Scotland. *BMJ*, 1, 556–8.

35 Bloom, A., Hayes, T.M. and Gamble, D.R. (1975) Registry for newly diagnosed diabetic children. *BMJ*, 3, 580–3.

36 Watkins, G.B., Sutcliffe, T., Pyke, D.A. and Watkins, P.J. (1980) Computerization of diabetic clinic records. *BMJ*, 281, 1402–3

37 Sonksen, P. and Williams, C. (1996) Information technology in diabetes care 'Diabeta': 23 years of development and use of a computer-based record for diabetes care. *Int J Biomed Comput.*, July, 42(1–2), 67–77.

38 Jones, R.B., Hedley, A.J., Peacock, I., Allison, S.P. and Tattersall, R.B. (1983) A computer-assisted register and information system for diabetes. *Methods of Information in Medicine*, 22, 4–14.

39 Gordon, M., Venn, J.C., Gower, P. and De Wardener, H.E. (1983) Experience in the computer handling of clinical data for dialysis and transplantation units. *Kidney International*, 24, 455–63.

40 Ezedum, S. and Kerr, D.N.S. (1977) Collaborative care of hypertensives, using a shared record. *BMJ*, 26, 1402–3.

41 Hetzel, M.R., Williams, I.P. and Shakespeare, R.M. (1979) Can patients keep their own peak-flow records reliably? *The Lancet*, 1, 597–8.

42 Koscel, J.M. (1974) How patients use a log-book. *Am. J. Nursing*, 74, 1307.

43 Jones, R.B., Hedley, A.J. and Peacock, I. (1980) A patient-held record and new methods of long-term care for patients with diabetes mellitus. In Velthoven, J.J. (ed.) *The 8th International Congress on Health Records*, pre-papers, vol. 1, pp. 357–72. Red Book, The Hague.

44 Shenkin, B.N. and Warner, D.C. (1973) Sounding Board. Giving the patient his medical record: a proposal to improve the system. *New Eng. J. Med.*, 289, 688–92.

45 Murray, F.A. and Topley, L. (1974) Patients as record holders. *Health and Social Services Journal*, July, 27, 1675.

46 Golodetz, A., Ruess, J. and Milhous, R. (1976) The right to know: giving the patient his medical record. *Arch. Phys. Med. Rehabil.*, 57, 78–81.

47 Simonton, M.J., Neuffer, C.H. and Stein, E.J. (1977) The open-medical record – an educational tool. *J. Psychiatr. Nurs.*, 15, 25–30.

48 Jones, R.B. and Hedley, A.J. (1987) Patient-held records: censoring of information by doctors. *J. Roy. Coll. Phys.*, 21, 35–8.

49 Etzwiler, D.D. (1973) Current status of patient education. *JAMA*, 220, 583.

50 Wise, P., Farrant, S., Dowlatshahi, D., Fromson, B. and Meadows, K. (1985) Computer-based learning and prescriptive feedback improves diabetes knowledge and control (Abstract). *Diabetic Medicine*, 2, 302a.

51 Homer, G.R. (1985) Computer-aided training in a renal dialysis ward – a unique approach. In Bryant, J. and Kostrewski, B. (eds) *Current Perspectives in Health Computing*, pp. 113–27. BJHC for BCS, Weybridge.

52 Slack, W.V., *et al.* (1966) A computer-based medical history system. *New England Journal of Medicine*, 274, 194–8.

53 Slack, W.V. and Van Cura, L.J. (1968) Patient reaction to computer-based medical interviewing. *Comput.Biomed.Res.*, 1, 527–31.

54 Evans, C.R. and Wilson, J. (1972) A program to allow computer based history-taking in cases of suspected gastric ulcer. *Natl. Phys. Lab. Rep. Com.Sci.*, 49.

55 Evans, C.R., Price, H.C. and Wilson, J. (1973) Computer interrogation of patients with respiratory complaints in a London hospital. *Natl. Phys. Lab. Rep. Com. Sci.*, 69.

56 Card, W.I., Nicholson, M., Crean, G.P., Watkinson, G., Evans, C.R., Wilson, J. and Russell, D. (1974) A comparison of doctor and computer interrogation of patients. *Int. J. Biomed. Comput.*, 5, 175.

57 Lucas, R.W., Card, W.I., Knill-Jones, R.P., Watkinson, G. and Green, G.P. (1976) Computer interrogation of patients. *BMJ*, 2, 623–5.

58 Jones, R.B. and Knill-Jones, R.P. (1994) *Electronic Patient Record Project: Direct Patient Input to the Record*. Report for the Strategy Division of the Information Management Group of the NHS ME, April. (Available from ray.jones@plymouth.ac.uk).

59 Al Barwani Al-Barwani, F., Jones, R.B., Cawsey, A. and Knill-Jones, R.P. (1997) A randomized trial of different patient–computer interactions. In Richards, B. (ed.) *Current Perspectives in Healthcare Computing*, pp. 77–84. BJHC for BCS, Weybridge.

60 Webref (accessed May 2007) Mayo Clinic Proceedings. www.mayoclinicproceedings.com/Abstract.asp?AID=263&Abst=Abstract&UID=

61 Webref (accessed May 2007) Instant Medical History. www.medicalhistory.com (see also a presentation by John Bachman in March 2005 conference www.chiirup.org.uk/conference_presentations.html).

62 Ley, P. (1982) Satisfaction, compliance and communication. *British Journal of Clinical Psychology*, 21(4), 241–54.

63 Ley, P. and Spelman, M.S. (1965) Communications in an outpatient setting. *British Journal of Social & Clinical Psychology*, 4(2), 114–16, June.

64 Webref (accessed May 2007) Gray, M. and Rutter, H. The resourceful patient. Chapter 1: The rise and fall of the medical empire – the evolution of medical power. www.resourcefulpatient.org/text/sec1.htm

65 Balint, M. (1955) The doctor, his patient and the illness. *The Lancet*, 268 (6866), 683–8.

66 Starfield, B., Wray, C., Hess, K., Gross, R., Birk, P.S. and Dlugoff, B.C. (1981) The influence of patient–practitioner agreement on outcome of care. *American Journal of Public Health*, 71(2), 127–31.

67 Charles, C., Gafni, A. and Whelan, T. (1997) Shared decision-making in the medical encounter: what does it mean? (Or it takes at least two to tango). *Soc. Sci. Med.*, 44, 681–92.

68 Marinker, M. (1997) Writing prescriptions is easy. *BMJ*, 314, 747–8.

69   Britten, N., Stevenson, F.A., Barry, C.A., Barber, N. and Bradley, C.P. (2000) Misunderstandings in prescribing decisions in general practice: qualitative study. *BMJ*, 320, 484–8.

70   (Author unknown) (1997) In focus – John Wennberg, renegade father of the outcomes movement. *Healthcare International* (The Economist Intelligence Unit), Chapter 2, 1st quarter.

71   Coulter, A., Entwistle, V. and Gilbert, D. (1998) *Informing Patients: An Assessment of the Quality of Patient Information Materials.* King's Fund, London.

72   Webref (accessed May 2007) www.hfht.org/chiq

73   DoH (1996) *Patient Partnership: Building a Collaborative Strategy.* Department of Health, Leeds, June.

74   DoH (1999) *Patient and Public Involvement in the New NHS.* Department of Health, Leeds, September. www.healthinpartnership.org (URL could not be accessed September 2007).

75   Coulter, A. (1997) Partnerships with patients: the pros and cons of shared clinical decision-making. *J. Health Serv. Res. Policy*, 2, 2, April.

76   Coulter, A (ed.) (1999) Paternalism or partnership? *BMJ*, 319, 719–20, 18 September.

77   Eaton, L. (2003) Saddle up, partner. *BMJ*, 326, 1275, 14 June.

78   BMJ (2003) From compliance to concordance. *BMJ*, 327, 7419, 11 October.

79   Jones, R. (1999) Consumer health informatics: the need to integrate with clinical practice. Keynote paper. In *Proceedings of the Primary Health Care Specialist Group of the BCS*, pp. 4–9. PHCSG, Cambridge, September.

80   Jones, R., Tweddle, S., Hampshire, M., Hill, A., Moult, B. and McGregor, S. (2000) *Patient-led Learning for the Clinical Professions in Fulfilling the Information Needs of Patients.* NHS Information Authority, Report ref. 2000-IA-280, June.

81   Kohner, N. and Hill, A. (2000) *Help! Does my Patient Know More Than Me?* King's Fund, London.

82   Levy, R.P., Cammarn, M.R. and Smith, M.J. (1964) Computer handling of ambulatory clinic records. *JAMA*, 190, 1033–7

83   Kennedy, F., Cleary, J.J., Roy, A.D. and Kay, A.W. (1968) 'SWITCH': a system producing full hospital case-history on computer. *The Lancet*, 2, 1230–3.

84   Jones, R.B., Nutt, R.A. and Hedley, A.J. (1984) Improving the quality of data in a computerized patient master index: implications for costs and patient care. *Effective Health Care*, 3, 97–103.

85   Bronson, D.L., Rubin, A.S. and Tufo, H.M. (1978) Patient education through record sharing. *Quality Review Bulletin*, 4, 12, 2–4.

86 HMSO (1984) *Data Protection Act 1984*. HMSO.

87 Baldry, M., Cheal, C., Fisher, B., Gillet, M. and Huet, V. (1986) Giving patients their own records in general practice: experience of patients and staff. *BMJ Clinical Research*, 292, 596–8.

88 Webref (accessed May 2007) www.dh.gov.uk/Policyandguidance/ Organisationpolicy/PatientsAndPublicInvolvement/Copyinglettersto patients/index.htm

89 DHSS (1982) *Steering Group on Health Services Information* (Chair Körner, E.). HMSO.

90 DoH (1998) *Information For Health: An Information Strategy for the Modern NHS 1998–2005*. DoH

91 NHS/DHSS Health Services Information Steering Group (1984) *Making Data Credible*. Proposals made by workshop held in February. (Chair Mason, A.). King's Fund, London.

92 Cited by Kay, S. (1998) Of babies and bath water. *British Journal of Healthcare Computing and Information Management*, 15, 21–2.

93 Jones, R. (2001) *Evaluation of NHS Direct Kiosks*. Final Report to NHS Direct. University of Glasgow.

94 Howe, A., Owen-Smith, V. and Richardson, J. (2002) The impact of a television soap opera on the NHS Cervical Screening Programme in the North West of England. *Journal of Public Health Medicine*, 24, 299–304.

95 Karpf, A. (1988) *Doctoring the Media: The Reporting of Health and Medicine*. Routledge, London.

96 British Broadcasting Corporation (1976) *The BBC's Medical Programmes and Their Effects on Lay Audiences*. BBC, Broadwater Press Ltd, Welywn Garden City.

97 Anonymous (1975) Medicine on television. (Editorial). *BMJ*, 1, 539 (March 8).

98 Anonymous (1980) An appalling *Panorama*. (Editorial). *BMJ*, 281, 1028.

99 Impicciatore, P., Pandolfini, C., Casella, N. and Bonati, M. (1997) Reliability of health information for the public on the World Wide Web: a systematic survey of advice on managing fever in children at home. *BMJ*, 314, 1875–8.

100 Jones, R.B. and McGhee, S.M. (1990) *Patient Online Access to Medical Records*. Scottish Home & Health Dept, Final Report to Health Service Research Committee.

101 Jones, R.B., McGhee, S.M. and McGhee, D. (1992) Patient online access to medical records in general practice. *Health Bulletin*, 50, 143–50.

102 Jones, R.B., Cawsey, A., Binsted, K., *et al.* (Date unknown) A Patient Workstation? In Hedley, A.J. (ed.) *MIC94: Third Hong Kong (Asia–Pacific) Medical Informatics Conference*. Hong Kong Computer Society, Hong Kong.

103 Jones, R.B., Cawsey, A., Al-Barwani, F., Reynolds, J. and Knill-Jones, R. (1996) Researching a patient workstation. In Richards, B. and de Glanville, H. (eds) *Current Perspectives in Healthcare Computing*, pp. 678–86. BJHC for BCS, Weybridge. ISBN 0 948198 24 9.

104 Cawsey, A.J., Binsted, K.A. and Jones, R.B. (1995) An online explanation of the medical record to patients via an artificial intelligence approach. In Richards, B., de Glanville, H. and MacOwan, H. (eds) *Current Perspectives in Healthcare Computing*, pp. 269–75. BJHC for BCS, Weybridge. ISBN 0 948198 20 6.

105 Webref (accessed May 2007) www.jiscmail.ac.uk/lists/consumer-health-informatics.html

106 Jones, R.B., Pearson, J., McGregor, S., Cawsey, A., Barrett, A., Atkinson, J.M., Craig, N., Gilmour, W.H. and McEwen, J. (1999) Randomized trial of personalized computer-based information for cancer patients. *BMJ*, 319, 1241–7.

107 Eysenbach, G. (2000) Consumer Health Informatics. *BMJ*, 320, 1713–16.

108 Webref (accessed May 2007) www.sdo.lshtm.ac.uk/index.html

109 Reiser, S.J. (1978) *Medicine and the Reign of Technology*. Cambridge University Press, Cambridge.

110 Smith, R. (1997) The future of health care systems. (Editorial). *BMJ*, 314, 1495 (May 24).

PART 5

Nursing

INTRODUCTION

The nursing, midwifery and health visiting staff comprise the largest group of professional staff within the NHS. The terms 'nursing' and 'nurses' have been used for brevity. Nurses provide a 24-hour, seven days a week service to patients and advice direct to the community. Initially it was nurses and midwives based in hospitals who had some opportunities to become involved in the use of 'computers in nursing'. They worked with their medical colleagues and others on clinical systems. The specific midwifery work is covered in Chapter 13.

Health visitors were early users of the call and recall systems developed for immunization schemes. Community-based nurses and midwives also had more access when the management structure changed at the end of the 1960s. Some of this work has been covered in Chapters 2 and 3.

Professional status within the NHS has always been a source of tension and consultant nurse positions are a relatively recent development in the UK. Individuals found it very difficult to get funding within the NHS. Nursing research was a new professional discipline and slowly grew within the 1980s, competing for regional funds with the GPs, as the teaching hospitals had an informal agreement to divide up the bulk of the annual funding between them. The ward sister's office of Florence Nightingale's time, her personal domain, had been taken over as a staff coffee area by the 1980s, so there was no location for a nursing research base for the ward sister or charge nurse. There were many other practical problems in developing new computer-based approaches to the nursing care of patients with specific conditions and diseases.

Managers were under pressure to justify the nursing 'resource', as it was a large part of the staff salaries. Although each individual salary was relatively small, there were simply a lot of people needed to provide a 24-hour service. Most of the early, big systems were focused on resources.

Teamwork, to provide a 24-hour service, had to be reflected in any computer-based system to provide individual patient care. As students were in apprentice-type roles in the first part of the period covered, their education was crucial to maintaining an adequate standard of patient care. Nurse education was then hospital based: this had an impact on the contact nurse teachers had with other academic disciplines. Medical resources were often protected from the encroachment of 'hoards of nurses'. Access to a medical library, even for senior managers, had to be negotiated with skill and charm. While clinical doctors might have access to more resources and to contacts within university departments, these were not accessible to the nurse, unless he or she left the bedside for higher education.

In the face of the difficulties, nurses have played an active role in the development of health informatics, both in the UK and internationally. Over the decades their contribution has often been more collegial than discipline based, so less apparent in the published literature. Nurses, midwives and health visitors have had a real impact on health informatics and will continue to do so.

The chapters in this part of the book bring together these and other external influences on the nursing professions over five decades. There are six chapters in all:

Chapter 17: Five decades of nursing informatics

Chapter 18: Influencing factors for nursing information systems

Chapter 19: Nursing events and organizations

Chapter 20: Nursing information system development

Chapter 21: Computers in nursing and patient education

Chapter 22: Enabling technology and the future

These chapters describe computers in nursing in the UK since their first introduction in the 1960s. The influence that environmental, social, economic, technological and political drivers had on nurses' use of computers is described and is illustrated in diagrammatic form in Figure I5.1. Examples of early projects and their pioneers are outlined. Also included are the British Computer Society's Nursing Specialist Group activities, and the part concludes with a summary of the state of nursing information in the 1990s, reflections on the past and thoughts on future possibilities.

The authors acknowledge that this is incomplete, reflecting only their knowledge of events. Some developments are more fully described elsewhere in the book, but they are included again here to complete the nursing perspective.

A part on nursing would be incomplete if it did not mention Florence Nightingale (1820–1910) who is credited with being the founder of modern nursing. Nightingale was also a statistician well known for collecting, analysing, organizing and reporting information about the health of the British population in order to improve it. This led to the proposal that Nightingale was the first nursing informatician.[1] Nightingale's focus on information management to support health improvement is still relevant today, as described in the recent definition of nursing informatics by the *International Medical Informatics Association* – Nursing Informatics Special Interest Group at the IMIA conference in Seoul, Korea, in 1998:

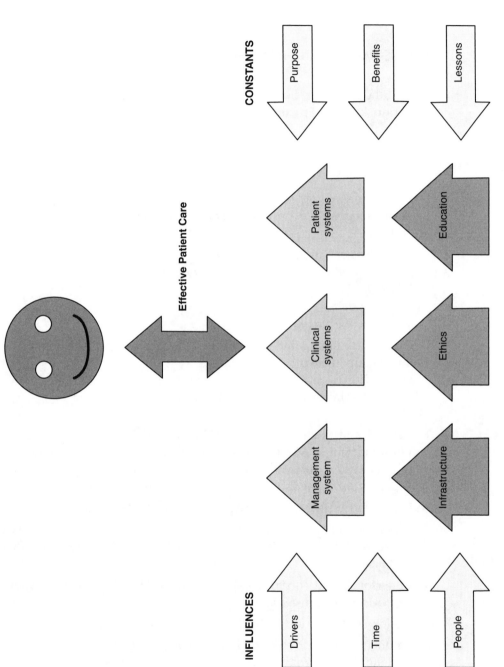

FIGURE I5.1 *Influences on nurses' use of computers*

Nursing Informatics is the integration of nursing, its information, and information management with information processing and communication technology, to support the health of people world wide.

REFERENCE 2

DENISE BARNETT AND HEATHER STRACHAN

The authors would like to acknowledge Nicholas Hardiker, David McKendrick, Paula Procter and Rod Ward for their contributions to this part.

REFERENCES

Five books are referred to extensively in the chapters in this section of the book, and together they provide the source for much of the early work. They are as follows:

The Impact of Computers on Nursing[3]
MEDINFO 74[4]
MEDCOMP 77[5]
MEDINFO 80[6]
International Nursing Informatics[7]

It is also worth noting that there is an extensive bibliography compiled by Yvonne Bryant in the first of these books.

Many of these references can be consulted in the Health Informatics Library of the University of Central Lancashire in Preston.

1   Betts, H. and Wright, G. (2003) Was Florence Nightingale the first nursing informatician? In Marin, H., Marques, E., Hovenga, E. and Goosen, W. (eds) *Proceedings of the 8th International Congress in Nursing Informatics.* E-papers Servicos Editorials Ltd, Rio de Janeiro. CD-Rom.

2   Webref (accessed May 2007) www.imia.org/ni/index.html

3   Scholes, M., Bryant, Y. and Barber, B. (eds) (1983) *The Impact of Computers on Nursing: An International Review.* North-Holland, Amsterdam. ISBN 0 444866 82 5.

4   Anderson, J. and Forsythe, J.M. (eds) (1975) *MEDINFO 74*, Stockholm. North-Holland, Amsterdam.

5   Online Conferences Ltd (eds) (1977) *MEDCOMP 77.* ISBN 0 903796 16 X.

6   Lindberg, D.A.B. and Kaihara, S. (eds) (1980) *MEDINFO 80*, Tokyo. North-Holland, Amsterdam.

7   Scholes, M., Tallberg, M. and Pluyter-Wenting, E. (2000) *International Nursing Informatics: A History of the First Forty Years 1960–2000.* British Computer Society, Swindon. ISBN 0 953542 72 6.

17 Five decades of nursing informatics

MAUREEN SCHOLES AND HEATHER STRACHAN

NURSING 'BEFORE COMPUTERS'

The 1950s brought an air of optimism. The Second World War was over and there was a new National Health Service offering free health care from the cradle to the grave. Child health and maternity services were seen as especially important and vaccination programmes and disease prevention stressed. Everywhere ideas flowed, expectations were high, yet resources were limited. To stretch these resources, particularly nursing time, there were initiatives such as the centralization of linen and equipment. Autoclaves were installed providing sterile packs to reduce the boiling of instruments at ward and theatre level. The introduction of a central diet kitchen was another effective initiative. Patients were sent to new recovery wards post-operatively or to intensive care units, resulting in specialized nursing and records. The laborious ruling of report books was replaced by the card index system, Kardex, for both individualized patient and nursing personnel records. Confidentiality of patient information and records was a high priority for nurses. Nurses' personnel information was less confidential. Their medical information was often shared with nursing managers, who were *in loco parentis* for those students under 21 years of age: the large majority.

A few State Registered Nurses, helped by a constant supply of student nurses, undertook the nursing care in general wards. While students' educational needs were considered, the ward staffing needs were paramount. The students were almost always resident, marked in a register as 'on duty', and had to request late passes, special days off and holidays.

The 1960s brought major advances in medicine, surgery and drug treatments. Infections were prevented or treated by a multiplicity of antibiotics. There seemed less need for scrupulous cleaning and the previous skilled wound care. Patient throughput was high and nurses became more involved with patient scheduling and records. Patient participation was important and, while verbal communication was vital, good individual records of observations, treatment ordered and given, were essential to continuity of care. Outcomes were difficult to record. Drug administration was more complex, error rates were found to be high. There were local and national initiatives to solve this.[1-4] The education of nurses was also pre-planned and the resulting national programmes were essential to State Registration and other post-registration qualifications. The allocation, use, quality and cost of resources to meet the needs of care and training were now centre stage.

THE PIONEERING SIXTIES

Nurses had, and continue to have, a pivotal role in the patients' care. Thus, any major changes have a profound effect on their professional practice, their systems and management and the education and research to support them. The information revolution, and the computer as a tool to support it, was just such a change. Thus, when the computer became available, there were many problems to be solved. At first, enthusiasts from different disciplines, including doctors, researchers, technologists and some nurses, grasped this new tool clumsily. The last of these were drawn into the sphere of interest because, being constantly with patients, they knew about the care systems, the clinics, wards and how a hospital 'ticked'. As early as 1965 the vision of what computers could do for nursing was being explored. Maureen Scholes, a nurse allocator at The London Hospital sought help with nurse allocation. She explored the issues with the computer and operational research specialists, but at that time the resources were not available. More preliminary groundwork was needed before significant help could be provided, but the issues raised at that time prepared the way for subsequent work. Moreover, in the late 1960s, she attended a handover session between nursing shifts, sitting round a ward computer terminal at King's College Hospital. Professor John Anderson and his ward sister, Jackie Streeter, were attempting to automate the patients' medical and nursing record. John Anderson remarked that:

> . . . if the Nursing Record was [to be] kept up to date, be accessed and available quickly, then online interactive systems were essential.

The results proved time-consuming for the junior doctors but successful for nurses. Interacting with visual display units may have been less of a problem for nurses than doctors due to better system design and the already formalized system of the Nursing Record.[5-7] The system therefore benefited nurses and the interaction with computer systems undoubtedly captured their imagination. This work resulted in a Nursing Record System being implemented at King's College Hospital in 1970: the first of its kind in the UK. Unfortunately the support from the DoH was withdrawn and the Nursing Record System discontinued, although the software was subsequently used for a successful banking system.

In 1965, at The London Hospital, enthusiasts Bud Abbott, from finance, and Barry Barber, from operational research, were exploring various ways in which the computer could benefit the hospital. The requirements of the finance systems were relatively clear although still difficult to implement in this early computer era. The various disciplines required for the successful implementation of systems were not yet fully established. However, the hospital gave them freedom to explore the scientific and medical uses of the computer whenever there were spare resources from the needs of the finance system. Much work was undertaken supporting the Physics department and many medical research surveys were undertaken. The crucial next step was the move from handling the hospital's medical statistical returns to the proposal, in the third annual report of the Computer department, that a full-scale Patient Administration System (PAS) should be developed (see Chapter 4).

In 1968, following encouraging comments from the DoH, the hospital embarked on the lengthy process of implementing a real-time PAS across the whole hospital.[8, 9] A multidisciplinary Executive of five senior members of staff representing, management, medicine, nursing, operational research and computing was established to manage the project. The members of the Executive retained their usual work on the staff, thus living with their own computing decisions. The London Hospital aimed to create a fully integrated system that would improve the patients' progress through their care, and at the same time, monitor the functioning of the organization and derive the necessary management statistics.[10] As each system was implemented hospital wide, with the clinically-sensitive systems last to ensure the computer was trusted, the manual systems were discontinued. As this was at a time when most potential users were unfamiliar with computers, a huge educational programme using a cascade approach was necessary. This involved the medical, nursing and administration staff as well as the relevant clerical staff. This project later formed part of the DoH Experimental Computer Programme (see The Developing Seventies). The early development of these ideas and an initial review of the key decisions were described by the Executive.[11, 12]

The Experimental Project system at Stoke-on-Trent was outlined by Lawton,[13] Peter Hills described the Hospital Information System at the Queen Elizabeth Hospital in Birmingham,[14] and Peter Hammersley *et al.* described the system for the coordinated projects at Addenbrooke's Hospital, Cambridge, and the three London Teaching Hospitals: Charing Cross, St Thomas' and University College.[15] It was in the context of some of these major developments that a number of important nursing systems were developed.

In 1969, at the Queen Elizabeth Hospital, Birmingham, Claire Ashton, a ward sister, joined the computer project as a project leader. She was responsible for designing a ward-based computer system including patient administration and drug administration, as well as nursing systems. Claire became a senior systems analyst in 1974, a significant achievement at that time for a woman and a nurse without computer qualifications. She published a report on computers and nursing in 1971,[16] and she described in detail her system for drug prescribing and its impact on drug administration in 1975.[17]

Much of this early work related to nursing and computers in the UK was documented in the next decade. This and related work going on in the USA is listed in Yvonne Bryant's bibliography.[18]

THE DEVELOPING SEVENTIES

Now that the possibilities for computers to support nursing practice were becoming a reality, the 1970s saw their development on a much wider scale. In 1968 the DoH had initiated a far-sighted research and development project, the DoH Experimental Computer Programme, to study the potential uses of computer technology in the NHS generally (see also Chapter 5). The programme was centrally funded and consisted of several projects with the objectives of:

- ensuring better patient care;
- increasing clinical and administrative efficiency; and
- improving management and research facilities.

The approach was described by Gedrych,[19] and in 1972 the DoH published a report on using computers to improve health services.[20] Within these objectives, each project had its own local objectives and a nurse working as an adviser, as a systems analyst or as a member of a multidisciplinary steering group. The projects developed and tested a range of systems including nurse management, intensive care, operating theatre, patient administration and nursing records.

A patient monitoring system for use by nurses was introduced at Wythenshawe Hospital in 1973. This intensive care system was based on a real-time, patient-data system that had been developed in 1969 at the Karolinska Hospital, Stockholm (see also Chapter 13).[21]

The North Staffordshire Health Authority was chosen, as a DoH Experimental Computer Project, to develop a real-time patient administration system (PAS) (see Chapter 4). It was, at the end of the 1960s, one of the largest authorities in the country, serving a population of nearly half a million. For nurses, who had always been involved in recording in-patient's information, the multiplicity of ward books became a memory with the implementation of the PAS.[22]

As the DoH Experimental Computer Programme and other initiatives developed around the country there emerged a need to share experience and information. The formation of the Computer Projects Nursing Group (CPNG) met this need and is described in Chapter 19. Many members of this group shared their work at international conferences. Of the five nursing papers at the first *MEDINFO* conference in 1974, four were from the UK and given by members of the CPNG,[23-26] another four were presented by CPNG members at *MEDCOMP 77*.[27-30] At this time UK nurses were very active internationally in promoting nursing involvement in the development of computer systems. They understood at an early stage that nurses were going to be crucial to the effective implementation and design of usable systems, even though most of their medical and administrative colleagues had not yet appreciated this fundamental truth. Beryl Warne and Joyce Wiseman addressed various issues in community care, and Jean Jarvis explored a number of management issues.[31–33] Ron Hoy and Brian Hambleton considered various aspects of nursing education and Ian Townsend provided a well-referenced paper on the introduction of new technologies in education.[34–36] Finally, Sheila Collins' paper, on the development of systems, described some of the practical work that was being undertaken in connection with the Princess Alexandra School of Nursing at The London Hospital.[37]

THE PROFESSIONAL EIGHTIES

The 1980s saw both nursing and the use of computers in nursing develop professionally, bring many nurses into contact with computers and the associated disciplines. It was essential for nursing to have good information on which to base clinical decisions as nurses strove for professional recognition. It was also essential for nurse management to have good information to manage the nursing resources efficiently, while assuring quality of care was not compromised. Nursing was the largest cost component of the health service and therefore provided a tempting target for reducing expenditure. The measures to achieve this ranged from reducing student nurse intakes to transferring tasks that were previously carried out by qualified nurses to unqualified or less costly nurses, or to less qualified auxiliaries. Many of these decisions were not based on good information. Short-term savings were made that would cause adverse effects later. The experimental projects of the 1970s had demonstrated the great potential that computers had to improve the provision of information in the NHS. In 1982, the National Staff Committee for Nurses and Midwives issued clear recommendations on computer training for nurses in middle-management posts.[38] Various Government inquiries into the NHS recognized that information to improve the efficiency, effectiveness and equity of health services across the NHS was inadequate. This eventually led to significant Government investment in computers, particularly nursing information systems, via the Resource Management Initiative (RMI – see Chapter 6).[39]

Within the growing field of medical informatics both nationally and internationally nurses were making a significant and professional contribution, but while some people considered that medical informatics applied to all aspects of health care, others saw it as pertaining just to medicine and doctors. In 1979, at the IMIA Working Conference on Hospital Information Systems, Barber highlighted the pivotal role of nurses in the implementation and use of these systems.[40] At *MEDINFO 80* there was a vastly increased interest in nurses' use of computers and the Japanese hosts arranged a specially enlarged nursing session. During the same congress the term 'nursing informatics' was used by Scholes and Barber.[41] They defined nursing informatics as 'the application of computer technology to all fields of nursing: nursing service, nurse education and nursing research'.

It was following this event that Bud Abbott and Barry Barber persuaded Maureen Scholes that an international conference on nursing should be hosted in London. In 1982, *The Impact of Computers on Nursing* conference was held at Church House, Westminster (the title echoing that of a paper by Barbara Hartmann given at *MEDINFO 74*). Enthusiasm was plentiful but money scarce, so a large conference was arranged with 550 paying delegates to fund a post-conference workshop of 59 invited international experts. The organisers were fortunate in being able to persuade Guy Barnett, MP for Greenwich, to host a dinner for international guests in the Palace of Westminster, following a tour of the Palace. This introduction to the conference was a welcome recognition of the value of computers in nursing for both nursing and BCS colleagues. The conference itself described current work around the world, while the post-conference workshop (held at the NHS Training Centre – The White Hart) discussed the impact of computers on nursing practice, education and research, and formulated future strategy.

Following this event, the CPNG, which had been totally involved, proposed the formation of a nursing specialist group within the BCS. It was also decided to request that a working group in IMIA should be allocated to nursing. The latter was agreed and *IMIA Working Group 8 (Nursing)* had its first meeting in Amsterdam in August 1983.[42] In addition to a strong nursing session within the main conference, special arrangements were made to hold a series of seminars in conjunction with *MEDINFO 83*. The nursing seminar included some historical highlights from Maureen Scholes,[43] and the context within which she contributed to The London Hospital Project in John Rowson's description of the overall system.[44] After this conference, programme committees were convinced of the importance of nursing sessions. This message was taken to medical colleagues as 'Methods of information in nursing as seen from the UK' within the prestigious *Methods of Information in Medicine* journal.[45] In 1984, a Government-backed Channel 4 documentary programme issued a series of booklets, including *This is IT and Your Health*, which included nursing systems as a matter of course.[46]

THE MATURING NINETIES

The constantly changing patterns of health, the shift of emphasis from an acute to a community setting, and increasing specialization, continually focused attention on the need for efficiency. Together with the growth of information available to both clinicians and patients, prompted by evidence-based practice and consumerism, the need for a different kind of nurse training was essential. Nurse education moved from an apprentice-type system into an academic environment in the higher education sector.

Advances in technology enabled early visions to be realized more easily and economically. The computer and its applications became more user-friendly and more suitable for use in health care situations.

The 1990s were a time for reflection and some of these were outlined by Abbott and Scholes in their paper on 'The development of nursing informatics in the UK 1963–1991' at the Melbourne Nursing Informatics Conference in 1991.[47] The investments in computers and nursing information systems had not brought all the promised savings and benefits expected, but much experience had been gained from the early projects. Introducing computers into health care impacted on all aspects of the organization from culture to structure, and had proved a significant challenge in an organization as big and complex as the NHS. The importance of ensuring education in information management and technology was recognized, but it had become quite clear that developing good information systems required significant investment and user involvement. Ultimately, if efficiencies were to be made and quality was to be improved, it was health care professionals who had to grasp the opportunities by changing the way they did things. For this they needed:

- information to support the health care process;
- tools to support communication of that information across the multidisciplinary team; and
- a patient-focused approach to health care rather than one based on different professional boundaries.

The future development of computers was refocused to support knowledge management and communication for clinicians and patients.

REFERENCES

1 Hill, P.A. and Wigmore, H.M. (1967) Measurement and control of drug administration incidents. *The Lancet*, 25 March, 671–4.

2 Sykes, C.H. and Oakes, A.E.M. (1968) Drug administration at The London Hospital. *The Pharmaceutical Journal*, 3 February, 117–18.

3 Oakes, A.E.M. and Wigmore, H.M. (1968) The London Hospital prescription sheet. *J. Hospital Pharmacy*, July, 177–81.

4 Crooks, J., Clark, C.G., Caie, H.B. and Mawson, W.B. (1965) Prescribing and administration of drugs in hospital. *The Lancet*, 1, 373.

5 Anderson, J. (1974) (Title unknown) In Collen, M.F. (ed.) *Hospital Computer Systems*, pp. 457–516. John Wiley & Sons, New York.

6 Knight, J.E. and Streeter, J. (1970) The Computer as an aid to nursing records. *Nursing Times*, 19 February, 66, 233–5.

7 Anderson, J. (1975) Informatics and clinical nursing records. In Anderson, J. and Forsythe, J.M. (eds) *MEDINFO 74*, Stockholm, pp. 126–32. North-Holland, Amsterdam.

8 The London Hospital Computer Executive (1968) *Preliminary Study for the Installation of a New Computer.* Internal document: The London Hospital, 9 October.

9 The London Hospital Computer Executive (1969) *Invitation to tender: The Board of Governors of The London Hospital Invite Tenders for a Real-time Computer System, Installation of a New Computer.* Internal document: The London Hospital, 10 October.

10 Fairey, M.J. (1969) Information systems in hospital administration. In Abrams, M. (ed.) *Medical Computing: Progress and Problems*, pp. 384–89. Chatto & Windus for BCS, London.

11 Barber, B. and Abbott, W.A. (1972) *Computing and Operational Research at The London Hospital*; 'Computers in Medicine' series. Butterworths, London. ISBN 0 407517 00 6.

12 Barber, B., Cohen, R.D. and Scholes, M. (1976) A review of The London Hospital computer project. In Laudet, M., Anderson, J. and Begon, E. (eds) *Proceedings of Medical Data Processing*, Toulouse, pp. 327–38. Taylor & Francis, London. ISBN 0 850661 06 4.

13 Lawton, M.D. (1970) Systems design for a management-oriented hospital information system. In Abrams, M.E. (ed.) *Medical Computing: Progress and Problems*, pp. 358–73. Chatto & Windus for BCS, London.

14 Hills, P.M. (1969) The objectives and design philosophy of the Real-Time Computer Project at The Queen Elizabeth Medical Centre. In Abrams, M.E. (ed.) *Medical Computing: Progress and Problems*, pp. 375–83. Chatto & Windus for BCS, London.

15 Hammersley, P., Roach, M.E., Robertson, V.S., Campbell, T. and Terry, H. (1977) Planning for transferability. In Online Conferences Ltd (eds) *MEDCOMP 77*, pp. 259–70. ISBN 0 903796 16 X.

16 Ashton, C.C. (1983) Caring for patients within a computer environment. In Scholes, M., Bryant, Y. and Barber, B. (eds) *The Impact of Computers on Nursing: An International Review*, pp. 105–14. North-Holland, Amsterdam. ISBN 0 444866 82 5.

17 Ashton, C. (1975) A computer system for drug prescribing and its impact on drug administration. In Anderson, J. and Forsythe, J.M. (eds) *MEDINFO 74*, Stockholm, pp. 175–96. North-Holland, Amsterdam.

18 Scholes, M., Bryant, Y. and Barber, B. (eds) (1983) *The Impact of Computers on Nursing: An International Review*. North-Holland, Amsterdam. ISBN 0 444866 82 5.

19 Gedrych, D.A. (1969) The department's policy and objectives. In Abrams, M. (ed.) *Medical Computing: Progress and Problems*, pp. 345–52. Chatto & Windus for BCS, London.

20 DoH (1972) *Using Computers to Improve Health Services*. DHSS, London, November.

21 Martin, J. (1983) Computers – help or hindrance. In Scholes, M., Bryant, Y. and Barber, B. (eds) *The Impact of Computers on Nursing: An International Review*, pp. 156–62. North-Holland, Amsterdam. ISBN 0 444866 82 5.

22 Gossington, D. (1983) Amending and extending an existing patient administration system. In Scholes, M., Bryant, Y. and Barber, B. (eds) *The Impact of Computers on Nursing: An International Review*, pp. 406–10. North-Holland, Amsterdam. ISBN 0 444866 82 5.

23 Ashcroft, J.M. and Berry, J.L. (1975) The introduction of a real-time patient display system into the cardio-thoracic department at Wythenshawe Hospital. In Anderson, J. and Forsythe, J.M. (eds) *MEDINFO 74*, Stockholm, pp. 101–7. North-Holland, Amsterdam.

24 Scholes, M. (1975) Education of health staff in computing. In Anderson, J. and Forsythe, J.M. (eds) *MEDINFO 74*, Stockholm, pp. 213–15. North-Holland, Amsterdam.

25 Henney, C.R., Brodlie, P. and Crooks, J. (1975) The administration of drugs in hospital – how a computer can be used to improve the quality of patient care. In Anderson, J. and Forsythe, J.M. (eds) *MEDINFO 74*, Stockholm, pp. 271–6. North-Holland, Amsterdam.

26 Hartmann, B. (1975) The impact of computers on nursing. In Anderson, J. and Forsythe, J.M. (eds) *MEDINFO 74*, Stockholm, pp. 305–8. North-Holland, Amsterdam.

27 Scholes, M., Forster, K.V. and Gregg, T. (1977) Continuing education of health service staff in computing. In Online Conferences Ltd (eds) *MEDCOMP 77*, pp. 639–48. ISBN 0 903796 16 X.

28 Bryant, Y.M. and Bryant, J.R. (1977) Towards an integrated system for nursing administration. In Online Conferences Ltd (eds) *MEDCOMP 77*, pp. 439–51. ISBN 0 903796 16 X.

29 Butler, E.A. and Hay, B.J. (1977) The fervent statistician: a computerized system to record and analyse nursing sickness and absence. In Online Conferences Ltd (eds) *MEDCOMP 77*, pp. 453–67. ISBN 0 903796 16 X.

30 Head, A. (1977) Maintaining the nursing record with the aid of a computer. In Online Conferences Ltd (eds) *MEDCOMP 77*, pp. 469–83. ISBN 0 903796 16 X.

31 Warne, B.E.M. (1983) Community nursing information systems. In Scholes, M., Bryant, Y. and Barber, B. (eds) *The Impact of Computers on Nursing: An International Review*, pp. 200–6. North-Holland, Amsterdam. ISBN 0 444866 82 5.

32 Wiseman, J. (1983) Statistics, computer forms and health visitor service planning. In Scholes, M., Bryant, Y. and Barber, B. (eds) *The Impact of Computers on Nursing: An International Review*, pp. 215–21. North-Holland, Amsterdam. ISBN 0 444866 82 5.

33 Jarvis, J.G. (1983) The computer: a tool for nurse managers to improve standards of care. In Scholes, M., Bryant, Y. and Barber, B. (eds) *The Impact of Computers on Nursing: An International Review*, pp. 445–56. North-Holland, Amsterdam. ISBN 0 444866 82 5.

34 Hoy, R. (1983) Nurse education and computers: time for change. In Scholes, M., Bryant, Y. and Barber, B. (eds) *The Impact of Computers on Nursing: An International Review*, pp. 257–64. North-Holland, Amsterdam. ISBN 0 444866 82 5.

35 Hambleton, B. (1983) Training nurses in computing in the UK. In Scholes, M., Bryant, Y. and Barber, B. (eds) *The Impact of Computers on Nursing: An International Review*, pp. 265–8. North-Holland, Amsterdam. ISBN 0 444866 82 5.

36 Townsend, I. (1983) The second coming: resurrection or reservation? In Scholes, M., Bryant, Y. and Barber, B. (eds) *The Impact of Computers on Nursing: An International Review*, pp. 334–46. North-Holland, Amsterdam. ISBN 0 444866 82 5.

37 Collins, S. (1983) Computer-based systems for professional education and training. In Scholes, M., Bryant, Y. and Barber, B. (eds) *The Impact of Computers on Nursing: An International Review*, pp. 350–5. North-Holland, Amsterdam. ISBN 0 444866 82 5.

38 Nurses and Midwives National Staff Committee (1982) *Recommendations on Computer Training for Nurses in Middle Management Posts*. September.

39 NHS Management Board (1986) Health Notice HN(86)34. November.

40 Barber, B. (1979) Patients' perspectives in health information systems. In Shannon, R.H. (ed.) *Hospital Information Systems: An International Perspective on Problems and Prospects*, pp. 31–9. North Holland, Amsterdam.

41 Scholes, M. and Barber, B. (1980) Towards nursing informatics. In Lindberg, D.A.B. and Kaihara, S. (eds) *MEDINFO 80*, Tokyo, pp. 70–3. North-Holland, Amsterdam.

42 Scholes, M., Tallberg, M. and Pluyter-Wenting, E. (2000) *International Nursing Informatics: A History of the First Forty Years 1960–2000*, pp. 24–30. BCS, Swindon. ISBN 0 953542 72 6.

43 Scholes, M. (1983) The use of computers in nursing: historical perspective. In van Bemmel, J.H.V., Ball, M.J. and Wigertz, O. (eds) *MEDINFO 83*, Amsterdam, pp. 310–12. North-Holland, Amsterdam.

44 Rowson, J.E.M. (1983) Implementation of a hospital information system. In van Bemmel, J.H.V., Ball, M.J. and Wigertz, O. (eds) *MEDINFO 83*, Amsterdam, pp. 94–101. North-Holland, Amsterdam.

45 Barber, B. (1984) Methods of information in nursing as seen from the UK. *Methods of Information in Medicine*, 23, 173–5.

46 Barber, B. (1984) *This Is IT and Your Health.* Channel 4 Television Broadcasting Support Services. ISBN 0 906965 21 7.

47 Scholes, M. and Abbott, W.A. (1991) The development of nursing informatics in the UK 1963–1991. In Reinhoff, O. and Lindberg, D. (eds) *Nursing Informatics 91: Proceedings of the 4th International Conference on Nursing Use of Computer and Information Science*, pp. 29–34. Springer Verlag, Berlin.

18 Influencing factors for nursing information systems

MAUREEN SCHOLES AND HEATHER STRACHAN

In the early days of computers, nurse pioneers were keen to explore how the computer could help nurses do what they did better and more efficiently and hence to improve patient care. The drivers to promote greater efficiency and effectiveness are described here as environmental, social and economic. Many of these drivers were reflected in the political imperatives of the time. Together with technological advances, the development of computers in nursing allowed the visions of the early pioneers to begin the journey towards fulfilment.

ENVIRONMENTAL DRIVERS

Changing health care needs

Nursing is shaped by the health care needs of the population. These needs change according to the age of the population, lifestyles and disease patterns. Together with an increase in the range of health care technologies available for treatment, the changing needs have influenced the relationship between secondary, primary, community care and public health. It has also changed what nurses do, the way they do it, as well as when, where, and why they do it. New nursing roles have emerged with increasing specialization. Care is provided by a wider array of health care professionals in an increasingly complex environment. This has moved from a hospital-based, professionally focused, and disease orientated approach to one where the emphasis is on supporting the patient within their community to maintain health, to prevent sickness and to enhance self-help activities. The computer provides an essential tool to support clinical communication and decision-making in this multifaceted situation.

SOCIAL DRIVERS

The nursing profession and consumerism

As nursing matured as a profession it became important to be able to describe the elements of nursing, not just in terms of what nurses do but in terms of their decision-making process of assessment, planning, implementation and evaluation. This became known in the 1980s as the 'nursing process'. These intellectual activities and observable skills of nursing, while not new, had not previously been described in quite this way. The articulation of the elements of nursing led to significant changes in documentation from simply describing what care was given to a more proactive goal setting, care planning and evaluation approach. Nursing began to strive for recognition as a research-based profession, with nursing care based on evidence and best practice. Nurses became accountable for their own actions and omissions. Their Professional Code of Conduct supported the idea that the scope of nursing would continue to evolve according to patient needs and the skills required in the multidisciplinary health team. If nurses were competent they could perform a wide range of roles. The drive for recognition was also influenced by the role of women in society and their rising career expectations.

In addition to the changes in the nurses' role, there were changes in the role of the patient. The influence of consumerism had reached the NHS and by working in partnership with the patient it was recognized that concordance with treatment and patient satisfaction could be improved. The computer had as much potential to provide patients with information about services and health care issues as the nurse. The nurse is no longer simply an information provider, but a knowledge broker for the patient.

TECHNOLOGICAL ADVANCES

Power and flexibility

Technological advances made possible the early vision of using the computer to collect and communicate information about the patient – their health problems and their treatment options – to the multidisciplinary health care team, in a timely manner, thus supporting clinical decision-making and patient care. These advances provide opportunities to add value to that information and improve patient safety through decision-support processes such as alerting, interpreting, critiquing and diagnosing. Nurses are with patients 24 hours a day in hospital settings and are coordinators of care in the community setting. These advances enable that information to be collected, retrieved and analysed in ways that were previously cumbersome, opening up even more uses for the technology to support nurses in delivering high quality, individualized patient care at the point of need. These advances were enabled by real-time computing, touch-screen data entry and telemedicine.

ECONOMIC FACTORS

Fewer nurses, fewer resources

Economic factors continued to demand that the best use was made of all health care resources. Nursing, as one of the biggest cost components of health care, has always been a key target when looking for greater efficiency. While limited resources, changes in disease patterns and advances in health care technologies resulted in demand exceeding supply, other factors also had the effect of producing a nursing shortage.

Better working conditions with shorter hours of duty and more holidays resulted in fewer nursing hours available. This brought about extra staff categories, some of which had little training. The method of organizing nursing changed from being task orientated to a team approach and resulted in increasing communication needs. There was a general shortage of nursing students entering the profession as population changes resulted in less school leavers and as there were wider opportunities available to female school leavers. Project 2000, a new university-based system of nurse education, brought about a change from a traditional apprentice-type system, where learner nurses made a substantial contribution to the workforce, to a system were nurse education was based in colleges with close links to higher education. Students became supernumerary in the hospital until their third year of education. As needs increased, but the ability to meet them lessened, other factors such as increased patient expectations strained the limited resources available. This created the need to examine in detail what nurses did, how much they cost and their impact on the patients' length of stay and outcomes. The computers' ability to analyse large amounts of data provided an opportunity to improve access to information and to support manpower planning.

POLITICAL DRIVERS

Quantify to improve quality and efficiency

These environmental, societal and economic factors, together with technological advances, led to a number of political drivers that had far-reaching implications for all types of health care information and the subsequent computerization to support data capture and analysis. The main programmes of legislation, official reports, consultative papers and recommendations significant to nursing are highlighted in the following paragraphs.

In 1979, the Royal Commission on the NHS reported on the best use and management of the financial and manpower resources of the NHS, having noted both social inequalities and geographical variations in the provision of health care resources. Among the recommendations were the need for medical audit, more involvement of medical staff in management and a review of nurse education.[1] In 1974–5, Doncaster Hospital Management Committee was one of the first to introduce nursing care audits, in parallel with a management audit for the nursing services and for teaching. These had antecedents in work done by the King Edward's Hospital Fund for London with origins as far back as 1960. Effective communication was emphasized.[2–4]

The Körner Reports, commissioned in 1979, recognized the inadequacy of information in the NHS.[5] A key concept of the reports was the identification of minimum data-sets to improve information management. Seven main reports were published between 1982 and 1984, dealing with aspects of activity and workload measures, manpower and financial information. The data-sets were designed to be collected largely as a by-product of operational procedures. However, they addressed the needs of managers rather than the information requirements of health care professionals.

The NHS Management Inquiry, led by Sir Roy Griffiths,[6] identified the need to improve the management of the NHS and set objectives. It recognized the importance of information, which would be detailed enough to enable episodes of care for individual patients to be costed. The Inquiry led to the introduction of general management in 1984 and provided the foundation for the resource management initiative in 1986.

Resource management (see Chapter 6), as described in a Health Notice, was about:

- *stimulating, encouraging and developing a hospital management process involving doctors, nurses and other clinical and managerial staff in strategic and operational policy decision-making;*
- *ensuring that such a process was underpinned by a patient-based information system, which was timely, accessible and credible to all participants.*

REFERENCE 7

It brought a significant investment in nursing information systems. Of the four computerized systems considered essential, one was a nurse management system that included nurse care planning and nursing records, patient dependency and workload calculation, rostering/scheduling and personnel management.

The Information Management and Information Technology Strategy for the NHS in Wales supported 22 projects concerned with a range of IM&IT issues from systems procurement, system requirements, improving National Health Service information and its use, organizational development and training, data quality and standards, and telecommunications. All were geared to the effective use of resources and improvements in services to the patient. Nursing was one of the key strategic systems identified in the strategy. The core specification for a Nursing Information System (NIS) in Wales included the following functionality: patient/nursing administration, care planning, pre-discharge planning, workload, rostering and scheduling, costing, quality assurance and decision support.[8]

In Scotland a formal policy and a strategy for a NIS was agreed in 1989. The key to this strategy was a Ward NIS that would support the following integrated functions: care planning, staffing and rostering, workload estimation, patient costing and clinical audit.[9]

The recognition of the need to change the way health care was provided, and to control the cost of health care at the same time as ensuring the best possible outcomes for patients, culminated in the *Strategy for Nursing*, published in 1989.[10] The strategy set various targets for nursing across England and Wales: in practice, manpower, education, leadership and management. Among other targets the strategy expected nurses to:

- set nursing standards;
- monitor patient outcomes;
- use systemic methods to agree the number and deployment of staff in all health care settings;
- create opportunities to continue post-registration education.

Most pertinent to computers in nursing practice was the recommendation that information technology, of proven suitability, should be installed in all health care settings: all procedures – practice, organizational and managerial – should be examined and, where desirable, computerized.

Changes in management practice initiated by Sir Roy Griffiths, were further developed and expanded in subsequent white papers. However, the next major NHS review came about as a result of the NHS and Community Care Act 1990. This review combined responsibility for primary health care and hospital services, at regional or health board level. It introduced a more direct system of funding and cross boundary charging, as well as the extension of medical audit and the resource management initiative. From a community perspective it signalled closer collaboration between the NHS and Social Services – which had already been established in Northern Ireland. Key themes that influenced information systems were the importance of planning and monitoring services to ensure cost-effectiveness, and the need to improve the quality of services.

The beginnings of consumerism in the NHS started as far back as 1974 when Community Health Councils were established to represent local communities' interest in the NHS to those responsible for managing them. Better informed users of the health service were helped by the media, and an increasing number of self-help and voluntary groups emerged. Consumerism in the NHS reached a new level with the launch of the Patients' Charter in 1991.[11] This gave patients the right, among other rights, to information on local health services from health authorities or health boards, GPs or Community Health Councils. Further legislation including the Data Protection Act 1984, Access to Medical Reports Act 1988, Access to Health Records Act 1990, and white papers recognized the rights of patients to information, either in the form of their health records or information about ranges of treatment options.

A strategic direction for computers in nursing in the UK was created and Government commitment provided the resources. Many projects sprang up around the country, some of which are described in these nursing chapters. Many of these early projects did not always reap benefits for those institutions that initiated or worked on them, as they were experimental in nature and were testing the possibilities. However, their contribution to the future cannot be underestimated.

REFERENCES

1 HMSO (1979) *Royal Commission on the NHS Report.* HMSO, London.

2 Doncaster Hospital Management Committee (1974) *Nursing Care Audit.* Doncaster Hospital Management Committee. January.

3 Doncaster Hospital Management Committee (1974) *Management Audit Teaching Division.* Doncaster Hospital Management Committee. March.

4 Doncaster Area Health Authority (1975) *Management Audit for the Nursing Services.* Doncaster Area Health Authority. August.

5 DHSS (1982) *Converting Data into Information.* Steering Group on Health Services Information. King's Fund, London.

6 DoH (1983) *Report of the NHS Management Inquiry.* Chairman Roy Griffiths. DHSS, London.

7 NHS Management Board (1986) Health Notice HN(86)34. November.

8 Roberts, R. and Melvin, B. (1994) Benefits assessment. In Grobe, S. and Pluyter-Wenting, E.S.P. (eds) *Nursing Informatics: An International Overview for Nursing in a Technological Era,* pp. 122–6. Elsevier, Amsterdam.

9 Hoy, D., Hyslop, A. and Wojcik, E. (1994) Rapid prototyping in the development of nursing information systems: creating and using databases to improve patient care. In Grobe, S. and Pluyter-Wenting, E.S.P. (eds) *Nursing Informatics: An International Overview for Nursing in a Technological Era,* pp. 188–91. Elsevier, Amsterdam.

10 DoH (1989) *Strategy for Nursing.* HMSO, London.

11 DoH (1991) The Patients' Charter – Raising the Standard. HMSO, London.

19 Nursing events and organizations

MAUREEN SCHOLES AND HEATHER STRACHAN

The Computer Projects Nurses' Group (CPNG) was formed in 1974 and chaired by Maureen Scholes, Senior Nursing Officer at The London Hospital. The CPNG were effectively the first 'voice' of nurses in this field in the UK. The original members of this group are listed at the end of this chapter.

Dame Phyllis Friend, then Chief Nursing Officer for England and Wales, had fortunately been Matron at The London Hospital during the DoH Experimental Computer Programme, and recognized the importance of computers to nursing. She gave her active encouragement to this rather isolated group at a crucial time and provided further support via DoH Nursing Officers Barbara Rivett and Doreen Redmond, who had also been associated with the development of the DoH Experimental Computer Programme.

The CPNG met regularly to clarify, describe and advise on the role of the nurse in computing and to discuss matters of common concern. They disseminated information and recommended publications of use or of interest to nurses. They considered the objectives for computer-based nursing systems, methods of achieving them and their effect on health care. Interestingly the lessons learned at that time are still appropriate today.[1]

At the Group's meetings members discussed conferences, courses and projects covering a wide variety of the topics that reflected the individuals who attended, and computers and their application to:

- education;
- research and development;
- nursing records;
- nurse allocation and manpower planning.

Also discussed were confidentiality, microcomputers, the roles of the nurses in informatics, future developments and specific national and international projects. The CNPG formed links with the British Computer Society's Medical Specialist groups and the Royal College of Nursing.

The CNPG were invited by the BCS to help organize the International Conference *The Impact of Computers on Nursing*, held in London in 1982 (see Chapter 17).[2]

A number of other groups and events emerged following the challenges and ideas that arose from this event. A proposal was made by Maureen

411

Scholes to the IMIA to start a nursing informatics working group (see Chapter 17). Maureen became its first chair and first Honorary Member. The Network of Users of Microcomputers in Nurse Education (NUMINE) was created, later inspiring the Open Software Library (OSL) in 1984. These groups' activities are described in Chapter 21. The CNPG also formulated proposals for a NHS Nursing Computer Advisory Group and a BCS Nursing Specialist Group.

Formation of the BCS NSG began with a small steering group, chaired by Claire Ashton, then chair of the CNPG, who set out the terms of reference, membership and finance. The BCS approved the new group, which held its inaugural meeting and first Annual General Meeting on 16 May 1983. A programme of activities was planned for 1983, publicized by the BCS Medical Specialist Group's newsletter, edited at that time by Jean Roberts. Under Claire Ashton, with Yvonne Bryant as secretary and Brian Layzell as treasurer, the Group flourished. By 1985 there were 431 members of which 56 were full BCS members. Key activities of the Group were: publications, an annual conference, other meetings, and the Dame Phyllis Friend Award. Executive committee members from 1983 are listed in Chapter 1.

The first regular publication, edited by Ron Hoy OBE, was *Computers in Nursing News* (CINEWS), which was published twice a year in *The Nursing Times*. A newsletter, sponsored by a computer firm, was produced three times a year. Later these efforts were channelled into an official journal (*Information Technology in Nursing* – ITIN), which was first published in 1989 and is now listed on the CINAHL database. The journal editors are listed at the end of the chapter. Other publications included two videos. The first, *Introduction to Computers in Nursing*, was launched in January 1985.[3] The accompanying booklet was a collaboration between the BCS NSG, the DoH, the Training Aids Centre of the NHS Training Authority and the National Staff Committee of Nurses and Midwives. The second video, *Beyond the Theory*, described the uses and benefits of a nursing information system during the patients' journey through care.[4] It won the *British Medical Journal* Certificate of Educational Merit for 1992 Film and Video Awards. A care planning system demonstration called AGNIS, devised by Derek Hoy, was issued to all NSG members on a disk in 1993.[5] The NSG website became a key source of information and, with Rod Ward as its webmaster, it won the BCS specialist group best website award in 2003 and 2004.

The first annual NSG conference (*Microcomputers in Nursing*) was held in Leicester in 1984. There was an exhibition of microcomputer-based nursing systems and *The Nursing Times* sponsored a lecture and subsequently published it. The Group took part in the BCS Health Informatics Specialist Group annual conference *Current Perspectives in Health Computing* held in Brighton. An annual conference was re-established in 1990.

The Dame Phyllis Friend Award was established by the Group and sponsored by Dame Phyllis in 1989. This award, which consisted of £500 and an accompanying scroll, was given to a NSG member who, through the use of computers, advanced nursing practice, education, research or management. The first award, in 1990 was to Derek Laren and colleagues at Bamber Bridge for *Creating a Community Mental Health Information System.*

Initially the NSG had a number of local branches. The largest branch, set up in April 1985, was the Northern Ireland section, chaired by Joan Thompson, Nursing Officer, DoH Northern Ireland. It was inaugurated in the Great Hall, Queen's University, Belfast. Another group was set up in South West Thames. The purpose of these groups was to provide a local focus and support on a geographical basis. Later, under the chairmanship of Graham Wright, focus groups became subject based. Topics included primary care, nursing terminology, mental health, education and multimedia, multidisciplinary patient-held records and midwifery. The NSG also provided a nurse representative on many national and international nursing informatics groups. These included the *IMIA Working Group 8 (Nursing)* (this later became *IMIA Special Interest Group on Nursing Informatics* and at the time of writing is chaired by Heather Strachan). There was also the *European Federation for Medical Informatics Working Group 5 on Nursing Informatics.*

The European Summer School of Nursing Informatics came into being in 1990,[6] when a subgroup of an inter-university meeting, set up to look into the future of advanced nurse education, identified the paucity of informatics topics in the nursing curricula. This group included Paul Wainwright, Mary Chambers, Nicola Eaton, Jos Aarts and Derek Hoy. They formed the first organizing committee and the first summer school was held in 1991 in The Netherlands. It has continued as an annual event open to all nurses involved in any way with informatics, from practice to projects or education.

The Strategic Advisory Group for Nursing Information Systems (SAGNIS), chaired by Yvonne Moores, then Chief Nursing Officer for England, was established in December 1991. Its remit was:

- to advise the National Health Service Management Executive in England on the pace and direction of information systems; and
- the overall development of information technology to support nursing, midwifery and health visiting.

The membership comprised senior nurses, midwives, health visitors and information specialists mostly from England, but with representation from Scotland. Simon Old, from the English Nursing Division, was instrumental in making SAGNIS an influential group. The group had a vision, which was to support nurses, midwives and health visitors to provide quality, cost-effective, individualized patient/client care in a multidisciplinary environment through the use of computer technology and information management. The vision was based on four principles:

- Information should be properly managed.
- Nurses, midwives and health visitors should be appropriately trained to collect, analyse, interpret and use data, and keen to use technology and information.

413

- Nursing information strategies should be integrated within unit information strategies.
- Technology should be flexible to meet changing needs.

The group initiated a research project, which reported in March 1994, providing a comprehensive evaluation of the progress of Nursing Information Systems up to that time.[7] On the initiative of Yvonne Moores and Graham Wright, the Nursing Professions Information Group brought together representatives from four professional organizations: the Royal College of Nursing; the Royal College of Midwives; the Community Nursing and Health Visitors Association; and the BCS NSG. It was still active in 2006.

APPENDIX

Computer Projects Nursing Group (CPNG) – original/founder members

Claire Ashton	Project Leader, WMRHA Management Services (Computing), Queen Elizabeth Hospital, Birmingham
Charmaine Astbury	Nursing Officer, Royal Devon and Exeter Hospital
A. Brunton	Senior Nursing Officer, The Royal Liverpool Hospital
Yvonne Bryant	Nurse Systems Designer, Addenbrooke's Hospital, Cambridge
Elizabeth Butler	Senior Nursing Officer (Computing), St Thomas' Hospital, London
Deidre Gossington	Senior Nursing Officer, North Staffordshire Royal Infirmary, Stoke-on-Trent
Brenda Smith	Senior Nursing Officer, South West Thames RHA
Beryl Warne	Regional Nurse, Wessex RHA
Kathleen Whelan	Nursing Officer, Charing Cross Hospital
Paul Wright	Killingbeck Hospital, Leeds

In attendance were:

Christine Henney	Nursing Officer (Computing), Ninewells Hospital, Dundee
Jackie Streeter	Area Nursing Officer, Kensington, Chelsea and Westminster AHA
I. Adam	Principal Nursing Officer, DoH
Margaret Clark	Nursing Officer, Scottish Home and Health Department

Francis Harrison	Regional Nursing Officer, South West Thames RHA
N. Hill	Nursing Officer (Regional Group), DoH
Brian Layzell	Senior Executive Officer, MSC/3A DoH
B. Mulligan	Deputy Chief Nursing Officer, Welsh Office
Doreen Redmond	Nursing Officer (Computing), DoH
P. Wall	Nursing Officer (SDG), DoH
J. Whitehead	Deputy Chief Nursing Officer, DoH

Information Technology in Nursing **journal editors**

Ron Hoy	1989–1990
Chris Dowd	1991
Denise Barnett	1992–1997
Elaine Ballard	1998
Peter Murray	1999–2001
Denis Antony	2002–2007

REFERENCES

1 Redmond, D.T. (1975) An overview of the development of National Health Service computing. In ref. 2, pp. 59–69. (Note: this paper includes the Recommendations of the final CPNG Report.)

2 Scholes, M., Bryant, Y. and Barber, B. (eds) (1983) *The Impact of Computers on Nursing: An International Review*. North-Holland, Amsterdam. ISBN 0 444866 82 5.

3 BCS/DHSS (1985) *Introduction to Computers in Nursing*. Training video.

4 BCS/DHSS (1991) *Computers in Health Care: Beyond The Theory*. Training video. Acorn and ENB.

5 Hoy, D. (1993) *A Gnomic Nursing Information System* (AGNIS). A disk-based care planning system demonstration. *Information Technology in Nursing*, 5, 2, 4.

6 Paget, T. (1999) The European Summer School of Nursing Informatics. *Information Technology in Nursing*, 11, 4, 13–15.

7 Barnett, D. (1997) Advisory structures on information. *Information Technology in Nursing*, 9, 2, 10–11.

20 Nursing information system development

MAUREEN SCHOLES AND HEATHER STRACHAN

Nursing Information Systems cover nursing practice, management and education. Nursing practice and management systems were often interchangeable, as it was an aim that management information should be automatically derived from operational systems. A ward nursing management system often included, as core elements, care planning and evaluation, workload assessment and rostering, including nursing personnel and time-out analysis. Nurses also used specialist systems to support specific areas of practice or patient groups. For example, in intensive care, nurses used computers to support monitoring and charting of patients' vital signs. Nurses used general health care systems such as order entry and results reporting, medication administration, patient administration systems, and theatre systems. Christine Henney, a ward sister at Ninewells Hospital, Dundee, and Professor J. Crooks were involved in an early drug administration system following their work on a manual system.[1, 2]

Computers also supported nurse education and patient education. To underpin the use of nursing information systems, research and development has been undertaken in such areas as decision support and nursing terminology. This chapter provides an overview of the purpose and benefits of these systems and examples of early innovation.

COMPUTERIZED CARE PLANNING SYSTEMS

Computerized care planning systems (CCPS) provided support to the nurse through the stages of the nursing process including assessment, care planning, evaluation and discharge planning. The benefits of computerizing the nursing process were anticipated as saving time in planning and documenting care, improved completeness, timeliness, conciseness and legibility of nursing records. This would support nurses' accountability, as the resulting documentation became part of the 'legal' record of care. Improved accessibility of the nursing record also served to improve communication. There was potential to support quality assurance and audit, to provide decision support for qualified nurses, and educational support for student nurses. In addition, these systems were often linked to some sort of dependency or acuity calculation, which could support nursing workload management. CCPS were relevant to both the community and hospital settings. With the exception of three commercial systems from the USA, most of the early care planning systems used in the UK were developed in-house.

The system developed as part of the Experimental Computer Programme in the 1970s owed much to the first nursing records system in the UK developed at King's College Hospital. They were also being developed at a time when a more individual approach to nursing care was replacing the traditional task-orientated approach.

The nursing records system at the Queen Elizabeth Hospital, Birmingham, was designed for care planning: to enter and cancel nursing orders directly into the computer using visual display units, which were placed in the ward sisters' office or on the nurses' station in the ward. Nursing care was selected from screens that listed the most common nursing orders: basic and technical, grouped, for example, into mobility, dressings, fluids, etc. They reflected the specialty of the ward concerned. A date was inserted indicating when the care could begin. Daily lists for the total care in each ward were printed to allow the ward sister to manage care in the wards either by task allocation or team nursing and yet identify an individual's care.[3]

At The Royal Devon and Exeter Hospital, Wonford was developing an online interactive system, covering the patient record from GP through to hospital admission to the ward. Alison Head joined the project in 1970 as a senior systems analyst. She had a mathematics background and worked with the nurses, including Paula Procter, to develop the computerized nursing record. They developed a patient-orientated record, which interestingly pre-dated the use of the 'nursing process' in the UK. Screens were designed to display all the relevant care for a particular condition or intervention (e.g. tracheostomy), and information collected on admission regarding special needs and precautions (e.g. diabetes). In addition to care plans and nursing orders, care given was reported by 'ticking' when it had been completed. More detailed progress notes were still handwritten on the Kardex.[4, 5]

As part of the Scottish Home and Health Department's Ward Computer Project at Ninewells Hospital, Dundee, Christine Henney designed a system similar to those at Birmingham and Exeter.[6] Printers were available in the wards. An individual summary was produced, on discharge, for insertion into a patient's medical record and for use by the community nursing service.

The evaluation of these pioneering projects was interesting and useful to nurses. The benefits for nurses were noted as:

- care plans were more accurate, precise and complete than the previous manual records;

- the use of non-approved abbreviations had been virtually eliminated;

- care plans were legible and more consistent through the use of standard terminology;

- the identity of the author of all entries was achieved;
- where reporting existed, a complete record of care given was provided.

As a consequence of computerizing nursing records, patient safety was improved. Nurses were able to assess a patient's clinical condition more easily, therefore reducing the possibility of a nursing order being misinterpreted. The computer systems took slightly longer than the previous manual ones, but this was compensated for by the much more complete record. Communication, coordination and continuity of care had been enhanced with the improvement of the quality, quantity and availability of individualized patient information. There was a basis for a patient dependency or workload information system as a means to measuring care without additional nursing input.

In Scotland, Morag Harrow later enhanced the nursing system developed in Ninewells Hospital and developed the Computer-Assisted Nursing Orders (CANO) system.[7] CANO provided daily nursing orders, including specimen collections and tests. Its aim was to reduce administration and improve communication among nurses. As the nursing process became the required method of organizing nursing care, it was decided to modify CANO, away from a task-based system, to support the nursing process. The result was the Computer-Assisted Nursing Information System (CANIS), which included nursing assessment and evaluation. The system's success was reported by user acceptance, however, there were still difficulties. The nursing process was introduced alongside the new CANIS and meant a change in the way nurses worked, which was not always perceived as beneficial. The charge nurse no longer recorded all the nursing orders and other staff were required to update patient data. CANIS was also introduced into Crosshouse Hospital, Kilmarnock, in 1986, led by Nan Luney.[8] Although successful, development was discontinued and it was replaced by a commercial system, as recommended by the Scottish Home and Health Department as part of the Nursing Information Systems Strategy.

In the 1980s, a DoH initiative in England developed a care planning system in partnership with Bloomsbury Health Authority at The Elizabeth Garrett Anderson Hospital. The Clinical Nursing Computer System project was managed by Kathie Woods, and supported by Barbara Rivett from the DoH. The project's aim was to computerize the nursing Kardex and the patient care plan, using the nursing process and an 'Activities of Living' model as its framework. The primary purpose of the nursing record was recognized as a communication tool between nurses and other disciplines and it was the 'legal' record of the nursing care given. An additional objective of this project was to realize the benefits that computers could offer, in analysing the large amount of data held in the nursing record, to support quality assurance. This involved the creation of a database of nursing practice. To reduce duplication of data entry it created interfaces between nursing systems and other systems such as PAS, laboratory services and the district-nursing network. Developing the data-set proved one of the greatest challenges and the extent of the work and skills required to do this became apparent during the development of the project.[9]

The Welsh Computer Strategy Committee devised a National Strategy that included the development of a nursing care planning system. The project was led by Ruth Roberts and implemented in Llanelli General Hospital, Dyfed. This system had a noticeable feature: it supported nurse's decision-making with an extensive help facility which used a knowledge base of information on appropriate care for common problems.[8]

In early 1987, one of the first commercial care planning systems to cross the Atlantic was a system called Excel Care™, marketed by Price Waterhouse. Project managed by Heather Strachan, for the Riverside Health Authority, the intention was to Anglicize the system. Excel Care™ used predefined standard care plans called 'Units of Care'. These defined nursing practice for specific nursing diagnoses or patient problems and were developed locally. In addition to care planning, Excel Care™ included a manpower component, which costed nursing care per patient, based on the timing of each nursing activity within the units of care.

Unfortunately, the Riverside Health Authority had little success implementing the system. After an initial pilot on one surgical ward it was concluded that the proposed benefits were unlikely without a major redesign of the software. The system did not support key aspects of the nursing process, including assessment, evaluation and recording of actual care given. Individualization of the standard care plans on the computer was only possible by recording additional free text. Identification and retrieval of standard care plans was time-consuming in the absence of controlled nursing terminology. The computer did little to support quality assurance, as only the manpower information from the care plans was analysed. Paper was produced in vast quantities.[10] The manpower component of the model used a mathematical model, which it has been suggested was fundamentally flawed when applied to nursing.[11] The system assumed that each nurse, whatever grade or experience, did one thing at a time and took the same length of time to do it. As the standard care plan could not be individualized other than by the use of free text additions, further inaccuracies were built into the final calculation.

The need to share Riverside's experience was at odds with the commercial goals of the suppliers and, with the exception of Price Waterhouse, there was no formal communication of the findings outside the Riverside Health Authority. Later the system was introduced as part of a research project in Weymouth and District General Hospital in West Dorset in November 1987,[12, 13] and as part of the Resource Management Initiative in Huddersfield Royal Infirmary in November 1988. Considerable work was undertaken in Huddersfield in relation to activity timings to support the manpower component. Weymouth invested heavily in developing and validating units of nursing care.

Other companies were working with the health service to develop, pilot and evaluate nursing care planning systems, including ICL who aimed to build on the DoH Experimental Computer Programme and had designed a care planning system in consultation with nurses across the UK.[14] The system used the 'Activities of Living' nursing model and the 'Criteria for Care' workload estimation. It was designed to integrate with a third-party scheduling system. Its prototype was piloted with limited success at The Westminster Hospital and at St Thomas' Hospital in London.

The first wave of the resource management initiative (RMI – see Chapter 6) sites provided significant opportunity to develop computerized care planning systems. Each site took an individual approach. In 1989, The Royal Hampshire Hospital, Winchester, had chosen the TDS Hospital Information System as part of the initiative. However, the care planning software was deemed unsuitable and therefore the Doctors Personal Orders software was adapted. The system allowed the hospital's own acuity system to be used for workload estimation, although it was not linked to care planning in any way.[8] The benefits to documentation of the nursing process were evaluated in a research study supported by Southampton University. The results showed that after three months of using the computerized care planning system, the increase in the number of patients for whom care plans were made and the increase in quality scores for care planning and evaluation were statistically significant. There were also noticeable improvements in assessments. After one year and after two years of use the improvements were further increased.[15]

The TDS Hospital Information System was also used at Arrowe Park Hospital, Wirral (another RMI site). The system, which was very successfully implemented, was linked to the ANSOS computerized nurse scheduling system. This system was the only automatic scheduling system, of the time, that was based on criteria entered by the ward manager. A review of the system, conducted by an external consultant, concluded that it delivered major benefits to patients and staff. In particular, nurses benefited from order-communication and results reporting. The full nursing benefits did not occur while nurses duplicated the information on traditional paper-based notes, or viewed computer activities as clerical or management functions. The problem was the nurse–computer interface, which did not have the flexibility, speed or ease of paper and pen.[16]

COMPUTERIZED CARE PLANNING SYSTEMS IN THE COMMUNITY

A CCPS that also produced management information was the Wiseman system. The Körner Report for Community Information Systems made recommendations for three data-sets: population registers, clinical activity and staff individual-activity data. The DoH in England commissioned a project, led by Joyce Wiseman, to address these and other issues facing community nurses. Key problems for community nurses included loss of the patient record, poor communication, excessive time spent recording activity and statistics, and duplication of information.[17] It was hoped to address these issues by using memory cards and portable microcomputers, which linked to the central computer and recorded patient details, assessment, care plans, visit records, nursing actions, mileage and future visits. Management reports were a by-product. The evaluation focused not only on anticipated benefits but also on usability, the patient as the guardian of their record in the form of the smart card, and the effect of the system on nursing and management practices.

Elsewhere in the UK there were other examples of Community Nursing Systems (CNS) that supported the collection of Körner data as a by-product of operational management systems. Savill and Goldberg from East Dyfed Health Authority reported one in the BCS NSG Newsletter.[18] The COMCARE Computer Health Information System, which ran on ICL's DRS 300 'super-microcomputers', was a patient-based system that covered the whole range of staff working in the community. A National User Group funded enhancements to the basic system. One such development was a mini-version of COMCARE to allow nurses to hold their own caseloads on a hand-held Psion computer.

By the early 1990s CNS were being influenced by the NHS and Community Care Act 1990, which had brought about the separation of the purchasing of services from the providing of services. The IM&T strategy promoted person-based operational information systems. Community nursing services were purchased in what was known as block contracts, with indicative volumes of activity specified. This approach did not take account of patient need, resource utilization or outcomes. It was seen as a disincentive to developing and improving services. Work to develop community data-sets to address this problem was being undertaken in both England and Scotland. This built on work undertaken as part of the Financial Information Project (FIP – see Chapter 11), and from the Blackpool, Wyre and Fylde District Health Authority.[19] This work identified two descriptions of care:

- Care Objective or Aims, which defined why something was being done.
- Care Programmes or Profiles, which defined the actual or potential problem being addressed.

In England, this work was handled under the auspices of the Information Management Group's Community Information Systems for Providers initiative. Scotland used this experience to build their own project, Effective Purchasing and Providing in the Community (EPPIC), which was project-managed by Alan Hyslop from the Health Systems Division.[20]

In 1989, Yvonne Moores, then Chief Nursing Officer for the Scottish Home and Health Department, commissioned Derek Hoy to undertake a review of the professional issues arising from the current use of computer-assisted nursing care planning in the UK. The review identified that commercial systems required extensive adaptation. While, in some cases, this was because they only recorded parts of the nursing process, it was also appreciated that the nursing process was in various stages of implementation across the UK and therefore local variations existed. The systems in use at the time were being introduced on an experimental, small-scale basis led by pioneers who had worked hard to win local acceptance. Extension of these systems might not receive the same commitment. Where hospital information systems were in use, nursing would be under pressure to accept the nursing components of these systems. Problems of support had the impact of slowing development and, with commercial systems, the high level of dependency on the supplier was noted. There was little evidence of evaluation of systems. Although there was a general feeling that computers could meet the demands for more thorough documentation, it was too early to evaluate the benefits at that time. It was recognized that more development work was needed and that this required realistic resources. It was necessary to find the best way of using the technology, to help nurses decide what they did and did not want, and to develop their skills appropriately.[8]

The various shortcomings of these early computerized care planning systems were identified by the Audit Commission a few years later. The time and effort required to create the nursing practice elements, or standard care plans, for use in the system was considerable. Often only parts of the nursing process were computerized. Care plans could not be individualized. Screens and reports were poorly designed and difficult to read or interpret. Hardware constraints, and systems not being linked, meant duplication of data entry and inflexibility; information could not be transferred from ward to ward with the patient.[21] The investment in the development of these new systems for both the industry and the users was both time-consuming and often unprofitable, but the pressure to implement was intense.

COMPUTERIZED NURSE SCHEDULING SYSTEMS

Developing the duty rosters every month and providing manpower reports to meet the Körner requirements was a time-consuming task for all ward managers. Although an administrative task in itself, it required a detailed knowledge of the ward routine, patient characteristics and staff skills. The computerization of this task was therefore very attractive. The anticipated benefits of computerized nurse scheduling systems included: saving time formulating a duty roster; producing more accurate manpower, payroll and personnel reports; more accurate costing of nursing care; and with the potential to improve the monitoring and control of nursing cost through better information and rostering practice.

Early nursing systems supported the reporting of nursing manpower without the scheduling component. In 1973, at St Thomas' Hospital, London, Elizabeth Butler was responsible for designing a sickness and absence system for nursing staff to achieve better use of nursing resources.[22] She also designed an operating theatre management system.[23] It is interesting to note that a century earlier at St Thomas' Hospital, Florence Nightingale, the 'passionate statistician', had designed a model statistical form to help in management. Nightingale had persuaded four other large hospitals to use the same form.

The concept of a ward nursing management system with all the nursing system's elements linked together was in development as early as the 1970s. At Addenbrooke's Hospital, Cambridge, Yvonne Bryant, Nursing Adviser, and John Bryant, Senior Systems Designer, developed another sickness and absence system as part of an integrated nursing management system. This was linked with a nursing dependency system, a training programme evaluation and an outline nursing record system.[24]

In 1982, the Regional Nurse Manpower Group from the DoH in Northern Ireland realized that the quality and quantity of nursing care was dependent on successful manpower planning. It made early strides in developing a nursing information management system. Joan Thompson, an active member of the BCS NSG was involved in this project. The aims of the system were in line with the expected benefits of Computerized Nurse Scheduling Systems (CNSS). The system was developed by the Information Technology Unit in Northern Ireland and was operational by November 1987. It had personnel, management and learner modules, with links to existing finance and personnel systems. It was introduced across most of Northern Ireland. The benefits were described modestly as 'producing standard reports which had previously required a lengthy manual process' and 'locating information is quicker and potentially easier than searching through a paper filing system'.[25]

The North West Thames RHA Hospital Nursing Project piloted two commercially available CNSS with a view to recommending one for implementation across the region. The evaluation from one pilot, which commenced in 1988 at Westminster Hospital, was positive. The system saved ward managers' time in rostering and report production. In addition, information was obtained that had not previously been available. Following various amendments to cope with hospital-wide implementation, which included a rewrite in the UNIX operating system and the purchase of a more powerful server, the system was fully implemented across Westminster and Charing Cross Hospitals. Work continued to add a patient dependency module to the system. However, following full implementation many problems occurred, few benefits were being gained and the system was eventually withdrawn.[26]

The conclusions arising from this project were that:

- the system required all users to be up to date and their returns centrally processed before reports could be produced;
- central processing was time-consuming and often failed;
- the system did not link to the manpower/payroll system; and
- the company never fulfilled its promises to undertake many of the required enhancements, including adding the dependency module.

Pilot projects do not necessarily reflect the implications of implementing a system throughout an organization. The enthusiasm of the pilot staff and the additional support can tend to be underestimated (the 'Hawthorne Effect'). In addition, much hinges on the continuing support of the company supplying the system.

Many of these early CNSS did not address the newer concepts in nursing, such as primary or team nursing. A prototype computer-based, hospital ward duty planning system developed by Nicholas Hardiker won the BCS NSG Dame Phyllis Friend Award in 1992. Hardiker approached the nurse-scheduling problem by examining the underlying scheduling method. He found that the systems available at the time took into account, to varying degrees, grade and number of nurses covering a shift, quality of the roster for the individual nurse and efficiency of the schedule. He designed a system that was able to prioritize at multiple levels and with sufficient flexibility to ensure a roster that was satisfactory to both the charge nurse and the individual nurse, and saved time compared with manual rostering.[27]

A number of problems were encountered with CNSS. Time was not always saved as the ward manager often did the roster manually at home and entered it on the computer later. Where auto-rostering was a feature, the results were often not satisfactory to either the ward manager or the staff. The systems could not always cope with work organization issues such as team or primary nursing, mentorship of students or shift patterns.[21] While benefits were gained by many, the problems and effort involved often outweighed these.

Lancaster Royal Infirmary also did nurse allocation and nursing workload measurement.[28–30] Greenhalgh *et al.* compiled an independent directory of comparative nursing systems and another of operating theatre systems,[31–32] sponsored by the suppliers themselves over a five-year period.

WORKLOAD ASSESSMENT

Workload assessment systems aimed to provide data to improve the match between staff numbers and skill mix available to meet patients' needs and quality of care, in order to improve nurse scheduling, work organization, establishment setting and manpower planning. The computerization of workload assessment was believed to provide benefits by enabling nursing activity data (i.e. what nurses do and how long it takes) to be collected and analysed in more detail by individual patient, on a shift-by-shift basis, and for long-term trends. The data could also be accessed easily and information disseminated widely. It was also anticipated that if workload could be automatically derived, from a patient's care plan, this could also contribute to the evaluation of a patient's progress to independence. Links with nurse scheduling would enable the actual staffing levels to be identified and areas of shortage highlighted.

Two enthusiasts, not themselves nurses, developed computerized nursing dependency systems in the 1970s. Catherine Rhys Hearn and Alex Barr from the Oxford Regional Hospital Board produced a program for nurse allocation on the Elliott 803 computer.[33, 34] Alex Barr's program was developed in the mid-1960s and considered for use at The London Hospital, but finally rejected in favour of developing a Nurse Training Database which could provide the input for such a program at a later stage. (See also Chapter 4.)

In the Riverside Health Authority the search for a workload monitoring system continued as part of the North West Thames RHA Hospital Nursing Project. Criteria to evaluate existing patient dependency and nurse activity methods were developed. These included the ability of the method to accurately reflect patients' needs, indirect care and associated ward activities, the effect of support services and ward layout on workload, ward standards, and skill mix requirements. Also identified were the method's ability to reflect changes in patterns of workload in a 24-hour period, to predict workload for 24 hours ahead, and its inter-rater reliability. No method available met all those criteria. However, the method chosen was easy to use and supported the intended use of the data. Nurses and managers believed it was a useful tool to support professional and managerial decisions with regard to daily scheduling of staff and workload analysis.[35] The method used a Lotus® spreadsheet to support data entry and analysis. It was planned to integrate it into a future, computerized, scheduling system.

The importance of validity and reliability of the methodology used to calculate the patient dependency or nursing activity, as well as the data collection method itself, is crucial to ensure that it represents the complexity of what nurses do, and the skill and time required to do it. Rimmer *et al.* identified these issues and designed a program that saved time collecting activity data and improved that data's reliability.[36]

The Freeman Hospital, Newcastle upon Tyne, a first-wave RMI site (see Chapter 6), had a nursing strategy that appreciated the interrelationship between individual patient care, quality of services to patients and the required manpower. As a consequence 'Criteria for Care', which was a tool to monitor patient dependency and workload, was computerized. 'Monitor' was chosen to cope with the issue of quality. Computerized care planning and rostering were also planned. Reported benefits arising from the information derived from the nursing management information system were outlined at a BCS NSG conference on Resource Management in Nursing in 1990. The use of information derived from the system had led to a skill-mix overhaul and grade-mix adjustment: reviews of what was done, when and by whom, focusing around the patient days and flexible scheduling. As a result of these changes there was a sustained reduction in sickness rates among nurses.[37]

The Freeman Hospital's experience proved that workload assessment could increase efficiency, but this was not universal. However valid the information, it has to be used to provide benefits and it was not always possible to use the information in the way intended. Rather than releasing savings, it often demonstrated the need for additional staff, yet there was no money to increase establishments. If workload information was intended to be used for forward planning, it was often not possible to alter the roster without considerable impact on either the remaining roster or on the staff.[21] Collecting workload information on a daily basis was recognized as a costly exercise with little benefit. Nurse establishments remained at their historic levels because few resources were available for increases even if they were indicated.

REFERENCES

1   Henney, C.R., Brodlie, P. and Crooks, J. (1975) The administration of drugs in hospital – how a computer can be used to improve the quality of patient care. In Anderson, J. and Forsythe, J.M. (eds) *MEDINFO 74*, Stockholm, pp. 271–6. North-Holland, Amsterdam. ISBN 0 444107 71 1.

2   Johnston, S.V., Henney, C.R., Bosworth, R., Brown, N. and Crooks, J. (1976) The doctor, the pharmacist, the computer and the nurse: the prescription, supply, distribution and administration of drugs in hospital. In Laudet, M., Anderson, J. and Regon, E. (eds) *Proceedings of Medical Data Processing*, Toulouse, pp. 137–48. Taylor & Francis, London. ISBN 0 850661 06 4.

3   Ashton, C.C. (1983) Caring for patients within a computer environment. In Scholes, M., Bryant, Y. and Barber, B. (eds) *The Impact of Computers on Nursing: An International Review*, pp. 105–14. North-Holland, Amsterdam. ISBN 0 444866 82 5.

4   Head, A. (1977) Maintaining the nursing record with the aid of a computer. In Online Conferences Ltd (eds) *MEDCOMP 77*, pp. 469–83. ISBN 0 903796 16 X.

5   Head, A. (1983) Planning and controlling patient care with the Exeter System. In Scholes, M., Bryant, Y. and Barber, B. (eds) *The Impact of Computers on Nursing: An International Review*, pp. 115–19. North-Holland, Amsterdam. ISBN 0 444866 82 5.

6   Henney, C.R. and Stewart, L.H. (1983) Data capture – from a real-time computerized nursing system. In Scholes, M., Bryant, Y. and Barber, B. (eds) *The Impact of Computers on Nursing: An International Review*, pp. 147–53. North-Holland, Amsterdam. ISBN 0 444866 82 5.

7   Harrow, M. (1988) Computer-assisted nursing information systems: a clinical nursing tool. In Daly, N. and Hannah, K.J. (eds) *Proceedings of Nursing and Computers, 3rd International Symposium on Nursing Use of Computers and Information Science*, Dublin, pp. 524–30. The C.V. Mosby Co, St Louis. ISBN 0 801632 35 8.

8 Hoy, D. (1989) *Computer-assisted Nursing Care Planning Systems in the United Kingdom: A Review of Professional Issues.* Nursing Division, Scottish Home and Health Department.

9 Woods, K. (1989) The development of a clinical nursing computer system. In Barber, B., Cao, D., Qin, D. and Wagner, G. (eds) *MEDINFO 89*, Singapore and Beijing, pp. 658–62. North-Holland, Amsterdam.

10 Strachan, H. (1987) *Excel Care™ Project Report.* Department of Computing, Riverside Health Authority. December.

11 Hyslop, A. and Jones, B. (1988) A cognitive model as a nursing expert system – potential for decision support and training in patient assessment. In Daly, N. and Hannah, K.J. (eds) *Proceedings of Nursing and Computers, 3rd International Symposium on Nursing Use of Computers and Information Science*, Dublin, pp. 565–74. The C.V. Mosby Co, St Louis. ISBN 0 801632 35 8.

12 Mason, E. (1988) Case study: implementing the Excel Care™ nursing system in West Dorset. In Daly, N. and Hannah, K.J. (eds) *Proceedings of Nursing and Computers, 3rd International Symposium on Nursing Use of Computers and Information Science*, Dublin, pp. 603–5. The C.V. Mosby Co, St Louis. ISBN 0 801632 35 8.

13 Sale, D. (1988) The Weymouth Case Study. In Daly, N. and Hannah, K.J. (eds) *Proceedings of Nursing and Computers, 3rd International Symposium on Nursing Use of Computers and Information Science*, Dublin, pp. 644–50. The C.V. Mosby Co, St Louis. ISBN 0 801632 35 8.

14 Palmer, B. (1988) The implications of nursing practice and how to improve improve the quality of care with the introduction of computerized care planning. In Daly, N. and Hannah, K.J. (eds) *Proceedings of Nursing and Computers, 3rd International Symposium on Nursing Use of Computers and Information Science*, Dublin, pp. 644–50. The C.V. Mosby Co, St Louis. ISBN 0 801632 35 8.

15 Newton, C. (1995) Examples of good practice – Royal Hampshire County Hospital. In Eaves, D. (ed.) *Benefits Realisation Monograph on Nursing Information Systems*, pp. 21–3. NHS Executive Information Management Group. June.

16 Richardson, K. (1995) Examples of good practice – Wirral Hospital Trust. In Eaves, D. (ed.) *Benefits Realisation Monograph on Nursing Information Systems*, pp. 33–5. NHS Executive Information Management Group. June.

17 Wiseman, J. (1989) Memory cards, portable microcomputers and community nurse recording. In Barber, B., Cao, D., Qin, D. and Wagner, G. (eds) *MEDINFO 89*, Singapore and Beijing, pp. 647–9. North-Holland, Amsterdam.

18 Savill, A. and Goldberg, C. (1988) Use of handheld computers: computers in the community. *BCS Nursing Specialist Group Newsletter*, 3, 4, 10–13.

19   Blackpool, Wyre and Fylde District Health Authority (1991) *District Nursing Dependency and Care Objective Taxonomy – Final Report.*

20   Hyslop, A. (1997) EPPIC – a community information systems project in Scotland. In Barnett, D. (ed.) *Sharing Information: Key Issues for the Nursing Profession* (Keynote presentations from NSG Conference in 1994 and European Summer School in Informatics 1994), pp. 41–3. BCS, Swindon.

21   Audit Commission (1992) *Caring Systems: A Handbook for Managers of Nursing and Project Managers.* HMSO, London.

22   Butler, E.A. and Hay, B.J.(1977) The fervent statistician: a computerized system to record and analyse nursing sickness and absence. In Online Conferences Ltd (eds) *MEDCOMP 77*, pp. 453–67. ISBN 0 903796 16 X.

23   Butler, E.A. (1983) A computerized operating theatre management system. In Scholes, M., Bryant, Y. and Barber, B. (eds) *The Impact of Computers on Nursing: An International Review*, pp. 421–9. North-Holland, Amsterdam. ISBN 0 444866 82 5.

24   Bryant, Y.M. and Bryant, J.R. (1977) Towards an integrated system for nursing administration. In Online Conferences Ltd (eds) *MEDCOMP 77*, pp. 439–51. ISBN 0 903796 16 X.

25   Halligan, M., McCormack, R., Thompson, J. and Waddel, M. (1988) Nursing information management system. In Daly, N. and Hannah, K.J. (eds) *Proceedings of Nursing and Computers, 3rd International Symposium on Nursing Use of Computers and Information Science*, Dublin. The C.V. Mosby Co, St Louis. ISBN 0 801632 35 8.

26   Strachan, H., Cogan, M., Kelly, K. and Webber, A. (1994) A benefits realization 'rescue plan' for a nursing information system. In Grobe, S. and Pluyter-Wenting, E.S.P. (eds) *Nursing Informatics: An International Overview for Nursing in a Technological Era*, pp. 108–12. Elsevier, Amsterdam.

27   Hardiker, N. (1992) Duty rostering in a primary nursing environment. *Information Technology in Nursing*, 4, 4, 8–9.

28   Maguire, G. and Roberts, J. (1982) Nurse allocation with computer assistance – extension to a manpower planner. In O'Moore, R.R., Barber, B., Reichertz, P.L. and Roger, F. (eds) *MIE, Dublin 82: Lecture Notes 16*, (pages unknown). Springer Verlag, Berlin. ISBN 0 387112 08 1.

29   Roberts, J. (1983) Computerized nurse allocation and the identification of related service tools. In Scholes, M., Bryant, Y. and Barber, B. (eds) *The Impact of Computers on Nursing: An International Review*, pp. 364–7. North-Holland, Amsterdam. ISBN 0 444866 82 5.

30   Greenhalgh, C. and Roberts, J. (1991) Nursing workload measurement – computer support. In Adlassnig, K.P., Grabner, G., Bengtsson, S. and Hansen, R. (eds) *MIE, Vienna 91: Lecture Notes 45*, pp. 829–32. Springer Verlag, Berlin. ISBN 3 540543 92 9.

31 (authors unknown) (1992) *Nurse Management Systems: A Guide to Existing and Potential Products.* Greenhalgh and Co. Ltd.

32 (authors unknown) (1992) *The Operating Theatre Systems: A Guide to Existing and Potential Products.* Greenhalgh and Co. Ltd.

33 Rhys Hearn, C. (1971) The more fruitful uses of hospital resources. In Abrams, M.E. (ed.) *Spectrum 71: BCS Conference on Medical Computing*, pp. 57–74. Butterworths for BCS, London. (See also D18 in Chapter 17 Reference 18.)

34 Barr, A. (1967) *Measurement of Nursing Care.* Operational Research Unit, Report 9, Oxford Regional Hospital Board.

35 Strachan, H. (1989) *Patient Dependency Pilot Study Evaluation.* Riverside Health Authority Nursing Information System Project. November.

36 Rimmer, P., Jones, B., Munro, S. and McMahon, J. (1994) The reliability of summary descriptions of monitored nurse activities within a care of the elderly ward in a psychiatric hospital. In Grobe, S. and Pluyter-Wenting, E.S.P. (eds) *Nursing Informatics: An International Overview for Nursing in a Technological Era*, pp. 383–7. Elsevier, Amsterdam.

37 Beeston, A. (1990) The resource management project at the Freeman Hospital. *Information Technology in Nursing*, 1, 4, 70–1.

21 Computers in nursing and patient education

MAUREEN SCHOLES AND HEATHER STRACHAN

Computers are used in a number of ways, in both pre- and post-registration nurse education, to support the management of nurse education as well as learning.

In 1975, The London Hospital introduced a computerized Learner Nurse Allocation and Nursing Personnel System. The staffing of the wards and departments was still largely dependent on learners but training requirements were now much more stringent. There were seven basic nursing courses and seven sites for clinical experience. The students had planned experience, but were essential to the ward staffing levels throughout the 24 hours. Annual leave and sickness were recorded for both learners and trained staff, and wastage statistics could be produced, saving clerical time. Responsibility for training and the learners' personnel records were now in the hands of the head of the School of Nursing, Dr Sheila Collins, who outlined the change from the manual system to a computer-based system.[1] It was initially a batch-processing system and later went online for 1000 students. This work was supported by Ashwin Shah from the Operational Research Unit at The London Hospital and subsequently continued when the unit was transferred to the North East Thames Region, extending over a long period of time as the various participants gradually understood what the information systems could do and what the nursing administrators and educators needed. At a later stage an interactive allocation program was produced for newly qualified nurses.[2] In 1978, in Manchester, Professor Brian Moores produced a formulation of the student nurse allocation problem.[3, 4]

The General Nursing Council for England and Wales had developed a computerized record system to maintain a register for UK nurses. Maintaining students' training records on a computer was common in most Schools of Nursing by the end of the 1980s.

Computer-assisted learning (CAL) was already well established in higher education and in nurse education in the USA and Canada by the early 1980s. However, it was still new to nurse education in the UK. A national survey was undertaken by the English National Board (ENB) in June 1986 to determine the amount of hardware and software required, and its potential use within Schools of Nursing and Midwifery. This was in preparation for a proposal for an organized approach to computer and IT development

431

in nurse education. There was a response rate of 85.5 per cent ($n = 327$). Only 37.5 per cent of educational establishments had microcomputers for educational use. The number ranged between one and 29 machines. Of those schools that had computers, 28 per cent had no educational software available for use by teachers or students, only 23.5 per cent ($n = 2,345$) had attended some sort of computer appreciation course and only a few had attended more advanced computer courses.[5]

CAL supported nurses to learn about specific topics, initially using the drill and practice techniques. However, this approach changed increasingly towards educating the user by moving them from the unknown to the known in a more cognitive manner. Computer simulation and robotics were, for example, used to teach about cardiac arrhythmias and resuscitation. Multimedia interactive-disk and video were also used in educational packages.

Computers provided students with access to a wide range of information as well as improved communication tools. CAL benefited students and teachers, allowing students to set their own pace and repeat the lesson as often as they required. Teachers were able to set standards and track students' learning. Above all, CAL generated self-study principles and enhanced distance and open learning opportunities. CAL was seen to be well placed to meet the needs of future nurses: to complement the UKCC Project 2000 proposals for greater emphasis on promoting self-study skills in nurses,[6] and in its proposed mandatory periodic refreshment for re-registration purposes.[7]

The growth in the number of CAL packages available between 1982 and 1987 was mainly due to development by nurse teachers, who were skilled in computer programming in BASIC or in using authoring tools. At a computer club at Warrington General Hospital, set up by Graham Wright, then a nurse tutor and member of NUMINE, computer enthusiasts from the health service met to share and demonstrate applications. This led to the development of a small library of teaching programs. Graham Wright conceived the formation of the Open Software Library (OSL) to gather similar innovative material from others in the health service and to make such programs more widely available.[8] OSL began in 1984 as a voluntary organization formed by several club members, including David McKendrick, Peter Annets, Colin McMurchie, Mike Wilkinson and Tom Newton. In this way the library acted as a clearing-house and repository for programs, which could be purchased at minimal cost with the author receiving a royalty. Additionally, it acted as a resource to promote the use of the new technology in the health service. An advertisement was placed in a nursing magazine and programs began to be collected. The group also demonstrated programs at nursing conferences to offer practical experience in using computers.

Initially, the teaching programs were written on a range of machines – Spectrums, Amstrads, Commodores and BBC Micros. Authors were mainly nurses or doctors, and amongst the most prolific were Sandie Massie,

Richard Butler, Peter Annets, Steve Ward, Dr Roger Snook, Steve Wheeler, Peter Jones, Elizabeth Fish, Sydney Chellen, Emrys Jenkins and Steve Jones. They were writing programs on such subjects as drug dose calculation, blood groups, ward management, clinical observations, patient assessment, the Mental Health Act, cardiopulmonary resuscitation and schizophrenia. As the BBC microcomputer became the recommended machine for use in Schools of Nursing, programs were mainly written in BBC BASIC and later in BBC Microtext, a programming tool offering relatively fast program development. By the mid-1980s the volume of material was too large for the group to continue on a voluntary basis, so it was restructured and, in 1986, it was incorporated as a limited company.

In 1986, the NHS Training Authority (NHSTA) met members of NUMINE and OSL to discuss a means of fostering the use of computers amongst health professionals. OSL and others, following this, undertook a number of funded projects.[9] Two national conferences to foster the use of computers in healthcare education and training were funded by the NHSTA.[10, 11]

From 1987, the IBM PC computer began to be gradually adopted as the machine of choice and the library of mainly BBC material, which OSL held, became obsolete.

Today, OSL continues to distribute a small number of titles produced by health practitioners, but the emphasis is now on reselling a wide range of medical and nursing titles produced by the major publishers and providing community websites for non-profit organizations.

One particular CAL project, led by Sue Norman, focused on the use of CAL in post-registration education. The Nightingale School CAL project ran from 1982 to 1988. It was run in collaboration with the University of Surrey and explored the development, application and impact of CAL in a nurse education context.[12] It produced an 'Intravenous Drug Administration Package' that was reported to be well received. The users thought the package, content and activity of information gathering was relevant to practice, contributed to their understanding of the subject and would change their practice. This early innovation was not extensively used, as it exceeded the knowledge of most Schools of Nursing at the time. However, the project did a lot to foster the recognition that the standards of development and the appropriate integration of CAL into the curricula were essential to the success of CAL.

The move of nurse education from hospital sites into universities facilitated the wider use of internet technologies. Computers were also used in libraries to support literature searches using a CD-Rom. The Joint Academic Network (JANET), supported communications, such as the email and bulletin boards, and provided opportunities for one-to-one or group discussion and tutorial help, independent of time and place. Electronic conferencing was also possible. Students increasingly used general software such as word processors, databases and spreadsheets in all areas of the curricula.

In the 1980s, a range of communication technologies was in use at Queen Margaret College, Edinburgh, which had one of the largest health care student populations in UK higher education. The electronic publishing and computer-mediated communication system, developed at the college, was called JIMMY. It allowed multiple access to information in an electronic form as viewed on a VDU, from a database, via a network. Its purpose was to support easy access to a range of electronic publications from departmental documents, past examination papers, assignment specifications, minutes of meetings, abstracts of periodicals and lecture handouts. JIMMY was designed in such a way that each department could manage its own file space and edit any of its own documents through its own bulletin board. The benefits were seen as providing nurses with the skills to create, send, receive and analyse electronically stored information. In addition, the college introduced students to systems in use in the health service in Scotland such as CANIS.[13]

Other early projects that supported students included the computerization of the library accession register in Romford College of Nursing and Midwifery by Chris Dowd. A network of Apple® Macintosh® machines using the *Ominis* database was developed into a library ordering system and database of all library contents to enable users to find a particular book quickly and with ease.[14]

However, for computers in education to flourish and support learner nurses, it was necessary for the nurse tutors to be computer literate. Following the success of the Wessex CAL project, led by Derek Morrison, which aimed to increase the level of computer awareness among nurse tutors in the eight Schools of Nursing in the Wessex Region, a national project was initiated.[15] The ENB, the statutory educational body for nursing in England and Wales, knew that if this integration was to occur there was a need to support educators of nurses to develop their understanding of information technology and its application to nursing and the curricula. The three-year ENB Computer Assisted Learning Project, led by Paula Procter, involved supporting the education of nurse tutors in educational computing, software and technical understanding. The intention of the project was to broaden the educator's perspectives with regard to computers and information technology, in order that they could effectively implement such technology in the curricula. The participating tutors then returned to their educational establishments and continued to build and consolidate their knowledge through distance learning, using multimedia and the Times Network System, before cascading that knowledge to their colleagues.[16, 17] The Times Network continued to be used as a tool for the management of education, enabling wide dissemination of information such as an educational media resource database, developed by the ENB resource and careers services. This held information on nationally available support materials for teaching and learning. The project achieved its objectives of removing the fear of computers and demonstrating their use in nurse education,[17] but still many tutors felt that it had failed them.[18]

In 1990, the statutory educational body, the National Board for Scotland, carried out a survey that identified that only 20 per cent of nurse teachers in Scotland used CAL and that it was not integrated into the curricula.[19] It was suggested that the reasons for this rather poor situation included limited hardware, limited availability and quality of software, and poor computer literacy skills among students and tutors.

In September 1993, the NHS Executive sponsored the first national study conducted in England to assess the readiness of nurse education to provide programmes that addressed the emerging Information Management and Technology (IM&T) initiatives, promoted by the Executive's strategy. The study reviewed the IM&T provision within pre- and post-registration nurse training. The results indicated 41 per cent of respondents included computer skills in the curricula and 18 per cent included information management, while eight per cent included research skills. From these points came a number of recommendations. Among them were developing guidance on:

- how IM&T can be included in the pre- and post-registration curricula;
- education of nurse teachers; and
- sources of advice.

It was recommended that existing products, materials and services should be mapped against curricula guidelines, and examples of good practice should be identified, developed and disseminated. It also advised that a feasibility project should explore the need for specialists in IM&T and the need for the establishment of research and development centres.[20] These recommendations were similar to the SAGNIS proposals of the same time, which resulted in a number of products.

The Institute of Health and Care Development and the English National Board produced *Information for Caring – Framework for Including Health Informatics in Programmes of Learning for Nurses, Midwives and Health Visitors and Other Clinical Professionals* in November 1997.[20] Other products included a video (*A Patient's View*) and learning packages on 'Security for the nursing profession' and 'Terms, records and information, and The Electronic Patient Record'.[21] Other recommendations were met through such projects as 'Dearing Skills',[22] and the Benefits Realization Monographs produced by the BCS NSG.[23, 24]

Probably the most well-known product to emerge was the Rainbow Pack, which was created to support nurses' learning about information management. The Rainbow 1 Pack learning programme, *Using Information in Managing the Nursing Resource*,[25] was commissioned by the NHS Management Executive, England and Wales. It was developed by a consortium of five regions in conjunction with Greenhalgh and Company Limited. Graham Wright, Christine Greenhalgh and Helen Jackson were instrumental in its development, and the system was marketed by Jean

Roberts. It was designed to be used on its own or alongside other training material in a number of ways: as an in-house training programme; as a self-study activity; as reference material; or as a higher education course. Rainbow was accredited in a number of ways. In Scotland it was used as part of the National Board's Charge Nurse Diploma. It was given four credits towards the diploma, and two towards the certificate, by the Open University. The Rainbow Pack was used as the basis for a Diploma in Using Information in Nursing (DUIM) at the University of Manchester, run by Graham Wright and John Kelly. The course trained assessors who could then run the Certificate Programme within hospital departments. Southmead Hospital, Bristol, and the Radcliffe Infirmary, Oxford, took a number of cohorts through this programme. The Certificate Programme was franchised to the University of Wales, which ran it in Cardiff and Glan Cwyld.

Reports on the use of the Rainbow Pack in England identified it as an excellent developmental programme for nurses and relevant to the NHS agenda, providing a flexible approach to learning.[26] An evaluation of the use of the Rainbow Pack in Scotland found that Rainbow 1 was used by over half the Colleges of Nursing and Trusts or directly Managed Units for post-registration training, and in a few cases for pre-registered nurses. It concluded that Rainbow 1 provided a flexible approach to ward managers' training in information and resource management. The new knowledge gained by participants was applied to practice, and in a number of cases led to changes.[27]

Subsequent surveys undertaken in Scotland and Wales in 1999 and England in 1998 have shown a marked improvement in the integration and use of learning technology in nurse education, but it is suggested that it still falls short of the expected future needs of the nursing profession.[20, 28]

In 1994, Manchester College of Midwifery and Nursing undertook a training-needs analysis. They devised a framework to integrate information management and technology into the nursing diploma course that prepared students for registration as nurses. Starting with the Common Foundation programme this provided nine hours of study each term. The associated skills training for nurse teachers and clinical assessors used the National Vocational Qualifications scheme.[29]

A pilot course at Avon and Gloucester College of Health focused on the management of information, rather than computer technology. It was evaluated, between 1 April 1994 and 31 October 1995, by Matthew Hughes. While the content indirectly built on the pre-registration course by Manchester College of Nursing, the core could be used by other health care disciplines. The Avon course was validated by the ENB as N32: Nursing Informatics. The evaluation showed the content could be applied in the daily work of course students and some gained promotion as a result.[30, 31]

PATIENT EDUCATION AND INVOLVEMENT

While CAL was seen as being beneficial to nurses for their own education, it was recognized that patients could also gain from a flexible, convenient and effective way of learning about specific conditions and treatments. CAL could be used as an adjunct to nurses providing individualized information and the teaching of clinical skills needed by patients to manage their own care.

A three-year, DoH-funded, study involving the development and evaluation of CAL material for patients about to undertake continuous, ambulatory, peritoneal dialysis was well received by the patients. The evaluation reported that patients had benefited from their experience of using the computer. It had provided them with useful information, an opportunity to learn in private, at their own pace, and in an uncritical manner without having to take up the time of busy nurses. CAL could meet a wide range of learning needs, as it was able to give choice to patients over what they would learn and how often they could repeat the material. The only negative comments were the lack of humanity in the computer.[32] This lack of humanity has been used to the benefit of patients elsewhere. It was found that patients using computers as an assessment tool to measure motivation for recovery in problem drinking were highly reliable: it appeared that patients disclosed more sensitive information using this method than with a paper-based approach.[33]

There were many other opportunities, for the public and patients, arising from the direct use of computers, for example:

- to get general information about health;
- to get personal information from their own clinical record;
- to communicate with their clinician either directly via email by completing a standard online interview, or through self-help group email lists.

Health consumers could access these resources from public terminals, from home or in libraries and health centres. See Chapter 16 for more information on this.

REFERENCES

1   Collins, S. (1983) Computer-based systems for professional education and training. In Scholes, M., Bryant, Y. and Barber, B. (eds) *The Impact of Computers on Nursing: An International Review*, pp. 350–5. North-Holland, Amsterdam. ISBN 0 444866 82 5.

2   Shah, A.R. and Hollowell, J.A. (1981) An optimal allocation of newly qualified nurses to wards. In Grémy, F., Degoulet, P., Barber, B. and Salamon, R. (eds) *MIE, Toulouse 81: Lecture Notes 11*, pp. 389–96. Springer Verlag, Berlin. ISBN 3 540105 68 9.

3   Moores, B., Garwood, N.H. and Briggs, G.H. (1978) The student nurse allocation problem: a formulation. *Omega*, 6, 1.

4   Moores, B. and Wood, I. (1977) Nursing allocation using a time-shared computer. *Nursing Times*, August.

5   Procter, P. (1991) *English National Board for Nursing, Midwifery and Health Visiting, Computer Assisted Learning Project 1988–1991*. Internal Report.

6   United Kingdom Central Council for Nursing, Midwifery and Health Visiting (1986) *Project 2000: A New Preparation for Practice*. UKCC, London.

7   United Kingdom Central Council for Nursing, Midwifery and Health Visiting (1987) *Mandatory Periodic Refreshment for Nurses and Health Visitors – A Discussion Paper*. UKCC, London.

8   Webref (accessed May 2007) www.opensoftwarelibrary.co.uk

9   Massie, S. (1988) Oak trees from little acorns grow. In (eds unknown) *Proceedings of the 2nd National Conference on the Use of Computers in Healthcare Education and Training*. NHSTA.

10 NHS Training Authority (1986) CBT'86. *Proceedings of the 1st National Conference on the Use of Computers in Healthcare Education and Training*. NHSTA.

11 NHS Training Authority (1988) CBT'88. *Proceedings of the 2nd National Conference on the Use of Computers in Healthcare Education and Training*. NHSTA.

12 Norman, S. (1988) Using computer-assisted learning to promote safe practice: an evaluation of the medium and the message. In Daly, N. and Hannah, K.J. (eds) *Proceedings of Nursing and Computers, 3rd International Symposium on Nursing Use of Computers and Information Science*, Dublin, pp. 676–81. The C.V. Mosby Co, St Louis.

13 Wyatt, M., Taylor, A., McMurdo, G. and Thomson, B. (1989) Computer-mediated communication and the teaching of nursing studies. In Barber, B., Cao, D., Qin, D. and Wagner, G. (eds) *MEDINFO 89*, Singapore and Beijing, pp. 663–7. North-Holland, Amsterdam.

14 Dowd, C. (1987) *The British Computer Society Nursing Specialist Group Newsletter*, 3, 3 17–19.

15 Morrison, D. and Betts, H. (1990) The Wessex Computer Assisted Learning Project. *Information Technology in Nursing*, 2, 2, 19–23.

16 Procter, P. (1988) We have the technology to rebuild . . . . In Daly, N. and Hannah, K.J. (eds) *Proceedings of Nursing and Computers, 3rd International Symposium on Nursing Use of Computers and Information Science*, Dublin, pp. 308–8. The C.V. Mosby Co, St Louis.

17   Procter, P. (1989) Network building with technology: the English nursing experience. In Barber, B., Cao, D., Qin, D. and Wagner, G. (eds) *MEDINFO 89*, Singapore and Beijing, pp. 683–937. North-Holland, Amsterdam.

18   James, M. and Turner, P. (1992) Computer-assisted learning – Any progress? *Information Technology in Nursing*, 4, 4, 10–11.

19   National Board for Nursing, Midwifery and Health Visiting for Scotland (1995) Using computers to teach nurses and midwives. *Information Technology in Nursing*, 7, 2, 5–7.

20   Wright, G., Kelly, J., McMurchie, C., *et al.* (1994) *A Review of the Current IM&T Provision Within Pre- and Post-Registration Nurse Training*. University of Manchester, Health Services Management Unit, Centre for Health Informatics.

21   Betts, H., Wright, B. and Olsson, G. (2000) One Step Back? A follow-up Review of IM&T Provision in Pre-registration Nurse Education in the UK. *Information Technology in Nursing*, 12, 3, 4–8.

22   Dearing, R. (1997) *Report of the National Committee of Inquiry into Higher Education*. HMSO.

23   Eaves D. (ed.) (1995) *Benefits Realisation Monograph on Nursing Information Systems*. NHS Executive Information Management Group.

24   Betts, H. (ed.) (1996) *Benefits Realisation Monograph on Maternity Information Systems*. NHS Executive Information Management Group in conjunction with BCS NSG.

25   Greenhalgh and Co. Ltd in conjunction with a consortium of five RHAs (1994) *'The Rainbow Project': Using Information in Managing the Nursing Resource*. Charlesworth, Huddersfield.

26   NHS Training Directorate (1993) *Information Management and Technology Programme – Evaluation of Rainbow 1, Final Report*. December.

27   Strachan, H., Waugh, J. and Watt, S. (1994) *Using Information in Managing the Nursing Resource. Rainbow Project – Final Report*. The NHS in Scotland, Management Development Group Report. November.

28   Topp, H. and Kinn, S. (1990) The use of learning technology in nurse education – a survey of Scottish and Welsh nurse educators. *Information Technology in Nursing*, 11, 4, 6–9.

29   Prestwich, M. (1994) *A Training Needs Analysis: Information Management & Technology in the Pre-registration Nursing Curriculum*. North Western RHA.

30   Barnett, D. (1996) *Information Management and Technology in Post-registration Education – Report of the Avon and Gloucester Project 1994–95*. NHS Training Division. March.

31 Hughes, M. (1996) A post-registration course in nursing informatics. In Richards, B. and de Glanville, H. (eds) *Current Perspectives in Healthcare Computing*, pp. 721–2. BJHC for BCS, Weybridge.

32 Luker, K., Ackrill, P. and Caress, A. (1994) Computer-assisted learning for use in the education of renal patients on continuous ambulatory peritoneal dialysis – results of an evaluation study. In Grobe, S. and Pluyter-Wenting, E.S.P. (eds) *Nursing Informatics: An International Overview for Nursing in a Technological Era*, pp. 419–22. Elsevier, Amsterdam.

33 McMahon, J. and Jones, B. (1994) Computer assessment by nurse therapists of motivation to quit in problem drinkers: the virtual page is treated differently to the real page. In Grobe, S. and Pluyter-Wenting, E.S.P. (eds) *Nursing Informatics: An International Overview for Nursing in a Technological Era*, pp. 476–7. Elsevier, Amsterdam.

22 Enabling technology and the future

MAUREEN SCHOLES AND HEATHER STRACHAN

ENABLING TECHNOLOGY

In addition to nursing information systems that support clinical practice, management and education, significant work was undertaken by nurses to underpin these systems. This work included research and development in the areas of decision-support tools, terminology and standards. The first part of this chapter describes some of this work and its application to nursing informatics.

Decision-support research

As its name implies, a Clinical Decision Support System (CDSS) can provide information to help in making clinical decisions. This can range from the simple task of alerting the clinician to an immediate problem that requires prompt action, to more complex decision support, such as critiquing, reviewing information and proposing actions. A CDSS, if used in the right situation, has the potential to support effective clinical reasoning and patient safety.

The 1988 IMIA International Conference on *Nursing Use and Information Science: Where Caring and Technology Meet* was followed by an exciting workshop on decision-support systems in nursing that contained a great deal of useful reference material.[1, 2]

Hyslop, Jones and Ritchie undertook early work on decision support in nursing while at Glasgow University. This research identified that the value of decision support in nursing was not necessarily in providing the decision, which is the nurse's obligation, but in providing a set of decision paths that represent alternatives to the decision path being either contemplated or followed. The user then handles these alternatives in whatever way they see fit. The work recognised that, while mathematical modelling could be a reliable and precise process when applied to a domain in which numbers had been applied, in nursing where the patient would equate to the number, the modelling could not necessarily be reliable and precise. Hyslop observed that the early expert systems were not reliable or flexible enough to replace the professional decision-maker. While optimistic about the future, he noted that research into nurses' decision-making was a time-consuming activity.[3, 4]

Nursing terminology

The use of computers in nursing provided an opportunity to collect and record large amounts of data to support the delivery of effective and efficient patient care. However, in order to express, retrieve, communicate, analyse and compare that data in a meaningful way, nursing terminology must be defined. A lack of agreed nursing terms created many of the difficulties in the early computerized care planning systems. A large workshop was held by the BCS NSG in Eastbourne in November 1992, and The Nursing, Midwifery and Health Visiting Terms Project, led by Anne Casey, addressed this problem.[5] The latter, a major national initiative in England, began in April 1993 and was part of the Clinical Terms Project based at the NHS Centre for Coding and Classification. The project involved over 120 nurses, midwives and health visitors throughout the UK, to produce a comprehensive set of terms used to record and communicate patient care. The project worked in close liaison with the medical and allied health profession terms project working groups to share terms and reduce duplication of effort.[5] Apart from producing an agreed set of nursing, midwifery and health visiting terms, there were patient and shared terms (which were relevant to social workers in community settings). The project also produced educational material about the terms and their use in information systems. These terms have now been incorporated into the Systematised Nomenclature of Medicine and Clinical Terms (SNOMED CT®) and must be used by all clinical information systems in the UK.

One of the most influential European clinical terminology initiatives has been the GALEN programme, which started in 1992.[6] It sought to provide a suitable technological foundation for the next generation of clinical information systems. The seeds of the GALEN programme were sown by the PEN&PAD programme (see Chapter 14), which was led by the Medical Informatics Group at the University of Manchester. The PEN&PAD (General Practice) project (1989–91) carried out early research into clinical workstations, focusing particularly on structured data entry. The tools and techniques developed were adopted by the PEN&PAD (Elderly Care) project (1992–94).[7] This project took the technology into a hospital environment to determine its applicability in other domains, specifically nursing, and to investigate issues around collaborative record-keeping.[8, 9] Some of this work fed into the TELENURSE programme (1996–2000), coordinated by the Danish Institute for Health and Nursing Research.[10] TELENURSE focused on refining the International Classification for Nursing Practice (ICNP® – an ongoing programme within the International Council of Nurses), fostering its use across Europe and assessing its applicability in operational nursing information systems. Before this, the GALEN project (1992–95), again coordinated by the University of Manchester, had worked towards further enhancing the emerging technology. A series of European demonstrator and application projects followed. The thinking behind the GALEN programme – compositionality, concept-orientation, description logic representation, terminology services, etc. – has had an impact on many clinical terminology initiatives, including the Read Codes and NHS Clinical Terms. It continues to play a major role in more recent developments, such as SNOMED CT® INCP® Version 1 and the ISO standard for nursing terminology systems ISO 18104, *Integration of a reference terminology model for nursing.*

The internet

The internet is a gateway to a wealth of information. During the internet's early development there was little or no nursing involvement, but when academics started using it, during the early 1980s, nursing-related information, largely from the USA, started to appear. Communication amongst nurses started with newsgroups (sci.med.nursing and uk.sci.med.nursing) set up by Denis Anthony, and mailing lists (nursenet, in the USA, and nurse-uk, in the UK). In 1992, the fledgling World Wide Web made it easier to publish and find relevant information. Four nursing-related websites emerged, three in the USA and the *NURSE* site, again created by Denis Anthony, in the UK.

Throughout the rest of the 1990s the number of nurses able to access the internet and the relevant information available grew massively. Various attempts were made to provide indexes of these including, in the UK, *Nursing and Healthcare Resource on the Net*, created by Rod Ward. This site listed over 2000 websites, mailing lists, etc. before it closed in 2001 to be replaced by NMAP,[11] which provides a greater level of quality assurance and improved search and browse capabilities.

The internet has now become a major information source and medium for nurses throughout the world, with the Google™ search engine identifying several million websites for nursing with almost every hospital, university department, publisher, organization and many individuals providing information. The key issues today are quality, amount and access to the information. Nurses need to ensure accuracy and relevance of the information available and, as patients themselves have access to this information, the relationship between the patient and the nurse is changing.

The internet has allowed greater access to knowledge for both staff and patients. Staff now have access to elibraries. The development of the National Electronic Library for Health in England and Wales (NeLH) has been based on the principle that it would complement and supplement the skills of NHS librarians and the resources in NHS libraries.[12] The NeLH has not sought to overlap the traditional functions of NHS libraries in providing books, journals or databases such as CINAHL. Its collection focuses on the new types of quality-assured knowledge produced by organizations such as The Cochrane Collaboration and the Centre for Reviews and Dissemination. The main source of advice on know-how it provides is from National Institute of Clinical Excellence.

Scotland has taken a different approach by providing an elibrary (www.elib.scot.nhs.uk) with access to an ejournal, ebooks and databases, as well as news, discussion groups and other knowledge services.

Patient information can be obtained from a wide range of sources. A key source from the UK is NHS Direct (www.nhsdirect.nhs.uk). It provides a health encyclopaedia and information on best treatment and self-help guides (see Chapter 16 for more details). NHS 24 (www.nhs24.com) provides a similar service in Scotland.

REFLECTIONS AND THE FUTURE

The state of affairs by the 1990s

The significant growth in the use of computers in nursing in the late 1980s and early 1990s followed the investment made during the Resource Management Initiatives, and a number of national reviews on usage and the impact of computers in nursing.

One survey, in 1993, identified that 128 of the 248 acute hospital providers in England claimed to have Nursing Systems either for scheduling or care planning. While sites had acquired a system, many were still at the pilot stage and did not cover all the hospital wards. The acceptance level was not high, with nine per cent of sites describing their systems as minimally acceptable. A further 23 per cent deemed their systems only partially acceptable. Where details were submitted, 53 per cent of sites reported that their nursing systems linked to other applications.[13]

A research study for the NHSME Information Management Group and SAGNIS concluded that:[14]

- more needed to be done to encourage sites to have a clear Nursing Information System strategy;

- progress toward NIS implementation was slow – only 51 per cent of the 180 sites that responded had completed their implementation;

- of the four phases of NIS introduction (preparation, procurement, implementation and evaluation), the last mentioned was the least successful;

- more guidance was required on benefits realization;

- only 26 per cent of hospitals with over 100 beds had integrated their NIS with a PAS or HISS;

- there was a clear need to review basic nurse training and post-registration training to identify where and when the use of information should be taught.

A number of recommendations included:

- the need for research and development in clinical information systems;

- education in information skills;

- the importance of leadership from the top;
- the need for a strategy for NIS; and
- benefits assessment and realization plans.

Despite the large investment in NIS, none of the hospitals studied in the Audit Commission report of 1992 had undertaken a full evaluation of the cost and benefits of NIS.[15] Reports compiled by Derek Eaves and by Helen Betts for the BCS NSG and the Information Management Group of the NHS Executive, [16, 17] attempted to address this and encourage more benefit realization. Information systems used by midwives were often adapted from nursing, community systems or hospital information systems with a few stand-alone systems designed specifically to meet midwifery needs. An impressive list of benefits was reported as follows:

- Care Planning Systems
  - More complete and improved quality of assessments, care plans and evaluation.
  - Reduction in time spent on care planning.
  - Improved quality of care.
  - Facilitation of audit.
  - Named nurse concept.
  - Multi-professional communication.
  - Discharge planning.

- Workload Systems
  - More accurate and up-to-date workload information.
  - Better use of staff.
  - Greater productivity.
  - More time spent on direct care.
  - Less use of temporary staff.
  - More appropriate skill mix.
  - Less time spent on nurse handover.
  - Reduction in clerical work by qualified staff.

- Scheduling Systems
  - Less time spent on designing and amending rosters.
  - More accurate timesheets and personnel data.
  - Quicker retrieval of personnel information.
  - Less time spent on administration by managers.

- Drug Administration Systems
  - Saved time on drug administration.
  - Drugs more likely to be administered at prescribed times.

- ◆ Improvement in legibility of prescriptions.
- ◆ Improvement in quality of care.
- Departmental Communication Systems
  - ◆ More efficient and effective communication.
  - ◆ Time saved on telephone calls.
  - ◆ Supplies ordering more efficient.
- Maternity Information Systems
  - ◆ Audit facilities.
  - ◆ Easy retrieval of statistics.
  - ◆ No duplication of data entry.
  - ◆ Printouts of records.
  - ◆ Improved standards of record keeping and documentation.
  - ◆ Highlighting risk factors for clients.
  - ◆ Save time.
  - ◆ Online access to results.
  - ◆ More accurate clinical information.
  - ◆ Better managed services and continuity of care.

Issues for future success

In the 1960s and 1970s there were expectations as to what computers could do for nurses and for patients. There was a belief that the computer could:

- save nursing time, thereby allowing more time for direct patient care;
- collect and analyse information that previously was too time-consuming and costly to do, in order to improve the quality of nursing care; and
- add to nursing's body of knowledge for the benefit of the public.

These beliefs were tested in early experimental projects.

A Delphi study undertaken in 1993/94 asked members of the BCS NSG, mostly project nurses from computing and resource management, and nurse educators, 'What issues need to be addressed to ensure their [Nursing Information Systems] successful usage?' Many of the issues identified were people issues, such as education and involving staff in the procurement and implementation phases. Others were across a wide range of topics, including management of change, hardware and software issues, use of information, and evaluation.[18] Many of these issues have been reflected in this part of the book.

The 1980s and 1990s were a time of setbacks, reflection and refocusing. Nursing saw more projects emerge across the UK. For most of the nurses involved, whether in developing, introducing or using these systems, the work was still exploratory in nature. Nursing systems came and went, some aspects were successful, others were not, but ideas and experience from these early projects were not lost. Neither was the belief in the common purpose and potential benefits computers could provide to patients' health. To consider the future it is useful to examine why there were failures, what was successful, what has now changed and what has remained constant.

Redesign not just rework

Early nursing information systems, like most of those in other disciplines, tended to computerize existing manual processes. They did not exploit the full potential of the new tool. In addition, the benefits were not perceived as great enough for the investment in the time required to develop and implement them. Where the computer was used as a vehicle for changing working practices these problems were compounded and resistance or failure could result.

An example of this is the lack of sustainability or success of care planning systems. A system that supported a single discipline became less appropriate as the emphasis moved towards individual patient care being supported by a multidisciplinary team. This requires clinical information systems that support the patient rather than a specific discipline. Before computerization it is essential to review information needs and existing work practice in order to optimize the use of the computer. This requires detailed measurement and analysis. Most importantly, user education and participation in the change processes and validation of these changes in work practice is essential alongside the computerization.

Home-grown versus commercial?

Nursing Information Systems were either developed or funded by the NHS itself or by commercial companies. The locally developed systems tended to have more success than commercial systems. Locally developed systems had more user involvement in the design and development stages, and so a greater use of local terminology and sense of ownership. In addition, the experimental nature of these early systems, which inevitably required frequent changes as users became more familiar with the possibilities offered by the new technology, did not fit well with commercial goals.

However, not all organizations have the necessary skills or capacity to develop their own systems. It would not make sense to create hundreds of 'computer cottage industries reinventing the wheel' and not sharing experience. The NHS is in the business of patient care, not computing, so a partnership with the computing industry is essential for development purposes. Technology has also moved on and rapid prototyping enables iterative development with users, analysts and designers working closely together addressing the problems. Time spent in getting the specification of a system right will later realize the benefits and avoid delays. Experience of early development has also helped demonstrate what works well and what works less well.

The weakest link

Many of the early projects were pilots that involved small numbers of enthusiastic staff, often determined to make the system work for them. Problems arose when introducing a system organization wide: these could be technical, organizational or cultural. The reasons range from a lack of experience caused by being first, to a lack of risk assessment, or simply enthusiasm and pressure to get systems working quickly and providing benefits for users. A key problem, at first, was the fact that nursing systems were not linked to the PAS or to manpower and payroll systems. Alternatively these organizational systems did not exist, and this led to the need for duplication of data collection and entry rather than saving time and effort. Essentially there were two possibilities for implementation:

- The big bang (horizontal) approach – involving the spreading of a single application throughout an organization and then subsequently adding further applications when the initial system had bedded in.
- The incremental (vertical) approach – involving the piloting of a cluster of applications in one or two experimental areas.

The advantages and disadvantages of these approaches are often debated. The key to implementation must be good project management and identification of critical success factors using risk assessment.

Lack of controlled terminology

A major problem with the early systems was their lack of controlled terminology. The ability of computers to store, analyse and retrieve large amounts of nursing information is one of its main benefits. However, to do this, it is first necessary to define these data. Defining nursing terminology is as challenging as defining nursing itself. Nursing care is provided in a variety of settings, and has a range of care objectives, delivered within a range of models, taking into account the views of individual patients and the medical possibilities. Nursing is one part of a multidisciplinary team with different terminologies and overlapping roles. The unavailability of defined nursing terminology in early care planning systems meant users had difficulty in collecting and entering data that often needed to be supplemented with free text. Appropriate analysis and retrieval of information from the computer was not always achieved. Ultimately this made nursing information systems time-consuming to use and provided little that was useful. This issue has now been addressed and the resulting clinical terms thesaurus that includes nursing, midwifery and health visiting terms (part of SNOMED CT®) is intended for use in all UK clinical information systems, allowing international comparisons and research projects.

Management versus clinical needs

Early computerization supported the provision of management information rather than supporting the clinical process. The environmental, social, economic and political drivers that influence the use of computers by nurses tended to focus on the need to improve the efficiency of nursing, the greatest cost component in the NHS. Nevertheless, it is important that nursing costs should be balanced with the optimum quality of nursing care delivered. When a system was proposed via a business case it was expected that the cost of the system would be recouped by greater efficiency. In reality, this rarely happened due to other changes that had an effect on the demand and availability of resources. The emphasis on cost is hardly surprising, as until the advent of clinical governance in the late 1990s, chief executives and boards were not legally accountable for the quality of clinical care provided. This has now changed and clinical governance, patient safety and corporate governance are the responsibility of all NHS staff. The resulting emphasis on the organizations' responsibility to support nurses and other clinicians in delivering high quality, safe care has refocused the requirements for nursing systems.

The constants

In 1999, Maureen Scholes described these fundamental threads running through the processes of delivering care despite a changing background as:

- resources are always limited, therefore optimum use of them is required to maintain or improve health care;
- nurses are essential to the understanding of health care systems;
- information and documentation needs to be accurate, up to date, relevant, legible and complete;
- needs are not finite and more are uncovered as health care advances;
- training has to meet the future needs; and
- technology is available to help.[19]

Above all nurses need to be valued, not only by adequate financial rewards, but by enabling them to give good quality care leading to personal and professional satisfaction.

A scenario for the future

The new millennium will, perhaps, realize some of the visions of the early pioneers. New generations of nursing systems are being introduced that support knowledge management and communication for multidisciplinary teams along a persons' health journey from the cradle to the grave. Patients will also be much more involved and have appropriate access to these communications. The solutions are just around the corner.

It is not easy to predict the future accurately but, like the pioneers, it is possible to imagine the potential. In 1993/94, Heather Strachan asked members of the BCS NSG, 'What information systems should nurses be using in the next five to ten years?' Respondents believed future systems should support the communication of patient-focused information and supported the move toward a common patient record that integrated with other systems and crossed institutional, as well as disciplinary, boundaries.[17] This is now becoming a reality 10 years later. Few ideas are completely new but often resurface when viewed from a different angle. Our present vision of how computers will have an impact on nursing is not new. The information and communication technology and systems are available now. They may not yet be universal but they are moving in the right direction.

We have a dream

In the future every UK resident will have their own NHS personal computerized health record, which they can access from home. When they have a health problem, or need, they will access this record and search for information from a national database. They can choose whether to have the information they requested as a multimedia interactive tutorial or simply as information. The information will be individualized according to their circumstances, as identified in their record, and support self care. When the support of a health care professional is required they will request an appointment or visit, via an email or secure web-based form to link to their local health centre. The triage nurse receiving the message will ensure that the right service is accessed as near as possible to the patient's home. The personal computerized health record will be updated at each and every contact made with the health service, including self-care advice as relevant, after each contact or episode. The patients themselves will be able to view their record at any time and identify who else has viewed or changed it. They will add their own information about their care and medication. They will be alerted to any routine screening automatically due according to their age, sex and heath history and an appointment will be arranged at a mutually convenient time. They can contact the specialist nurse for a consultation, via email, if they have a non-urgent problem. In emergencies their record will be available to authorized health professionals to ensure that decisions can be made on the best information available. In other cases, the patient's consent will be required to access the health record.

The nurse who is dealing with the patient will use a decision-support system to prompt questions about the nature of the patient's symptoms. If treatment is planned or medication prescribed the system will check that it is compatible with the individual patient's circumstances or other medications.

On confirmation of the nursing diagnosis, the nurse will be able to access the latest evidence and local clinical guidelines to include in the individual's care plan. The nurse will alert the patient's local primary care team to any changes in treatment. During the nurse's continuous professional development session, she/he will use the health care records database to audit a group of patients with the same diagnosis. Following the introduction of a new care regime she/he can audit the outcomes. The nurse will receive an alert about a new patient safety issue and search the database to identify its relevance to his/her patients.

Challenges for the future

- Developing clinical information systems that support the patient rather than specific staff groups.
- Designing sufficient flexibility to provide care that reflects the individual patient's physiological responses, wishes and desired outcomes.
- Frequently reviewing work practices and information needs.
- Achieving active participation of nurses in all clinical system designs and developments that will have an impact on patient care.
- Validating changes in work practices.
- Regularly reviewing and proposing revisions to nursing terminology to reflect current clinical practice.
- Assessing the risks and finding accurate, reliable ways to measure and monitor them.
- Prototyping and testing again and again to get effective, efficient systems.
- Developing accurate, reliable measurements.
- Developing new, effective ways to share the experiences of those working on or using new computer-based systems.
- Integrating health informatics competencies into all staff assessment and development programmes.
- Integrating SNOMED CT® approved terms into all clinical nursing activities.
- Developing nursing knowledge based on clinical records as part of the professional practice of all registered nurses and midwives.
- Achieving access to reliable decision-support and evidence-based options for clinical care beside every location where patients receive care, treatment or advice, be it bed, couch or consulting room.
- Developing new processes for providing personal advice and information at the request of the patient, with related quality assurance and audit mechanisms.

In the future people will have easy access to the most appropriate advice and services. The nurse will have this access:
- to choose relevant patient information and treatment options;
- to support good clinical decision-making;
- to communicate with other members of the multidisciplinary team; and
- to communicate with the patient.

These communication facilities will function easily and support continuous quality improvement and patient safety. Regular audits will inform the public whether nurses are doing their best with the resources available to them. All these elements have been developed. Extending their availability and integrating them is already under way.

Technology and health care will continue to advance and will, no doubt, stretch the resources even in this utopia. Unless the nursing practice matches the accumulated knowledge and capacity of our nursing systems our patients will not see any benefits and utopia will become cloud-cuckoo-land! It has been said that:

> *The epigram for the future should be 'patient heal thyself' – take charge of your own destiny but with information technology tools linked into a health care knowledge base and support from your professional health care advisors.*

REFERENCE 20

REFERENCES

1 Daly, N. and Hannah, K.J. (eds) In *Proceedings of Nursing and Computers, 3rd International Symposium on Nursing Use of Computers and Information Science*, Dublin. The C.V. Mosby Co, St Louis. ISBN 0 801632 35 8.

2 Osbolt, J.G., Vandewal, D. and Hannah, K.J. (eds) (1990) *Decision Support Systems*. The C.V. Mosby Co. ISBN 0 8016 3236 6.

3 Jones, B., Hyslop, A. and Richie, I. (1988) Computers as decision-support tools: wither expert systems? In Daly, N. and Hannah, K.J. (eds) *Proceedings of Nursing and Computers, 3rd International Symposium on Nursing Use of Computers and Information Science*, Dublin, pp. 330–6. The C.V. Mosby Co, St Louis. ISBN 0 801632 35 8.

4 Hyslop, A. and Jones, B. (1988) A cognitive model as a nursing expert system – potential for decision support and training in patient assessment. In Daly, N. and Hannah, K.J. (eds) *Proceedings of Nursing and Computers, 3rd International Symposium on Nursing Use of Computers and Information Science*, Dublin, pp. 565–74. The C.V. Mosby Co, St Louis. ISBN 0 801632 35 8.

5 Casey, A. (1994) Nursing, midwifery and health visiting terms project. In Grobe, S. and Pluyter-Wenting, E.S.P. (eds) *Nursing Informatics: An International Overview for Nursing in a Technological Era*, pp. 639–42. Elsevier, Amsterdam.

6 Rector, A., Bechofer, S., Goble, C.A., Horrocks, I., Nowlan, W.A. and Solomon, W.D. (1997) The GRAIL concept modelling language for medical terminology. *Artif. Intell. Med.*, 9, 139–71.

7 Heathfield, H.A., Hardiker, N., Kirby, J., Tallis, R. and Gonsalkarale, M. (1994) The Pen and Pad medical record model: development of a nursing record for hospital-based care of the elderly. *Meth. Inform. Med.*, 33, 464–72.

8   Hardiker, N. and Rector, A.L. (1998) Modelling nursing terminology using the GRAIL representation language. *J. Am. Med. Inform. Assoc.*, 5, 120–8.

9   Hardiker, N., Bakken, S., Casey, A. and Hoy, D. (2002) Formal nursing terminology systems: a means to an end. *J. Biomed. Inform.*, 35, 298–395.

10 Mortensen, R. (ed.) (1997) *ICNP in Europe: TELENURSE*. IOS Press, Amsterdam.

11 Webref (accessed September 2007) www.intute.ac.uk/healthandlifesciences/nursing/

12 Webref (accessed September 2007) National Library for Health. www.library.nhs.uk

13 Thorpe, J. (1994) NHS Computer developments and hospital nurses. *Information Technology in Nursing*, 6, 1, 12–13.

14 (authors unknown) (1994) *Nursing Information Systems: A Research Study for the NHSME Information Management Group and SAGNIS on Nursing Information Systems*. Greenhalgh & Co. Ltd.

15 Audit Commission (1992) *Caring Systems: A Handbook for Managers of Nursing and Project Managers*. HMSO, London.

16 Eaves, D. (ed.) (1995) *Benefits Realisation Monograph on Nursing Information Systems*. NHS Executive Information Management Group.

17 Betts, H. (ed.) (1996) *Benefits Realisation Monograph on Maternity Information Systems*. NHS Executive Information Management Group in conjunction with BCS NSG.

18 Strachan, H. (1996) Nursing informatics: a Delphi Study. In Brender, J., *et al.* (eds) *Human Facets in Information Technology*, pp. 867–71. IOS Press, Amsterdam.

19 Scholes, M. (1999) Reflections. *Information Technology in Nursing*, 11, 2, 2–4.

20 Scholes, M., Abbott, W. and Barber, B. (1989) Into the next millennium – healthcare and information management. In Barber, B., Cao, D., Qin, D. and Wagner, G. (eds) *MEDINFO 89*, Singapore and Beijing, pp. 8–13. North-Holland, Amsterdam.

PART 6
Common issues

INTRODUCTION

Early health informatics was concerned with issues of whether a computer system could be made to work in terms of:

- clinicians' willingness to have their patients' data computerized;
- whether the data could be captured adequately or in time;
- whether the patients' data were held confidentially;
- whether the technical problems of using particular equipment could be overcome.

It was accepted that there would be technical problems, but were the technical staff competent enough to sort them out? In many cases development involved the understanding of unexpected and new kinds of technical activities that were required to support the new technologies. However, time after time it was the people and overarching system problems that required attention if the computer systems were to be used successfully. A 'successful' program or suite of programs was totally useless if:

- the clinical or clerical staff could not use it;
- they refused to use it;
- it did not fit into their working practices;
- the staff did not trust it;
- it gave inaccurate information.

Devising a new computer system could expose other organizational problems that needed managerial attention. The temptation to replicate the manual system on the computer had to be resisted. It took until the 1980s before re-engineering the patient's pathway through the care process became a major focus.

Initial work at The Royal London Hospital on the finance systems and on small-scale survey or research projects helped to build up local confidence in computer systems. The hospital staff expected and received urgent attention whenever computer problems arose, whether this involved computer staff dropping other planned work, cancelling meetings outside the hospital, or late night or weekend working to fix the problem. Immediate solutions were not always possible but immediate attention was.

The development of a great deal of scientific and medical computing provided a valuable learning mechanism for a lot of staff as it helped them in their own activities. It is worth remembering that, at this time, there were no personal computers and no experience of computers in general education or common use. The staff needed to become computer literate, so that they could understand what was possible with the facilities available. As systems became available, CAL became a reality. This enabled the processes of continuing professional education to be accessible whenever required, as opposed to when the staff had a training day. Advanced simulators helped professionals to learn complex body systems and monitor their reactions to differing types of stress or disease.

This developed into decision-support systems to help with the interpretation of complex patterns of clinical data. Originally, decision-support systems were thought of in terms of the Japanese Fifth Generation of computers: as rule-based 'expert' systems. Since then they have become much more widely interpreted as systems that can support decision-making, however configured.

As with medical record systems, the 'holy grail' has been seen as systems that include the accumulated 'knowledge' based on data from the patients' care. This could help professionals in the diagnostic process by assessing an individual patient's record in the context of that 'knowledge'. In time this led to wider use of evidence-based practice: a professional expectation by the start of the twenty-first century.

The issue of confidentiality was ever present and in the early days at The Royal London Hospital it was thought that:

1 *the new computer system shall be no less confidential than the present manual system;*

2 *the confidentiality procedures shall not be so cumbersome as to detract from the acceptability and usefulness of the system.*

REFERENCE 1

The initial fear was that the computer record could be made a great deal more confidential but that this might render it a more difficult to use. As familiarity with systems increased, and there were many more computer systems around, data protection legislation was enacted to ensure that systems were not misused. In due course, the security issues raised moved up the agenda, so that the risks of accidental or malicious interference with health information systems were seriously addressed.

In developing the Experimental Computer Programme (see Chapter 5), the DoH established ideas about evaluating computer applications so that successful Experimental systems would be moved into a Development phase and then onto widespread Implementation (the EDI concept). A great deal of effort was put into developing these ideas over a decade or more. Initially, a 'before and after' approach was thought possible, but this suffered from the immense amount of other change that happened in hospitals during the long process of implementing these early systems. These other changes confounded the interpretation of the computer system's effect. Also, perhaps the most important question was whether a proposed system could be made to work at all. Some testing was undertaken on simulation models to explore the behaviour of systems under differing loads. The DoH developed 'performance criteria' to assess the impact of various computer systems that could be compared with manual systems in similar conditions. Unfortunately, the overall hospital systems were extremely varied. The range of performance measures that emerged was extremely large and detailed. In the end it was difficult to make serious sense of the final and very detailed performance criteria report – although the DoH went on to apply these concepts to the wider field of hospital efficiency with multiple targets and performance criteria for health care activities.

FUTURE CHALLENGES

Common challenges for the future include:

- defining the competencies in information use and management for every level of professional education and practice;
- attaining a consistent level of computer and information use across the NHS;
- developing ways for all professionals to contribute to the collection of data and its interpretation in order to develop evidence-based clinical practice as new drugs and technologies impact on daily patient care and treatment;
- maintaining patient privacy in the face of Government desires for integrating data-collection systems and the temptations of 'hackers' to seek unauthorized routes into patient data;
- finding robust ways to evaluate the impact of new information systems on patient care and treatment when change is rapid, ever present or difficult to monitor;
- retaining the focus on the individual patient and their choices when there is a financial and technical pressure to 'standardize'.

BARRY BARBER

REFERENCES

1 Barber, B., Cohen, R.D. and Scholes, M. (1976) (title unknown). *Medical Informatics*, 1, 1, 61–72.

23 Education in health care

BERNARD RICHARDS

This chapter covers the development of computer-assisted learning (CAL) and other IT routes to education in health care. Chapter 21 discusses how these were applied in nursing.

The Open University (OU) was an important influence on the field of distance learning for higher education and also the use of computers to reach a scattered student population. Initially it used television programmes with text guides, and filmstrips with hand-held viewers for students working at home, or with scientific computer use through other university and higher education premises particularly during the OU's intensive summer schools. As early as 1978 it was promoting *Computer Science: BASIC Language Programming*, as a set book.[1] The Joint Academic Network (JANET) provided an early communication route between students and academic staff. As the technology gave way to personal computers so the home computer became important, later complemented by the use of the internet and then the World Wide Web. Hypertext provided a means of creating different routes through the content of a subject. By the 1990s the use of interactive video instruction (IVI) and the CD-Rom provided more flexible approaches: especially useful for health care staff who worked shifts, particularly at night.

CAL OR CBT OR CAT?

When people are expanding their areas of knowledge they often need definitions. So it will be appropriate to introduce several of the terms used in this area of computer learning. 'Computer-Assisted Learning' (CAL) or 'Computer-Based Training' (CBT), the most common terms, mean using the computer to enhance the user's knowledge. A less used term, 'Computer-Assisted Testing' (CAT), means that the computer is merely used to test the student's knowledge on topics studied by other means. This chapter gives examples of both CAL and CAT. Some computer programs comprise integrated modules comprising CAL followed by CAT.

The concept of 'learning' followed by 'testing' in the one medium is not new. In 1962, Crowder produced a book using the then new concept of 'Programmed Instruction' (PI).[2] In this book, the reader traversed his/her way, reading pages that were definitely not in consecutive or ascending order. The reader's route through the book was determined by the choice of answer given to the Test Question at the end of each topic. From the chosen answer, the reader may have been:

- deemed to be in need of further guidance for not fully understanding the question;

- in need of correction for clear misunderstandings; or

- demonstrated a sufficient grasp of the subject matter to be allowed to proceed to the next topic.

In the first two of the cases above, the reader was instructed to re-read certain numbered pages, which should then have allowed the correct answer to be given to the test question.

One enterprising company produced an electromechanical device to perform the same operations. It was not until the mid-1960s that computer systems, such as the PLATO system in the USA, took over this concept of programmed interaction. Bitzer, at this time, reported on his CAL program for nurses.[3, 4] Some years passed before computers became readily available. Then Katherine Hannah in Canada,[5, 6] and Pam Tymchyshyn in the USA,[7, 8] started using CAL in nursing. It was through the international collaboration of the nurses and their interaction at World Meetings that the nurses in the UK became interested.[e.g. 9] The founding of the British Computer Society's Nursing Specialist Group (BCS NSG) in 1983 helped to introduced CAL ideas into UK nursing. Sue Norman was one of the UK pioneers.[10, 11]

Notwithstanding the BCS NSG's initiative, the emergence of CAL across the UK was limited. Various efforts were made by the Open Software Library (OSL) and by Graham Wright (see also Chapter 21). In 1986, the OSL ran *The First National Conference on the Use of Computers in Healthcare Education and Training* with support and sponsorship of the NHS Training Authority.[12]

MEDICAL SECRETARIES

At a time when computers were about to become more available it was felt that people training to become medical secretaries should have some training in the use of, and knowledge of, the potential of computers.

In 1991, Richards, Webster and Vincent wrote a program to train medical secretaries.[13] It comprised a set of CAL modules that instructed the user on various aspects of computing, such as how the computer was used, its potential and the value to a medical secretary. These modules each had a set of questions to test how much of the material had been absorbed and understood. This pioneering suite was one of the first in the UK to provide training in computer literacy.

DOCTORS

One of the first systems for training doctors was that devised by Dr Gavin Kenny in Glasgow, for training paediatric anaesthetists. A program called 'Blue Baby' displayed a newborn baby to the user with a diagram of the baby in which the face was the dominant feature. The doctor had to prescribe appropriate gases (oxygen) and drugs so as to maintain the baby's PO_2 at the correct level and also to maintain the baby's PCO_2 at the appropriate level.

In a separate exercise, Professor David Ingram *et al.* devised a program called MACPUFF to demonstrate the physiology of the lungs. This work contained a rigorous mathematical analysis of the physiology involved. His book describes the processes admirably.[14] Subsequently, the ideas caught on and MACMAN was produced for Cardiovascular Simulation, and MACPEE for Body Fluid Simulation.[15]

In 1996, Colman, Lewis, and Richards wrote a program for training doctors to carry out bronchoscopies.[16] After logging on, the trainee doctor was shown a menu on the PC screen. The first few menu items described the various types of bronchoscope in current use, the justification for carrying out a bronchoscopy, and the likely findings. All purely factual, but nevertheless useful information, often filling gaps in the doctor's knowledge. The major item on the menu was the virtual reality 'bronchoscopy' itself. In this, a diagram of the bronchial tree was displayed on the left of the screen with the computer's cursor sitting at the patient's lips. On the right of the of the screen was a display of the view that the tip of the bronchoscope (and hence the eye of the doctor) would see if the tip were in the position currently occupied by the cursor (i.e. at the lips of the patient). The doctor, using the mouse, could then drive the cursor down the trachea into the main bronchus. As the cursor was moved down the bronchial tree, the right screen would then show the bifurcation into the right bronchus and left bronchus. As the cursor was moved further down the bronchial tree, the screen depicted an ever-narrowing orifice. In this way the trainee could experience the journey down the tree. Eventually the cursor could traverse the three lobes of the right lung and the two left lobes.

Views of the bronchial tree could be selected to show that of a healthy patient or that of a patient with cancer or other defects. The system was acclaimed for its ease of use but more importantly for the quality and clarity of the views down the bronchial tree. While accepting that this program was a prime example of CAL, others felt that, in addition, it was almost an example of 'virtual reality'.

A CAL system for training anaesthetists was produced by Doran and Richards.[17] It took the junior doctor through the various procedures necessary before the start of an operation.

RADIOGRAPHERS

Nurses had the BCS NSG in which to sound out their ideas and seek collaboration, however, the radiographer had no such enthusiastic group. It was left to Alan Watson, the Principal of the School of Radiography in Glasgow, to see the potential of CAL and he sought my help at UMIST to bring his dreams to fruition. A series of programs were written covering computer literacy, anatomy and physiology.[18, 19] All these modules contained a teaching/learning package followed by a testing module, in which the learner was quizzed on their understanding of what they had just seen and read on the computer screen.

The program on computer literacy was similar to that provided for the medical secretaries.[13] The anatomy module displayed the bones in the various limbs and torso and asked the user to name them. As with secretaries and nurses, one of the new benefits of such a system was that students could do the learning and take the tests in their own time, and that the teacher could produce a list of student marks when it was required.

The physiology module was quite innovative. One section illustrated fertilization of the ovum by the sperm during the process of conception by means of animation: a first in computing in the field of radiography. There were the obligatory questions at the end of the module.

Overall the exercise was judged to be a worthwhile innovation for the students and the programs were considered to be very successful. No subsequent programs for CAL in radiography are known about.

DENTISTRY

In comparison with the other sectors of health, for example primary care, dentistry was for long the poor relation, and was slow to get off the ground. This comment applies both to the uses of CAL in dentistry as well as the use of the computer for patient records.

One of the earliest uses of CAL in dentistry, or in this case, more exactly, dental health, was a request by the Regional Dental Officer in Liverpool to provide: (a) a suite of programs for use in the schools to teach the children about their teeth and dental hygiene; and (b) to let them do this using computer games. The latter requirement was timely because computers were new in schools: the BBC-model B having just been introduced. A partnership comprising Joanna Morris and Bernard Richards produced several computer games in which the objective was to win by demonstrating knowledge of dental care and a knowledge of which foods contained sugar.[20]

The Dental Charting Program was a more academic program written by Khoury and Richards for use in the University Dental School to train dental students.[21] The program comprised two aspects, namely a Patient Database and a Charting Module. The former did what every good patient database should do, namely it effectively stored patient records and profiled the database. The Charting Module was innovative and pioneering. With a patient seated in the chair, mouth open, the dentist would first bring up on the touch screen a screen showing an array of 32 teeth. He or she would then touch a screen tooth, say Left Upper 8 (a molar), and this would then zoom to occupy the whole screen, showing clearly the five surfaces. The dentist would then examine the 'real' tooth noting existing fillings and new cavities. The dentist would then go back to the computer and, using the handle of the dental mirror (or his finger), could select from a list of icons a 'state' of the 'real' surface, such as a buccal filling or a lingual cavity; and then by touching the screen surface display the charted result. Hence the whole mouth could be checked and charted without the need for an assistant or having to put down the dental instruments. On completion of the chart, a record could be printed out and the Electronic Patient Record was updated.

PATIENTS

In the early 1980s, patients were beginning to realize that medical knowledge was not the exclusive property of health care professionals. They were becoming better informed and people were wanting to know more and to have their questions answered. In the USA, patient education was becoming a mandatory part of health care. It was felt that a basic health education was needed so that the patient would achieve an optimum level of health, and it was recognized that it would be negligent to assume that patients would learn unaided. Some writers even went as far as saying that health education was more cost-effective than illness care. The work described in this section was apposite as computers were becoming more widespread – people were keen to use them. Research generally shows that people retain 20 per cent of what they hear, but 50 per cent of what they hear and see. Finally, self-education using CAL relieved medical staff of having to spend time explaining treatment to the patients.

This section distinguishes between a person with a 'medical condition' who is in contact with a doctor and seeks more information on that 'condition', and its medical treatment.

In 1995, Colman produced a suite of CAL programs to educate patients with rheumatoid arthritis.[22] The purpose was to supplement the information provided by the clinician, but also, more importantly, to try to maximize compliance with the drugs prescribed for the treatment. Text and graphics were used in the programs that ran on a standard PC. The disease was described, pictures of the joints were displayed and the drugs used were identified, and their actions described. Users marvelled at the easily understood diagrams, and at the navigation keys that gave easy access to the lesson.

In 1989, Hayes and Richards, assisted by Leong, produced a CAL system that gave advice to pregnant women.[23–25] The program started by making the patient familiar with the computer and its keyboard and then proceeded to go through the stages of pregnancy: from conception to labour and delivery. This system first underwent field trials in the waiting area of an ante-natal clinic. Women found it more exciting using the computer than reading the usual magazines. After their period at the computer the patients filled in a field-trial questionnaire. They were asked about its useability, the suitability of the colours, and whether they had learnt anything new.

In 1990, Kathleen Wong and Bernard Richards wrote a much enlarged program that entered into a dialogue with the user.[26] The user was either someone contemplating pregnancy (male or female) or a woman who was already pregnant. This program asked the user to introduce themselves by name and then offered either the pre-conception route or the current pregnancy route. The computer screen then offered two options:

1 I want to have general information on my present situation.
2 I have a particular problem to discuss.

In the general advice (route 1), the computer sought information from the user in the same way that a doctor or nurse would by asking for personal details such as age and weight, smoking and drinking habits, education level and social status. If the patient revealed smoking and/or drinking habits, the program would point out the damage that could be done to a fetus if these habits continued into or throughout pregnancy. Furthermore the computer would then indicate the savings in monetary terms of giving up smoking and/or drinks, usually of the order of £500 to £1000: enough money to furnish and equip the nursery. The content of this system was of a quality sufficient for it to be called 'an expert system'.

Chapter 16 gives further information on the use of computers by patients.

In 1991, hospitals were using computers with young people admitted to hospital. The intentions included: to stimulate interest when a child was depressed, withdrawn or negative; to distract or divert a child from the hospital experience; to encourage the reluctant child to sit up; to develop manipulative skills; as well as for more conventional education. Nelson-Smith identified seven programs for the BBC B computer and three groups of different games for two to 16 year olds.[27]

NURSING

Some of the DoH Experimental Computer Programme projects in the 1960s included a clinical nursing system with elements of what was to become CAL in its broadest sense.[28] The learning elements supported practice, generally in the form of textual notes (see Chapter 21).

In the late 1970s, two nursing groups were to become influential in nursing's ongoing development with CAL. These groups were NUMINE and the BCS NSG. Sadly the former no longer exists but it was responsible in collaboration with the Open Software Library and the NHS Training Authority for two significant CAL conferences in 1986 and 1988.[29, 30]

The 1980s saw both individual and national development of CAL in nursing. Individuals were exploring the possibilities of using basic programming techniques to deliver multiple-choice questions to students with varying degrees of success. Koch and Rankin noted some criticisms about expense, incompatibility between hardware and available programs, and concerns that computers would usurp the teacher's authority.[31]

In 1986, the United Kingdom Central Council for Nursing, Midwifery and Health Visiting (UKCC) included the following statement in its guidelines for nurse education under the new Project 2000 course structure:

The use of technological aids that will allow students to experience at a distance, and not to infringe unnecessarily upon privacy, will be crucial.

REFERENCE 32

This was a significant turning point in nursing's development and use of CAL. Nationally, two major projects were started in the mid-1980s: the English National Board's for Nursing, Computer Assisted Learning Project (ENB CAL),[33] and the Wessex Regional CAL Project.[34] Both of these projects addressed the growing need for teacher instruction in order to move CAL forward. In Scotland, CAL in nursing was also moving forward through the work of the National Board for Nursing, Midwifery and Health Visiting in Scotland (NBS) on the Scottish Microelectronics Development Project (SMDP) established in 1984.[35] This project was concerned with identifying opportunities for CAL program development.

The ENB CAL Project instructed around 750 nurse teachers, across England, in the use of computers to advance nurse education. The preparation included three days of face-to-face instruction at one of seven centres, followed by online continued instruction. This required the teacher to download material, formulate a response and return this to their centre using a computer network (Campus 2000).[36] The teachers using this network also had access to discussion forums, which allowed exchanges of information between geographically distanced individuals. Many of these discussions were based around the development and implementation of CAL materials in the nursing curriculum. National development ceased in 1991 and once again individual enterprise took control.

By 1995 nursing education had migrated into higher education. This became a Government-directed move that even today is still going through a positional stage. In this uncertain climate the opportunity arose within the Computers in Teaching Initiative (CTI) for a new centre for nursing. Proposals were put forward and the centre was awarded to the University of Sheffield. The CTI formed a collection of 24 centres across the UK with a mission:

... to maintain and enhance the quality of learning and increase the effectiveness of teaching through the application of appropriate learning technologies.

REFERENCE 37

The subject centre primarily assessed and disseminated CAL information across nursing education. This was achieved through a paper-based newsletter and the emerging Web. The CTI worked closely with the Teaching and Learning Technology Programme (TLTP) whose remit was:

> *To make teaching and learning more productive by harnessing modern technology.*
>
> REFERENCE 38

These two initiatives moved forward nursing use of CAL to support the curriculum. Most nursing schools installed computer rooms for students and most teachers were provided with computers to use both educationally and in administration. The Web provided a new dimension to education with access to online graphics, text and people. Although there remained a place for CAL programs, migration towards the Web continued. An example of CAL produced during this time was that of the CONCEPT Project.[39] This produced five CD-Rom-based programs, covering different surgical specialties, which were designed to meet the educational needs of health care students in higher education institutions and on clinical placement.

By 1999 nursing education had embraced technology to support education alongside other subject disciplines. There remained much work to do and avenues to explore, but the grounding was solid and from early tentative steps nursing education was ready to move forward.

PUBLIC ACCESS LEARNING AND INFORMATION SYSTEMS

Dr Ray Jones, at the University of Glasgow, was a leader in public learning systems. He produced information software to which the public had access in kiosks. These became known as 'Healthpoints' and their use and value was extensively studied in the 1990s.[40, 41] This topic is dealt with in greater detail in Chapter 16.

EDUCATION

Downie and Basford identified a number of advantages of using computer-based training in health care. These included:

- Learners working without constant supervision – trainers have more time to assess needs and for remedial work.

- No waiting for a course to start – it can be used at any time to suit the individual or the organization.

- It suits learners with different levels of ability and individual learning paths.

- It can present situations and give users the opportunity to practise in a safe environment without direct risk to patients.

- It is cost-effective and materials are inflation proof while their content remains current.

The range of available programs included leadership, supervising and interviewing skills, assertiveness and stress management, financial and resource management.[42]

Steve Wheeler, a Learning Resources Officer from South West College of Health Studies, identified some of the problems encountered by 66 teachers when using computers.[43] He quoted three reasons posited by Suppes in 1980:[44]

- Computers cannot listen to students talking nor observe non-verbal reactions – vital functions for a teacher.

- Computers cannot communicate to students in the many ways a human teacher can.

- Teaching is more than just 'information giving' – it is the facilitation of learning (although CAL packages can do this to a limited extent).

In 1991, the *Multimedia Show* held at Olympia provided an opportunity to check how other major organizations were making computers work for them. By this time Basford was claiming that there was sufficient evidence to demonstrate students learnt much faster and retained knowledge longer through CBT. Authoring systems were available commercially to help individual professionals develop their own materials.[45] Some came with a resource library that included animations, graphics, movies and sound, and allowed the user to store their own images and video sequences.

WIDE-SCALE COMPUTER-BASED TRAINING ACTIVITIES

Introducing computers into the daily work of increasing numbers of health care staff in the NHS led to a drive, in 1989, to train one million health workers in computer and information skills. The NHS Training Authority published its 'Information Management and Technology Training Strategy' and £4.5 million were allocated to fund 20 projects to develop training courses and materials.[46] Martin Southeran, IM&T Training Programme Manager at the NHS Training Division (NHS TD), produced a list of training products to support the strategy, which included brief outlines of the four medical, four nursing and two PAM-oriented projects in the IM&T Training and Development for Clinicians Programme. There was also a list of IM&T Regional Training Coordinators for the UK.[47] By 1996 the Institute of Health and Care Development (formerly the NHS TD) was running an IM&T Programme for clinicians that covered:

- pre-qualification education;
- post-qualification education;
- continuing education; and
- training and development for practitioners.[48]

Kostrewski examined the provision of education and training in IT, reporting the seven locations, in 1991, that were providing some form of post-graduate course for health professionals. The benefits and disadvantages of using IT in medicine, dentistry, pharmacy and nursing were described, and those of CAL were ranked. She also reviewed training provision at a local level.[49]

In 1995, the Centre for Health Informatics and Multiprofessional Education (CHIME) was established, based at The Whittington Hospital in London. Jeanette Murphy and Professor David Ingram were actively involved.[50] The centre set up the Champions database of enthusiasts and activists in informatics teaching. The rapid pace of change in medical schools by the 1990s was already such that recording details of informatics courses and equipment was not feasible.[51]

The Teaching and Learning Technology Programme, an initiative funded by four UK higher education funding bodies, was allocated £22.5 million over three years for its first phase starting in January 1996. It funded 76 projects in a wide range of disciplines. Those for the medical sciences included:

- PCCAL – Pharmacy Consortium (University of Bath).
- ProCARE – shared curriculum material for the caring professions (University of Southampton).
- Applying behavioural sciences to the teaching and training of health professionals (Queen's Medical Centre, Nottingham).
- Computer-based courseware for public health medicine (Ninewells Hospital, Dundee).
- Pharma-CAL-ology – 17 CAL packages (University of Leeds).
- Technology-based learning in medicine – beyond the courseware (Ninewells Hospital, Dundee).
- Bio-net: changing the patterns of teaching in biology (University of Leeds).[52]

In 1998, Procter and Hible provided a resource guide that included details of 19 publications and 35 reviews of software for use in nursing, midwifery and other health disciplines, many drawn from Computers in Teaching Initiative centres. The list of reviewers reflected the current 'movers and shakers' in health care IM&T education.[53]

EVALUATION OF CAL AND TELEMATICS FOR EDUCATION

Peter Jones of Bolton explored the role of simulation in health care and identified factors affecting the production of effective CAL applications and whether commercialism should dictate what was produced.[54]

In 1996, the evaluation of teaching and learning with CAL had developed to a stage where a variety of instruments were being applied by the Teaching and Learning Technology Support Network (TLTSN), these included:

- a computer experience questionnaire;
- a task experience questionnaire;
- observations using an evaluator or video;
- pre-task/post-task questionnaires;

- focus groups;
- learning resource questionnaires;
- student confidence logs;
- knowledge quizzes;
- an examination or assessment of performance;
- a post-course questionnaire;
- interviews.[55]

Adrian Vranch and Steve Wheeler of the University of Plymouth examined how the new technologies, including video conferencing, computer-mediated communications, digital satellite TV and data transmission, could be used. The university provided live, interactive broadcasts, from its own TV studio and a satellite uplink transmission facility, on loan from the European Space Agency, to four main sites and a number of telelearning centres in hospitals across Cornwall, Devon and Somerset. They used a Benefits Analysis Map to quantify the parameters: high quality, wide access and low cost, in order to identify which technologies to use to deliver particular distance learning courses.[56]

The advent of the internet, and then the Web, provided new routes to distribute distance learning materials, and UK-based organizations, such as South Bank University, began to explore the provision of multimedia materials to overseas institutions and individuals.[57]

CONCLUSION

While much early work was done in the USA and Canada (as reported at the beginning of this chapter), many good CAL developments also came from the UK in those early years. Today, the greater resources in the USA and Canada are being used, but the level of clinical detail in the work of Ingram *et al.*,[14] Coleman *et al.*[16] and Doran *et al.*[17] has not been surpassed.

As Professor Mary Chambers noted in 1994,[58] many of the teachers in health care had received their own education 'in an era when IT was confined more to science fiction texts than nurse education and practice'.

Health care educators, particularly those in university settings, rose to the challenge of becoming adept at using the new technologies to educate fellow and future health care professionals.

ACKNOWLEDGEMENTS

I gratefully acknowledge Paula Procter and Denise Barnett for information on the nursing use of CAL and on the education section, respectively.

REFERENCES

1 Forsythe, A.I., Keenan, T.A., Organick, E.I. and Stenberg, W. (1974) *Computer Science: BASIC Language Programming*. John Wiley & Son, London. ISBN 0 471266 77 9.

2 Crowder, N.A. (1962) *The Arithmetic of Computers*. EUP, London.

3 Bitzer, M.D. (1963) *Self-directed Inquiry in Clinical Nursing Interactions by means of the PLATO Simulation Laboratory*. Report R-184. University of Illinois, Urbana.

4 Bitzer, M.D. (1966) Clinical Nursing Information via the PLATO simulator laboratory. *Nursing Research*, 15, 2, 144–50.

5 Hannah, K.J. (1978) An overview of computer-assisted learning in nurse education at the University of Calgary. In Stanton, M.C. (ed.) *Perspectives: Nursing Education, Practice and Research*, pp. 143–56. University of Calgary.

6 Hannah, K.J. (1983) Computer-assisted learning in nurse education: a macroscopic analysis. In Scholes, M., Bryant, Y. and Barber, B. (eds) *The Impact of Computers on Nursing: An International Review*, pp. 280–7. North-Holland, Amsterdam. ISBN 0 444866 82 5.

7 Tymchyschyn, P. (1983) Nurse education in the computer age in retrospect and prospect. In Scholes, M., Bryant, Y. and Barber, B. (eds) *The Impact of Computers on Nursing: An International Review*, pp. 300–6. North-Holland, Amsterdam. ISBN 0 444866 82 5.

8 Tymchyschyn, P. (1982) *An Evaluation of the Adoption of an Innovation (CAI via PLATO) into a Nursing Program*. Unpublished doctoral dissertation. University of Illinois, Urbana. June.

9 IMIA Workshop (1982) In Scholes, M., Bryant, Y. and Barber, B. (eds) (1983) *The Impact of Computers on Nursing: An International Review*. North-Holland, Amsterdam. ISBN 0 444866 82 5.

10 Norman, S. (1982) Computer-assisted learning: its potential in nurse education. *Nursing Times*, 1 September, 1467–8.

11 Norman, S. (1988) *The Nightingale School Computer Assisted Learning Project*. University of Surrey. April.

12 Wright, G. (ed.) (1986) *Proceedings of the First National Conference on the Use of Computers in Health Education and Training*, Keele University. NHS Training Authority.

13 Richards, B., Vincent, V. and Webster, A.W. (1991) Teaching medical informatics to medical secretaries. In van Bemmel, J.H. and Zvarova, J. (eds) *Proceedings of Knowledge Information and Medical Education*, Prague, pp. 203–12. North-Holland, Amsterdam.

14 Ingram, D., Dickinson, C.J. and Ahmed, K.A. (1987) *A Simulation of Human Respiration, Gas Exchange and Control*. Oxford University Press.

15 Dickinson, C.J., Ingram, D. and Shephard, E.P. (1971) A digital computer model for teaching the principles of systemic haemodynamics. *J. Physiol.*, 216, 9.

16 Colman, A., Richards, B. and Lewis, R. (1996) A hypermedia computer-aided learning system for bronchoscopy. In Brender, J., Christensen, J.P., Scherrer, J-R. and McNair, P. (eds) *MIE, Copenhagen 96*, pp. 808–12. IOS Press, Amsterdam. ISBN 9 051992 78 5.

17 Richards, B. and Doran, B. (1982) Training anaesthetists to cope with an emergency in the operating theatre: the use of a microcomputer. In O'Moore, R.R., Barber, B., Reichertz, P.L. and Roger, F. (eds) *MIE, Dublin 82: Lecture Notes 16*, p. 673. Springer Verlag, Berlin. ISBN 0 387112 08 1.

18 Richards, B., Watson, A. and Wardale, K. (1988) Computer-assisted learning for radiographers. In Hansen, R., Solheim, B.G., O'Moore, R.R. and Roger, F.H. (eds) *MIE, Oslo 88: Lecture Notes 35*, pp. 303–7. Springer Verlag, Berlin. ISBN 3 540501 38 X.

19 Richards, B., Turfrey, M., Wardale, K. and Webster, A. (1989) Computer-aided learning for radiographers. In Salmon, R., Prottie, D. and Mohr, J. (eds) *Proceedings of the International Symposium on Medical Informatics and Education*, Victoria, Canada, pp. 471–4. School of Health Sciences, University of Victoria, Canada.

20 Richards, B. and Morris, J. (1987) Computer-aided education in dental health. In Serio, A., O'Moore, R., Tardini, A. and Roger, F.H. (eds) *MIE, Rome 87*, pp. 756–61. Springer Verlag, Berlin.

21 Richards, B. and Khoury, D. (1991) The use of a touch-screen computer for dental charting. In Adlassnig, K.P., Grabner, G., Bengtsson, S. and Hansen, R. (eds) *MIE, Vienna 91: Lecture Notes 45*, pp. 145–9. Springer Verlag, Berlin. ISBN 3 540543 92 9.

22 Colman, A. and Richards, B. (1995) The use of CAL and multimedia in patient education. In Maurer, H. (ed.) *Proceedings of ED_MEDIA 95*, Graz, Austria, June, pp. 163–8. AACE, Virginia, USA.

23 Hayes, G. and Richards, B. (1989) Health education while waiting to see the GP. In Duncan, J. (ed.) *Current Perspectives in Healthcare Computing*, pp. 32–4. BJHC for BCS, Weybridge.

24 Richards, B. and Hayes, G. (1989) Pregnancy: an electronic textbook and knowledge-based interactive system for use by patients. *Journal of Perinatal Medicine*, 17, 1, 156.

25 Richards, B., Hayes, G. and Leong, F. (1989) A knowledge-based system for giving expert advice to patients in all matters relating to conception, pregnancy, and childbirth. In Barber, B., Cao, D., Qin, D. and Wagner, G. (eds) *MEDINFO 89*, Singapore and Beijing, p. 1183. North-Holland, Amsterdam.

26 Richards, B. and Wong, K. (1990) An expert advisory system for pregnancy related matters. In O'Moore, R.O., Bengtsson, S., Bryant, J. and Bryden, J.S. (eds) *MIE, Glasgow 90: Lecture Notes 40*, pp. 544–8. Springer Verlag, Berlin. ISBN 3 540529 36 5.

27 Nelson-Smith, J. (1991) Programs for play. *Nursing Times*, 887, 42, 55–7.

28 Ashton, C.C. (1983) Caring for patients within a computer environment. In Scholes, M., Bryant, Y. and Barber, B. (eds) *The Impact of Computers on Nursing: An International Review*, pp. 105–14. North-Holland, Amsterdam. ISBN 0 444866 82 5.

29 Wright, G. and Brookman, P. (eds) (1986) *Proceedings of the First National Conference on the Use of Computers in Health Care Education and Training*. NHS Training Authority.

30 Snook, R., Wright, G. and Brookman, P. (eds) (1988) *Proceedings of the Second National Conference on the Use of Computers in Health Care Education and Training*. NHS Training Authority.

31 Koch, B. and Rankin, J. (1984) Swift idiots in the classroom. *Senior Nurse*, 1, 32, 16–18.

32 United Kingdom Central Council for Nursing, Midwifery and Health Visiting (1986) *Project 2000: A New Preparation for Practice*. UKCC, London.

33 Procter, P.M. (1986) To boldly go: CAL in nursing education. In Wright, G. and Brookman, P. (eds) (1986) *Proceedings of the First National Conference on the Use of Computers in Health Care Education and Training*, pp. 49–55. NHS Training Authority.

34 Morrison, D. (1986) Computers in nurse education: a focus for curriculum development. In Wright, G. and Brookman, P. (eds) (1986) *Proceedings of the First National Conference on the Use of Computers in Health Care Education and Training*, pp. 129–35. NHS Training Authority.

35 Rankin, J. and Koch, B. (1987) Application of computers in nurse education. In Koch, B. and Rankin, J. (eds) *Computers and their Applications in Nursing*, p. 35. Lippincott Nursing Series.

36 Procter, P.M. (1988) Network building with technology. In Snook, R., Wright, G. and Brookman, P. (eds) (1988) *Proceedings of the Second National Conference on the Use of Computers in Health Care Education and Training*, pp. 55–60. NHS Training Authority.

37 HEFCE (date unknown) *Computers in Teaching Initiative*. The Higher Education Funding Council for England.

38 Atkins, M. J., Hasywood, J., Homes, P., Layzell-Ward, P., Rigby, W. and Saunders, D.M. (1988) *Report of the CTI-TLTSN Review Group*. Higher Education Funding Council for England.

39 The Higher Education Funding Council for England (1998) HEFCE TLTP Phase 3: Outcome of bids for funding under the Teaching and Learning Technology Programme.

40   Jones, R.B., Navin, L.M. and Murray, K.J. (1993) Use of a community-based touch-screen public-access health information system. *Health Bulletin*, 51, 34–42.

41   Campbell, G.M. and Jones, R.B. (1991) A computer-based patient education system in the radiology department. In Richards, B. and MacOwan, H. (eds) *Current Perspectives in Healthcare Computing*, pp. 113–20. BJHC for BCS, Weybridge. ISBN 0 948198 12 5.

42   Downie, C. and Basford, P. (1991) Use computer-based training. *Nursing Times*, 87, 37, 63.

43   Wheeler, S. (1992) Managing technological change in nurse education. *Information Technology in Nursing*, 4, 2, 5–7.

44   Suppes, P. (1980) The future of computers in education. In Taylor, R. (ed.) *The Computer in the School*. Teacher College Press.

45   Basford, P. (1991) Software review: Multimedia show. *Nursing Times*, 87, 33, 58.

46   DoH (1989) *New drive to train one million NHS staff in information skills*. Press release 89/269: 28 June. (Includes list of centrally funded projects).

47   NHS Executive Information Management Group (undated) *Information Management and Technology: Training Products*. NHS Training Division publication 3110194.

48   Pearson, C. (1996) Information proficiency in clinical practice and service management – an integration challenge for education. In Procter, P. and Hible, G. (eds) *Computers in Nursing & Midwifery Education – the Vision: Conference Proceedings*, pp. 37–8. CTINM Publications, Sheffield. ISBN 0 952847 10 8.

49   Kostrewski, B.J. (1991) The provision of education and training for IT applications in healthcare. In Richards, B. and MacOwan, H. (eds) *Current Perspectives in Healthcare Computing*, pp. 105–16. BJHC for BCS, Weybridge. ISBN 0 948198 11 7.

50   Webref (accessed June 2007) CHIME. www.chime.ucl.ac.uk

51   Murphy, J. and Stramer, K. (1996) The Champions database: identifying enthusiasts and activists in informatics teaching. In Richards, B. and de Glanville, H. (eds) *Current Perspectives in Healthcare Computing*, pp. 144–50. BJHC for BCS, Weybridge.

52   Turpin, S. (date unknown) The Teaching and Learning Technology Programme – what have we learned and where do we go from here? In *New drive to train one million NHS staff in information skills*, pp. 11–24. DoH Press release 89/269: 28 June.

53   Procter, P. and Hible, G. (1998) *CTI Centre for Nursing and Midwifery – Resource Guide*. University of Sheffield CTINM Publications, Sheffield. ISBN 0 952847 11 6.

54 Jones, P. (1994) An enthusiast's view of CAL. *Information Technology in Nursing*, 6, 2, 4–6.

55 Glasgow University (1996) *Managing the Adoption of Technology for Learning*. Teaching and Learning Technology Support Network Case Studies 11, pp. 10–12.

56 Vranch, A. and Wheeler, S. (1997) Distance education and convergent technologies: the benefits of telematics. *Information Technology in Nursing*, 9, 6, 7–11.

57 Tatlow, M.P., Dowd, C.D.P. and Rainbow-Towers, D. (1996) Incorporating and exploiting emerging information technology for the delivery of distance-learning courses to multidisciplinary health care professionals. In Richards, B. and de Glanville, H. (eds) *Current Perspectives in Healthcare Computing*, pp. 151–6. BJHC for BCS, Weybridge.

58 Chambers, M. (1994) Information technology and the curriculum. In Wainwright, P. (ed.) *Nursing Informatics*, pp. 139–58. Churchill Livingstone, Edinburgh.

24 Decision-support and expert systems

BERNARD RICHARDS

Some explanation of the terms 'decision-support system' (DSS) and 'expert system' is required as an introduction to this chapter. The former is a computer program that helps the user to make the appropriate, some would say the correct, decision. It does not imply that it is a substitute for the skill of the user. The latter are programs that contain so much data that they can often be a substitute for the 'expert' user's knowledge and experience.

One of the first decision-support systems was a program for analysis of the electrocardiogram (ECG) developed by Hubert Pipberger in 1961.[1, 2] He fed the ECG derived from a 12-lead system into the computer, carried out a signal analysis and then examined the waveform, noting the salient characteristics: namely the start of the waveform as indicated by the P-wave, the height of the R-wave, the time-length of the QRS complex, the height (positive/negative) of the t-wave and the heart rate. Taking all the factors into account, such as a lengthened QRS complex, a diagnosis was made in the computer and printed out. Later, in 1968, Bonner and Schwetman produced their version of an analysis system.[3]

In 1969, Howard Bleich produced what was probably the first DSS for use in critical care situations.[4] It was a simple computer program that analysed a patient's electrolytes and blood gases and recommended an amount of bicarbonate to correct the imbalance in the patient's acid–base status. His (simplified) solution was to use a multiplier of three to correct the acid–base imbalance. A more comprehensive approach to this problem resulted in a program by Goh, Doran and myself.[5]

Also, in 1969, Jennings and colleagues analysed the work load of on-call anaesthetists and produced a 'predictor' for the number of likely calls over a 24-hour period. The system was very accurate in its forecasts. While this was a predictor system, it was used to plan the staffing levels for the on-call anaesthetic team. In that same year, Jennings also produced a DSS as an aid to planning operating theatre lists.

In 1972, Dr Tim de Dombal produced his first DSS for the diagnosis of the cause of abdominal pain (see Medicine).[6]

In 1973, Edward Shortliffe produced a, now famous, expert system called MYCIN.[7, 8] Its role was to analyse all the biochemical and bacteriological information about an infectious patient and to recommend an appropriate antibiotic.

In 1974, Howard, Longson and myself produced a DSS for the Programmed Investigation Unit.[9]

Dr John Fox, who was to become one of the advocates and producers of DSS in the UK appeared on the scene with his paper on rule-based systems (see Oncology).[10]

Fired with enthusiasm and with the knowledge of what was happening in the USA, several pioneering programs were written in the UK. One of the most popular machines was the DEC PDP-8. It came to the UK in the late 1960s and was then available in some hospital departments. It was ideal for signal processing.

MEDICINE

In 1972, Dr Tim de Dombal produced a DSS for computer-aided diagnosis of abdominal pain.[6, 11, 12] This program took in all the non-invasive information obtained by the doctor on the circumstances and site of pain in the abdomen, the duration of the pain, tenderness, etc., and then made a diagnosis. Of particular importance was which of the four quadrants of the abdomen had the most severe pain. The computer had to cover such diagnoses as appendicitis, gallstones, severe indigestion and pancreatitis. As with many such DSS, its role as an expert soon became recognized and it was a most valuable tool to aid junior doctors: having the computer on hand was as good as, if not better than, having the consultant alongside. The value of this program was recognized by the US Navy Department, which installed it in all its submarines. Also, this program was actively used in the Royal Infirmary in Leeds.

This work of de Dombal was actively taken up by the North East Thames RHA and was also put into trial use at Whipps Cross Hospital in London. Dr de Dombal's work was used as a base for work done at Airedale District General Hospital by W.A.F. Adams, a consultant surgeon.

In 1979, surgeons in Glasgow sought to increase the accuracy of their diagnoses in the accident and emergency department, for those patients attending with abdominal pain, by using the de Dombal system. Dr Gavin Kenny, an anaesthetist there, was one of the pioneers of computer usage.[13]

Scheduling

In 1974, the team of Longson, Richards and Howard produced a DSS for use in the Programmed Investigation Unit (PIU) at the Manchester Royal Infirmary.[9] The PIU was also known as 'The Five-day Ward' since it only functioned from Monday through to Friday. Patients were admitted to this ward for periods of one to five days to undergo a pre-scheduled series of investigations. The scheduling of each admission and the matching of patient-test against availability of resources in the paramedical service departments had previously been done manually.

The problem of precisely defining the various constraints was a necessary task and the sequencing was demanding. For example: 'Test B must not follow Test A closer than eight hours to avoid interactions but it must precede Test C; however, Test C can follow Test B after one hour, but must wait 12 hours if it is to follow Test A'. It was subsequently proved that the computer program, in fact an expert system, was a valuable tool on the ward, as the Ward Sister had been the only person who could regularly do the sequencing job.

To make possible the decision-making process, a matrix was constructed and stored in the computer. It showed how soon one test could be done after another. For instance, a lumbar puncture can be done immediately after a red cell survival test, but staff must wait 12 hours after a lumbar puncture before doing a red cell survival test. Also some tests cannot be carried out on the same admission, for example a cholecystogram and an intravenous cholangiogram. Again some tests requested by the doctors in the wards were part of a much larger investigation done in the PIU. Thus a 12-hour fasting blood sugar test was the first part of an insulin tolerance test with cortisols. The full matrix of tests comprised some 180 tests resulting in 32,400 interactions and this proved to be beyond human capabilities. The other advantage of using this DSS was that the computer could minimize the length of stay for a patient in the PIU: an important factor when the demand exceeded the 26 beds available.

Gastroenterology

A computer program to aid the gastroenterologist was produced to work both as an expert system and also as a training program for junior doctors.[14] The aim was to create a model for the training of the junior doctors in the techniques of interviewing patients, obtaining their medical histories and examination procedures. The expert system aimed to be as good as a consultant in the field. The strategy used in this program differed from that used by de Dombal,[6] but it initially followed the ideas of Richards.[15] It used a knowledge base that included the experiences of several well-qualified doctors and an expert in the field.[16] Details of the patient were obtained by structured interview and then a diagnosis was made. As with many expert systems, the user asked 'Why?' and the computer then illustrated the logical pathways used to arrive at that diagnosis. Many junior doctors regarded this as one of the more helpful parts of this package.

INTENSIVE CARE

In the late 1970s, Dr Dave Morrison, Director of Intensive Care in North Manchester General Hospital, collaborated with Dr Powner in the Electrical Engineering department at UMIST. Eddie Powner built a special purpose computer for displaying the results of laboratory tests carried out on patients in the ICU. Their first idea was to display four electrolyte values (namely sodium, chloride, potassium and bicarbonate) on four axes: $+x$, $-x$, $+y$, $-y$. On joining the four points, a diamond was produced. The four values were scaled so that, at normality, a perfect square was seen. At any abnormality, the square would appear a misshapen. The thinking here was that a person could more easily see the changed pattern than read and interpret a set of four numerical values. The idea caught on and was expanded to displaying eight values whose perimeter would be a perfect octagon (the corners representing blood sodium, urine sodium, blood urea, urine urea, in addition to the four previously mentioned electrolytes). This display system was the front end to a second module that then prescribed the treatment. In 1980, Morrison was also the originator and the prime mover in introducing a database for ICU patients, and a national manual register, so that ICUs across the UK could compare problems and performances. In 1984, he wrote a series of articles describing work in the ICU.[17]

Biochemical status

In 1977, the team of Richards, Doran and Goh took Bleich's work much further. They produced a DSS to take in all the laboratory results from a patient in the ICU to make a diagnosis (such as damaged spleen, heart defect, infection, electrolyte disturbance) and to suggest a possible treatment to correct any acid–base imbalance or electrolyte imbalance.

A further module was added to the above program in 1996.[18] This took in data from the patient's blood pressure, heart rate and central venous pressure, diagnosed any abnormalities and directed a course of treatment to correct any imbalance, such as drug administration, or adjustment of the ventilator. This program was recognized as being an expert system as it provided the junior staff with a 'virtual consultant'. Such was the accuracy of this program that it was used in the ICU in both Poland and in the Czech Republic, as well as in the UK.

Dr David Price, working in the Neurological Intensive Care Unit in Pinderfields Hospital in Yorkshire, developed a closed-loop system for perfusing Mannitol into a patient who had suffered a head injury. He incorporated modeling techniques into a conventional rule-based artificial intelligence shell in order to monitor the relationship between intra-cranial pressure and intra-cranial volume, and thereby select the optimum dose of Mannitol to control the intra-cranial pressure.[19] This was the first closed-loop system in the UK.

Geoffrey Shearing, working with colleagues in the ICU at Newcastle upon Tyne, and with Peter Harvey in Lancaster, produced a DSS program running on a DEC PDP-8 to calculate the exact contents of the intravenous infusions for feeding an individual patient during the next 24 hours.[20]

SURGERY

Although there were advances being made in robotics to help surgeons, there was once thought by some to be little need for computer software in surgery. Computer-controlled robotics has improved operations such a hip-replacement and brain surgery and will continue to develop, but what of software developements in other areas of surgery?

Cardiovascular surgery

In 1977, Doran, Goh and Richards produced a DSS to aid the anaesthetist during cardiovascular surgery (often called open-heart surgery).[21] This, and heart transplants, are difficult areas for the surgeon and the anaesthetist as many organs are involved and mastering the physiological demands involves a good deal of complicated mathematics.

The principle involved in open-heart surgery is that of cardiac-bypass, wherein the heart is stopped and its function is taken over by the heart-lung machine because the absence of a beating heart means that the lungs are not able to perfuse the blood with oxygen. So, before the surgeon made the first cut, the above expert system requested data on the patient. These included the current values of blood-gases and electrolytes as well as other data that allowed the computer to calculate the patient's blood-volume. Knowing the volume of blood in the body enables the computer to calculate how much blood and saline should be put into the bypass pump before the start of the operation. During the operation it is not unusual for the patient's blood to be diluted by a factor of two to reduce viscosity and so avoid complications. The computer then indicated what chemicals needed to be added to the bypass pump. These included potassium and magnesium in addition to the more usual electrolytes and dextrose. During the operation the computer controlled all the materials (oxygen, carbon dioxide, electrolytes and fluids) put into the patient. As the heart was restarted after the operation, the patient resumed breathing using a ventilator. Again the computer indicated the settings on the ventilator.

Controlled studies using the above system have shown that the patient's physiology, as measured by many variables, is in a much better state than otherwise would have been the case without computer intervention. At least two hospitals in Manchester used the system.

The program ran on a Commodore PET computer and on the much larger Cyber 70 mainframe. It was subsequently transferred to a desktop computer.

Anaesthetics

Another DSS was developed by Richards to make available the skills of the consultant anaesthetist to the junior anaesthetist working alone in the operating theatre, possibly at night.[22, 23] The concept was simple. Before staring the operation, the duty anaesthetist would switch on the computer and feed into it certain information, such as the patient's ID, height and weight, clinical condition (e.g. peritonitis) and the planned operation (such as 'repair of a perforated duodenal ulcer'). The patient's blood pressure and electrolytes data were also fed in. The computer would then require the anaesthetist to test the equipment, especially doing the 'one-hose test'. If, during the operation, a problem arose, the concept was that the anaesthetist would tell the computer the area of difficulty and the machine would reply with a series of steps to be carried out, for example checking various patient values. Typical emergency situations would be cardiac arrest, cyanosis and anaesthetic machine failure.

Neural networks

As computing systems became more complex, clinical databases were used to build knowledge acquisition systems. Gerard Conroy, in the Computation department at UMIST, used genetic algorithms (neural networks) to advise the surgeons. In 1990, he used his DSS algorithms to help the renal transplant surgeons.[24] In 1999, he used similar techniques when collaborating with his colleague Dr Tim Ritchings to provide a diagnostic DSS for patients with cancer of the larynx.[25] His system analysed the response from the vocal chords and thus advised the surgeon of the degree of damage the cancer had done to the chords: this information being vital in decisions as to the amount of surgery or chemotherapy to be undertaken.

CARDIOLOGY

In the USA, in 1963, Caceres developed a computer program for analysing the 12-lead system.[26] Much later, in 1971, Macfarlane used a three-lead approach to make a significantly better diagnostic system.[27] Not only was the three-lead system (sometimes called 'the orthogonal system') easier to administer, it also suffered no loss of essential information. One of the first major successes of this DSS was that it was able to show significant improvements in the diagnosis of myocardial infarctions by using a careful timewise analysis of the QRS complex. Macfarlane's claim to fame must be based on the work he did in the analysis of arrythmias.[28] That this program was able to work was due to previous efforts by the same team at Glasgow Royal Infirmary, where they designed both digital filtering techniques and smoothing algorithms. For the diagnosis of the arrythmias, decision-tree logic was used. Macfarlane and colleagues were also responsible for carrying out a comparison of the three-lead and 12-lead systems, and for showing that the three-lead system proved faster to analyse by the computer.[29] The breakthrough in being able to diagnose myocardial infarctions was due to the ability of the computer to analyse 'serial electrocardiograms' collected over several days.[30] This work was widely acclaimed and the team went on to make significant contributions to worldwide standards in the recording and analysis of ECGs using the computer.

PAEDIATRICS

The computer can help in two ways in the field of paediatrics. Firstly, in providing an expert system for the treatment of babies in neonatal ICUs where a process analogous to that described in the adult ICU (see Intensive Care), but with appropriate modifications, can be used to control the neonate's blood gases and electrolytes. Secondly, the computer offers great potential in diagnosing congenital abnormalities. In 1980, Richards and co-workers, including Dimitropoulous, produced such an expert system.[15] The paediatrician would carefully examine all the features of the child, noting the face, the limbs and where possible the fingerprints. No invasive procedures were required. Every detail was noted and entered into the computer. Using its 'knowledge' the computer program would then produce a list of likely diagnoses ranking these in descending order of the probability of the given diagnosis in that child, based on the reported characteristics. The number of possibilities was enormous and so provided a good role for a computer using a knowledge-based system. The program was much admired when it was presented at conferences such as the *Genetics Conference* at Lviv, in the Ukraine, and at *MEDINFO 80*.[15]

The 'intelligence' in the system came about through the use of maximum-likelihood methods, but with normalization and appropriate weighting. The mathematics produced a 'points score' – the likelihood of that syndrome being present in the child. The computer then printed out the five most likely syndromes, together with a 'probability' of that syndrome being present in that particular child. This program was very valuable as a source of guidance to the non-expert clinician.

So what was the value of the computer program? While one might argue that it could not compete with the best consultant in medical genetics, it was recognized that it was a very useful tool for the GP who was the first port of call for a sick baby. The GP would be able to make a first-line assessment and reassure the parents before making an appropriate decision as to a referral to hospital.

PREGNANCY

Pioneering work was done by obstetricians Lilford and Chard, in London. They set up a computer system to aid the obstetrician in prescribing treatment for the expectant mother. It started as a simple 'booking system', but was enhanced with rules as to when, and for what tests, the mother should be seen.

In 1981, the British Obstetric Computer Society was formed. Members included amongst others, Drs Kevin Dalton, Rupert Fawdry, Howard

Jenkins, Richard Lilford, Mike Maresh and Bernard Richards, who working both collectively and independently, produced databases (Richards in 1971 and Maresh in 1985), computer-aided history-taking systems (Lilford in 1984) and telemedicine systems (Dalton in 1986).

Labour

Howard Jenkins, working in the maternity unit at Queen's Medical Centre, in Nottingham, realized that the fetal electrocardiogram (FECG), recorded from a fetal scalp electrode during labour, was a signal often marred by 'noise'. He produced a computerized system to clean up the signal. But he then went further and produced a DSS that analysed the FECG and was able to predict a drop in the fetal heart rate some 20 minutes in advance of the event. This was a much better accuracy than the midwife could attain, and hence the staff were forewarned and able to take remedial action.[31]

At the end of the pregnancy, when the mother was due, by dates, to give birth, often she would be induced by an oxytocin infusion. Getting the infusion rate correct so as to maintain the contractions, but so as not to incur distress in the fetus, required much skill and was a job for a computer system. The team of Lieberman, Richards, Cadman and Levit produced an expert system, which controlled the flow of the oxytocin, taking into account these two factors.[32] In a control study, the results showed that using this system required less than half the oxytocin normally given in an unaided regime, and that labour was not lengthened (if anything it was shortened slightly). Of critical importance was the fact that, using this DSS regime, fewer babies, 10 per cent as compared with 20 per cent, were born with clinical jaundice. This improvement may be because, as fluid crosses the placenta, excessive introduction of the dextrose can make the baby water-logged, a decidedly unhealthy state for the newborn.[33]

These systems are also discussed in Chapter 13.

Support for women

Richards, Hayes and Leong produced a DSS for use in pregnancy.[34] The *Electronic Textbook of Pregnancy* was interactive and personal. It asked the user relevant questions as to smoking habits and alcohol consumption, and gave help and advice based on the information gleaned (see Chapter 23). The system was able to tell the pregnant woman what to expect and give advice at each stage of pregnancy (e.g. to take folic acid before conception), and to indicate an appropriate course of action when anything untoward was occurring. The program was a generalized information system as it comprised four subsections. These were: (1) The pregnancy textbook *From An Idea To A Birth*; (2) a pregnancy encyclopaedia; (3) a medical dictionary devoted to matters relating to pregnancy; and (4) a selection of 'previously answered questions'. The somewhat curious title in (1) above was explained by the fact that the early chapters dealt with the situation before pregnancy and gave advice on how to prepare for pregnancy and conception, and there was a chapter covering labour through to delivery and dealing with the newborn baby.

NURSING

One of the earliest uses of the computer in the field of nursing was to involve it in the tricky task of nurse scheduling (see also Chapter 20), which meant planning the shift-rotas for the ward. Professor Brian Moores was one of the people who made a significant contribution to this field.

In 1974, when the NHS was still very dependent upon the contribution of trainee nurses, the DoH funded a team at UMIST, led by Moores, to undertake an investigation of nurse-activity patterns in 30 wards in three acute hospitals. Qualified nurse observers made over 200,000 recordings at three-minute intervals of what all nursing staff on the ward were doing over a seven-day period. It was found that, across all these wards, 64 per cent of the available nursing staff hours were contributed by either pupil or student nurses, who were following two- and three-year courses respectively.

Direct patient care was found to account for 49 per cent of the work of all the nursing staff during the day shifts and, of the hands-on nursing, 75 per cent was contributed by trainees.[35] The pressure of work being experienced each day was determined using patient dependency data and staffing figures collected over 13-week periods. In two of the three hospitals, the proportion of the considerable variation in that index, explained by the difference between weeks, was 46 per cent, whereas, variation between different days of the week accounted for only three per cent.[36] This suggested that, in both hospitals, this might be reduced considerably were trainee availabilities better matched with patient needs. In the third hospital, these two figures were 18 per cent and 22 per cent, respectively, implying that the production of off-duty rotas also warranted serious attention. This prompted Moores to consider providing a DSS to do that job.

This traditional, but surprisingly high, reliance on trainees reflected, to some extent, a widely held belief that trainees represented a cost-effective form of labour. A study in 1979 revealed that this was not the case as the substitution of trainee nurse labour by an appropriate mix of qualified and auxiliary staff would be significantly less expensive.[37] That said, the prevailing situation demanded that the nurse training programmes had, of necessity, to take into account the demands of the wards for pairs of hands. Each of the programmes was structured around a specified number of weeks of ward experience and blocks of classroom study in the nurse training school. Typically, the classroom blocks were from two to four weeks duration and ward experiences ranged between 15 and 20 weeks. The General Nursing Council had to approve all training programmes.

The production of supposedly well-balanced nurse allocation programmes occupied many man-hours of trial-and-error effort, much use of erasers, and numerous sheets of graph paper. When microcomputers appeared on the scene in the 1970s, it seemed obvious that this highly labour-intensive activity might lend itself to some form of computerization. This was indeed the case. The initial efforts of the UMIST team focused on producing readily interpreted computer-generated schematic representations of the consequences of any specified training programme. The planner would specify the date on which any one of any number of trainee intakes would commence and the lengths of each period of classroom teaching and assignment to wards. Given a set of specified parameters, the computer would then determine the number of trainee nurse intakes that would be in the classroom on any one given week, and the associated number available in the wards. Clearly, this constituted a significant breakthrough as these results were displayed in a user-friendly format within seconds. The specifications could then be modified and the impact of these changes evaluated. An accomplished planner could use this fairly rudimentary computational device to help in the production of reasonably balanced training programmes in a relatively short time, and several planners used that first prototype in this iterative manner.

The next stage was to incorporate optimization into this planning process. It proved possible to formulate the problem of determining balanced training programmes where the conflicting goals were the numbers of trainee intakes in the classroom and in the wards.[38] The model was tested with the aid of an experienced planner who was attempting to generate programmes for three student nurse intakes and two pupil nurse intakes each year, over a three-year planning cycle. This generated a problem with 219 binary variables (zero-one) representing the possible starting dates for the holidays, the classroom blocks and the ward assignments, and 150 goals corresponding to 52 desired ward staffing levels, 52 to school capacity, and 46 to the times between holidays and blocks. This necessitated the development of an algorithm for solving zero-one goal programming problems.[39]

At that time, a linear programming (LP) problem of this magnitude could only be solved on a mainframe computer. A planner produced a first attempt and provided weightings for the conflicting goals, which it was felt reflected the appropriate priorities in the hospital. The computer produced a solution, which the planner then examined. If there was too much variability in the number of intakes in the school, for example, the weightings were adjusted accordingly and another solution was obtained. It took just four iterations of this process to produce a final acceptable version and which was indeed implemented. Whereas the original plan would have resulted in the number of trainees on ward duty ranging from 222 to 346, by the fourth iteration, this range had fallen to 248 to 300. Likewise, under the original plan, there would have been five intakes in the school of nursing in one week.

The final piece of the jigsaw did, to a large extent, represent a return from this esoteric mathematical programming formulation to the use of the elegant pictorial representation of the programmes. By 1981 the enhanced processing speed of microcomputers meant that these could be generated in seconds. It had been found that an accomplished planner could use them to produce balanced programmes in a heuristic manner. Indeed, the aforementioned programme, achieved after four iterations of the goal programming algorithm, was arrived at heuristically in seven iterations. Also by this time, UMIST was receiving a steady flow of enquiries from Schools of Nursing anxious to implement a computer-based solution. Several workshops were organized at which the software was demonstrated.[40] Feedback obtained from these workshops indicated that the software had to be expanded in such a way that it also served as an information storage and retrieval system. Accordingly, the software was developed along these lines in association with an enthusiastic School of Nursing. For example, information on each trainee was readily retrievable and a record of her or his training could be routinely produced for submission to the General Nursing Council.

Another clerical activity that occupied a great deal of time was the production of Change Lists, which informed the wards and departments of the identities of the trainees they could expect to appear on the Unit. The production of these and other reports were incorporated into the package.

It was felt appropriate to solicit the cooperation of more Schools of Nursing in the further development of the software. Accordingly six schools signed up to become members of a user group. In addition to implementing the package, members of this group met on a regular basis with the UMIST-based project leader to swap experiences and help shape the future direction of the software. It was this group that suggested incorporating estimates of the impact of wastage into the software. Previous research had revealed that if a trainee did not complete the training, then that trainee's departure from the programme occurred early, rather than later. This pattern had obvious ramifications as specialties towards the end of a trainee's programme would inevitably be allocated fewer trainees than those that appeared earlier.

The user group remained in existence for several years. Perhaps not surprisingly, a software company recognized the commercial potential of the approach. It was not overly difficult for any reasonably efficient programmer to write code that produced virtually identical input and output screens to those featured in the UMIST package. In truth, the company was able to invest substantially more staff-hours on their version in anticipation of a substantial revenue stream. Both packages were, of course, eventually rendered obsolete by the complete reorganization of nurse training stemming from the Project 2000 proposals, which now see students no longer counted in a ward or department's establishment.

As nursing practice changed, the concept of 'named-nurse' came to the fore, Nick Hardiker produced a nurse scheduling system,[41] which enabled the ward to act in accordance with the new guidelines. The expert system was well received by the UK nursing fraternity and widely acclaimed. The full significance of this work is revealed in Chapter 20.

OPERATING THEATRE SCHEDULING AND THE ANAESTHETISTS WORKLOAD

One of the vexing problems associated with an operating theatre is the overrunning of the scheduled list that supposedly is to occupy three hours and no more. It was widely known that the lists almost invariably overran. In Northampton, Dr Alan Jennings produced a DSS that significantly increased the chances of the list finishing on time. This was achieved by noting the surgeon, the specific operation and its complexity. This DSS was found to be most useful.

Dr Jennings with colleagues, at the Northampton General Hospital, The Royal London Hospital and at the Alder Hey Children's Hospital, Liverpool, also did a multi-site study of the workload of on-call anaesthetists. He was able to produce a system that would accurately plot their likely workload over a 24-hour period, taking into account the number of anaesthetists on-call on a given occasion.[42, 43] The system could then aid the planners in constructing the duty rotas.

ONCOLOGY

One of the first DSS in oncology was ONCOCINA, a rule-based expert system for oncology protocols developed at Stanford, USA. Reference has already been made to the work of Conroy in producing a DSS for detecting cancer in the larynx.[25] But there is no doubt that one of the chief producers of expert systems in oncology was Dr (now Professor) John Fox, now at Cancer UK. Alone and with others he has produced a steady series of decision-support systems, including the LISA System,[44, 45] and culminating in his latest system called Proforma.[46]

CONCLUSION

There is one significant area that is not covered in this chapter, that of radiation treatment planning. DSS for this were in use in Manchester and in London. Further information can be found in Chapter 9.

In summing up, it is sufficient to say that, while some early DSS were developed in the USA, the work done in the UK, by those pioneers mentioned, really advanced the techniques and furthered development of that subject.

REFERENCES

1 Pipberger, H.V., Arms, R.J. and Stallmann, F.W. (1961) Automatic screening of normal and abnormal electrocardiograms by means of a digital computer. *Proc. Soc. Exp. Biol. Med.*, 106, 130–2.

2 Pipberger, H.V. (1965) Computer analysis of the electrocardiogram. In Waxman, B. and Stacy, R. (eds) *Computers and Biomedical Research*, p. 377.

3   Bonner, R.E. and Schwetman, H.D. (1968) Computer diagnosis of electrocardiograms II: a computer program for EKG analysis measurements. *Computers in Biomed. Research*, Feb., 1, 4, 366–86.

4   Bleich, H.L. (1969) Computer evaluation of acid–base disorders. *Journal of Clinical Investigation*, 48, 1689.

5   Richards, B., Goh, A. and Doran, B. (1977) Computer assistance in the treatment of patients with acid–base and electrolyte disturbances. In Shires, D.B. and Wolf, H. (eds) *MEDINFO 77*, Toronto, p. 407. North-Holland, Amsterdam.

6   de Dombal, T., Leaper, D.J., Staniland, J.R., McCann, A.P. and Horrocks, J.C. (1972) Computer-aided diagnosis of acute abdominal pain. *BMJ*. 2, 5804, pp. 9–13.

7   Shortliffe, E.H., Axline, S.G., Buchanan, B.G., Merigan, T.C. and Cohen, S.N. (1973) An artificial intelligence program to advise physicians regarding anti-microbial therapy. *Computers in Biomedical Research*, 6, 544–60.

8   Shortliffe, E. (1976) Computer-based medical consultations: MYCIN. *Artificial Intelligence Series*. Elsevier Computer Science Library, New York.

9   Richards, B., Howard, P. and Longson, D. (1974) Computerization of a Programmed Investigation Unit. In Anderson, J. and Forsythe, M. (eds) *MEDINFO 74*, Stockholm, pp. 529–32. North-Holland, Amsterdam.

10 Fox, J., Barber, D.C. and Bardhan, K.D. (1980) Alternative to Bayes: a quantitative comparison with rule-based diagnosis. *Meth. Infrm. Med.*, 19, 210–15.

11 de Dombal, F.T., Dallos, V. and McAdam, W.A.F. (1991) Can computer-aided teaching packages improve clinical care in patients with acute abdominal pain? *BMJ*, 302, 149–57.

12 de Dombal, F.T. (1984) Clinical decision-making and the computer: consultant, expert or just another test? *British Journal of Healthcare Computing*, 1 (Spring), 7–13.

13 Graham, D.F., Kenny, G. and Wright, R. (1979) Computer diagnosis of abdominal pain. In Barber, B., Grémy, F., Überla, K. and Wagner, G. (eds) *MIE, Berlin 79: Lecture Notes 5*, (pages unknown). Springer Verlag, Berlin. ISBN 3 540095 49 7.

14 Richards, B. and Lugovkina, T. (1991) The use of the computer in doctor training and patient diagnosis. In Adlassnig, K.P., Grabner, G., Bengtsson, S. and Hansen, R. (eds) *MIE, Vienna 91: Lecture Notes 45*, p. 620. Springer Verlag, Berlin. ISBN 3 540543 92 9.

15 Richards, B. and Jeffery, C. (1980) The computer diagnosis of congenital malformations. In Lindberg, D.A.B. and Kaihara, S. (eds) *MEDINFO 80*, Tokyo, p. 779. North-Holland, Amsterdam.

16 Lugovkina, T. (1990) An expert system in gastroenterology. *International Conference on Medical Informatics and Medical Education*, Prague.

17 Morrison, D. (1984) The saga of a simple physician in the world of high technology. *British Journal of Healthcare Computing*, 1, 1, 38.

18 Hollingsworth, R. and Richards, B. (1996) An expert system for ITU treatments. In Brender, J., Christensen, J.P., Scherrer, J-R. and McNair, P. (eds) *MIE, Copenhagen 96*, pp. 594–8. IOS Press, Amsterdam. ISBN 9 051992 78 5.

19 Price, D.J. and Mason, J. (1984) An attempt to automate control of cerebral oedema. In Go, K.G. and Baethmann, A. (eds) *Recent Progress in the Study and Therapy of Brain Edema*. Plenum Press, London, pp. 597–607.

20 Shearing, G., Wright, P.D., James, R.M., Roberts, J.M., Cooper, R.T., Bellis, J.D., Harvey, P.W. and Rich, A.J. (1978) A computer program for calculating intravenous feeding regimes and its application in a teaching hospital and a district general hospital. In Anderson, J. (ed.) *MIE, Cambridge 78: Lecture Notes 1*, pp. 641–50. Springer Verlag, Berlin. ISBN 3 540089 16 0.

21 Richards, B., Goh, A. and Doran, B. (1979) The use of computers in open-heart surgery. In Barber, B., Grémy, F., Überla, K. and Wagner, G. (eds) *MIE, Berlin 79: Lecture Notes 5*, pp. 340–44. Springer Verlag, Berlin. ISBN 3 540095 49 7.

22 Richards, B. (1982) Training anaesthetists to cope with an emergency in the operating theatre: the use of a microcomputer. In O'Moore, R.R., Barber, B., Reichertz, P.L. and Roger, F. (eds) *MIE, Dublin 82: Lecture Notes 16*, p. 673. Springer Verlag, Berlin. ISBN 0 387112 08 1.

23 Richards, B. (1992) An expert system providing aid to the anaesthetist in an emergency or in routine work. *Fifth International Conference on System Science in Health Care*, p. 1321. Omni Press, Prague.

24 Cameron, C.A., Conroy, G.V. and Kangavari, M.D. (1990) Machine learning techniques for patient and program management in renal replacement/transplantation therapy. In O'Moore, R.O., Bengtsson, S., Bryant, J. and Bryden, J.S. (eds) *MIE, Glasgow 90: Lecture Notes 40*, pp. 286–91. Springer Verlag, Berlin. ISBN 3 540529 36 5.

25 Ritchings, T. and Conroy, G. (1999) A neural network-based tool for speech therapists: post therapy. In Bryant, J. (ed.) *Current Perspectives in Healthcare Computing*, (pages unknown). BJHC for BCS, Weybridge. ISBN 0 948198 11 7.

26 Caceres, C.A. (1963) Electrocardiographic analysis by a computer system. *Arch. Inter. Med.*, 111, 196.

27 Macfarlane, P.W. (1971) ECG waveform identification by digital computer. *Cardiovascular Research*, 5, 141.

28   Macfarlane, P.W. (1974) Arrythmia interpretation by digital computer. Schubert (ed.) *Nue Ergebnisse der Elektrokardioologie II*, pp. 243–5. Berlin.

29   Macfarlane, P.W. (1971) Three- and 12-lead electrocardiogram interpretation by computer. *British Heart Journal*, 33, 266.

30   Macfarlane, P.W., Carwood, H.T. and Lawrie, T.D.V. (1975) A basis for computer interpretation of serial electrocardiograms. *Computers in Biomed. Res.*, 8, 189–209.

31   Jenkins, H.M.L., Kirk, D.L. and Symonds, E.M. (1982) Continuous intrapartum fetal electrocardiography using a real-time computer. In O'Moore, R.R., Barber, B., Reichertz, P.L. and Roger, F. (eds) *MIE, Dublin 82: Lecture Notes 16*, pp. 136–42. Springer Verlag, Berlin. ISBN 0 387112 08 1.

32   Richards, B., Cadman, J., Levitt, J. and Liebermann, B. (1991) Control of oxytocin in labour: a proven algorithm. In Richards, B. and MacOwan, H. (eds) *Current Perspectives in Healthcare Computing*, p. 300. BJHC for BCS, Weybridge. ISBN 0 948198 11 7.

33   D'Souza, S. and Richards, B. (1986) Oxytocinon induction of labour: hyponatraemia and neonatal jaundice. *European Journal of Obstetrics, Gynaecology and Reproductive Biology*, 22, 309–17.

34   Richards, B., Hayes, G. and Leong, F. (1989) A knowledge-based system for giving expert advice to patients in all matters relating to conception, pregnancy, and childbirth. In Barber, B., Cao, D., Qin, D. and Wagner, G. (eds) *MEDINFO 89*, Singapore and Beijing, p. 1183. North-Holland, Amsterdam.

35   Moores, B. and Moult, A. (1979) Patterns of nurse activity. *Journal of Advanced Nursing*, 4, 137–49.

36   Moores, B. and Moult, A. (1977) Sources of variation in the pressure of work index in 30 hospital wards. *Journal of Advanced Nursing*, 2, 271–9.

37   Moores, B. (1979) The cost-effectiveness of nurse education. *Nursing Times*, June, 65–72.

38   Briggs, G.H., Garrod, N.W. and Moores, B. (1978) A zero-one formulation of the student nurse allocation problem: a formulation. *Omega*, 6, 1, 93–6.

39   Garrod, N.W. and Moores, B. (1978) An implicit enumeration method for solving zero-one goal programming problems. *Omega*, 6, 4, 374–7.

40   Gwinnett, A.E. (1984) *The Development of Microcomputer-based Models for the Production of More Effective Nurse Training Programmes.* Unpublished doctoral dissertation, UMIST.

41   Hardiker, N. (1993) Rostering for primary nursing with quality control. In Reichet, A., *et al.* (eds) *MIE, Jerusalem 93*, pp. 218–21. Freund Publishing House, Jerusalem.

42 Taylor, T.H., Jennings, A.M.C., Nightingale, D.A., Barber, B., Lievers, D., Styles, M. and Magner, J. (1969) A study of anaesthetic emergency work. *Br. J. Anaesth.* Paper I: The method of study and introduction to queuing theory, 40, 70–5; Paper II: The workload of the three hospitals, 41, 76–83; Paper III: The application of queuing theory to anaesthetic emergency work, 41, 167–75; Paper IV: The practical application of queuing theory to staffing arrangements, 41, 357–62; Paper V: The recorded service failures and general conclusions, 41, 362–70.

43 Jennings, A.M.C. (1969) A statistical study of anaesthetic emergency work. In Abrams, M. (ed.) *Medical Computing: Progress and Problems,* pp. 321–31. Chatto & Windus for BCS, London.

44 Fox, J. and Thomson, R. (1998) Decision support and disease management: a logic engineering approach. *IEEE Transactions on Information Technology, Biomedicine,* December, 2, 4, 217–28.

45 Bury, J., Hurt, C., Roy, A., Bradburn, M., Cross, S., Fox, J. and Saha, V. (2004) A quantitative and qualitative evaluation of LISA: a decision support system for chemotherapy dosing in childhood lymphoblastic leukaemia. In Fieschi, M., Coiera, E. and Li, Y.J. (eds) *MEDINFO 04,* San Francisco, pp. 197–201. North-Holland, Amsterdam.

46 Fox, J., Johns, N. and Rahmanzadeh, A. (1998) Disseminating medical knowledge: the *Proforma* approach. *Artificial Intelligence in Medicine,* 14, 157–81.

25 Confidentiality and security

BARRY BARBER

When personal health information first began being included in Health Information Systems, it was normally in collaboration with particular clinicians who would set out their clinical requirements in terms of access to these various items of data. In this way the early work was clearly under the control of health professionals who were subject to the professional rules and discipline of clinical confidentiality. The early concerns about the confidentiality of personal health information were reflected in the Sweden Data Protection Act of 1973 and other subsequent adopters of data protection legislation. These concerns led in due course to the adoption of the Council of Europe Convention *For the Protection of Individuals with Regard to Automatic Processing of Personal Data*[1] in 1981 and the associated Recommendation *On Automated Medical Data Banks.*[2] This Convention and its various Recommendations seek to implement the requirements of Article 8 of the Convention *For the Protection of Human Rights and Fundamental Freedoms* in the context of the special problems arising from the handling of computer-held information.[3] It is worth noting that The French National Assembly adopted a charter of Human Rights in 1789, *Des Droits de l'homme et du Citoyen*, as a means of protecting their citizens even though the outside world thought that this was just a paper exercise.

In addressing these issues at The London Hospital, before international guidelines became available, the Real-Time Computer Project Executive decided that:

- in terms of confidentiality, the computer system should be at least as effective as the previous manual system;
- the measures employed to further security should not be so cumbersome as to destroy the advantages of the system.

This approach was described by Barber, *et al.*[4, 5] The approach worked very effectively during the early stages of the development of the project, where the key issues were those of finding practical ways of securing the benefits of the computer's facilities in a convenient fashion. The computer terminals were under the same administrative controls as a patient's case notes and unauthorized access was dealt with administratively in the same fashion. It was only in later years that the access control systems had to be changed to come in line with modern security practice: with an identifier and a password for authentication purposes. The initial concerns were to ensure the development of useful working systems that would support and benefit the wide variety of clinical and administrative processes involved in the delivery of patient care.

THE CHILD HEALTH SYSTEM

The development of the Child Health Computer System in the early 1970s, which built on the successful vaccination and immunization computer systems of Galloway at West Sussex,[6] raised confidentiality issues in the sense that the system necessarily crossed a number of organizational boundaries.[7, 8] The strength of the requirement for confidentiality had not been appreciated by the DoH. This issue delayed the implementation of the Child Health System for a considerable time while a Confidentiality Protocol was developed in conjunction with John Dawson of the British Medical Association (BMA). Amongst these requirements was the need for a 'Data Custodian' conversant with the ethical requirements for data protection. This was before the first Data Protection Act in 1984 and the local arrangements for child care. The requirement also preceded the establishment of the still developing role of Caldicott Guardians by some 20 years.[9]

The Confidentiality Standards Working Party of the Child Health Computing Committee developed guidelines that were endorsed by the full committee in May 1979, and amended in the light of further experience in May 1983. These guidelines addressed the complex issues of data protection in the context of the child care arrangements, which involved a wide variety of adults, with a wide variety of areas of responsibility and authority in respect of particular children. The guidelines were incorporated directly into the NHS Data Handbook under the guidance of Dianna Osborn.[10] She had been involved in the process of developing acceptable solutions. The approach to data protection was showcased in one of the Child Health Computing Committee's conferences in 1987.[11] The basic material was subsequently incorporated into the South East Thames RHA Guidance.[12]

The delays associated with the introduction of systems across organizational boundaries have echoes with the clinical concerns expressed about the National Programme for Information Technology (NPfIT).[13] These matters can usually be resolved, but care is required to get the systems right and to convince the users that they have been sorted out ethically, legally and practically. These professional concerns always take time for their satisfactory resolution and, generally, the systems development cannot be undertaken until this has happened.

PRIVACY AND DATA PROTECTION IN THE UK

The issues of privacy had been exercising Parliament for many years and, of course, the UK had been involved in the development of the European Convention on Human Rights and Fundamental Freedoms. A draft Privacy

Bill was introduced in the House of Lords as early as 1961 and there have been many other attempts since that time. The Younger, and Lindop, Committee Reports in 1972 and 1978,[14, 15] respectively, together with the White Paper in 1975,[16] looked as if the issues of Data Protection and Privacy were beginning to be taken seriously. The 'Ten Younger Principles' were already recognizing the key issues that became the 'Data Protection Principles', but they included a requirement that 'a monitoring system should be provided to facilitate the detection of any violation of the security system': a rather more explicit requirement than was included subsequently. The Data Protection Act was finally enacted in 1984 so that the UK could sign up to the 1981 Council of Europe Convention. The Act was the barest minimum that would allow a UK signature indicating compliance but the opportunity was taken, unlike the Irish Data Protection Act 1988, to use non-standard terminology, which then had to be changed when the European Directive was implemented in the Data Protection Act 1998.[17]

George Knox edited a European Review of the issues of medical confidentiality across the European Economic Community in 1984.[18] He looked at specific issues arising in particular countries and developed a 'Model Code of Practice', paying particular attention to the issues of medical research. The concept of 'guardianship' of medical records was developed, which, like the work on child health, foreshadowed the ideas of the Caldicott Guardians a decade later, and the report began to consider some of the aspects of computer security.

David Kenny and the Körner Report

The Health Services Information Group, under the chairmanship of Edith Körner, was a landmark in the use of personal health information for purposes other than the direct care of individual patients. Within that context, and under the chairmanship of David Kenny, a report from the Confidentiality Working Party was produced in relation to the protection and maintenance of confidentiality of patient and employee data.[19] This report was developed in the context of the imminent passage of the Data Protection Act 1984, but when its details were still not completely clear. It ante-dated the project to develop a strategy for implementing the Act within the NHS but it set out clearly, in a 102-page report, a number of recommendations in a 'Policy for Protecting and Keeping Confidential Patient Data' and a 'Code on Confidentiality of Personal Health Data'. The working group examined the implications of holding personal health data on computers, and they initiated consideration of the security issues involved. The computer subgroup included two Regional Computer Services Officers, Jack Sparrow and Peter Lock, as well as Professor F. Flynn, who was running his computerized biochemical laboratory systems. The working group listed the legislation, at that time, that required disclosure of personal health data despite the common assumption by patients that their data were held confidentially.

The Körner Working Group carried out a one-month survey in six District Health Authorities, which had established policies on the confidentiality of patient information, in order to assess the size of the traffic in patient information. During that month they found that 1273 requests had been made for information. In 59 per cent of cases the consent was implied within the request. Written consent was secured in 10 per cent of cases, in four per cent of cases oral consent was obtained and in 27 per cent of cases consent was sought and there had been no response. The largest number of requests, 436, came from doctors outside the hospital, the next was 314 requests from the DoH and 89 from other Government departments. In these pre-Data Protection Act days, 34 requests came from patients, 32 from relatives and friends and 144 from non-medical representatives. A further 100 came from voluntary bodies, 35 from insurance companies, 30 from local authorities and 20 from researchers. From the organizational point of view, 70 per cent of requests were covered by the Confidentiality Policy. In 17 per cent of cases written consent was obtained from the relevant doctor and in 12 per cent of cases oral consent was given. Consent was refused in one per cent of cases. Before this inquiry, little concern appeared to have been expressed about the large and growing volume of requests to access patient data for purposes that were not directly connected with their treatment and care.

Although there has been no published repeat of this type of survey, it would appear that the demands for access to personal health data have increased over the years. The Caldicott Report found,[9] in 1997, extensive data flows that were 'partially justified' and the present emphasis on 'data sharing' can only have contributed to the problem of access without consent to personal data, despite the requirements of the EU Data Protection Directive. Since that time, there has been a dramatic increase in these types of disclosure. Legislation giving much wider legal access to personal health data and the advent of Government-backed enthusiasm for 'data sharing' and 'anti-terrorist security monitoring' leaves one much less sure about the confidentiality of our medical information.[20-23] The scheme for genetic profiling of criminals has been extended beyond those convicted of serious criminal offences without the detailed protections that were afforded by an earlier generation with respect to the fingerprinting of criminals. In addition, proposals for building up a DNA register for all newborn infants, to provide support for their subsequent medical treatment, is only acceptable if it can only be used for this purpose and not for a wide variety of other Government or private sector purposes. Similarly, the legislation currently proposed for a 'voluntary' identity card scheme, that might be widely used throughout the public and private sectors, can only erode public confidence in the protection provided by the data protection legislation.

The DoH eventually was forced to the conclusion that access to personal health information should normally be by consent. It would have saved itself a lot of recent trouble if it had reached this conclusion earlier and bothered to act on this working group's recommendations in respect of:

- confidentiality standards in patients' leaflets;
- staff awareness and training;
- confidentiality clauses in staff contracts and handbooks;
- publication of statistics on the number and nature of exceptional disclosures at least annually;
- developing guidance on the access to patient information in the context of the developing NHS.

At about the same time an Inter-Professional Working Group (IPWG) had been set up to examine access to personal health information. This group eventually produced a code of practice that was, curiously, published in an early version of the *Encyclopaedia of Data Protection* by Chalton and Gaskill,[24] but it was never finally adopted and put into effect.

CIT and the Data Protection Act 1984

The Centre for Information Technology (CIT) established a working party, chaired by Jack Waldron, with representatives from the RHAs and the home countries, the DoH, the NHS Training Authority and the Office of the Data Protection Registrar. This work developed an initial set of guidelines for the implementation of the Data Protection Act 1984 within the NHS. The guidelines set the scene for the work to follow and included not only the text, but overhead transparencies with which these concepts could be presented to colleagues and at regional gatherings. This material was distributed widely throughout the NHS to support a training and awareness programme. Similar material was provided by the National Computer Centre, which had collaborated with the NHS Training Authority to develop it.

This initial work was further developed into a central Data Protection Project to address the issues of subject access within the NHS and I was seconded from North East Thames RHA as the project manager. The first step was to develop a Data Protection Handbook for the NHS,[9] and to explore the complexities required to handle the 'subject access' within the NHS. The DoH was helpful in providing draft Statutory Instruments concerning the exemptions to subject access in health care, so that the key issues could be explored, guidance developed and presentations prepared. Version 3 of the handbook was issued in June 1987 to help prepare for the first 'subject access requests' that could be lodged with Health Authorities on 11 November 1987.

At the time there was considerable apprehension about the possibility of having large numbers of requests that would place a large administrative burden on health authorities. There were, also, worries that trade unions might generate a mass of requests in order to bring pressure on authorities. In the event, there were relatively few 'subject access requests' during the first decade of the operation of the 1984 Act. This might have been because the personal information about patients held on the NHS computer systems was still very much less informative than that held in the conventional case notes. The 1984 Act related only to automatically processed 'personal data'. The Access to Health Records Act 1990 later provided a more useful approach to securing information about an individual's care. Subsequently, of course, the implementation of the European Data Protection Directive resulted in the Data Protection Act 1998,[17] which combined access rights to both automatically processed information and personal data held in conventional paper records. It also included personal data as images and sounds. The 'subject access exemption' relating to information that might cause 'serious harm' or identify a third party who had not consented appeared to be needed in a small proportion of cases (probably less than 10 per cent, although reliable statistics were hard to come by). The other problem that arose was that of 'enforced subject access requests' in which the 'data subject' was required to make a request and then supply the results to a third party. This issue has largely been addressed by legislation providing certain direct rights of access to some criminal records by employers.

Another issue that arose was the consideration of the Council of Europe *Recommendation On Automated Medical Data Banks*,[2] which was not well known or observed much within the UK. However, it did incorporate some useful ideas. In particular the notion that medical data banks should be subject to published regulations (which would be called a 'security policy' in modern terminology). This recommendation was updated subsequently to become the Council's *Recommendation On the Protection of Medical Data* in 1997.[25] Although the Home Office representative had not registered an interest, I was able to represent the NHS during the initial phases of the work that led to this Recommendation, alongside John Woulds, and then David Smith from the Office of the Data Protection Registrar.

The NHS Security and Data Protection Programme

Following the work on data protection, the Information Management Centre (the amalgamation of the CIT and the Centre for Data Administration) carried out some security work utilizing the CCTA Risk Analysis and Management Methodology (CRAMM).[26] This involved the development of some 'worst case scenarios' arising from a breach of security in the health care environment. Since the analysis was carried out using the Government's own method of risk assessment, this provided compatibility between NHS systems and other national Government systems. The first conclusions were that breaches of integrity or availability could have much more damaging results than breaches of confidentiality – which had been the key worry up until then. It was also clear that quite severe security countermeasures were called into play by virtue of the fact that serious integrity and availability breaches could lead to contraindicated, damaging or life-threatening treatment. Or, contrariwise, the damage or life-threatening situations arising from treatment that was withheld. These ideas about 'worst case scenarios' and the use of CRAMM were published at *HC92*[27] and at *MEDINFO 92*.[28]

Papers on the development of a security policy[29] and on training[30] were published at BCS health care conferences. Since that time the issues of safety in health care has become more and more important as they have been investigated more and more thoroughly.

The Security and Data Protection (S&DP) programme was initiated to address the wider issues in a variety of ways:

- CRAMM was used in a number of locations and in respect of a number of central initiatives in order to examine the risks and ensure that adequate security measures were put in place.

- A top-level security policy was developed as the basic policy of the NHS on which local policies could be based.

- A security handbook was developed to provide basic advice for the NHS.[31]

- A wide variety of training seminars were developed on various aspects of security and data protection within the context of the NHS.

- Scientific and technical papers and presentations were produced for NHS and IT conferences and journals.

- The Data Protection Handbook was revised and issued in the light of practical experience.[32]

- After the European Directive had been adopted on 24 October 1995, the S&DP programme was involved in trying to assess the implications for the NHS.

- Two videos were produced to help with security awareness and training programmes. They were *Threat to the System*,[33] which used a hospital-based scenario, and *Chain Reaction*.[34] The latter arose from the collaborative AGNIS (Awareness among General Practitioners of the Need for Information Security) project,[35, 36] set up in conjunction with the Joint Computing Group of the BMA and the Royal College of General Practice Physicians led by Dr Tim O'Connor. This project provided a detailed handbook for the practice manager; a small introductory booklet to the key security issues facing general practitioners; and a set of slides for an introductory talk about the issues. The introductory booklet illustrated ways in which patients' data could be lost, interfered with or protected.

The advice given by the S&DP programme was much less permissive than that offered by the DoH,[37] so it is satisfying to note that the more recent advice has become much more realistic (and closer to the views of the Information Commissioner) in respect of informed patient consent[38, 39] and the use of encryption for transmitting personal health information. The development of training videos was a major and interesting departure into a different area of technology that provided useful training material to support more formal lectures.

The S&DP team was a small one that never included more than Dave Garwood, Paul Skerman and myself, supported by Alison Treacher and a secretary, together with outside consultancy.

Encryption of personal health information

There was a good deal of controversy between the BMA and the NHS Executive about the confidentiality of personal health information in the context of developing an increasingly networked information strategy. This debate was initiated by Fleur Fisher, of the BMA, with Ross Anderson as her technical security adviser. Following the implementation of the Data Protection Act 1984 and the Körner Report on confidentiality,[19] there was some concern about establishing a code of practice on the confidentiality of personal health information. A draft code was in fact developed by an Interprofessional Working Group (IPWG) and it was published in the *Encyclopaedia of Data Protection*.[24] However, for some reason, it was never implemented and the BMA returned to this task with a Multi-Disciplinary Professional Working Group. This developed some draft legislation that would establish legal safeguards for personal health information[40] and an explanatory handbook of associated guidance.[41] Fleur Fisher explained the thinking behind this work at a conference of the BCS NSG.[42] During this period the DoH had been working on these issues,[37] as well as developing its NHS networking activities. The Information Commissioner issued somewhat more stringent guidance a bit later in 1998.[38] These various activities left the BMA distinctly uneasy about the ethical safety of the developments and Ross Anderson was asked to provide his technical advice. He spent a great deal of effort elucidating the various issues and the BMA published his advice as *Security in Clinical Information Systems* during which it was noted that:[43]

> . . . the advice of the British Medical Association to its members is that exposing unprotected patient identifiable clinical information to the NHS-wide network (or indeed to any other insecure network), or even sending it in encrypted form to an untrustworthy system, is imprudent to the point of being unethical.

Interim guidelines were published in the *BMJ*,[44] and the health professions were alerted to these ideas by his article 'NHS-wide networking and patient confidentiality'.[45] Ross Anderson also spoke at the *HC96* conference under the title 'Patient confidentiality – at risk from NHS-wide networking'.[46] The implication of this work was the DoH had to readdress these issues if it was to hope to secure clinical support for its major programme of IT activities. Article 17 on the Security of Processing, the EU Directive[17] requires that:

> Member States shall provide that the controller must implement appropriate technical and organizational measures against accidental or unlawful destruction, or accidental loss, alteration, unauthorized disclosure or access, in particular where the processing involves transmission of data over a network, and against all other unlawful forms of processing.

Having regard to the state of the art and the cost of their implementation, such measures shall ensure a level of security appropriate to the risks represented by the processing and the nature of the data to be processed.

Since personal medical data are special category data requiring special treatment, and since the risk arising from the loss of, or malicious modification of, these data might include inappropriate or unsafe care, suitable encryption services are virtually mandated unless they are disproportionately costly. Curiously, the DoH only began to address these issues seriously in 1996.[47]

EU COLLABORATIVE PROJECTS

Following the clarification of some of the data protection and security issues within the UK, it was felt that it was worth exploring some of these issues in other European countries. This became possible in the context of the EU programme of Advanced Informatics in Medicine (AIM) because:

- of the collaboration with colleagues within the European Federation for Medical Informatics (EFMI), of which the BCS UK was a founder member;
- Barry Barber was successively Secretary, Vice Chairman and IMIA Vice President [Europe]; and
- these activities made it possible to find suitable colleagues from across Europe to participate in a project.

The first attempt at a European Data Protection Project was named HOPE but it failed to be selected for inclusion into the AIM programme. However, the interest shown by the European Parliament in these topics made it possible to help the Commission develop a valuable conference, addressing the issues in these pre-Directive days, which helped to explore the way forward.[48] Also, I was awarded a contract by the AIM programme, along with many other colleagues across Europe, to participate in the AIM Impact and Forecast project to explore the way forward for the programme. During this period the issue of patient safety surfaced and a subgroup of the project developed the six 'Safety First Principles of European Health Information Systems', which were accepted by the whole group and provided a basis for the future.[49-52] The principles required that systems should provide:

- a safe environment for patients and users;
- a secure environment for patients, users and others;
- a convenient environment for users;
- a legally satisfactory environment across Europe for users and suppliers;
- a legal protection of software products; and
- multilingual systems.

The issue of safety was a surprise to the group, although it was an obvious requirement when the issue was considered in any detail. Similarly, the possibility of multilingual systems becomes clear, although such arrangements are only just becoming available on DVDs. Both these issues have some way to go before they can be considered as having been achieved.

Following this work the SEISMED project (a Secure Environment for Information Systems in MEDicine) led to the development of a European High Level Security Policy and enabled some comparative work to be carried out using CRAMM in a continental environment. The project developed a series of security guidelines for managers, technical staff and users,[53-55] as well as producing a monograph on current issues,[56] and examining the then current data protection law in a number of European countries. The follow-on project ISHTAR (Implementing Secure Healthcare Telematics Applications in euRope) included a wider group of verification centres. It produced a White Paper examining the situation as well as exploring architectural issues in health information systems. Considerable effort was put into exploring the legal implications of the design, development, use and maintenance of health information systems, and an attempt was made to develop a usable incident monitoring system for recording and reporting on security failures in health information systems.[57] An ISHTAR website was developed to provide security information for users of health information systems, and this has been taken over by l'Association Internationale pour la Sécurité des Systèmes d'Information de Santé (ASSIS). A training course was developed and the paper security guidelines were developed into a set of electronic security guidelines, which were intended to be developed into a commercially available tool. The S&DP programme was also involved with the SYNAPSES and EUROMED-ETS projects. The Good European Health Record (GEHR), managed by St Bartholomew's Hospital Medical College, provided a useful window onto the issues of developing an electronic health record, paying particular attention to the legal and ethical issues. These various European projects, and of course many others, helped to explore UK health care information systems activities in the context of developments in the rest of Europe.

STANDARDS ACTIVITIES

After the first group of AIM projects had been completed it was clear that there was a need for a health informatics standardization body to take up the results of the work that was being generated. Negotiations led to the formation of TC 251 within the European standards body, CEN (see Chapter 1). Initially it had seven working groups (WG) with WG6 as the Data Protection, Security and Safety group, but a later reorganization changed the numbering to WGIII. This group has produced a number of health care security standards. The most obvious, and simplest, is the password standard ENV 12551. In addition, RSA was adopted as the algorithm for handling digital signatures and ENV 12924 was developed, using CRAMM, into a categorization of health information systems from a security point of view, together with a set of security measures for the six categories of system.

In due course, the International Organization for Standardization (ISO) established a health information standards group, ISO 215, and WG4 is the current security group. The UK counterpart of TC 251 was IST/35 and it maintained a Mirror Panel III to track and advise on standards developed within both the European and world contexts.

After retiring from the NHS, I was involved with the MEDSEC project,[58] which was part of the ISIS programme, as a subcontractor to Professor François Allaert's CENBIOTECH company. The project involved an examination of the standards available for the security of health information systems. It was possible to validate some of the standards and to provide a standards training course as an introduction to them. The validation included the SEISMED High Level Security Policy and the CEN standard ENV 129924. A draft was developed of a standard for secure medical databases and Bernd Blobel helped develop a standard guide for EDI (HL7) communications security, which was adopted as an HL7 informative document soon after the end of the two-year project.

Spreading the information

The first *MEDINFO* congress in 1974 included five papers on the topics of data protection and security.[59] The IMIA Working Group in this area, WG4, was one of the first groups to be formed initially by IFIP TC4 under the chairmanship of Professor Gerd Griesser, but then transferred as IMIA Working Group 4 after the establishment of IMIA in 1979 in Paris. This working group developed a series of books and papers arising from their conferences under the chairmanship of Gerd Griesser, David Kenny, myself, Ab Bakker, Jochen Moehr and François Allaert.[60-66] These conferences continue to provide a snapshot of the many issues that have preoccupied the medical informatics community over the years, in more detail than can be addressed by the sessions allocated at *MEDINFO* and *MIE* congresses.

Reflections

Since retiring from the NHS, many issues of confidentiality and security have been taken on board under the driving force of the National Programme for Information Technology. The control of our Personal Data that was enshrined in the European Data Protection Directive appears to have withered away under this onslaught. However, it must be noted that the events of the early twenty-first century and the associated terrorist legislation make it difficult to believe that anyone's health records are secret from prying eyes – except possibly from those of a clinician who has lost his or her access arrangements. Nationwide access and data sharing are now widespread across many sectors as Government policy in the guise of 'modernization'. This brings tremendous security risks that have hardly begun to be addressed and this was emphasized in a report on identity theft on an industrial scale from a Government department.[67] In any case, the US Patriot Act appears to give access to any information held by American organizations and their subsidiaries wherever these subsidiaries are geographically located, or whatever security and confidentiality agreements have been agreed to protect another country's national databases.

The key issue remaining from the programme is that of safety. It surfaced briefly in the AIM programme and arrived in the standards work programme. However, the IEC 61508 standard took some time to emerge. By then the work of the TC 251 working group (WG6/WGIII) had run out of funding and it was not possible to follow up the original plans to develop guidelines for the use of this standard in health care informatics systems. However, the problem has resurfaced more recently in the context of the level of 'adverse incidents' in health care. The report of some work in the USA[68] and Sir Brian Jarman's Harveian Lecture,[69] which caught the attention of the media, encouraged the *BMJ* to devote a whole issue to the subject.[70] The work looks set to continue,[71] and it is hoped that the health informatics community will work to develop standards to ensure that their systems do not introduce additional errors into the processes of delivering health care and will preferably help in reducing errors from other causes. IST/35 MPIII has made a start by working on the development of an initial standard, categorizing types of health information system according to the risks associated with its use. Another encouraging sign is the enthusiasm with which these matters were discussed at a conference of the Royal College of Physicians held in London on 21–22 October 2004.[72] Safety in health care is, at last, coming out of the closet and beginning to be addressed in a systematic fashion, even as so many issues of confidentiality are in regression.

REFERENCES

1   Council of Europe Convention (1981) *For the Protection of Individuals with Regard to Automatic Processing of Personal Data.* Convention. 108, January.

2   Council of Europe Recommendation (1981) *On Automated Medical Data Banks.* R(81)1, January.

3   Council of Europe Convention (1950) *For the Protection of Human Rights and Fundamental Freedoms.* November 1950 (subsequently developed by a variety of additional protocols and interpreted by various judgements of the European Court of Human Rights, which was established by the Convention).

4   Barber, B., Cohen, R.D. and Scholes, M. (date unknown) A Review of The London Hospital Computer Project. *MED INFORM*, 1, 1, 61–72.

5   Barber, B., Cohen, R.D., Kenny, D.J., Rowson, J. and Scholes, M. (1976) Some problems of confidentiality in medical computing. *Journal of Medical Ethics*, 2, 71–3.

6   Galloway, T.McL. (1963) (title unknown). *The Medical Officer*, 109, 232.

7   Saunders, J. and Snaith, A.H. (1967) (title unknown). *The Medical Officer*, 113, 209– 303.

8   Davies, C.K. (1970) Recall procedures – local authority applications. In Abrams, M. (ed.) *Medical Computing: Progress and Problems*, pp. 283–8. Chatto & Windus for BCS, London.

9   Caldicott, F. (1997) *Report on the Review of Patient-Identifiable Information*. DoH, London, December.

10 NHS Centre for Information Technology (1987) *Child Health System*. NHS Data Protection Handbook, chapter 7.

11 Child Health Computing Committee (1987) *Child Health Matters*. Child Health Computing Committee Conference Report. Welsh Health Common Services Agency, Cardiff. September.

12 South East Thames RHA (1993) *Access to Data: The Child Health System*. S.E. Thames, June.

13 DoH (2002) *Delivering 21st Century IT Support for the NHS*. DoH, London.

14 Younger, K. (chair.) (1972) *Report on Privacy*. Cmnd 50121972. The Stationery Office.

15 Lindop, N. (chair.) (1978) *Report of the Committee on Data Protection*. Cmnd 7341. HMSO, London.

16 White Paper (1975) *Computers and Privacy* and its supplement: *Computers: Safeguards for Privacy*. Cmnd 6353 & 6354. HMSO, London.

17 European Union (1995) *On the Protection of Individuals with Regard to the Processing of Personal Data and on the Free Movement of such Data*. Directive 95/46/EC. OJEC L281/31 – 50, 24 October.

18 Knox, E.G. (ed.) (1984) *The Confidentiality of Medical Records: The Principles and Practice of Protection in a Research-Dependent Environment*. EUR9471 EN. European Commission, Brussels.

19 Kenny, D.J. (1984) *The Protection and Maintenance of Confidentiality of Patient and Employee Data*. Confidentiality Working Party report for the NHS/DHSS Steering group on Health Services Information chaired by Edith Körner.

20 Cabinet Office (2002) *Privacy and Data Sharing: The Way Forward for Public Services*. Performance and Innovation Unit. April.

21 HMSO (2000) *Regulation of Investigatory Powers Act 2000*. The Stationery Office, London.

22 HMSO (2001) *Health & Social Care Act 2001.* The Stationery Office, London.

23 HMSO (2001) *Anti-Terrorism, Crime and Security Act 2001.* The Stationery Office, London.

24 Chalton, S., Gaskill, S. and Sterling, J.A.L. (eds) (updatable text) *Encyclopaedia of Data Protection.* Draft IPWG Code of Confidentiality of Personal Health Information. Sweet and Maxwell, London.

25 Council of Europe (1997) *Recommendation On the Protection of Medical Data.* R(97)5, February.

26 CCTA (1991) *Risk Analysis and Management Methodology (CRAMM) User Manual.* CCTA, IT Security & Privacy Group, 157–61 Millbank, London. (The software has been frequently updated and has now been made available in a commercial context from Insight Consulting at Walton on Thames.)

27 Barber, B., Vincent, R. and Scholes, M. (1992) Worst case scenarios: the legal consequences. In Richards, B. and MacOwan, H. (eds) *Current Perspectives in Healthcare Computing,* pp. 282–8. BJHC for BCS, Weybridge. ISBN 0 948198 12 5.

28 Barber, B. and Davey, J. (1992) The use of the CCTA risk analysis and management methodology (CRAMM) in health information systems. In Lun, K.C., Degoulet, P., Piemme, T.E. and Rienhoff, O. (eds) *MEDINFO 92,* Geneva, pp.1589–93. North-Holland, Amsterdam. ISBN 0 444896 68 6.

29 Barber, B. (1991) Towards a security policy. In Richards, B. and MacOwan, H. (eds) *Current Perspectives in Healthcare Computing,* pp. 345–51. BJHC for BCS, Weybridge. ISBN 0 948198 11 7.

30 Treacher, A., Barber, B. and Osborne, D. (1991) Training for health information security: an approach to cultural change. In Richards, B. and MacOwan, H. (eds) *Current Perspectives in Healthcare Computing,* pp. 80–7. BJHC for BCS, Weybridge. ISBN 0 948198 11 7.

31 NHS Executive Information Management Group (1996) *The NHS IM&T Security Manual.* DoH, London. February.

32 NHS Executive Information Management Group (1994) *Introduction to Data Protection in the NHS.* Birmingham, ref. E5127.

33 NHS Executive (1994) *Threat to the System: Information System Security and Data Protection.* Video. DoH.

34 NHS Executive (1995) *Chain Reaction: Information Security in General Practice.* Video. DoH, ref. E5212, May.

35 NHS Executive (1995) *Handbook of Information Security: Information Security for General Practice.* DoH, London.

36 NHS Executive (1995) *Play IT Safe: A Practical Guide to IT Security for Everyone Working in General Practice.* DoH, London, ref. E5209.

37 DoH (1994) *Confidentiality, Use and Disclosure of Personal Health Information: Draft Guidance for Consultation.* DoH, London.

38 Information Commissioner (2002) *Use and Disclosure of Health Data: Guidance on the Application of the Data Protection Act 1998.* May.

39 DoH (2003) *Confidentiality: NHS Code of Practice.* DoH, London, gateway ref. 1656, November.

40 BMA (1994) *A Bill Governing Use and Disclosure of Personal Health Information.* BMA, London. July.

41 BMA (1994) *An Explanatory Handbook of Guidance Governing Use and Disclosure of Personal Health Information.* BMA, London. July.

42 Fisher, F. (1995) Saga or soap opera? Drafting a Bill governing the use and disclosure of personal health information. In Barnett, D. (ed.) *Patient Privacy, Confidentiality and Data Security.* The INFOrmed Touch Series, vol. 3, pp. 10–18. BCS Nursing Specialist Group Conference.

43 Anderson, R.J. (1996) *Security in Clinical Information Systems.* BMA, London. January.

44 Anderson, R.J. (1996) Clinical system security: interim guidelines. *BMJ,* 312, 7023, 109–11. 13 January.

45 Anderson, R.J. (1996) NHS-wide networking and patient confidentiality. *BMJ,* 310, 6996, 5–6. 1 July.

46 Anderson, R.J. (1996) Patient confidentiality – at risk from NHS-wide networking. In Richards, B. and de Glanville, H. (eds) *Current Perspectives in Healthcare Computing,* pp. 687–92. BJHC for BCS, Weybridge.

47 NHS Executive (1996) *The Use of Encryption and Related Services with the NHSnet.* DoH, London, ref. E5254. April.

48 CEC AIM programme (1991) Data protection and confidentiality in health informatics: handling health data in Europe in the future. *Studies in Health Technology & Informatics,* vol. 1. IOS Press, Amsterdam.

49 Barber, B., Jensen, O.A., Lamberts, H., Roger, F., de Schouwer, P. and Zöllner, H. (1991) Six safety first principles of health information systems. In Richards, B. and MacOwan, H. (eds) *Current Perspectives in Healthcare Computing,* pp. 296–314. BJHC for BCS, Weybridge. ISBN 0 948198 11 7.

50 Barber, B., Jensen, O.A., Lamberts, H., Roger, F., de Schouwer, P. and Zöllner, H. (1990) The six safety first principles of health information systems – a programme of implementation: part 1 safety and security. In O'Moore, R.O., Bengtsson, S., Bryant, J. and Bryden, J.S. (eds) *MIE, Glasgow 90: Lecture Notes 40,* pp. 608–13. Springer Verlag, Berlin. ISBN 3 540529 36 5.

51 Barber, B., Jensen, O.A., Lamberts, H., Roger, F., de Schouwer, P. and Zöllner, H. (1990) The six safety first principles of health information systems – a programme of implementation: part 2 the environment,

convenience and legal issues. In O'Moore, R.O., Bengtsson, S., Bryant, J. and Bryden, J.S. (eds) *MIE, Glasgow 90: Lecture Notes 40*, pp. 614–19. Springer Verlag, Berlin. ISBN 3 540529 36 5.

52 Barber, B., Jensen, O.A., Lamberts, H., Roger, F., de Schouwer, P. and Zöllner, H. (1991) Report for the AIM secretariat of the European Commission. In Roger-France, F.H. and Santucci, G. (eds) *Perspectives of Information Processing in Medical Applications*, pp. 235–50. Springer Verlag, Berlin.

53 The SEISMED consortium (1996) Data security for health care. Vol. 1: Management guidelines. *Studies in Health Technology & Informatics*, vol. 31. IOS Press, Amsterdam. ISBN 9 051992 63 7.

54 The SEISMED consortium (1996) Data security for health care. Vol. 2: Technical guidelines. *Studies in Health Technology & Informatics*, vol. 32. IOS Press, Amsterdam.

55 The SEISMED consortium (1996) Data security for health care. Vol. 3: User guidelines. *Studies in Health Technology & Informatics*, vol. 33. IOS Press, Amsterdam.

56 Barber, B., Treacher, A. and Louwerse, C.P. (1996) Towards security in medical telematics. *Studies in Health Technology & Informatics*, vol. 27. The SEISMED consortium. IOS Press, Amsterdam.

57 The ISHTAR consortium (2001) Implementing secure healthcare telematics applications across Europe. *Studies in Health Technology & Informatics*, vol. 66, IOS Press, Amsterdam.

58 Allaert, F-A., Blobel, B., Louwerse, K. and Barber, B. (2002) Security standards for healthcare information systems: the MEDSEC consortium. *Studies in Health Technology & Informatics*, vol. 69. IOS Press, Amsterdam.

59 Anderson, J. and Forsythe, J.M. (eds) (1975) *MEDINFO 74*, Stockholm. North-Holland, Amsterdam. ISBN 0 444107 71 1.

60 Griesser, G.G., *et al.* (1980) *Data Protection in Health Information Systems: Considerations and Guidelines*. North-Holland, Amsterdam.

61 Griesser, G.G., *et al.* (1983) *Data Protection in Health Information Systems: Where Do We Stand?* North-Holland, Amsterdam.

62 Barber, B., *et al.* (1994) *Caring for Health Information: Safety, Security and Secrecy*. Elsevier, Amsterdam.

63 Bakker, A., *et al.* (1996) Common security solutions for communicating patient data. *Int. J. Biomed. Comp.*, 43, special issue.

64 Bakker, A., *et al.* (1998) Communicating health information in an insecure world. *Int. J. Biomed. Comp.*, 49, special issue.

65 Bakker, A. *et al.* (2000) Security of the distributed patient record. *Int .J. Biomed. Comp.*, 60, special issue.

66 Barber, B., Gritzalis, D., Louwerse, K. and Pincirol, F. (2004) Realizing security into the electronic health record. *Int. J. Biomed. Comp.*, 73, 3, 215–331.

67 The Independent on Sunday (2006) *Revealed: The cash-for-fake-ID scandal at the heart of the Government.* 14 May, p. 5.

68 Kohn, L.T., Corrigan, J.M. and Donaldson, M.S. (1999) *To Err is Human: Building A Safer Health System.* National Academy Press, Washington, DC.

69 Jarman, B. (2000) The quality of care in hospitals. *Journal Royal College of Physicians*, 34, 1, 75–91.

70 British Medical Journal (2000) Reducing error and improving safety. *BMJ*, 7237, 18 March.

71 Shaw, R. (2004) Patient safety: the need for an open and fair culture. *Clinical Medicine*, March/April, 128–31.

72 Royal College of Physicians (2004) *Making Health Care Safer 2004.* Conference at the Royal College of Physicians, supported by the BMA.

26 Computer systems evaluation (1969–1980)

ALARIC CUNDY

In these early years of the twenty-first century we take the widespread use of computers in the NHS for granted – indeed without them the present day NHS just could not function. Through the National Programme for IT (NPfIT) we are on the verge of the largest investment in computing that the NHS has ever seen.[1]

The contrast with the early days of NHS computing is stark. In the1960s, there were very few computers in use within the NHS, and those that were in place were mainly used for commercial applications, such as payroll and accounting.[2]

During the 1960s, however, initial developments were taking place with computers to support immunization recall services, though at that time Local Authorities operated such services, rather than the NHS. Davies reported on early developments in this area in West Sussex,[3] and the paper also included what perhaps amounted to the first successful computer evaluation study for health care services in England. It showed the impact on immunization rates achieved as a result of the introduction of the computer system, leading to wider implementation.

In the late 1960s, the DoH established and financed the Experimental Computer Programme.[4] This programme was aimed at:

- discovering whether, how and where computers can help to:
  - improve patient care;
  - improve clinical and administrative efficiency;
  - provide facilities for management and research.
- giving some NHS staff practical experience with computers.

In a major review published in 1972, the Experimental Computer Programme was refocused to follow a simple three-phase framework, which was referred to as the EDI process:

- Experiment – to determine whether an application is useful, desirable, and feasible.
- Development – in which applications that promise good cost/benefit returns are developed into fully operational model systems.
- Implementation – in which developed model systems are put into widespread operation.[4]

Through this programme, the DoH funded 14 experimental uses of computers to support patient administration. The 'experiments' needed to be evaluated in order to decide which application areas and approaches were successful, and hence should be 'developed' for wider 'implementation' throughout the NHS. Chapter 5 covers this in more detail.

OVERVIEW OF EVALUATION WORK

The evaluation of the projects within the Experimental Computer Programme was viewed as an essential component of the overall programme. For example, in 1977, it was reported that collectively at 13 of the experimental projects,[5] a total of 35 staff were engaged full time on computer evaluation activity funded by the DoH, and they were coordinated and advised by a central team within its Computers and Research Division. This activity eventually spanned a period of 12 years, starting in 1969.[6]

In the 1972 review,[4] it was concluded that assessment criteria should be of one of three types:

- Resource allocation – helping to decide how the funds available to the NHS should be directed. It was noted that the impact of projects within the Experimental Computer Programme would be minimal and indirect on this issue.

- Resource utilization – by carrying out some routine tasks, the computer might free up the time of doctors, nurses or others, so that more patients could be seen in the same time, or the same number of patients in less time.

- Quality of care – helping doctors, in particular, to deliver better patient care.

There are three identifiable 'eras' of this evaluation work, which, though there was some overlap of timing, in retrospect could be described as:

- Initial phase – innovation and exploration.

- Improvement objectives – first attempt at a nationwide structured approach.

- Performance criteria – second attempt at a nationwide structured approach.

During the initial phase, each evaluation team was encouraged to develop and report on its own evaluation methodologies. For example, at The London Hospital, a major review of plans and achievements to date was published by Barber,[7] and later detailed in an internal report issued in November 1975,[8] followed by a discussion of future plans and prospects in February 1976.[9]

DoH staff carried out the 1972 review,[4] clarifying their interest in progressing systems through the experimental and developmental phases to that of widespread implementation, using an unspecified economic evaluation process to determine the progression of particular computer applications. The work completed up to then did not provide evidence that identified candidates for 'development' and 'implementation'. The DoH introduced a new structured approach based on the concept of 'improvement objectives', as described by Sharpe in 1974.[10] During 1975/6 attempts were made to summarize all the material available from the partly implemented projects within this evaluation format, but it became clear that the material could not readily be put together, and that it did not add up to a coherent and usable picture of the various projects.

An Evaluation Working Group, which was set up under the chairmanship of Professor R.D. Cohen to review the situation, reached two key conclusions:

- Firstly that 'before and after' evaluation did not provide reliable information to other potential system users because any improvement found at the original site was in part due to the efficiency or otherwise before implementation.

- Secondly, the concept of a 'shop window' was developed. It was thought that the material to be made available to other potential users should comprise a description of the system and a set of performance information that described the achievements at the originating site.[11]

Thus, the DoH adopted an alternative structured approach – based on the measurement of a set of 'performance criteria'.[6, 12]

This approach was applied to 16 computer sites and five manual sites within England. With the completion of the review of this work early in 1981,[13] the DoH funding for the computer evaluation work ceased, and the level of documented activity in the field in England reduced considerably.

Of course, throughout this period, evaluation of the use of computers to support health care was being explored throughout the world, and these works have been summarized by Barber and Cundy.[14] Particularly noteworthy was the work of Norwood, who developed the concept of 'benefits realization' at the El Camino Hospital in the USA.[15] This terminology, even if not some of the associated actions, has carried through to the present day, and most modern 'business cases' will include proposals about 'benefits realization'.

Throughout the rest of this chapter, the successes and disappointments of these three 'eras' have been examined in more detail, and the links forward into more recent developments have then been analysed. Other chapters in the book also cover some of the topics detailed here.

THE INITIAL PHASE (1969–1974)

An evaluation team was set up at each of the projects included within the Experimental Computer Programme. Their general terms of reference were recorded as:

... to identify, measure and, wherever possible, value the differences between the manual system and the computer-based system for a particular health care function (i.e. a computer application) and to assist in the comparison of the value of the benefits with the additional costs from using computer equipment.

REFERENCE 10

Initially, no standard approach for the evaluation work was set out, and this situation left the way open for each project to develop its own methods. It was noted by Barber, Abbott and Cundy that hitherto no serious attempt had ever been made to evaluate the performance of any health care system, and thus in parallel with the pioneering development of the computer projects, the development of the measurement methods was also pioneering.[16]

In the USA, Fanshel and Bush developed a Health Status Index in the 1960s.[17] This idea had potential applications in the arena of computer system evaluation. However, there is no evidence that this development was pursued in the context of the DoH Experimental Computer Programme.

Rosser and Watts developed some important ideas concerning the measurement of the effectiveness of health services in the 1970s.[18] They sought to address the question: 'to what extent is the health of the individual improved by the activity of the health service?' The work focused on attempting to measure the level of sickness, and the value of improvements brought about through exposure to health care services. Rachel Rosser and Tim Benson made reference to the application of these ideas to the evaluation of the Experimental Computer Programme,[19] though there is no clear evidence that the ideas were adopted beyond Charing Cross Hospital.

Rosser and Benson used a scaled classification of states of illness, originally developed by Rosser and Watts. This defined two dimensions of illness: disability, which was assessed on an 8-point scale, ranging from 1 = no disability, to 8 = unconscious; and distress, which was graded between 1 = no distress and 4 = severe. They developed a relative valuation of each feasible combination of disability and distress, derived through psychometric methods. Then, for example, each patient could be assigned a valuation at the start of their period of hospitalization, at points through their stay, and at the end, and hence a numeric assessment of the health gain could be achieved. By comparing scores for similar patients passing through different treatment regimes, for example with or without computer support, an estimate of the impact of the computer could be deduced.

It may be the case that this idea did not 'catch on' at the time because it was 'ahead of its time'. Certainly in the 1970s, there was not a recognized method of 'grouping' similarly profiled patients, such as the Healthcare Resource Groups (HRGs) that have been developed during the 1990s.[20] The American-originated scheme of Diagnostic Related Groups (DRGs) had been deemed to be inappropriate for use in England. (See also Chapter 6.)

Barber, Abbott and Cundy recorded that the main requirement to support the evaluation of the computer projects was the accurate perception of the objectives, and a clear link between them and the measurements undertaken.[16] This requirement involved some reconciliation of those participating in the programme, and thus one valid framework for evaluation could be summarized as:

- Project evaluation – an assessment of the key decisions regarding applications selected, overall approach adopted, and project management of each project. Could non-technical staff actually use these systems?

- Technical evaluation – an assessment of the detailed technical approach adopted.

- Application evaluation – an assessment of particular application subsystems for subsequent widespread adoption.

- Programme evaluation – an assessment of the programme as a whole.

These ideas were extended further in a subsequent evaluation of evaluation activity to include:

- Technical evaluation – multi-project.

- Systems evaluation – multi-project.

- Policy evaluation – multi-project.[14]

In practice, most attention was focused on Application evaluation, though an outline of the Project evaluation at The London Hospital was published.[21] The key headings for the evaluation of the programme as a whole were considered to be:

- the implementation of prototype operational real-time systems – it had been found that such systems were acceptable and that they were used in a multidisciplinary environment;

- computing expertise – each project had developed a nucleus of technical expertise in advanced computing systems, from which later systems could be developed;

- education of health service staff – the systems were capable of being used on a day-to-day basis by non-specialist staff. Not only that but, through the design and evaluation studies, such staff had also become accustomed to having their practices scrutinized by system developers and researchers;

- emphasis on the need for evaluation – the work associated with the evaluation of computer systems had opened opportunities for wider exposure of health care systems to evaluation and scrutiny;

- examination of system costs – the costs of the prototype systems would not necessarily carry over into wider implementation. Large-scale replication of the prototype systems would either require a high degree of standardization, or the availability of development environments in which individual tailoring of systems would be straightforward and rapid;

- examination of system benefits – since most of the early systems supported administrative processes, rather than providing direct clinical support, it could not be expected to find key clinical benefits from them. The types of benefits actually recorded included:

  - improved access to information – available faster and to more people;
  - improved completeness, accuracy, and legibility of information;
  - reduced transcription;
  - reduced clerical effort.[16]

Numerous detailed studies have been reported to explore the benefits in more detail. The comments below provide an overview and references to more detailed published material.

The level of use of the system

A significant measure of the success of a system is its level of use. If it is not used, it cannot be judged to be a success. Systems in the 1970s were relatively primitive in terms of the availability of audit trails, and the journal records required careful interpretation before they could be used to measure the use of the system. Cundy and Goldstein reported on measurements of the use of the system at The London Hospital at a point in time when the system was partly developed and implemented.[22]

A key pointer was the ratio of 'inspect' transactions to 'update' transactions: one of the principles of the system was that data would be entered once and then used many times. The absolute level of use was also a factor of interest. These studies showed that the system was well used, but they also showed the limitations of the long planned downtimes overnight every night, to enable housekeeping, batch processing, preventative maintenance and the development of new modules. Initially, the system was only available to users each day from 8 am to midnight, with the level of use peaking between 11 am and noon and 2 pm to 4 pm. In 1972, a second computer was purchased at The London Hospital in order to allow new developments to continue in parallel with the routine operation of the early modules on the main machine.

Simulation models

One of the issues facing computer system evaluators was that it took a while to develop and implement particular modules, and sometimes the evaluation studies themselves could take a significant time. Thus it could be the case that a significant time gap existed between the 'before' and 'after' measurements, and it was not necessarily straightforward to conclude that any observed benefit was solely due to the implementation of the computer system or a result of service changes in the meantime.

One solution to this issue was to seek to model the situation under investigation. For example, Cundy developed a simulation model of the administrative processes that supported the processing of laboratory requests and reports in the manual setting. The volume of activity in the laboratories was growing steadily year-on-year.[23] Using the simulation model it was possible to deduce the resources that would have been required – people and equipment – to continue to operate the manual system at the workload volumes later passing through the computer system. The study showed that, though the manual system was viable just before the introduction of the computerized requesting and reporting system, it was in fact on the point of collapse and could only have continued to operate with the injection of resources. Indeed, it later became self-evident that no Head of Biochemistry in particular could contemplate running an effective department without computer support.

System response

If the response times experienced by end-users becomes unacceptable, then a system will not be used, and hence no benefits will ensue.

At The London Hospital, this area was explored in two ways:

- Measurement of responses experienced by end-users, using measuring devices provided by the hospital's Physics department.
- A simulation model of the system itself.[23]

The simulation model was designed to help predict and prevent problems with response times as the use of the system increased. It also provided a means of apportioning use between different applications, which in turn helped to apportion costs between the various applications.

Patient states

Michael Fairey first promoted the use of industrial process control ideas to the passage of a patient through his/her period of care as an in-patient in a hospital (see Figure 3.5).[24] He hypothesized that a patient's journey from admission to discharge required the circumnavigation of a number of barriers – often waiting for something to happen, such as waiting for a slot in theatres or waiting for the results of an X-ray or a laboratory result to decide on the next appropriate care step. This concept became known as the 'Fairey Tube' (see Chapter 3).

Barber developed these early ideas into a computer evaluation methodology called 'Patient States'.[7] A series of 'states' to which a patient could be assigned at any specific moment in time were identified, including, as examples, the categories listed below:

- awaiting
  - results
  - decision
  - treatment
  - discharge

- under observation
- having treatment
- having terminal care.

A considerable amount of effort was expended during the lifetime of The London Hospital Project in trying to get estimates of the delays to patients through their in-patient stay. The final method adopted involved the house officers categorizing their patients at predetermined random times. The reported results showed that 45 per cent of patients were awaiting some form of activity.

Attitude surveys

Users' perceptions offer an important measure of the usefulness of any system – computer or manual – and this technique was used extensively within the Experimental Computer Programme, for example, at The London Hospital.[7]

The surveys developed were tested through pilot studies to confirm that the terminology was appropriate to the staff group under observation. The questions were designed to complement other forms of evaluation measure, and so they were grouped under headings such as: admissions office, A&E department, beds, staffing, theatres, clinical laboratories, discharge, after care and computers.

Desired sample sizes for each staff group were calculated so that changes in staff attitudes through the various phases of implementation could be detected. In practice, the published results were restricted to comments about the overall acceptability of the computer system; those comments were favourable.

Request to report intervals – 'transit times'

The interval between the lodging of a request for a laboratory result and the return of the report to the originating doctor was seen as an important measure of the efficiency of the system. The delayed return of a report could affect crucial decisions about a patient's care, and if the report was unduly delayed, it could result in unnecessary repeat requests.[7] The measurement of the 'transit time' for a computerized system was trivial, as the computer logged all the various stages in the processing of a request. The equivalent method of measuring 'request to report' intervals in the prior manual system would be labour intensive, and potentially could affect the actual time being measured, since all of the requests being tracked would have to be marked in some way, and hence laboratory staff would be alerted.

Some physical aspects of the process could not be speeded up by computerization, for example, even in the twenty-first century it is not possible to send blood samples for testing through the Local Area Network. Nevertheless, the computer was able to sort and schedule the samples to be collected for the benefit of the laboratory-based phlebotomists.

At The London Hospital, greater sophistication of measurement was developed using a Markov chain model of the laboratory requesting and reporting process (see Chapter 3), prompted by earlier work by Farrow, Fisher and Johnson that sought to analyse pathways of patients with renal failure (see Chapter 13).[25, 26] This laboratory model could be used to explore the effect on the end-to-end transit time of improving certain elements of the overall process.

The published results show that at The London Hospital the average transit times for common biochemistry and haematology reports were 2.27 days and 1.32 days, respectively, before the introduction of the computer system, and that savings of 35–45 per cent were achieved using the computer system.[22]

It was less easy to assess whether or not the computer had influenced the number of repeat requests. Cundy demonstrated that the physical number of request forms passing through the laboratory depended in part of the actual grouping of tests into individual requests.[23] Inevitably, the introduction of a newly designed computer system offered the scope to reassess the standard groupings, and hence in effect invalidated 'before and after' comparison of laboratory workloads.

Activity sampling

Activity sampling is a standard work measurement technique that was used to good effect in the field of computer evaluation, especially for ward staffing.[7] The method provided estimates of the amount of time spent by different staff groups on specific activities, and the concept was quite simple. The work area of interest was observed for a period of time, and a list of all the different types of activities was drawn up. The validity of that list could be verified with the subjects of the survey. Then, during the actual period of recording, an observer would shadow a subject for a period of time, typically a whole week, or complete work-cycle if different. At predetermined random times, the observer would note which of the listed activities the subject was engaged upon at that particular instant. The number of recording points could be established beforehand so as to ensure that the resulting statistical confidence intervals around the derived estimates were reasonably narrow.

By applying this technique using the same activity categories at various stages before, during and after the implementation of the computer system, the way the computer had changed the deployment of the subject staff groups could be assessed. The hope was that it would show that staff spent more time engaged on patient-care-related activities, and less on administrative activities.

Significant success was reported with this technique at The London Hospital, where it was used to measure the impact on nurses.[27] The time saving attributed to the introduction of the computer system was equivalent to 15 whole-time nurses, though this time was distributed around all of the hospital's wards. In practice, the study showed an increase in patient-related time, and a reduction in the time spent on activities not associated with patient care, such as administration. It was not reported that there had been an actual reduction in payroll costs.

This particular idea had been pursued rigorously by Norwood at El Camino Hospital in the USA, where a similar impact on staff time was noted, particularly for ward nurses.[15] The average time spent on clerical aspects amounted to 15–25 per cent of the total. Here, the term 'benefits realization' was used to describe those circumstances where management action was required to convert potential benefits into actual benefits, for example, reduced payroll costs. A total of 31.3 whole-time equivalents of staff savings were 'realized', about two thirds of which were savings from the nursing payroll.

Completeness, accuracy and legibility

Key benefits associated with the introduction of a computer system are gains made in terms of completeness, accuracy and legibility of records. Computer systems introduce a discipline that encourages complete recording, and certain data items may automatically be 'copied' from one subsystem to another. For example, the basic identification details recorded at the time of admission can be reused on laboratory requests. One of the key aims of administrative computer systems within the NHS is to reduce the number of occasions on which essentially the same information is captured, or copied from one document to another. Every act of 'transcription' introduces the scope for errors. Handwriting can be notoriously difficult to read and key instructions or records may be misinterpreted.

As an example, studies carried out at The London Hospital showed high occurrences of missing identification details on manual laboratory requests, with, for example, over 15 per cent missing the case notes number.[7]

Overall hospital efficiency

There could be considerable added value to the evaluation results if an impact on major measures of hospital performance could be demonstrated. For example, if it could be shown that improved administrative processes resulted in a reduced overall length of stay, then it could have been argued that the same patient caseload could be treated in fewer beds, with fewer staff, and hence leading to an overall savings in costs. On the other hand, the same number of beds and support staff could have been used to treat a greater number of patients, probably at a higher total cost, but a lower unit cost. Norwood argued that such a link could not be proved.[15] Work at The London Hospital sought, but failed to find, such a link.

But in the 1970s, inflation was high and there were a number of changes within the NHS – including the major administrative reorganization effected in April 1974.[28] Against a background of generally falling lengths of stay and increasing costs, there is no evidence available that any evaluation team within the UK actually carried these ideas through to a convincing end-point.

Management information

Now, in this first decade of the twenty-first century, the concept of 'management information' and its contribution to both operational and strategic planning and day-to-day tactical operations is assumed, but in the 1970s it was virtually an unknown concept. There is some evidence of early attempts to improve management processes,[24] but one of the great expectations of the introduction of computerized administration systems was that they would provide a wealth of knowledge as to how the hospital was actually functioning, and thus automatically they would lead to improved management efficiency – one of the previously identified key objectives of the Experimental Computer Programme.[4] Barber discussed the potential of these concepts,[29] but arguably the ideas were at least 10 years ahead of their time. They did not begin to gain a hold until after the major review of NHS data collection undertaken by the Steering Group on Health Services Information in the 1980s – often referred to as the Körner committee.[30]

Review of laboratory systems

Tim Benson and Eric Thwaites recorded an appraisal of hospital laboratory computing across several hospitals, drawing out the impact on:

- entry of request details;
- request by clinician;
- specimen collection;
- laboratory reception;
- production of worksheets;
- laboratory records.[31]

Their paper included an example of a seven-year discounted cash flow (DCF) calculation, which followed the emerging DoH guidance. They concluded that, assuming a DCF rate of return of 20 per cent, a typical laboratory system would have a 'net present value' of around £2,000 for an initial outlay of £60,000 – that is there was a modest return on investment in parallel with the achievement of benefits against the headings listed above.

Drawing the programme results together

In the early 1970s, each of the projects within the Experimental Computer Programme was required to provide a report and a presentation to the DoH Chief Scientist, including an overall description of the progress with the project and a summary of the evaluation results. The DoH drew these experiences together into a major review report, which:

- examined the investment to date;
- established an inventory of working systems, and those planned or under development;
- provided an assessment of the benefits that had hitherto been identified by the evaluation projects.[5]

One of the major recommendations from this review was that a new and structured approach to computer evaluation was to be adopted by all of the projects within the remit of the Experimental Computer Programme.

THE IMPROVEMENT OF OBJECTIVES PHASE (1974–1977)

Sharpe had set out the framework for the new approach, which was to be based on Improvement objectives.[10] Each project was viewed as a set of individual applications, or groups of applications. For each such entity the local project team was expected to draw up a list of expected 'improvement objectives'. For each objective there would be an associated tangible measurement that could be used to gauge the impact of the introduction of the computer application.

Application-level improvement objectives were categorized under three main headings:

- Resource allocation
- Resource utilization
- Quality of health care.

Some published examples of such improvement objectives are:

- Resource allocation:

  - No examples identified (not unexpected).

- Resource utilization:

  - To reduce the amount of time spent by clinicians on admission procedures.
  - To reduce the amount of time spent by nurses on admission procedures.
  - To reduce the amount of time spent on locating in-patients.
  - To reduce the amount of time spent on document production.[10]

- Quality of health care:

  - To shorten the length of time taken to register emergency admissions.
  - To reduce the numbers of errors and omissions in checking registration at time of admission.
  - To reduce the number of errors in identifying and locating patients.
  - To reduce the number of errors in admission procedure documentation.[10]

Alongside the development of improvement objectives and their associated measurements, attention was devoted to establishing a standard methodology for valuing recorded improvements. Three key points were:

- The main focus of attention was on individual applications.[10] It was recognized that some improvements resulted from the complementary effects of more than one application, but in such cases attempts were to be made to apportion the value of the benefits to individual applications.

- Any additional costs associated with the operation of the computer system were to be documented.[10]

- Costs and benefits were to be calculated on a discounted cash flow basis for individual applications, and amortized over a seven-year period.[22]

It would be wrong to imply that there was a sudden conversion to the new 'improvement objectives' approach to computer evaluation. Many of the innovative approaches already in progress would have been completely wasted had they suddenly been terminated. Some observers viewed the new approach to be unnecessarily narrowly focused.[27] Nevertheless, the improvement objective approach increased in importance, and each project was expected to develop an appropriate framework and to report progress to the DoH on a regular basis.

The whole idea of the Experimental Computer Programme was that a number of 'applications' would be developed in a number of different experimental settings, and the success or otherwise would be judged. Those that were deemed to be successful would be developed and then implemented more widely: the 'Experiment, Develop, Implement' (EDI) approach.[4]

The idea was that wherever possible appropriate measures related to the improvement objectives would be taken before and after the implementation of the computer module. Hence the degree of 'improvement' could be assessed, and, wherever possible, valued in monetary terms. Successful applications could thus be distinguished from less profitable ones, and then further developed for implementation throughout the NHS.

There were some obvious problems associated with this approach, which Cundy and Goldstein listed as:

- In some cases the application had already been implemented before the 'improvement objective' approach was introduced.

- Applications took a while to program and implement, and thus the time gap between the 'before' and 'after' measurements could be substantial. It could not necessarily be deduced that any improvement was due solely to the introduction of the computer system. Indeed, the positive effects of the computer system may have been *dampened* by the passage of time.

- The improvement objective and EDI concepts would have had more validity had the Experimental Computer Programme been established as a proper, designed, statistical experiment. In practice there was no overall design to the programme, and hence there could be no allowance for random effects.

- The amount of measured improvement in part depended on the efficiency of the prior manual system.

- The system analysis and the associated process design that preceded the design and implementation of the new computer system in themselves brought about improvements, irrespective of the merits of the computer system. Further system and process scrutiny through the evaluation activity compounded this effect.

- There was temptation to make the lists of 'improvements' look as impressive (i.e. as long) as possible, and thus they included fine details such as the saving of floor space through the elimination of the need for filing cabinets, as well as more substantive measures, such as turnover and length of stay. The minute detail tended to be easier to assess than the higher level measures.[22]

Despite these difficulties, attempts were made to value the benefits and assess the costs of computer systems in accordance with the DoH guidance. Cundy and Goldstein,[22] and Cundy,[32] reported a comparison of projected costs against minimum and maximum predictions of the value of benefits over a seven-year period. The variability in the forward projections was in part due to a distinction being drawn between financial and non-financial benefits. In practice, the projected costs almost exactly bisected the maximum and minimum financial value of projected benefits.

It came as no great surprise that, when the results of the improvement objectives era were drawn together in a major review,[4] no major firm conclusions about the success of any of the elements of the Experimental Computer Programme could be drawn out. A working group considered the review findings, and another new standard approach was proposed.[11]

THE PERFORMANCE CRITERIA ERA (1977–1980)

The 'Performance criteria' era of computer evaluation can be viewed as comprising four phases:

- Development of the theory.
- Development of lists of performance criteria.
- The measurement of the criteria in appropriate settings, including in 'non-computer' institutions.
- Overall reporting and review.

Development of the theory

The development stage was led by Professor R.D. Cohen, a consultant physician at The London Hospital. A seminar was held in May 1977 at which the Evaluation Working Group's review was considered,[11] and the 'Performance criteria' approach to computer evaluation was commissioned.

Previously, it had been inferred that, because a successful computer project ought to improve the working of the hospital or health centre, measurements taken before and after the introduction of the computer system should yield a predictive statement about the improvements that could be achieved elsewhere by the installation of a computer system. The attempts to use this approach had met with the following difficulties:[11, 12]

- During the considerable time required to develop computer systems, many other causes of change operate. It is impossible to isolate changes caused solely by the computer, except in matters that are only a means to an end, such as the timely and accurate presentation of information.

- Even if the effect of a computer in one hospital could be precisely identified, there are good reasons to doubt any simple prediction of the same results elsewhere. No two hospitals are alike. Improvements achieved in the first hospital with the help of a computer may have been achieved already in the second, before a computer has been considered.

- The improvement objectives were lists of possible improvements drawn up logically and chronologically after the prior decision to implement a computer system. They were rather long lists of desirable improvements, which were unlikely to match the priorities for change that might have been derived from a prior analysis of any major problems facing the hospital or health centre before the computer project was proposed. Nor was there any certainty that the particular improvement objectives were of interest outside the hospital to other prospective users of the system.

It was argued that the key issue of interest was 'the level of performance that could be achieved by a hospital or health centre with the support of a computer system'.[12]

Development of lists of performance criteria

A project to develop an appropriate set of performance criteria was established, under the leadership of Brian Molteno, Regional Management Services Officer for the Trent Regional Health Authority.[13]

The Evaluation Working Group Report gave the project team the following objective:

> *To determine performance criteria for appropriate areas of activity in the NHS where computers have been used or are planned. Performance Criteria are salient pieces of information which indicate the efficiency of areas of NHS activity, and which can be measured to help in making decisions on the effectiveness of particular systems (computer-based or not).*

REFERENCE 11

This overriding objective was further clarified in the detailed work of the project team, as follows:

- *The project objective was to propose an aid to help clients decide whether a computer system was worth having in a hospital, clinic, or health centre, and to help the DHSS make judgements about the experimental computer programme in those fields.*

- *Credibility was vital for the project. Success would be measured by the extent to which clients judged that the criteria identified were useful in the decision-making that existed. Clients could be encouraged to strive towards objectivity, but it could not be expected to achieve more than to bring people up to the best level that already existed in the client's world.*

- *The criteria were to measure activities, not computer systems. However, the work was focused on activities that had been served by the experimental computer programme somewhere.*

- *The pilot study needed to demonstrate to what extent clients thought that useful criteria could be produced, and then to extend the study as appropriate.*

REFERENCE 33

The scope of the project was determined by the scope of the then current projects within the Experimental Computer Programme,[6, 33] and nine topic areas were identified:

- in-patient administration
- outpatient administration
- waiting list management
- medical records
- nursing records
- pathology services
- pharmacy services
- X-ray services
- health centre records.

A decision was made that the lists of performance criteria should be based on consultation with 'key decision makers' (i.e. senior managers) and with 'users' (i.e. doctors and nurses). A number of NHS institutions were approached and they agreed to provide a sample of appropriate individuals to participate in the project, covering large and small hospitals, health centres, teaching hospitals, and institutions with and without computers.

The project was conducted in two phases:

1 Structured, but open, interview with a randomly selected group of 'decision makers'. The object of this stage was to generate lists of potential candidate performance criteria, but without any attempt to rank them in any way.

2 Market research techniques were used to ask a large number of participants to rank the criteria within an individual topic area, and by type of respondent (hospital doctor, GP, nurse, etc.).[6]

The study showed that agreement within respondent groups was greater than agreement between respondent groups. That is doctors tended to agree with each other, but disagree with administrators on the relative importance of suggested performance criteria as a useful measure of efficiency. Potentially, this observation led to difficulties in combining the lists into something manageable.

Within each topic area, the criteria were placed into 'bands' depending on the highest ranking ascribed by any one respondent group. Thus a band-one criterion was one that was judged to be the most useful by any one or more respondent groups, and band two comprised those that were ranked second most important by any respondent group, so long as they were not already in band one.

It had originally been hoped/expected that a clear discrimination between 'useful' and 'not useful' potential criteria would be established. This outcome proved not to be the case, and thus the project yielded nine lengthy lists of criteria, which were individually ranked in decreasing order of importance. The project team thus commended all of the criteria developed through the project, but with a recommendation that measurement should be focused on those appearing in higher bands.

This project was written up and published in a handbook by the DoH.[34, 35] As an illustrative example, the band-one performance criteria for in-patient administration were listed as:

- Bed usage: bed usage measured by one or more of the following:

  - turnover interval (average time a bed remains empty)
  - occupancy
  - patient throughput (discharges and deaths per available bed per year)
  - average length of stay.

- Timeliness of discharge letters: average delay between discharge and the discharge letter being despatched to the GP.

- Patient satisfaction: patients' and relatives' views concerning in-patient procedures.

- Missing case notes and X-rays.

The measurement of the criteria

The handbook also included recommended standard methods of measurement for each of the criteria,[34] and it was mandated for use within the projects comprising the Experimental Computer Programme, and a small selection of 'non-computer' hospitals.

There is very little published material relating specifically to the measurement of the performance criteria at individual hospitals and health centres, although there is evidence that internal working papers from each of the projects contributed to a major exercise that sought to bring all the results together into a coordinated report.[36]

Cundy and Nock reported on the application of the performance criteria technique to the evaluation of X-ray services at one hospital.[37] In discussion they observed problems in interpretation of the performance criteria because the lists for each of the nine topic areas were so extensive. They further proposed that the individual criteria could be grouped to contribute to answers to five key questions:

- Speed – how quickly does the service respond?
- Quality – how accurately does the service respond?
- Workload – how much workload can the service process for this level of response?
- Resource – how much resource (staff, equipment) does the service need to offer this level of service for this level of workload?
- Opinions – how acceptable is the level of response?

The complete set of hospitals and health centres at which the performance criteria were measured were:

- Computer supported:
  - Sunderland
  - Leeds
  - Nottingham
  - Cambridge
  - Charing Cross
  - Southend
  - The London
  - University College Hospital
  - St Thomas'
  - Oxford
  - Exeter
  - Birmingham
  - Stoke
  - Liverpool
  - Lancaster
  - Salford
- Non-computer:
  - Sheffield
  - Chelmsford
  - Basingstoke
  - Bath
  - Winchester
  - Oldham.[36]

Overall reporting and review

The results of the measurements undertaken at both the manual and computer-supported hospitals were drawn together in the final phase of the project by a team based within North East Thames RHA Management Services Division, and they were presented to the Evaluation Working Group.[11] The detailed results were published in a series of reports.[36, 38, 39]

The summary report acknowledged that the results were generally disappointing in that they did not show 'good news' with respect to the benefit and value of the use of computers in the NHS.[13] Obvious first-order results were demonstrable, for example improvements in legibility and reduction in transcription/copying effort, with consequential reduced errors. However, it was not possible to show any significant advantage in terms of key issues, such as bed or theatre usage, the management of waiting lists, or the availability of case notes at the point of doctor–patient contact. The more detailed supporting documents suggested that the lack of strong, positive conclusions was in part due to the relatively small number of data points for any one criterion, and that the results were additionally impacted by the relatively short timescales for the application of the performance criteria method.[36]

A small-scale study into the effects of computers on nursing record systems did, however, show stronger advantages of computerization.[38] A key effect was the ability of the computer to carry forward nursing orders automatically. A similar small-scale study into the effects of computers on primary care systems also yielded generally more encouraging results.[39] In particular, benefits were found in terms of information retrieval and information communicated to outsiders.

The performance criteria approach certainly had advantages compared with the improvement objective approach – it addressed most of the problems listed previously – although it also introduced its own difficulties. For example, it was noted that three of the four leading measures of the 'In-patients' application area concerned GP–hospital communications – yet none of the then computer experiments in that application area attempted to support such processes.[13] Thus it subsequently came as little surprise that the then in-patient applications could not be judged as major successes in terms of the standard performance criteria. Cundy and Goldstein subsequently noted that, if the object of the Experimental Computer Programme was to identify applications that could be implemented widely and successfully throughout the NHS, then it would have been an idea to develop the list of performance criteria before the earliest applications were even scoped.[22] Hindsight is a virtue.

MORE MODERN TIMES

In the early 1980s, the general conclusions that could be drawn from the attempts at evaluating the projects within the Experimental Computer Programme were:

- It had been shown that systems could be developed for routine use by non-technical staff. Working systems had been developed in a small number of locations.

- Generally, those systems were widely used and their users liked them.

- Some clear benefits had been identified, though generally the monetary value of them was dwarfed by the costs of the original developments and the ongoing maintenance.

- Costs were high, though in real terms they were falling, and that might tip the economic balance in favour of the systems within the near future.[14]

A purist economist would have been excused for concluding that the Experimental Computer Programme had been a monetary failure. Yet, less than five years later, the Körner review was published, in which sets of minimum data-sets were identified.[30] These sets described the minimum information that a competent District Management Team needed to manage its affairs effectively and efficiently. Already it was evident that the capture and processing of those minimum data-sets would not be feasible in the absence of computer support. Indeed, one of the key themes of the Körner review was that the necessary data-sets should be captured as a by-product of normal operational processes. Those operational processes needed the support of computer systems that very closely resembled those that had been developed as part of the Experimental Computer Programme.

In order to enable implementation of the Körner proposals, a majority of the then 14 RHAs subscribed to the development of the ICL-based Inter-Regional Collaboration (IRC) Patient Administration System (PAS) (see Chapter 7). This PAS was based on a fusion of the experimental projects developed at the Queen Elizabeth Hospital, Birmingham, and at North Staffordshire Hospitals, Stoke-on-Trent. The system was still in use in some Trusts within the NHS in the early years of the twenty-first century. The collaborative project colloquially known as 'The Four Horsemen' (Charing Cross Hospital, United Cambridge Hospitals, St Thomas' Hospital and University College Hospital), formed the basis of the system adapted for use at University College Hospital, and it remained in use there until 1992. A system developed at Southend inspired the development of the Remote Data Capture System in the early 1980s as used in many hospitals within North East Thames Region, and then the Remote PAS that supported the original Körner implementation in the same area. The longest surviving of the original projects is The London Hospital Computer Project, which was implemented in hospitals throughout the former City and East London Area Health Authority, and after a facelift in the 1980s, it was still operational in early 2008.[40]

The widespread growth in use of patient administration systems during the 1980s lent realism to the development of the Resource Management Initiative (see Chapter 6)[41] through which attempts were made to measure, compare and improve clinical performance at the level of individual doctors.

So, in effect, the aim was to evaluate health care delivery, and where it was found to fall short in some sense, to instigate change and improvement.

The performance criteria concept has also carried through in a modified form into the current NHS. The DoH has produced its national set of performance indicators since 1993/4,[42] and these were inspired by some earlier work by Yates.[43] These ideas can be traced forward into the performance assessment framework[44] and clinical indicators.[45] In September 2001, the Government published the first NHS performance ratings for NHS Trusts providing acute hospital services.[46] This was a step towards fulfilling the Government's commitment to provide patients and the general public with comprehensive, easily understandable information on the performance of their local health services. Dr Foster Intelligence® now publishes annual hospital comparative tables in the national press and on the internet.[47]

Benefits realization

Today, no proposal for the introduction of a major computer system would succeed unless it was supported by a rigorous business case. That business case would be expected to include a full evaluation of the merits and projected costs of each potential solution, a statement about the expected benefits of the preferred option and a statement as to how those benefits would be realized. Indeed, the term 'benefits realization' lives on – although not usually with the rigorous connotations developed at El Camino.[15] The concept has a key role to play in the modern day appraisals of the effectiveness of NHS computer systems.

Curiously, the position has been reached whereby, although there is no evidence that anyone judged formally that the Experimental Computer Programme was a success, nor that the associated evaluation activity made a positive contribution, it is certainly the case that both computer systems per se, and the evaluation of them, are now firmly engrained into the fabric of the NHS. The modern NHS just cannot survive without extensive investment in technology. Yet with many competing demands on a never sufficient supply of funds, this investment still needs to be justified on a case-by-case basis. It would be comforting to think that some of the pioneering evaluation work of the 1970s still has a role to play in such decision-making.

REFERENCES

1   Webref (2002) (accessed September 2007) *Delivering IT in the NHS: A Summary of the National Programme for IT.* DoH, London, 11 June. www.dh.gov.uk/en/Publicationsandstatistics/Publications/PublicationsPolicyAndGuidance/DH_4003250

2   Barber, B. and Abbott, W.A. (1972) *Computing and Operational Research at The London Hospital*; 'Computers in Medicine' series. Butterworths, London. ISBN 0 407517 00 6.

3   Davies, C.K. (1970) Recall procedures – local authority applications. In Abrams, M. (ed.) *Medical Computing: Progress and Problems*, pp. 283–8. Chatto & Windus for BCS, London.

4   DHSS (1972) *Using Computers to Improve Health Services*. DHSS, London, November.

5   DHSS (1977) *Interim Report on the Evaluation of the National Health Service Experimental Computer Programme*. DHSS, London.

6   Molteno, B.W.H. (1978) A new approach to evaluation in the NHS Experimental Computer Programme. In Anderson, J. (ed.) *MIE, Cambridge 78: Lecture Notes 1*, pp. 691–700. Springer Verlag, Berlin. ISBN 3 540089 16 0.

7   Barber, B. (1975) The approach to the evaluation of The London Hospital computer project. In Anderson, J. and Forsythe, J.M. (eds) *MEDINFO 74*, Stockholm, pp. 155–165 and pp. 1011–12. North-Holland, Amsterdam.

8   Computer Executive Committee (1975) *The Evaluation of The London Hospital Computer Project: An Interim Report*. The London Hospital internal document.

9   North East Thames Regional Health Authority, Management Services Division (1976) *The Future Direction and Progress of The London Hospital Computer Project*, report no 745. February, internal document.

10   Sharpe, J. (1974) Towards a methodology for evaluating new uses for computers. In Anderson, J. and Forsythe, J.M. (eds) *MEDINFO 74*, Stockholm, pp. 137–43. North-Holland, Amsterdam.

11   DHSS (1977) *Report of the Working Group on Computer Evaluation*. Computers and Research Division. May.

12   Cohen, R.D. (1979) Evaluation of computer systems in medicine. In Barber, B., Grémy, F., Überla, K. and Wagner, G. (eds) *MIE, Berlin 79: Lecture Notes 5*, pp. 931–7. Springer Verlag, Berlin.

13   North East Thames Regional Health Authority (1982) *The Evaluation of the NHS Computer R&D Programme: An Analysis of Performance Criteria Measurement Results*. Molteno, B. (ed. – Trent Regional Health Authority) Report number 1094, review paper on computer evaluation for the Evaluation Working Group (DHSS). February.

14   Barber, B. and Cundy, A. (1980) Evaluating evaluation: some thoughts on a decade of evaluation in the NHS. In Lindberg, D.A.B. and Kaihara, S. (eds) *MEDINFO 80*, Tokyo, pp. 602–6. North-Holland, Amsterdam.

15   Norwood, D. (1975) Economic evaluation of total hospital information systems. In Anderson, J. and Forsythe, J.M. (eds) *MEDINFO 74*, Stockholm, pp. 149–54. North-Holland, Amsterdam.

16 Barber, B., Abbott, W. and Cundy, A. (1977) An approach to the evaluation of the Experimental Computer Programme in England. In Shires, D.B. and Wolf, H. (eds) *MEDINFO 77*, Toronto, pp. 913–16. North-Holland, Amsterdam.

17 Fanshel, S. and Bush, J.W. (1970) A health status index and its application to health services outcomes. *Operations Research*, 18, 1021.

18 Rosser, R. and Watts, V. (1977) Measurement of the effectiveness of health services. In Shires, D.B. and Wolf, H. (eds) *MEDINFO 77*, Toronto, pp. 559–71. North-Holland, Amsterdam.

19 Rosser, R. and Benson, T. (1978) New tools for evaluation: their applications to computers. In Anderson, J. (ed.) *MIE, Cambridge 78: Lecture Notes 1*, pp. 701–10. Springer Verlag, Berlin. ISBN 3 540089 16 0.

20 Webref (accessed June 2007) *Healthcare Resource Groups*. NHS Information Authority. www.nhsia.nhs.uk/casemix/pages/hrg.asp

21 Barber, B., Cohen, R.D. and Scholes, M. (1976) A review of The London Hospital computer project. In Laudet, M., Anderson, J. and Begon, E. (eds) *Proceedings of Medical Data Processing*, Toulouse, pp. 327–38. Taylor & Francis, London. ISBN 0 850661 06 4.

22 Cundy, A.D. and Goldstein, M. (1976) The evaluation of the first phase of The London Hospital real-time computer system. In (eds unknown) *Proceedings of Computer Performance Evaluation*, London, pp. 209–24. Online Conferences, Uxbridge.

23 Cundy, A.D. (1971) A simulation model of a clinical laboratory office. In Abrams, M.E. (ed.) *Spectrum 71: BCS Conference on Medical Computing*, pp. 124–38. Butterworths for BCS, London.

24 Fairey, M.J. (1969) Information systems in hospital administration. In Abrams, M. (ed.) *Medical Computing: Progress and Problems*, pp. 384–89. Chatto & Windus for BCS, London.

25 Farrow, S.C., Fisher, D.J.H. and Johnson, D.B. (1971) Statistical approach to planning an integrated haemodialysis/transplantation programme. *BMJ*, 2, 671–6.

26 Davies, R., Johnson, D. and Farrow, S. (1975) Planning patient care with a Markov model. *Operational Research Quarterly*, 26.3ii, 599–607.

27 Computer Executive Committee, The London Hospital (1975) *Presentation to the DHSS Chief Scientific Officer*. Private communication from Dr B. Barber, Director of Operational Research, The London Hospital, to the Chief Scientific Officer, DHSS.

28 Rogers, P. (1972) *Management Arrangements for the Reorganized National Health Service*. DHSS, HMSO. ISBN 0 113204 85.

29 Barber, B. (1978) *Proposals for Data Analysis from The London Hospital Computer system*. North East Thames RHA, internal report.

30 NHS (1982) *A Report by the Steering Group on Health Services Information on the Collection and Use of Information About Hospital Clinical Activity in the National Health Service.* First report to the Secretary of State. HMSO, London.

31 Benson, T. and Thwaites, E. (date unknown) An appraisal of hospital laboratory computing based on NHS R&D experience in England. (Source unknown.)

32 Cundy, A.D. (1976) The London Hospital Computer Project: cost benefit analysis. *Univac Users Association Conference*, Europe, Rai Centre, Amsterdam.

33 DHSS (1977) *Report on the Pilot Stage to Evaluation Working Group.* Performance Criteria Project. September.

34 DHSS (1978) *Performance Criteria Project Report.* Computers and Research Division, DHSS. March.

35 DHSS (1979) *A Handbook for the Measurement of Performance Criteria.* DHSS, MSC4C Division. May.

36 DHSS (1980–81) *The evaluation of the NHS Computer R&D Programme: an analysis of performance criteria measurement results.* A series of reports published for the DHSS by North East Thames RHA Management Services Division: No. 997 (June 1980), No. 997A (July 1980), No. 1023 (September 1980), No. 1023A (November 1980), No. 1071 (March 1981), No. 1024 (September 1980), No. 1024A (September 1980), No. 1046 (December 1980), No. 1046A (April 1981), No. 1092 (June 1981), No. 1092A (June 1981).

37 Cundy, A. and Nock. J.D. (1979) An assessment of the use of performance criteria in the evaluation of the NHS Experimental Computer Programme. In Barber, B., Grémy, F., Überla, K. and Wagner, G. (eds) *MIE, Berlin 79: Lecture Notes 5*, pp. 117–130. Springer Verlag, Berlin.

38 Davis, A. and Kumpel, Z. (1981) *The Effects of Computerization on the Nursing Record.* West Midlands RHA. June.

39 Kumpel, Z. (1981) *The Evaluation of Primary Care Systems.* South Western RHA. June.

40 Fairey, M. (2000) The London Hospital Computer Project: a proto-HISS project. *BJHC&IM*, 17, 10, 31–4.

41 Battle, T., Cundy, A. and Rowson, J. (1995) The Resource Management Initiative: how has it helped at hospital level? *BJHC&IM*, 12, 4, 15–17.

42 Webref (accessed June 2007) *The NHS Performance Guide 1995–96.* DoH, London. www.performance.doh.gov.uk/tables96.htm

43 Yates, J. (1983) Inter-hospital and inter-health authority comparisons. *BURISA (British Urban and Regional Information Systems Association)*, 14–15 April.

44 Webref (accessed June 2007) *The NHS Performance Assessment Framework*. DoH, London, April 1999. Health Service Circular HSC 1999/078. www.dh.gov.uk/assetRoot/04/01/20/52/04012052.pdf

45 Webref (accessed June 2007) *Quality and Performance in the NHS: High Level Performance Indicators and Clinical Indicators*. DoH, London, November 2002. www.performance.doh.gov.uk/indicat/nhsci.htm

46 Webref (accessed June 2007) *NHS Performance Ratings, Acute Trusts 2000/01*. DoH, September. 2001. www.performance.doh.gov.uk/performanceratings/2001/index.html

47 Webref (accessed June 2007) *The Good Hospital Guide*. Dr Foster Intelligence Ltd. www.drfoster.co.uk/

27 A personal view from the private sector

ALAN W THOMAS

Over the last 30 years of 'health computing', as it used to be called, many people and companies in the private sector have devoted much of their working lives to significant developments in medical informatics in the NHS, even if many have never been NHS employees. They did it because they believed IT could contribute to a better health service and because they felt strongly that they and their companies had products, expertise and services that the NHS and the wider health care industry could use effectively.

Therein lies one of the basic problems from the private sector viewpoint. Do we go ahead and take the risk and develop products that we think the NHS should be using, or should we wait until the DoH and the NHS tells us what they want, and then we compete to provide them? We never really got an answer to that one. The reply usually was – if you can find a NHS organization that will take and use your product, then let us know, and we will think about what may happen next. Not a very useful basis for planning.

The companies involved are numerous, and they include the companies I was involved with: ICL, IBM, DEC, Honeywell, Hewlett Packard, Rank Xerox Data Systems, Sema and Sifo Ltd. There were, of course, many others, such as SMS, MDIS and Abies.

The people who came into computing in the 1960s were interesting. The received wisdom was that only mathematicians and electronic engineers could program computers, but it was soon realized that anyone with a logical, rigorous mind could master systems design, programming and computer operations. They included nurses, teachers, physicists, economists, clerks – one of the early heads of data processing at the National Coal Board was a former payroll clerk – and people of every background. But, they all shared an aptitude for logical thinking and enthusiasm for the development and application of computer systems. There was a greater equality of the sexes at that time before IT became more of a male profession; in gender terms, it was a time of almost equal opportunity.

My own early experience of computing, fresh from the London School of Economics (LSE), was as a National Coal Board economist working with such people as E.F. ('small is beautiful') Schumacher, programming an Elliott 803, and later being trained by International Computers and Tabulators (ICT) as a systems analyst/programmer on PLAN and the ICL 1900 range, working on public sector projects. This was where I joined the BCS.

I then gained experience with a multinational corporation, Rank Xerox (RX), where, with a very small group, we designed, programmed and implemented all the core business systems for 13 European RX companies, using standard data definitions, validations, code structures and programs. This led me, after Xerox bought Scientific Data Systems (SDS), to involvement with the DoH Experimental Project Programme 'Group of Four' project.

Exciting responsibilities were given to young people in the IT industry at that time, and we developed our own methodologies for systems analysis and design, and for project management. We were learning and making it up as we went along. There were very few UK university computer science departments. They were principally centred on Manchester and Cambridge, and focused on electronic hardware engineering and early operating system development, rather than 'systems'.

THE 'GROUP OF FOUR' PROJECT

The DoH Experimental Computer Programme 'Group of Four' Project (or 'The Four Horsemen') in 1970 required basic hospital systems that would be used at four hospitals (University College Hospital, St Thomas' Hospital and Charing Cross Hospital in London, and Addenbrooke's Hospital, Cambridge), all using the same computer hardware and operating systems software. The application systems for a patient index, in-patient management and summary medical records would be developed by the users themselves.

The Experimental Computer Programme was a laudable attempt to try to bring the benefits of IT to the NHS. It had been started in the late 1960s with a number of projects including:

- The London Hospital's ambitious hospital-wide set of systems, based on UNIVAC equipment;
- Dr John Anderson's ward-based system at King's College Hospital, using ICL equipment;
- North Staffordshire Hospital's outpatient system, again using ICL equipment;
- Exeter Community Health project;
- and several others.

See Chapter 4 for more details on these projects.

Rank Xerox Data Systems (RXDS) won the 'Group of Four' project with the versatile Sigma range of computers, designed by former Ferranti engineers, which allowed concurrent batch transaction processing and real-time connectivity to scientific instruments. The Sigma was also used extensively in the NASA space programme, and around the world in aircraft simulation and telemetry, the oil industry, and even in London's first dial-up online service for commercial services. RXDS was looking for new applications and markets that could use the flexibility of the Sigma, and hit on health care.

We did the sizing for the four configurations: the Sigma used a large Random Access Disk (RAD) as well as removable disks and tapes. We had to calculate the disk space required for all the hospitals' active and passive medical records, and the operational requirements for real-time connection of early laboratory instruments, with analogue-to-digital conversion. Sigma also had one of the first Committee of Data Systems Languages (CODASYL) approved database management systems – XDMS – which allowed complex data structures rather than simple index sequential, and the use of data base management systems in health care became a particular interest of mine.[1]

Taking a part-time Master of Technology degree in Computer Science at Brunel University, London, helped to bring some order and structure to my chaotic world of computing and information technology in the 1960s and early 1970s. The course included hardware design and logic, systems design, operating systems structure, and programming and testing methodologies. In the past, such techniques, methods and programs were developed pragmatically to meet the immediate requirements of the problems and human and business situations that were faced, rather than working to any predefined methodology. It was a stimulating environment and frequently kept you awake most of the night, but many people were often reinventing the same wheel, as it was difficult to find out what relevant work other companies were doing. The BCS and the USA Association of Computing Machinery (ACM) publications began to be helpful in that regard.

After surviving surgery and prolonged heavy radiotherapy treatment at the Royal Marsden Hospital, Fulham, it was time to become an academic, moving to the other side of the world, to New Zealand, to set up the first Department of Computer Systems and Quantitative Methods at the University of Otago, Dunedin.

ACADEMIA AND THE PATIENT INFORMATION SYSTEM

In 1960, there were only four universities in the UK teaching computer science or related subjects, but by 1970 there were approximately 50. New Zealand universities were hard on the UK's heels, having committed to a nationwide purchase of large-scale Burroughs mainframe computers for all its universities. It was a controversial decision.[2]

Preparing the curricula for new undergraduate and post-graduate computer systems courses from scratch caused me to rely initially on the work done in the USA, on information systems teaching in business schools, published in the ACM journals. It was pulling together a new science or art: attempting to trace its many roots back to logic, electronics and the social sciences. It was essential also to encourage sensitivity to all the human aspects of systems, with intellectual rigour and attention to detail.

The Otago Health Board, which contained two university hospitals, was trying to develop its own version of an NZ DoH 'experimental IT project'. An In-patient Management System for two hospitals in Dunedin, using an ICL

1902 with 8 k memory and magnetic tapes only, shared with the local city council. The project was based on teletype terminals and a local computing enthusiast had starting writing programs without first designing a system. We started again, and with the help of keen students. Working on this as practical project work as part of their studies, we designed, programmed and implemented a working system over the following 12 months.

As no software existed for handling online simultaneous access to the same patient record or file, we worked with ICL NZ to write our own software. These techniques were fed back to ICL UK to be incorporated in their next generation of operating systems. Other challenges included generating large statistical tables using only 8 k of memory and four magnetic tapes. This involved intricate computing and then repeated sorting of column and line records before a final printout could follow.

The implementation succeeded because we worked closely with the nursing and medical staff, including a chest surgeon who knew the intricacies of patient case notes and their structures, and with the clerical staff who understood the operational systems well. The patients were not consulted, but they were kept informed about the patient identification methods used and the data privacy implications, a long time before the UK Data Protection Acts.[3] The lessons learnt were:

- to keep the system simple for the users;
- to involve them in the system design and training; and
- to realise that a system is not working until it is successfully implemented.

These could not be achieved without detailed project management and cost control.[4, 5]

RADIOTHERAPY TREATMENT PLANNING AND NATIONAL HEART GENEALOGICAL DATABASES

Two other clinical projects at that time reflected the developments in medical information processing. The first was radiotherapy treatment planning. The Ontario Cancer Institute had pioneered Fortran programs that allowed consultants and technicians to propose sets of parameters for the radiotherapy machines for each patient treatment period, varying the intensity, angle and depth of the beams on to the human body. The programs calculated and illustrated, using shading on line printer paper (as there was no graphics output in those days), the potential effects of each set of parameters. This allowing the staff to select the optimum settings before commencing treatment on the patient. Ontario was contacted and they sent us, without charge, magnetic tapes with all their relevant programs and limited documentation. These were converted from IBM to Burroughs equipment, and soon the consultants were noting their requirements on paper forms at 8.30 am, and their possible settings for one day's group of patients. These were punched onto cards at the University Computer Centre, the programs run, and the output delivered back to the hospital by 9.30 am. The consultants then selected the optimum settings before the first patient arrived.

Previously it had taken two technical staff all morning to calculate, using hand calculators, just one set of parameters for each patient, with the attendant risk of initially calculating an incorrect set. Now the process was completed for multiple sets for all patients within an hour, with the ability to select the most advantageous treatment for each patient and to avoid the potential risks. This was one of the early applications of IT to health care that gave immediate qualitative and quantitative improvements for the patients and the clinical staff, and many of us have continued searching for that particular 'Holy Grail' ever since.

The other project was with the NZ National Heart Foundation. It was conducting research across New Zealand to identify members of the general public who had potential deficiencies in their lipoprotein metabolism, and as a result might be subject to unexpected heart attacks and possible death. This is a genetic disorder, and routine blood sample analysis may give only a six-week window for treatment before a potential attack. It was important therefore to trace those persons who might be subject to the condition, analyse their blood and treat them if necessary. Clinical researchers were located at all four NZ Clinical Schools, and a highly confidential database had to be built. We used the Burroughs Data Base Management System (BDMS), which was based on applied set theory: structuring a three-dimensional database allowing for the multiple 'unions' and 'matings' and any resultant offspring that might occur in the different familial cultures within New Zealand. Initially four separate databases were built and they were gradually linked like a jigsaw into a national database as the overlapping family links were identified. The system was of real value in reducing the death rate for this condition.[6]

New Zealand's first Medical Information Processing Conference was organized in 1974, with key note speakers from Australia and New Zealand. One of the papers was from Professor Ashley Aitken, of the Department of Community Medicine, at Otago University Medical School, and a former GP. He had spent many years categorizing the problems as presented by patients in primary care, and had classified them into a comprehensive set of codes.[7] This was a precursor of the primary care codes developed by Dr James Read in the UK in the 1980s.

After three years the computer systems courses were all in place and students were graduating into the growing NZ IT industry, including the growing medical informatics sector. The hospital systems were functioning as a normal part of the running of the hospitals. US consultants had begun advising the NZ Government on an expensive new approach to health computing. I decided to head back to the UK.

ICL AND HEALTH CARE IN THE MID-1970s

In the mid-1970s, ICL negotiated a 10-year agreement with a joint team of the DoH and NHS. ICL was established in a strong position in the UK public sector. After nearly a year of discussions, the agreement gave ICL a practical monopoly of the supply of mainframes and software to the 14 Regional Health Board Computer Centres and Wales.

Such preference for national computer suppliers was reasonable at the time. The UK IT industry had inherited from the 1950s' aerospace industry great, high-technology, innovative skills. The UK had eight independent computer manufacturers in the early 1960s and these had been gradually forced, by the new Minister of ('white hot') Technology, Anthony Wedgwood Benn, to metamorphose into ICL, with significant internal rivalries and, sadly, a resulting loss of competitiveness in the growing global IT industry.

The arguments in favour of the NHS–ICL agreement centred on a discount for ICL hardware and software for the NHS, which was not large; and the possibility of common working practices for NHS IT staff. 'Standardization' was a big theme of the NHS, following on the experience of corporate multinationals. The agreement allowed the framework for standard systems for payroll, stores, purchasing, financial ledgers, regional and national statistics (hospital activity analysis) and, in Wales, a comprehensive set of child health systems. However, despite considerable effort by the DoH, the NHS and ICL on the methodologies, development and implementation of standard systems,[8, 9] many RHAs and NHS Units continued to purchase and develop their own local systems, and the possible economies of scale were not achieved. The DoH did not appear to have the authority or will to insist on the standard systems approach and perhaps an opportunity was lost.

The association between ICL and the DoH gave the opportunity for the first UK joint marketing visit to the Middle East. I went with a senior DoH officer to Kuwait to propose, with the strong support of the British Embassy, ICL-based child health and stores systems for the Kuwait Ministry of Health; and to Dubai, to propose a PAS for the new Rashid Hospital. There was some initial success, which was not followed through.

ICL never developed effectively into the Area Health Authority or hospital systems markets despite its advantages.[10] Progress was made by IBM, DEC and other equipment suppliers and software houses. Unfortunately, the fate of a number of grandiose schemes in Wessex RHA and West Midlands RHA did little to improve confidence amongst the NHS and DoH managers, who had to fund and implement the IT projects, or the private sector suppliers.

During this period the DoH conducted a survey of the IT activity within the NHS and organized the application systems into a number of categories, including: general practice, community health, hospital patient-oriented systems, laboratory and scientific systems, etc., and regional and national requirements. It was against this background that the Körner committee started the movement towards standard data definitions and data-sets, the lack of which in the NHS had hampered many previous developments. Improved NHS general management and financial budgeting arrangements were also being introduced. The information systems needed to underpin such reforms were not in place, as they would have been in other large-scale national enterprises.

THE ENTREPRENEURIAL ASPECT

At this time a number of people were leaving the IT corporate environment to set up their own companies in a variety of sectors. Before starting my own health information systems company, Sifo Ltd, in 1977, I spent a year in Italy trying to develop health, government and scientific systems jointly with ICL, the European Union and the Italian national and regional governments. With Olivetti, we came close to confirming a major new computer centre based at the University of Rome, against the US competition of IBM, Control Data and UNIVAC. We were thwarted at the end by the activities of the *Brigati Rosse* in blowing up the Rome computer centre, the collapse of the Government with whom we had been negotiating and the assassination of the Prime Minister, Alberto Moro.

In 1978, the Riyadh and Al Kharj Hospitals (RKH) Programme in Saudi Arabia planned to implement a full range of patient-oriented and management systems at two Ministry of Defence and Aviation hospitals, in a short timescale. The project was awarded to a British health care company, Allied Management Group, who commissioned, staffed and managed the hospitals using British NHS approaches and procedures with many British former NHS staff. Initially the habits of planning and operational bureaucracy inherited from the NHS were in danger of stifling the project, but as staff realized that they were liberated from many of the previous constraints, the project moved ahead swiftly. The hospitals' buildings, built to a high standard by a German company, and the hospital equipment were commissioned. The clinical, nursing, technical and administrative staff were recruited and trained. Internationally high clinical standards were established, and the first patients were received within nine months of the signing of the contract.

Sifo was involved from the beginning in planning, selecting hardware and software, developing and modifying them to meet British health care requirements, testing, designing related administrative procedures, and implementing and managing the sets of systems in the two hospitals. The systems included:

- Patient registration and master index
- In-patient management
- Patient accounting and costing
- Medical records episode summaries
- Laboratory

- Pharmacy
- General ledger
- Payroll and personnel
- Purchasing and stock control
- Equipment maintenance scheduling.

The initial mainframe hardware had already been committed to Honeywell-General Electric in France, and sadly there were no sets of software systems available from any of the UK hospital IT projects. The St Thomas' Hospital, London, Honeywell-based project was in a very early stage and, to the disappointment of Honeywell UK, did not have a comprehensive, functioning set of systems. Urgently I trawled the USA for suitable systems and entered into agreements with Blue Cross/Blue Shield of Arizona and Becton Dickenson of Boston. Our UK- and Saudi-based staff modified these systems extensively to work in a British health care environment, cutting out approximately 50 per cent of the functions and adding another 50 per cent. The IT Production and Operations Manager was put in charge of all testing, rather than the systems development staff, as the final responsibility for the day-to-day running of the systems would be with his department.

The IT staff were organized into small teams for each set of application systems and infrastructure, with clear cut plans and tight timescales for each stage. We worked to tight cost constraints, comparable to the UK NHS information technology projects at that time. The systems were ready to go when the first patients arrived, and functioned effectively thereafter.

There were interesting problems with patient identification for Arab patients, with many people having similar phonetic names and no detailed date of birth. We identified them and gave them unique patient numbers on the basis by which they identified themselves and were known to each other, including their full names (of various spelling), father's names and place of origin. The method worked well. This came in useful when we later developed hospital systems in Egypt and other Arab countries.

The small systems teams worked closely with the clinical, nursing and administrative staff in developing the screens and outputs from the systems and, after initial scepticism, most of them participated well. Computer 'Moon landing' games were left ticking-over as background for the night nursing staff to encourage their keyboard skills, and they found 'discharging a patient' relatively easy after that.

The second stage of the RKH Programme developed its own outpatient management system based on Hewlett Packard 3000 equipment and linked it, using customized data communications, to the other systems. We had to allow for patients arriving for treatment a week early or late, and sophisticated airline booking overflow techniques were used. This system was later extended, working with the Programme Primary Care Director, Dr Bill Dodd, to become a comprehensive Primary Care Health Centre system, using problem-oriented records with 20 concurrent GPs. The primary care system acted as the filter for patients entering the main hospitals. A new large-scale laboratory system from Control Data, Medlab, was also implemented for all disciplines and linked to the other hospital systems. By the early 1980s there were scores of terminals connected to different systems linked to each other and accessing patient data where appropriate and allowed. This had been a dream of the DoH Experimental Project Programme of the 1960s/1970s.[11, 12]

LESSONS LEARNT

The lessons learnt for British health care IT were a recurring theme: clear specifications, tight contracts, clear specification of responsibilities, small working teams, close involvement with the clinical users, sound project management, effective use of technical alternatives, and steady concentration on implementation methods and reliable operational running.

One of the problems with hospital IT systems from the supplier's point of view is that you have to meet several, often contradictory, criteria for your clients. The customer wants to be sure that you have proven hardware and software and that you can demonstrate, from previous projects, the project management abilities to implement successfully on time and on budget. At the same time each hospital and NHS unit regards its situation as unique and wants a slightly different system from your previous installations. They will also have their own unique views on how the system should be implemented and organized at their sites. This was true both in the Middle East and in the UK. The fact that as a supplier you had achieved a large-scale success did not necessarily mean you would get the next business.

THE NEXT PROJECT

Sifo's next project in Saudi Arabia was to be in a very different context: a medium-sized hospital reserved for high-ranking Government staff, set in the mountains of Taif above Mecca, built by an Italian construction company, and managed by a thrusting Californian health care management company, National Medical Enterprises. It took a week of days and nights negotiating the contract, in Los Angeles, against a squad of US corporate lawyers. This time the full set of hospital patient and financial systems was based on three Hewlett Packard 3000 computers, linked to provide redundancy support for each application system. New systems were designed and developed for patient registration and master index, in-patient management, medical records and patient accounting; and HP-based systems for laboratory and pharmacy were brought in from external suppliers and linked.

The US-oriented internal management, standardized clerical procedures and operational manuals, and disciplined commercial environment, together

543

with the cooperation of the US and Saudi staff, made implementation less problematic. Concentration could be focused on improving the technical aspects of the systems while the US hospital management took greater responsibility for training, implementing and managing the systems, rather than leaving it to the supplier.

The project went well and Sifo went on to implement its HP-based systems in a variety of hospitals for different ministries in Jubail, Dhahran, Jeddah, and Riyadh, including the prestigious King Khalid Eye Hospital. The Sifo IT staff designing, developing, implementing, operating and maintaining the systems were nearly all UK nationals working for good salaries, six months on and one month off. Most of them stayed with Sifo for many years until the Middle East hospital tide began to ebb. Some went on to Australia and some returned to the UK. It is a pity that the experience and skills of this talented group were not used more effectively back in the UK's NHS, but there were some exceptions.

Sifo also agreed to project manage the implementation and support of other suppliers' systems and, using UK staff, implemented IBM's small-scale HIS in a private hospital in Cairo, and ICL's developing PAS in a military hospital in Egypt.

UK HEALTH CARE IN THE 1980s

With the development costs of the hospital systems software already paid for in the Middle East and the solid project management implementation experience in difficult circumstances, Sifo thought that the UK NHS systems market in the 1980s would come knocking on the door. New premises were set up near London, and exhibitions set up and papers presented at the BCS Medical Informatics conferences at Harrogate and elsewhere. But it was difficult to make progress. There were several competing hospital systems in the market including ICL's PAS, IBM's HIS, and DEC and SMS offerings.

The national IT Strategy was not clear between the various factions of the DoH, the Regional Health Boards and the Hospital Units themselves, and large parallel DoH and NHS IT planning bureaucracies had grown up in Birmingham and London.

From the viewpoint of the large and small private sector suppliers to the NHS the same set of problems and questions existed:

- What were the national, regional and local strategies?
- What standards had to be complied with?
- What application systems and related software were required?
- What should the supplier invest in?
- Who had written, and how good were the 'specifications of requirements' and proposed contract arrangements?
- What business was likely to come up over the next one, two or three years?

- Where was the contract likely to be located?
- Who were the final, real decision-makers?
- How long would the procurement cycle take?
- Would the contract end up being awarded, or simply run into the sand, as often happened?
- What was the cost of bidding?
- What was the potential value of the business?
- Who was funding it? Did the supplier or competition have the application software to meet the likely specifications or would it have to buy in from others?
- Did the supplier have the expertise and staff to project manage, develop, implement and maintain the systems?

Was it all worth the effort? Even the large companies were questioning whether it was financially viable to continue to do business with the NHS and the DoH.

This was further complicated by some large IT corporations seeking to enter the UK health care market, sometimes on the basis of experience in the USA or other countries' health care markets, and being prepared to take financial significant losses over several years in order to establish market share. This rarely worked to the advantage of the NHS in the medium term, as previous losses on undercutting to get the business were clawed back on support, development and maintenance contracts. Ironically, the overall result was that there remained significant underinvestment in IT for the health service, both by the NHS and the suppliers, and the application of information technology to health care fell further behind all other sectors of the UK economy, even local government.

Sifo bid for four Ministry of Defence hospitals' systems but was hugely underbid by a US company, presumably as a loss leader. No company should do business at a guaranteed loss; the customer is bound to suffer, either from poor service or escalating future support costs.

However, the Special Health Authority managing the National Hospital for Nervous Disease in Queen Square, London, and that for Great Ormond Street Hospital for Sick Children, plus the two children's hospitals in East London, selected Sifo's HP-based systems. Four patient-management systems, including outpatients, were implemented in four sites in four months, using fibre optic links for the first time in the UK. A module was added for matching the ICD-9 diagnostic codes with Great Ormond Street's own comprehensive child-oriented diagnostic codes. These had originally been devised when J.M. Barrie, the hospital's benefactor, was writing *Peter Pan*. The systems were also implemented in North Wales but further business did not materialize.

So, like other IT suppliers, Sifo looked to the private hospital market and participated in a joint venture, Modular Medical Systems, which developed and implemented patient accounting systems for private hospitals based on DEC/Systime equipment. Eleven systems were implemented over the next few years, including the Great Portland Street Hospital for Women and Children, in London.

At the same time ICL's PAS was being further developed as the principal system for the NHS in Wales and Sifo set up a new software development centre in Cardiff to work on it. An interesting clinical audit system was also developed jointly with Oxford RHA and several versions were implemented: Surgical Clinical Audit in the John Radcliffe Hospital, where the full range of potential complications were codified; Anaesthetics Clinical Audit in Northampton General Hospital; and Cardiology Clinical Audit, again in the John Radcliffe. The system was marketed across the UK and a number of systems were implemented.

At this time, the late 1980s, I was asked to develop an overall IT strategy for the NHS in Wales. Using the experience I had gained, I based it upon:

- clear core specifications of requirements in each application system area;

- tight procurement and contractual arrangements;

- defined basic data-sets; and

- a clear specification of how different hospital and primary care systems, even if from different suppliers, should communicate with each other.

This would allow national and local Health Authorities, Hospital Units and GP practices to select 'best of breed' systems from various suppliers if necessary, that best suited their needs, without having centrally purchased systems imposed upon them. They could be secure in the knowledge that the ability of the systems to communicate with each other with the defined data-sets would be built in by the supplier before implementation, and not as some later expensive modifications.[13]

The fashion for, and poor progress of, the NHS HISS had shown clearly that no single corporate IT supplier could supply and implement the full range of required systems, let alone provide links with GP and other external systems. However, the method of procurement tended to lock out the smaller companies. The strategic emphasis should have been on the standard intercommunication between systems, and the minimum data-sets, items and validation required, rather than on standard systems per se. However, this did not appear to be a high priority for the DoH. It is interesting to note that if the entire set of the Data Model Specifications prepared by the NHS Information team in Birmingham were printed out, they would have covered 22 football pitches. But there was still no definition of priorities, or definition of the key data items to be transferred between systems. The private sector trying to develop systems for the NHS scratched their heads in bewilderment.

DECISION TIME

For a number of private sector companies, including Sifo Ltd, it was decision time. Twelve years of running your own company with up to 50 staff can be hard work and one risks becoming a business school case study. Do you diversify into other sectors or systems products, looking for niche markets? Do you look for collaborations with other companies, which could lead to mergers or takeovers, in order to spread the management load, and build the company? A great deal of time and management effort can be taken up in discussions and negotiations with other companies, which can divert the management from running its own business properly. Sifo was involved in detailed discussions with ICL, Abies, Hewlett Packard and others. None proceeded to a satisfactory conclusion.

Also, the technology with which we were working was changing rapidly. Mainframes had become powerful minis; minis had become personal computers (PCs) with more power than a large mainframe of 10 years earlier, and these could be linked with fast, reliable data communications. The internet and email were born.

For the Sifo systems, even though the data items and functions that appeared on the screen would not change significantly, as we had followed an Apple® Mac®-type approach from the beginning, the hardware and software on which health care systems were based were changing radically. This would mean a complete redesign and rewriting of all the systems. There would have to be a significant investment of potentially non-productive time of one's staff as they developed new products for an uncertain market. The lack of any reasonably sure business from the NHS, with long procurement timescales and final fickle decisions, and negligible profit margins, caused Sifo to pull out of the active, health care systems market. The company moved into consultancy, project management and developing other larger companies' systems.

There still existed a huge market for health care systems but a thousand blossoms were not being allowed to bloom. There should be room for minnows as well as sharks in the health care ocean, as innovation is more likely to come from small groups. The NHS approach of relying primarily on large IT suppliers, influenced increasingly by corporate consultancy organizations, prevented this happening. Other European countries encouraged their smaller IT companies, but not the UK.

Personally I was also looking to spend more time on long-distance single-handed sailing (the antidote to medical informatics) and on politics.

MORE NHS ORGANIZATIONAL CHANGES

The late 1980s and early 1990s was another period of massive organizational change for the NHS in England and Wales. John Moore, Secretary of State for DoH, ironically who was at LSE at the same time as me, and Frank Dobson, a later Secretary of State, had been instructed by the Prime Minister, Margaret Thatcher, to 'do something about the NHS'. Moore failed to come up with sufficiently radical solutions and Kenneth Clarke, with cigars and 'Hush Puppies®', was drafted in to provide the structure for 'independent' NHS Trusts and GP fundholding. At the time it was thought that this would be a reasonable boost for IT within the NHS. It can be argued that unless one captures the basic data about individual patients throughout their health cycles (from diagnosis, GP prescription costs, through to detailed activity data on outpatient and in-patient episodes), then one could never properly define and manage case mix, which is essential for the proper management of a health care unit's quality and performance, or assess the realistic costs of treatment, individually or nationally. If systems were required to prepare a detailed itemized pseudo 'invoice' for all aspects of care for a patient episode, which could only be done with sound IT, then one could make major improvements in NHS delivery of health care and its management.

As a non-executive director of a first wave NHS Trust in Oxford in the early 1990s there was an opportunity to see the effects of the NHS reforms from the inside. Despite opposition from other parts of the NHS, the Trust Board and management made good use of its limited independence to create a hospital organization that was based on the involvement of the medical, nursing and administrative staff in the running and development of the hospital. The board paid close attention to the financial viability of the hospital and its relations with its GP 'customers'. The board members, led by the chair and chief executive, took overall responsibility for planning and management, but also concentrated on individual interests in specific areas of activity in which they had experience. In the hospital systems, procedures and IT area, the throughput and productivity of the hospital was improved significantly without lowering standards of clinical quality, and even initiated elective operations during the weekend to help reduce waiting lists. We implemented, with a small, keen and competent staff, a new hospital information system.

One major problem during this period was the proposed Nationwide Clearing Service, which would require data from every English hospital, covering all hospital episodes, to be sent to a new, large central computer system. It appeared that this massive sledgehammer had been proposed to try to crack the nut of a breakdown of viable statistics from the Hospital Activity Analysis (HAA) system, both locally and centrally. The widely differing estimates of NHS activity were causing embarrassment in Parliament. The intention apparently was to use the Clearing Service for the recording of all 'contract' episodes and use these to improve the statistics. Many Trusts, including ours, argued that as over 80 per cent of all activity was local, sending 100 per cent of our data to be cleared and relayed via a potentially unreliable and time-consuming central system was unnecessary, and that only 20 per cent of the 'cross-border' activity need be transmitted. As a member of an IT Advisory Committee to the Minister of State for Health, several papers were prepared advising against this solution.[14] But, despite long discussions with the NHS Chief Executive, the NHS Information Authority proceeded with the system. It was another example of an unsound, expensive IT strategy based upon expediency rather than sound planning, when other alternative methods were available. It did not appear to do much to improve the national statistics either.

CONSULTANCY

The DoH and the NHS have long been commissioners of consultancy studies and projects, and the corporate consultancy companies had assiduously developed significant shares of this very large market. Sifo had become a health care consultancy and project management company. The experience gained over the previous 20 years was used in preparing the NHS Training Authority Information Management and Technology (IMT) Training Strategy.[15, 16] It helped the Hong Kong Hospitals Authority to move from a disparate mix of state and private institutions to a single department in preparation for the British handover of Hong Kong to the Republic of China. The company helped the NHS Executive in setting up the systems and procedures for 'NHS Outposts', in a range of projects including performance indicators, case mix and contracting issues, care packages and cases, and local Health Authority IT strategies.

NHS SCOTLAND

In the 1990s, NHS Scotland invited tenders for the direction of a project for the radical overhaul of NHS Scotland's Scottish Morbidity Records (SMRs), and Sifo undertook the responsibility. Scotland had the vision to set up systems 30 years earlier for the collection, initially manual, of patient activity details covering in-patient, surgical, maternity, cancer, mental health and geriatric specialty episodes. The related IT systems had 'grown like Topsy' and the introduction of ICD-10 had led to the need for an appraisal of the data items to be captured and their processing. The project required the restructuring of all the records and data items, the redesign of the central statistical analysis systems, as well as changes to all the hospital and unit patient administration IT and manual systems throughout Scotland that supplied the original data. NHS Scotland had considered basing the new systems on Read Codes but this was found not to be viable given the nature of the Read Code structure and its relation to ICD-10.

The project took two years to develop but with a sound project structure, a sensitive project board, and the cooperation of the IT systems suppliers, it was successfully implemented in 1996.[17-19]

Outsourcing of IT systems had become fashionable in the NHS, as IT was not regarded as a 'core' business. One could, however, argue that IT is now simply part of the management and operational structure of any

NHS organization and having it supplied by an external organization could undermine commitment to the needs of the local organization. It could make as much sense as outsourcing nursing or laboratory services. Implementation of the SMR and other projects was complicated significantly by the large-scale outsourcing in 1995 of many of the national systems in Scotland to Computer Sciences Corporation (CSC). Approximately 500 NHS Scotland IT staff were transferred to CSC, and this number was reduced dramatically over the three and a half years of CSC's contract. Many of NHS Scotland's users perceived problems in the IT services being provided, and a complex contract structure may have caused problems of financial viability for all parties.

In the late 1990s, NHS Scotland decided to procure a new outsourcing contract. Sifo was given the responsibility for directing the procurement and subsequent management of the contract, which on this occasion was awarded to Sema. The IT systems and services involved all the national systems based on the Scottish unique personal identification number: the Community Health Index (CHI) number. This had originated in the Tayside Experimental Project in the 1960s. These systems included national, hospital and community patient identification, a wide range of child health and screening systems, national payroll, Compas PASs, data communications, and a range of financial and miscellaneous systems. Later the contract was extended to include a substantial part of the Scottish Care Index (SCI) programme.

The keys to successful outsourcing, if one decides to go for it, are the specification of the work, the structure of the contract, and the human relationships between the supplier and those who use the systems and direct their future development. For the procurement, which Sema won, a 'contract cost model' had been developed. It clearly defined the computer equipment, software, human resources and skills, and supporting components for each major area of NHS Scotland systems activity, combined with a strict procedure for tightly costed 'change control'. The contract gave the supplier an incentive to improve the quality of service whilst rationalizing its cost base. It gave NHS Scotland a clear controlled 'value for money' provision of its necessary IT services, and also allowed the development of new systems and technology approaches. The Accounts Commission, the equivalent of the National Audit Office in Scotland, examined the contract and its performance on several occasions and recommended the approach as a sound basis for public sector outsourcing contracts, not only in health. If such an approach had been adopted in England, it may have reduced the number of public sector IT contractual failures during the 1990s.

CONCLUSIONS

Medical informatics and its very wide range of applications were, and will continue to be, the most fascinating and potentially useful area of information technology applied in the service of humanity. The practitioners of 'health computing', whether those responsible for data capture and input, those who analyse the outputs, the terminal screen and computer operators, the programmers, systems designers, technical support advisers, data communications specialists, project managers, and even supervisory Board members, engage in their work with commitment, enthusiasm, creativity and sensitivity, for the benefit of the patients and the NHS staff.

Unfortunately the possibilities of medical informatics as they unfolded in the 1960s and 1970s were not realized in the 1980s and 1990s as they should have been. The investment and effective use of IT in the NHS and health services generally fell well behind its application in the private sector and other public sectors, including local government.

From the private sector viewpoint, the reason for this was, and maybe still is today, an excessive structural bureaucracy at the centre: a 'Civil Service' mindset of not being able to take a positive approach to the future; of being wary of all risk and not being able to take difficult decisions; a lack of confident, specialist IT knowledge and project management skills at senior management levels, because of the 'generalist' Whitehall approach; and a management inability to plan, direct, project manage and operationally run large-scale IT projects. Naturally, I think that the private sector, both large and smaller companies, should have been given a freer hand, under sound contractual control, to work directly with its NHS customers and users, to develop and implement innovative solutions making the best use of information technology. Paraphrasing Marlon Brando, 'UK health IT could have been great, could have been a contender', and Mao Tse-tung, 'but a thousand blossoms were not allowed to bloom'. There has been no cultural revolution.

Such innovation would naturally have to be within a framework of common data items, validation formulae, data-sets and data communications protocols. Not enough attention was paid to this and the work that was done was cumbersome, non-directed and practically unusable by the health IT industry. As a result there was a continual reinventing of the wheel by NHS users and IT suppliers: repeatedly analysing the same problems and developing new or marginally revised data-sets or procedures, for example in patient identification, in-patient management, outpatient management (still a wasteland), medical records, laboratories and so on, let alone payroll, financial records, stock-control and equipment maintenance, when many of the basic structures had been discovered many years before. Trying to adapt US and Australian health care systems to meet the NHS environment has not helped.

The real problems have been ineffective organization, indecisive planning, poor cost estimates and cost control, and using large-scale, top-down project structures, rehashing the PRINCE project management approach on every occasion.

Good medical informatics requires clear strategies and plans at national and local levels; real commercial awareness; effective, fast, procurement procedures; and careful anticipatory contract structures. It uses relatively small, committed working teams of users and dedicated IT professionals; the involvement of medical, nursing and all health care staff from project conception to full collaborative implementation; and strong, flexible project management with the total emphasis on successful implementation in the required timescales.

The health care systems applications targeted should be those that are of real value to NHS staff, with immediate short-term, as well as long-term benefits. The applications, with core common data-sets, should be progressively linked together in a comprehensive, economic telecommunications network, avoiding the top-down imposition of super large-scale systems. IT departments within NHS organizations should make reliable operational provision of services its prime objective, having undertaken rigorous testing of newly developed systems before they go live.

The large numbers of dedicated IT staff within the NHS and the supplying companies must be given the incentives to achieve the highest standards of professionalism, and allowed to develop their careers freely.

In summary, there should be better local planning and involvement, less central direction, and greater freedom for NHS organizations to engage with the numerous private sector companies and organizations that are keen to provide systems and expertise to the NHS.

That was the old, and recent, past. We may need some of these ideas today, with the NHS National Programme for Information Technology (NPfIT).

REFERENCES

1 Thomas, A.W. (1971) *Database Management Systems*. Management Centre Europe proceedings paper.

2 Thomas, A.W. (1973) New Zealand universities computer agreement. *Computing*, UK.

3 NZ Computer Society (1974) *Proposals for Privacy Legislation*. Social Implications Committee.

4 Thomas, A.W. (1973) The foundation and development of an online patient information system. *NZ Medical Information Processing Conference Proceedings*.

5 Thomas, A.W. (1974) Experience with an online patient information system. *European Computing Congress Proceedings*.

6 Thomas, A.W. (1974) A medical family register using generalized data base management software. In Anderson, J. and Forsythe, J.M. (eds) *MEDINFO 74*, Stockholm, (pages unknown). North-Holland, Amsterdam.

7 Aitkin, A. (1973) (title unknown) In Thomas, A.W. (ed.) *Proceedings of the NZ Medical Information Processing Conference*, University of Otago, (pages unknown).

8   Thomas, A.W. (1975) Standard computer systems and their characteristics. *NHS Computer Officers Workshop Proceedings*.

9   DHSS (1976) *Guidelines on the Documentation, Implementation and Maintenance of Standard Systems*. Dataskil.

10 Thomas, A., Sharp, A. and Parkinson, A.J. (1977) *Information Processing Requirements of an Area Health Authority*. Welsh Health Technical Services Organisation.

11 Thomas, A.W. (1979) The computer information systems of the Riyadh/Al Kharj Hospitals Programme. *First Saudi Arabian National Computer Conference Proceedings*, Dhahran.

12 Thomas, A.W. (1983) *Review of Health Care Computing in the Middle East*. British Computer Society Medical Specialist Seminar. BCS.

13 Thomas, A.W. (1987) Information systems strategy for the NHS in Wales. In (eds unknown) *BCS Health Computing Conference Proceedings*, (pages unknown). BCS.

14 Thomas, A.W. (1994) *NHS Nationwide Clearing Service: Implications*. Paper for Minister of State for Health. Unpublished/internal document.

15 Thomas, A.W. (1988) NHS Information Management and Technology Strategy for Training and Staff Development. In (eds unknown) *BCS Health Computing Conference Proceedings*, (pages unknown). BCS.

16 Thomas, A.W. (year unknown) The NHS Training Authority Information Management and Technology Training Strategy. *British Journal of Healthcare Computing*, (issue and pages unknown).

17 Thomas, A.W. and McKerron. J. (1994) *New Core Patient Profile Information*. Requirement Specification 1, Core Patient Profile Information in Scottish Hospitals (Coppish) Scottish Morbidity Records (SMRs). NHSiS Information and Statistics Division.

18 Thomas, A.W. (1994) *Implementation Implications for Systems*. Requirements Specification 2, Coppish SMR Project. NHSiS Information and Statistics Division.

19 McKerron, J. (1995) *Coppish SMR Data Manual*. NHSiS Information and Statistics Division.

28 Developing the health informatics professional*

JEAN ROBERTS

Since the early 1960s there have been few implementations of health informatics (HI) that have not fragmented in some way: by sector within heath care, by geographic location or by professional focus. The last has moved from unidisciplinary informatics to multidisciplinary informatics as the clinical team (tactically) and the management and policy people (strategically) progressively share informatics components and functionality. It now includes social care, lifestyle management and the citizen as the active subject of care.

At the beginning, computing in and for the health domain could not claim to be health informatics, the term used today, because the focus was on applications to address one functional area for one professional group in one location. The situation has changed immeasurably, as it has in other spheres such as commerce. Some aspects of the journey are reflected in personal careers and in a much wider involvement with professional informatics societies.

STARTING IN HEALTH INFORMATICS

In 1969/70, some of the operational areas that are now part of health informatics were within the purview of the local authorities; and it was for a local authority that I, as a computer science industrial year student, wrote and implemented an Immunization and Vaccination system. Responsibility for such activities now forms part of the community sector provision within the NHS.

In-house computing facilities in health authorities outside major conurbations were limited, when in 1977, Dr Peter W. Harvey and Dr John A. Farrar, pioneers at Royal Lancaster Infirmary (RLI), called on the facilities at the local University of Lancaster to process their first surveys of activity in the A&E unit. The RLI was a typical urban/rural district hospital. In the mid-1970s, there were some computing facilities at local district level in some parts of the country, such as central Birmingham, but nothing in peripheral

---

* This chapter is based on the author's Doctoral Thesis awarded at the University of Teesside (2006) and based on her published work entitled 'The contribution made to the support of health care delivery and management by health informatics at a local and national level over a period of thirty years'.

units like the RLI. Most local computing support was in the urban teaching hospitals and focused on one clinical specialty. Computer processing in the NHS at regional and national levels was on mainframes of considerable size but of limited processing power.

The university's link to large computing power through the universities' network, utilizing machines in both Lancaster and Manchester, and through them to those in the USA, allowed the use of an early version of the Statistical Processing for Social Sciences (SPSS) application.

In the initial stages, the clinicians at the RLI carried out their own data capture by transcription to self-designed forms, before a specialist data preparation department produced Hollerinth cards. Members of university staff helped external clients with their computing requirements. The additional resource was necessary because the clinicians were not computer-trained. At that time they understood what analyses they required but were not able to program the systems to do the investigations. SPSS was not as user-friendly in its configuration of queries as it is now.

Resources for local health organizations were frequently provided by centralized services or by secondment-funding from the Regional Health Board. Formal indications that challenged such outposted resources with relocation to another district necessitated the Lancaster Health Authority to declare 'UDI'. At that point, I became the first District Computer Services Officer in the country. I joined a still small community of diverse professionals designing, developing and delivering applications in-house and for the NHS.

The coding system used to translate data from the casualty records card, filled in by doctors when the patient presented, was locally defined.[1] The International Classification of Diseases (ICD) available at the time did not give the specificity of coding required to answer operational questions in a casualty department. Problems included expressing: 'are there any marked trends in the ailments of specific cohorts by age, gender, home location or presenting condition?' and 'are accidents caused by individuals doing tasks they are not trained for, for example DIY?'.

This period was also when the GPs introduced a formal appointment system locally and a survey was done on data-sets before and after their introduction. One hypothesis tested in the Lancaster research was whether GPs introducing appointments systems would make a difference to the profile of attendance at the casualty unit; the outcome being that they did not. The depth and range of the Delphic analysis of the data collected could not have been contemplated without informatics processing capability and a creative use of existing off-the-shelf software.

At the time, SPSS only produced tabular output. The mapping of data occurrences against geographic Enumeration Districts, to provide a picture of local activity, necessitated coloured crayons on an outline map. The correlation of data from the analysis with the General Household Survey and National Census entailed the same method with local maps to identify 'hot spots' and areas of clinical concern. One example was the identification of a distinct cohort of prisoners' relatives making particular demands on health facilities in Lancaster.

Publication of the results describing different aspects of the A&E activity study in various journals represented a very early example of locally based computing support to the NHS.[2-4] The study was available for local health management use and as part of national evaluation. The same data was subsequently used to validate selected tenets of the Black Report,[5] in terms of the types of emergency presentation at casualty from different population cohorts.

Direct results, through decision support to the District Community Physician (now more typically referred to as a Director of Public Health) and his team, were significant clinical and health management changes notably:

- changes to community support – in terms of outreach information units that pre-dated the NHS Direct telephone triage services by many years;
- changes to hospital working practices, in terms of staffing up peak times and with specific competencies;
- input to the national understanding of accident and emergency attendance profiles;
- triangulation of hypotheses of public health/population research done elsewhere.[5]

KÖRNER

In the 1980s, the NHS/DHSS Steering Group on Health Services Information (Körner initiatives) invited operational practitioners from a number of disciplines to help design the minimum data-set requirements for many aspects of information management in the NHS; in my case the non-hospital sector. The Körner definitions were successful and still form the basis for many items in a health informatics system. The Government did not allocate funding to take the definition phase forward into national implementation, preferring to indicate that cost reduction through change of work practice would cover costs. As this was not feasible the implementation of IT in the NHS continued in a fragmented way. The principles defined by the Körner Groups that distinguish data from information (being in the right format, at the right time and presented in a manner appropriate to the use to which it is going to be made) are still used. They can be identified in the (National) detailed Care Records Specification today.[6]

The idea of common minimum data-sets without which it was deemed impossible to be able to manage health services effectively went out of fashion when Körner was not allocated funding for implementation but are now re-emerging to support the National Strategic Frameworks.[7]

CLINICAL STUDIES

Local computing officers supported local operational practices on an ad hoc basis, for example Lancaster designed, developed and operated programs for calculating complex intravenous feeding regimes based on pathology test results.[8, 9] Interestingly, there were contradictory clinical views of the theoretical basis for the calculation, which resulted in its use by some clinical teams and not others. Some of this local programming was rolled out to other locations, but at that time many applications remained where they were first written. Other local bespoke programs were written for operational use, for example to evaluate the signal traces produced from electromyographic investigations to detect the differences between the profile produced by damaged or diseased muscle,[10] and another to support the complicated rosters required for allocating nurses in training to wards where the necessary high-level staff support was available.[11]

These local developments were carried out on a Ferranti Argus computer, our first major in-house computer, actually installed for the transfer and customization of a pathology management system, Delphi-Phoenix. The Phoenix system originated at the post-graduate Hammersmith Hospital and was re-versioned from a CTL base to the Ferranti Argus at Lancaster. Delphi was not intended to indicate we were unsure of the outcomes of the project, but it was indicative of the humour of the team leader (P.W. Harvey) that it locally meant 'Don't Ever Let Peter Harvey Interfere'.

Innovative design, re-versioning, development and implementation of a leading localized pathology system in the mid-1970s resulted in a system that was locally fit for purpose and was incorporated in the commercial portfolio of a then major departmental system producer: Ferranti Systems. The transfer from a large London teaching hospital to a district general hospital required extensive reconfiguration because the test range of the RLI was wider than that of a specialist-focused teaching unit: including regular lead levels testing for the local nuclear facility. In addition, the normal ranges of test results were required to be more flexible (including for non-human samples provided from veterinary 'patients'), and the professional mix of clinical staff who relied on outputs from the pathology systems was more extensive.

LOCAL HEALTH COMPUTING

The emphasis in 1983 was:

A   clinical services

B   general services

C   personnel

D   finance.

The proposal was that services:

> *. . . should have the objective of improving efficiency of resource use, while maintaining the same standard of service and where possible, achieving recurring savings.*
>
> REFERENCE 12

The general rule was that:

> *. . . any computerized departmental information system [should] be made of discrete modules, each of which is designed to tackle a specific operational task. A modular system can be expanded or modified and enhanced at comparatively low cost. It also allows a department the combination of modules which meet the specific user needs.*
>
> REFERENCE 12

This was the beginning of the 'best of breed' concept that has necessitated interface engines, rather than integration, in many cases.

It would be untrue to say that application linkage was not considered in 1983. It would 'depend on management requirements' and may be 'essential for clinical services' to transfer data files from one system to another, transmit messages, relate data held in one system with that in another and merge data from more than one system. The architecture at that time certainly did not support the real-time use of data from many sources on an open platform that is striven for today.

Pragmatism in the use of informatics in health was demonstrated by the statement that:

> *It is technically possible to arrange for all users of clinical systems to input data to the patient register as well as obtain information from it. However, the feasibility of the technology is not a justification for its implementation; of paramount importance is the need for organizational arrangements that ensure the quality of the data entered. It is because of the latter considerations that most districts with a patient register have limited the right to enter data on it to trained medical records staff.*
>
> REFERENCE 13

The roll-out of computing power into the local health areas outside large cities was slow in the early days. That is not to say that there were no 'islands of development'; for example Dr John Bryden, a community physician and epidemiologist from Glasgow, was analysing head injuries data, and Mike Garner was introducing computers for administration support in a District Health Authority in Huntingdon.

Overall, the phasing of computing usage was typically regional and national health computers for standard systems, such as payroll and governmental statistical returns, and third-party, usually academic, facilities for ad hoc surveys. This was followed by the introduction of stand-alone services to support operational health processes, for example computerization of the biochemical laboratory results, reporting administration, and then increasing the ongoing operation of ad hoc/ beacon solutions on a wider basis, and for routine use.

However, this migration indicated a need for computing power at local level. *Developing a District IT Policy* describes one of the next major steps in informatics to support health.[12] The information requirements for the NHS in England developed by the Körner Working Groups were, as stated by Edith Körner (in the report's Preface):

> ... to engender debate about and progress towards the introduction of information technology at the district level of the NHS.

REFERENCE 12

Notable about the Körner initiatives was the involvement of operational field professionals in strategic developments. The Körner initiatives were the first to explicitly recognize that:

> ... health authorities and their senior officers rely on the availability of timely, relevant information [for their] onerous and complex duties.

REFERENCE 12

It is interesting to see that the Körner initiative stated that:

> The ability to transmit electronically text, pictures and numbers can greatly improve communications between health professionals thus allowing them more time to care for patients.

REFERENCE 12

This objective is still a goal, not a total reality, today.

The realism was also necessary at that time because all staff that used computers for a substantial part of their duties were eligible for an additional (Whitley Council) salary allowance. The Körner Reports concluded that the next steps in the migration were not going to be easy.

> To the NHS Computer Policy Committee falls the difficult task of creating an environment in which staff from district, region and the commercial sector collaborate to produce the technological evolution in information and administration systems which is required urgently to help the performance of NHS managers.

REFERENCE 12

Echoes of these challenges still remain today.

WORKING AT A NHS REGION

By the 1980s the North Western Regional Health Authority, one of the nine English health regions, was in a position to commission and roll out both Patient Administration and Manpower Information Systems. Working with suppliers who were developing bespoke solutions for health, with the added complexity of customization requirements for each local district situation, was not unlike the challenges facing NHS *Connecting for Health* (NHS CFH) today. Nineteen districts were then receiving the systems in a staged implementation plan, each working from a different base with different priorities and staff competencies – and of course the programme attracted press interest even then.

In parallel to the PAS, the region was developing a manpower management solution with another vendor (EDS). Negotiation with the suppliers was, in pre-PRINCE project management times, a challenge, but the solutions produced were acceptable to the end-users and were still (mainly) in service until very recently. Some of them are still operational 20 years on – so much for the amortization of costs and replacement within seven years.

At this time (1987), the PAS reflected the activity within the hospital and did little to address the potential patients in the wider community who had received no acute health care locally. The statistical returns generated in the mid-1980s were basic and submitted to the local RHA system for processing. Errors were referred back with little explanation or guidance, and in extreme cases creative amendments were added to make the figures balance. Any enquiry on your data, whether relative to other similar districts or stand-alone, required a formal request to the RHA Computer Centre and an eventual response. The system was therefore very cumbersome and not responsive, so it was little used to inform local tactical decision-making.

Trying to create a PAS with a master index (PMI/MPI) prompted questions about data quality and the primacy of the content, especially names and addresses, held in each previous system. Pressure was gaining for data exchange, by pump priming and periodic (overnight) updates rather than sophisticated interfacing.

WORKING IN A VENDOR ENVIRONMENT

In the late 1980s, as today, many computing staff from the NHS, including myself, moved into working for the NHS, rather than within it.

Going to an established IT company in its early days of entry to the health domain was interesting and challenging. In 2007, UK Council for Health Informatics Professions (UKCHIP) registration has still to address the simplistic views of health informatics as 'the same' as any other sector IT. There were developments of new bespoke IT for health and consideration of re-versioning of solutions from other sectors into the health domain, such as travel agent booking and engineering queuing theory models. Companies participated in the Healthcare Computing congress in many ways, both showcasing product portfolios and by reflecting on where health and other sectors were the same and different. At a Harrogate meeting in the late 1980s, the then (MD) of AT&T ISTEL, John Leighfield (now Master of the Worshipful Company of Information Technologists), indicated that the health sector was typically 15 years behind industry in deployment of technologies. There was little outcry challenging his statement.

AT&T ISTEL then incorporated the Trent Regional Computer Centre (in line with a trend happening with other Regional Computing Centres and other commercial companies) and continued to successfully roll out a commercial customizable (and therefore slightly different in each installation) version of a PAS for a number of years.

Organizational changes and the innovative ability of systems to interwork reflected an emerging need in the NHS to have a clear picture of the local area on which to base plans and strategy. Very little progress had been made since the National Strategic Framework for Information Management in the hospital and community services was released in late 1986. The strategy claimed three facets:

- Regional information strategies deriving information requirements from service plans.

- A minimum set of common standards, for example, for data definitions and communication.

- Works plans relating to requirements for disseminating information and indices of performance.

The paper 'Towards an index of the community' drew heavily on market knowledge and a specific project being carried out in one area.[13] It flagged up the usefulness of an index of a geographic area, rather than the then current, stand-alone hospital, community and separate primary care systems. The concept of profiling a local health community holistically remains viable, and in practice is still not widely available, albeit it was a goal of the NHS CFH National Programme for IT.

Contractually, in 1987, general practice was under family practice groupings and it was legally not permissible to formally share data with hospitals, although it did sometimes happen in practice (for example the Salford joint administrative register). Community profiles were not as important to hospitals because they tended to 'body shop': dealing with patients as they were referred or presented. Hospitals based projected demands for staff and facilities on previous activity not local potential demands.

This fragmentation has now radically changed. A clearer picture of local needs, demands and previous activity in all areas of care delivery, disease prevalence and resource deployment is now recognized. The multifactored profiles could not have been delivered so easily without the more complex computer systems that are now available for epidemiological and resource demand analysis.

The NHS was moving towards an 'internal market' situation where patients did not necessarily have to be treated in the nearest hospital to their home, but could be directed to a unit that specialized in the treatment for specific clinical conditions. By 1988 there were local networks in place, albeit it with dumb terminals and some IBM, Amstrad and Atari personal computers. A DTI paper, in 1990, highlighted the 20 agencies and professions who could realize benefits from accessing an index of the whole community using, as examples, the 'islands of technology' seen frequently at this time in different organizations.[14] The paper advocated that:

> ... *the ability to transfer data from one system to another either to pump-prime a system or to share data should become a totally automated process.*

REFERENCE 14

In fact, the transition of data from existing systems to the NPfIT solutions was still, in 2005, proving problematical and expensive, and identifying many data-quality issues. The 1999 paper cited the needs of resource management (a national initiative), local performance indices and patient demographic and epidemiological profiles as some of the functions requiring automatic data transfer.

By 2004, digital TV and cable services were in use in some pilot areas, but were being devalued as a 'mainstream requirement for operational (support), rather than (for) citizens'.[14] The potential for digital services came to the fore within the National Programme for IT, the NHS Direct plan and as a contemporary research theme of e-health. The emergence of a broadband telecommunications capacity, which current Government plans indicate should be available across the country for business and public sector reasons, makes the digital solution more realistic and viable.

In 1989, the proposal was for a staged approach to integrated services – with access to knowledge management through a network link from Regional Health and local networks to a third-party network by dial-up connections. Patently this proposal has been subsequently superseded by direct access to the National Library of Medicine's Medline clinical databases and the generic Google®, etc. services of today. Other 'futuristic' suggestions made at that time and subsequently implemented in the NHS included:

- Store and forward and electronic data interchange (EDI) facilities to make supplies ordering more efficient.
- Appointment booking forms and legible discharge summaries.
- Customized screen enquiry formats for individual clinical teams.
- GP access directly to PAS through third-party gateways (little mention of secure gateway firewalls at this time).
- The introduction of Value-Added Data Services to the NHS as used by commercial organizations.
- Population registers as deployed in Scotland (GPASS and the common, national, community health index (CHI) number for each individual), which subsequently became conceptually the unique NHS number in England, now issued at birth to all babies.

CONSULTANCY – NICHE TO MULTINATIONAL MODELS

The late 1980s frequently saw product planning by individual application vendors, and strategic market analysis to design and develop new solutions. There was still a significant requirement for end-user organizational customization. There were now insufficient in-house resources and large and niche consultancies were addressing all aspects of health management, informatics being only one of these. For example, Greenhalgh and Company Limited, a niche consultancy of about 50 professional staff, covered technical assessment, evaluation and review of multinational research project bids, progress for the European Community (later Information Society and Media Directorate-General) research and development programmes, and a major three-year period in the IT EDUCTRA consortium.[15] This produced core educational materials about health informatics concepts. Transnational opportunities resulted in a widely circulated CD-Rom containing material in various European languages.

Independent, comparative analysis of market products in the areas of nursing systems and applications for operating theatre management were commissioned by suppliers and purchased by operational NHS units and others for reference in application specification, solution procurement and comparison of functionality. There were a number of substantial systems in operation at the time of these Greenhalgh reviews, demonstrating a vibrancy and dynamism of the market not prevalent today.

Education

Particularly wide recognition was gained by the *Using Information Effectively* series of printed open learning materials, referred to as the Rainbow series because of its construction as brightly coloured boxed sets addressing information use in various scenarios (see also Chapter 21):[16]

- *Using Information in Managing the Nursing Resource* (1992) was prompted as a response to the national RMI giving senior nurses a demanding new role in management.

- *Using Information in Contracting* (1993) was to support the fund-holding environment.

- *Using Information Effectively in Practice Management* (1994) was to address the information requirements of primary care.

- *Using Clinical Information in an Integrated Environment* (1996) addressed the full spectrum of care situations.

Retrospectively, academic credits and NHS recognition for the materials and related courses were sought. This material pre-dated the widespread use of CD-Rom and web-based materials for educational purposes, particularly in open learning, but is still frequently referred to and recognized in the health domain.

Systems procurement

During the period 1994/95, and related to work on procurement processes, product specification and contract assessment for both commercial and NHS end-users, a *Healthcare Computing* paper posited the effects of the concept of 'whole-life costs' of NHS solutions that resulted in almost all systems requirements being put out for Europe-wide commercial proposals.[17] This then required a more complex, open, auditable and structured assessment protocol for arriving at a solution choice.

The founding countries of the European Union signed the Treaty of Rome in 1957. The UK joined in 1973. By 1995 when Austria, Sweden and Finland joined there were 15 countries and accessibility to markets and procurement opportunities had been widened. European law was that contracts over a certain level for significant capital goods had to be advertised for international response. The scale of the involvement of the European Community in national procurements could also be construed as interference.

When large-scale computing power was being acquired for an increasing number of locations nationally the mass of documentation that was issued to guide potential purchasers appeared, in some cases, to complicate the process. In 1995, application systems, such as those for supporting the allocation of nurses within shift patterns, or systems to support the administration of general practices were typically felt to be 'low cost'. When the concept of 'whole-life cost', required by EC regulations, was brought to the attention of systems users and specifiers, it caused concern that such relatively small systems were also required to use the more complex procurement processes of the larger systems. Until then, only those who contractually procured goods and services for the NHS had known this information.

The European procurement process for locally funded applications and systems contrasted with the previous charitable trust, grant or ad hoc funding where decisions had been made locally. Under EC law, all procurements which had a projected whole-life cost of over 100,000 ECUs (approximately £70,000 at that time) had to go to international advertisement in the Official Journal of the European Community (OJEC). The costs of many factors that had a considerable impact on the success or failure of implementations in the NHS had been previously hidden. For example the costs to be covered should have included those of training staff to use the new functionality and the staff coverage necessary in the department whilst staff were receiving such training. The costs of modifications to the building and estate, where the application was to be housed, and the costs of additional utilities while the system was in operation were also sometimes omitted in the business plan.

The regulations about true costs still apply today although the value levels have changed. Early procurements after the regulations came in sometimes attempted to circumvent the processes by making purchases in tranches.

However, a ruling against a Scandinavian company for this, resulting in them being instructed to restart their procurement from the beginning, soon put paid to such creative procurement practices.

The NHS has since been through a period (1998–2002) where funding was ostensibly explicitly allocated (referred to as hypothecated funding) to IT developments. Then, because of competing priorities, the funding was taken away to address, for example, waiting list reduction and winter pressures. This caused many problems around commitments and plans that it was not then possible to implement. This caused frustration and loss of motivation among NHS staff at local levels.

The new model of procurement, described in the twenty-first century strategy,[18] has changed again, bringing in plans for the transfer of risk to commercial players and closer partnerships with applications and service providers. The model has yet to be proven in this domain, and, in fact, there are a number of case studies highlighting 'cracks' in the processes employed in this new millennium. It will still be necessary that, in order to satisfy European Community legislation, a wide procurement call be undertaken after the UK Treasury is satisfied with the business case(s) under Office of Government Commerce regulations.[19]

The NHS in England procurement method, used by the NHS Connecting for Health programme, is now being closely scrutinized and replicated in other Government and public sector procurements: in contract detail terms, performance requirements and on costs. As yet (2007) no health informatics procurement has reached full implementation, so there are still concerns about the overall costs and applicability when implementation, not just installation, is evaluated. The whole-life cost model should be reflected in the Office of Government Computing Risk Analysis Profile content, but no related detail can be identified in publically available documents to date.[19] In addition, re-versioning of NPfIT is currently reallocating cost implications from central to local levels, with unclear results. In terms of the niche/smaller systems solutions, ostensibly still procurable outside the national programme, the model can still apply, albeit many developments outside NPfIT remain to a greater or lesser extent 'blighted' by the focus on the major English initiative.

Transferring research into practice

Transferring Research to Practice for Health Care Improvement: Information Technology Strategies from the United States and the European Union,[20] addressed factors surrounding the transition of research solutions and concepts into the real market. Not all of the ideas in this book are followed today, although many universities do have proactive units that identify near-to-market and potential income generation ideas and take them forward, for example at the University of Ulster and at Leeds.

PERSONAL CONSULTANCY

The American mega-model of contracting for consultancy support in health informatics was tested in the UK in the late 1990s prior to roll out across Europe. Although some significant projects were secured, the market was not deemed to grow at a necessary rate so not all of those multinational corporate bodies are still active in the UK and mainland Europe.

In 2007, the model of niche consultants, both individually and collectively on a contract-by-contract basis, as associates on specific projects (such as scrutinizing the efficacy of e-learning modes and investigating the viability of health and social care informatics convergence) still operates viably. The development of learning materials for the Open University on knowledge, information and care for health and social service staff was produced by this model.[21]

ACADEMIA AGAIN

In 2006, market analyses indicated a considerable need for more domain sensitive and technology competent professionals. A significant number of undergraduate and post-graduate academic courses, and increasing vocational recognition, were emerging. As an example, this chapter's author is now a Senior Lecturer in Health Informatics at the University of Central Lancashire (Uclan) in the School of Health and Post-Graduate Medicine, having been an occasional lecturer over the years for various universities, and a founder member of the Centre for Health Informatics Research and Development, University of Winchester.

Design, development and accreditation of innovative Foundation Degrees in Health Informatics, and domain-specific short courses are also on the increase. Foundation degrees are a core plank in the Government target of over 50 per cent participation in higher education. In parallel to the Modern Apprenticeship and Cadet Scheme concepts, the Foundation Degree being pioneered at Uclan offers the optional opportunity to study part-time while still carrying out paid work. In a similar manner to the post-graduate qualifications, PG Certificate, PG Diploma and Masters degrees, students are being taught by a blended learning route.[22] Approaches like these can be seen in nearly 20 academic locations across the UK in 2007, and coverage has not yet reached a plateau.

BCS INVOLVEMENT

Professional and learned society involvement is a personal choice, sometimes encouraged by one's peers, and geographic contribution is patchy. From 1972, and under the visionary leadership of Dr Peter Harvey and Dr John Farrar, Lancaster played a part in the activities of the BCS health groups. Their participation was on a national and international basis, involving specialist groups in IMIA and EFMI. Such activities have been complemented by national initiatives for the BCS through the General Council and predominantly the BCS Health Informatics Specialist Groups Coordinating Committee (later called the BCS Health Informatics Forum (BCS HIF)). This 'gift' work provides excellent opportunities to validate one's work in a wider context, and to seek information about activities elsewhere that can frame local developments through others' experiences. Ongoing involvement brings rich rewards to both organizational development and to individuals over time.

An early major national meeting, a precursor of the *Current Perspectives* event, was held in a large 'shed' at Lancaster University in 1978, led by David Kenny of IMIA Working Group 4 just after their North-Holland book on Data Protection was issued.[23]

The *Healthcare Computing* event (previously called *Current Perspectives*) has since achieved status as the largest event of its kind in Europe with over 300 exhibiting organizations involved and a well-received five-stream scientific programme.

The BCS HIF has supported attendance for national representatives, specialist group chairs and individuals to share their findings and developments on a wider basis at many international events. The human network of experts generated by such activities is extensive and BCS HIF is now attempting to quantify its coverage.

The BCS has, since 1974, supported its members in attendance, presentation, publication and reporting of their findings at other international congresses. These include the American Medical Informatics Association Spring and Fall meetings, IMIA Regional events (such as regional APAMI, IMIA-LAC and meetings in the emerging African HELINA area) and specialist area events like the biennial Nursing Informatics meeting and the World Organization of National Colleges and Academics (WONCA) primary care meetings. The international knowledge exchange is aided day to day increasingly by initiatives like HIF-net@WHO and the Institute for Development Policy and Management.[24]

INTERNATIONAL IMPACT

There is a suspicion that the UK does not play as full a role as it might on the international stage. The recognition, in *Delivering 21st Century IT Support for the NHS*,[18] that a national strategic programme for the NHS in England cannot ignore health informatics originating outside England is to be welcomed. BCS activities include supporting European Health Telematics standards development, tracking international health IT strategy developments such as those in New Zealand and Australia as well as inviting external internationally respected experts to review progress. From 1986, with *MEDINFO* in Washington, DC, to 1995, with *MEDINFO* in Vancouver, the DoH, with the BCS, supported exhibitions as outward missions to health informatics congresses. These events contributed considerably to the networking of experts and specialists from the UK and abroad.

The strategic panel of the BCS HIF and its predecessors established a mechanism for coordinating collective responses to national consultations. Subjects included the NHS University, research management for the NHS, the National Detailed Care Records Service functionality, and the national Foresight direction for development. Consultations are now publically available through the BCS HIF website (www.bcshif.org).

Radical Steps

Many of the national statements represented top-down guidance that was supplemented by local documents. Most recently the *Delivering 21st Century IT Support for the NHS* description requires radical changes yet again, while building on the main themes, such as electronic patient records, e-booking of appointments and 24/7 access to records. It was because of the grass-roots unease and disquiet in operational health organizations, commercial players and relevant academia that the concept of the *Radical Steps* initiative was raised.

The *Radical Steps* action research/collaborative inquiry methodology has been used to tap into the experiences of practitioners in the field. The 2006 'Way Forward' statement generated considerable discussion and recommendations for action.[25] The results from this type of consultation and the mechanics of its processes appear to be of interest across the European Community. Interest has been shown internationally through the Information Society Directorate of the European Commission and at a *MEDINFO 2004* presentation.

It is interesting to contrast the recent apparent reluctance to involve experts in the field in developments within the NHSCFH with the stance taken by the Körner committees of the 1980s. Körner sought to 'engender debate and progress towards the introduction of IT'. A widely held view, often aired in the media, points to the difference being, in part, the character of those leading the management of the changes in each period.

The *Radical Steps* model has been applied in the area of open-source operation in the health domain,[26] and to address the educational laws, concepts and theories that constitute informatics in support of health.[27]

MATURING THE PROFESSION

Informatics in the health domain has gone from the early 'islands' to a much more embedded function in the fabric of health care delivery and management today. *The UK has been in IT for years!*, a short summary of the state of the health informatics market, described this distribution in 1990.[28] There was a growing interest at the time among the pharmaceutical and non-computer vendor industries in health informatics. This summary was an early attempt to inform those in the pharmacology business where health informatics was active, long before the exciting concept of the UK BioBank was established.[29]

In the late 1980s, the market horizon for UK health informatics was not just parochial. It had been opened up to more international contributions also. It was presented for an international audience in the journal *Medical Horizons*, read by the 'trade' and a major communication vehicle for the Association of British Healthcare Industries (ABHI). The ABHI also supported a UK showcase at the IMIA event held, in part, in Singapore in 1989.

The pharmaceutical focus has subsequently moved to bioinformatics and genomic informatics. An interest in operational informatics remains because the costs of identifying their own subject databases for random controlled trials are becoming prohibitive. Some research can be carried out, with the appropriate privacy permissions, on data already available in the local community indexes. In addition, the research themes were still valid in 2004 and reflected a potential cost-releasing area in e-prescriptions (NPfIT) and e-prescribing in the DoH e-health research priorities. Other members of the ABHI are involved in the commissioning of new hospitals and the like abroad. They make alliances direct with solution vendors if there is an informatics component within a large build.

The emerging service provision model, described in the *21st Century* strategy, had local service providers (LSPs) offering solutions to five clustered areas of health organizations in England,[6] bringing together a stable of application providers, addressing a full spectrum of health informatics applications that can be delivered as a service to the NHS. (Note that the cluster structure had by 2006 been revised, but the detail is still volatile so it is not addressed here.) Various major companies, some ABHI members, who currently are not providing informatics services to the NHS (but are in some cases to other UK public sector bodies) have put together teams to address this new requirement for holistic project management.

The encouragement of wider groups to participate in health informatics has frequently been inhibited by the inappropriate perceptions of those organizationally above the potential participant leading to a reluctance to invest in communication about their activities, whether to national or international meetings. *Image Of A Global Village* was intended to stimulate a wider attendance in international events,[30] both directly by enthusing others to write papers and to inform them of the types of worthwhile outputs that accrue, such as:

- knowledge of applications that are operational elsewhere;
- international research that could be of UK value and adaptable;
- access to extensive personal and professional contacts carrying out similar work in other countries.

It also sought to explain to managers of potential attendees that there could be considerable benefits to local projects if staff are allowed to attend, such as gaining a broader perspective that may result in a more effective local solution.

WHAT IS A HEALTH INFORMATICIAN?

Explicit informatics to support health had become more widespread in 1998 but tensions as to how to actually define a health informatician or the subject of health informatics itself kept re-emerging. Various analyses like that done by Stephen Kay on comparative qualities of academic and professional qualifications, in October 1995, had looked at the subject.[31] His work was carried out to identify synergy and differences between the BCS Groups and the newly formed ASSIST organization for information management and technology (IM&T) staff within the NHS, but did not resolve the problem of identity. No answers were forthcoming, either, from the international field, despite IMIA Working Group 1 (Education) repeatedly revisiting the issue. The challenge was to try to tease out some consistency from the eclectic experiences and prevalent qualifications.

The *MEDINFO* congress in 1998, held in Seoul, explored related issues in 'Health or unidisciplinary informatics initiatives – why are we where we are today?'[32] This presentation prompted considerable debate. The premise was that the development of informatics to support health (from the early 1960s in many countries) initially was clinical discipline-specific or purely administrative. Various factors have reduced the segregation between the health care professions, who now frequently work in multidisciplinary project management, implementation and operational teams. However, it is interesting to note that, even by 2002, where multidisciplinary projects existed, there were still indications of the lack of a typical mix. For example, few female nurses were leading projects even when there was a predominant nursing informatics component involved; also cross-sector projects were frequently led from the acute side; and projects predominantly have an informatics lead, not one drawn from the potential end-users.

The paper considered a five-phase emergence pattern for health informatics projects over time, namely:

- **Isolationist**, where computer processing was a 'support' function done by a separate profession, data being transferred, processed and returned sequentially with little professional interaction between users and technologists.

- **Selective involvement** of clinical or other relevant professionals in specifying system requirements.

- **Anarchic** 'free for all' where a wider range of professionals was involved and the potential for data duplication, parallel processing and inconsistencies between stand-alone systems were identified. This phase typically mirrored the introduction of minicomputers in individual departments where the computing functions could be carried out by technicians and interested parties within the department and not be specialist 'computer people'. Over time, this duplication was reduced through system interfacing or in some (well-blessed?) investment in 'big bang' new (more holistic) systems that addressed administrative processing and clinical functions in a closer coupled manner.

- **Increased synergy** created as the systems matured. For example data collected for one purpose such as staff operational management (rostering, pay and rations) was used for less direct functions such as strategic planning, skill mix analysis, audit and governance enquiries.

- **Integrated recording**, present since 1998. In a patient situation, an Integrated Care Records Service (IRCS) was outlined as a focus of one of the three pillars of the *Delivering 21st Century IT Support for the NHS*.[18] Back in 1998, the same philosophy was bringing nursing, paramedical and medical notes together to present a richer picture of patient status. The challenges then, as now, were mainly in the area of ensuring clarity in the content from the different sources.

Coding and classifications systems like the Read Coded Clinical Terms, International Classification of Diseases (ICD), North American Nursing Diagnosis Association (NANDA) and Systematized Nomenclature of Medical Terms (SNOMED) conventions were tools used to attempt to achieve comparability, consistency and compatibility of record content regardless of source. The challenge of integrating patient/client records from various sources will reoccur when health and social welfare organizations collaborate more closely in client/patient records. Solutions are not easy to identify or implement. Informatics is perceived as a lesser threat to professional domains than the working practices of other professionals and may conversely provide a stimulus to a more rapid introduction of multi-sector working.

UK Council for Health Informatics Professions

Ideas regarding a possible profession were mooted in 1987.[33] Exploration of a UK Council for Health Informatics Professions (UKCHIP) commenced in 2002.[34] The BCS HIF was one of the founding drivers. The experiences in other countries have proved useful input to the ongoing debate. UKCHIP was established as a body dedicated to increasing the professionalism of health informatics through accreditation, registration and possible regulation.[35]

UKCHIP will at last formalize the situation and define and position informaticians. It will use similar methods and models as those of the medical Royal Colleges and the Health Professions Council for medical physics technicians. It is interesting to note that in parallel, but slightly lagging, the UK Government agencies were considering a licensing procedure for all IT practitioners.[36]

As the operational standards for UKCHIP and the academic comparability and mapping of awards and vocational qualifications in health informatics become more stable, it will become easier to effect mobility. As international prima facie standards are formally adopted this can be across a wider transnational domain, in and out of the public sector, to and from commercial players. The domain definition is firming up in its traditional areas, but must remain capable of accommodating emerging practitioners: for example in genomics, bioinformatics, identity card technologies, multimedia applications, virtual realities and smart textiles.

These issues continued to be raised in publications and presentations,[37, 38] as did the NHS Information Authority's *Ways of Working with Information* National Health Informatics Competency Surveys in 2000, 2001and 2002.[39] In 2002, UKCHIP was formally established as a voluntary (regulation and) registration body.[35]

In 2005, it was still possible to identify projects and scenarios at each of the phases listed above, and to see health informaticians unclear about their roles. This might be expected from the diversity of thematic priorities and the roll-out plans for different projects within the NHS in England and internationally. As informatics to support health emerged more clearly, there were still cloudy issues relating to the status of the professionals providing informatics to health and social care bodies. These were noticeable in the NHS *Agenda for Change* initiative.[40]

International professional standards

The meaning of the term 'informatics' and the recognition of informatics as a profession have been considered since the mid-1990s,[40] the international debate about both continues.[41, 42] Many previous papers have looked solely at the academic components of health informatics. The UK does not have as many, explicit and per capita, health informatics courses at vocational, undergraduate and post-graduate levels as countries such as the USA and Germany. The international, invited participation, seminar in Madrid held in March 2000 was a move to determine a distinct identity for those who work in health informatics.

The UK does have many specialists who have come from different origins and work in high-level health informatics to support operational health care delivery and management. As with other professions who do not have direct patient responsibilities, the actions of informatics professionals do have an impact on patient outcomes.

Activity in the professional (regulation) and registration areas under the aegis of UKCHIP has attracted international interest, including exploration of UKCHIP principles and processes with the BCS HIF equivalent organization in Canada: COACH.

Informaticians in health service roles

The operation of early stand-alone applications was frequently by interested amateurs. Informatics specialists emerged as the field widened. Health informatics professional support for a local organization/area grew, also encouraging specialization. Thus the generalist health informatician is not as visible as previously but may be resurfacing with the demands that the NPfIT and NHS CFH are placing at Strategic Health Authority level, where its contracts were developed. Also at the 'cluster level', where the five (now distilled) geographic contracts with the LSPs on the industry side are lodged. The competencies needed by informaticians have not been well defined over time. People skills and roles have evolved to meet a localized need with little consistency and even less comparative positioning across organizations. This situation for health informaticians was altering by 2005 with various activities:

- The *Agenda for Change* initiative to place NHS staff into graded bandings.
- The Skills for Health (professional council) development of National Occupational Standards.
- The DoH publication of job evaluation handbooks, linking competencies to job roles.
- The establishment of UKCHIP to register fitness to practice.
- Liaison between professional groups representing many facets of informatics in the heath domain under the BCS HIF aegis.

Some of the above initiatives applied only within the NHS, but as Millen stated in NHS Information Authority presentations, the demand for HI staff is increasing exponentially: 23,000 needed by 2007. This demand will not all be met in-house and so external resources should be trained to meet the capacity requirement and have the competency to work at comparable levels to their NHS partners (and vice versa).

Various initiatives have moved the perception of a profession of informatics to support health from a dubious fix under Whitley Council rules, such as when I was classified as a medical physicist in the mid-1970s, to the current situation where:

- all HI staff have been evaluated and positioned against *Agenda for Change* descriptors (albeit some with personal protected positions due to a mismatch with formal areas);[40]
- UKCHIP had over 2,500 people in or through the process of voluntary registration, with further development taking account of the 2007 review of HI Services commissioned jointly by the Health and Social Care Information Centre, NHS CFH, and the Health Professions Council model regarding accreditation, regulation and registration;[35]
- the profession of 'health informatician' is recognized in TV adverts promoting NHS careers;
- UKCHIP registration has started to appear as a desirable prerequisite on job specifications and adverts;
- schools and undergraduate locations were being informed of career opportunities in health informatics, encouraging earlier entry into the domain to complement those staff who had made a sea change from initial clinical, management or technological professions into a health informatics specialism;[43]

- the UKCHIP model of regulation was being considered and assimilated into the BCS national professionalism agenda for 2006 and was acknowledged as consistent with the e-government intention to regulate all IT, ICT and similar staffs;[44]

- the UKCHIP entry criteria were being addressed by short courses to help move out-of-sector experts into health domain posts, such as the UCLAN Fast Track Fundamentals of the Health Domain (FTF). Pre-registration commitments to learn about the sensitivities of the domain were being sought from new entrants at all levels;

- pharmaceutical companies carried out research on their own data until such time as it became a legal and financial problem, then efforts were made to inform them of what health informatics could offer, both in the UK and around the world.[28]

Formal widespread recognition of HI as a profession requires completion of assessment, registration and continuing competence processes and their take-up by the increasing numbers actively involved in related informatics across the community within the next 10 years. Explicit procedures were being put in place in 2005; firstly, for the registration of existing practitioners by a fast track 'grandparent route'; then, for those new to the market by using extended development entry criteria, covering both qualifications and vocational experiences. In addition a formal process was under development for the withdrawal of an individual's registration and accreditation if they are no longer fit to practise.

International collaborative work

Since the early 1970s, the roll of countries involved in IMIA and its regional groups including EFMI increased to over 50. The same organizations were involved in standards and professional society activities. The European Commission has gained a greater influence over the direction of UK informatics. The vision of a global health informatics village[30] is not yet a ubiquitous reality, but there are some indicators of progress:

- IMIA is still gaining participating member countries.

- Implementations of commercial solutions are crossing continents.

- Telehealth is supportively contributing in isolated, post-war and disaster situations.

- Standards work (including open source) is aiding worldwide dialogue and interworking.

As yet, the sources of information about UK participation, personnel involvement and progress in multinational health projects are fragmented; work is needed to document the linkages and key contacts of members of the BCS in this arena. The overarching nature of European legislation is now much wider and requires the UK to contribute to research, design and development to avoid getting left behind. Members of the BCS HIF have been on the register for progress audit and bid evaluation for the CEC for over 10 years.

Heeks and Roberts put the case for dialogue between developed and (re-) emerging countries, and countries in transition in terms of their informatics experiences.[45] Subsequently, EFMI began a survey of both groups to identify patterns of positive and negative experiences that could bear on future informatics deployment.

In their 2006 agenda, the Board of UKCHIP focused on communicating the benefits of UKCHIP to employers and individuals, on developing more formalized continuing professional development (CPD) procedures and case studies, and on ensuring UKCHIP was consistent with e-government and other professional initiatives. These should stimulate mobility of 'fit to practise' professionals between sectors. However, there were still challenges of brand recognition and regulation of the emerging profession of health informatics that needed addressing.

Information standards for citizens

At the end of the twentieth century, the individual citizen was expressing a need to take a more involved role in his or her own health care. This was not just in the approximately five per cent of the population that had private medical insurance cover but also among NHS patients. The image of clinical professionals as 'doctor knows best' was frequently being questioned. The more open climate of care was expressed in the Government's 'Health of the Nation' targets. Progress towards targets was reported in the media. Health care was being commissioned against performance and quality criteria from locations outside the home area.

Inter-professional communication encouraged questions about what was happening to an individual patient. The challenges of common dialogues between the professionals active in the care of an individual were multiplied in complexity when lay persons, the subject or their informal carers, entered into the communication matrix. The increasing use of the internet required safeguards to provide effective communication of clinical topics to a range of audiences: lay and professional.[46–48]

Cultural changes were necessary to move away from a traditional clinical autocracy. It was historically usual for doctors to only tell patients what they as the professional felt the patient themselves 'needed' to know. Much of the evidence base and founts of knowledge, through teaching, paper-based media or increasingly internet-based sites were initially populated with professional concepts couched in profession-specific language. Whether the information to be communicated was about an individual or, for example, a disease group, the difficulties in interpretation were similar and covered:

- **Ensuring safety in the meaning of a message**, regardless of the profession or lay situation of the intended audience. This included unambiguous interpretation, a consistent level of seriousness/gravity implied by the message, and resultant repeatable actions and outcomes from any actions taken as a result of its receipt.

- **Communicating the provenance of the information content:** speculative suggestions, definitive conclusions and a bias from the source should be clearly distinguishable.

- **Reinterpreting content from the professional language** to offer optional consistent versions to suit other audiences. This required considerable resources and skilled effort, as criteria do not just include lay terminology but explicit concept explanation, regardless of the natural language presentation media or the English language competence of the recipient of that data or information.

It was at this time that the first of the codes addressing website content in the health domain, the Health on the Net (HON) Code, emerged.[49] Its use by new sites was wide, but for a site to award itself the HON rosette attribution after a retrospective review of existing content was still intermittent. Taking the requirement for 'fitness for purpose' further, also at this time, the French government threatened to take to court any website that was accessible in France that did not have a French language section in parallel to its base language. This requirement did not exist for long, due to its impracticality. Multi-language requirements were frequently met with parallel copies of publications in other languages, for example the Pan-American Health Organisation (PAHO) model, provided Spanish and English copies of key documents.[50] Another solution was integral translators and translations, for example on some European Community websites; while the Singapore Hospital One website provided Cantonese and Mandarin. Some stand-alone websites in their native language provided some translated material, such as the Danish ones: Health Telematics and Medcom.[51, 52]

E-health research priorities in 2004 recognized that research and reinforcement of the need for implementation of good practices were still required. UK research in this area informed and re-enforced the work of, for example, international colleagues Eysenbach and Jadad in promoting concepts of good practice.[53, 54] International codes of practice and guidance are still being developed and refined. This is an ongoing topic with legitimate revision being carried out because the domain of those interested in web-based information about health, care and lifestyle management constantly develops.

The principal challenges for clinicians dealing with more active, demanding patients are highlighted in concerns about the presentation and declared provenance of openly accessible clinical content on the web, and the caveats currently attached (or not) to its use.[49] Analysis of the usability of existing websites for lay users demonstrated the eclectic nature and value of current material. It emphasized the need for considerable investment in addressing fitness for a range of purposes. The subject is returned to frequently, including at the Asia–Pacific Association of Medical Informatics (APAMI) meetings.[55] Thankfully there were increasingly fewer residual 'page turner' sites reported where the content just replicated paper-based presentations.

Patients and their carers actively seek information about specific clinical conditions from the internet. They frequently use it to challenge and question primary health practitioners including GPs within the consultation. There is ongoing work to change working practices and cultural stereotypes to maximize the potential partnership from this dialogue, and also to enhance the quality and accessibility of health-related web-based information.

PREPARATION FOR A CHANGE OF PRACTICE: CONVERGING SECTORS

In the early 1980s, primary care was delivered by GPs under the previous Family Practitioner Authorities and was explicitly distinct from hospital and community care.[13] At this stage it was not politically feasible to produce a population index for a local health community so PASs, as described above, were implemented without a capacity for patient, and information, flows between these sectors.

The situation has changed over time. Since *Making IT Work: Implementing the IM&T Strategy in the NHS*,[56] and confirmed in *Information for Health*,[57] steps are being taken to focus on capturing data from a patient journey regardless of where the care is delivered or the information originates. However, there still is considerable debate and concern about the 'client–server' model of content, referring to the NHS Data Spine centrally and the application systems locally. The solutions procured by NHS CFH often do not reflect the selected content in each location that general practitioners think should be there to support them. There is little immediately available in the phased implementations for community staff or the nursing professions. The constructive critical commentary is typified in the 2005 *Radical Steps* position statement:

> ... *many of the benefits of health informatics will not surface until professional and sectoral inter-working are more widespread.*

> REFERENCE 58

Political changes had to be accommodated to ensure the independent GPs, the Primary Care Trust service commissioners and the traditional hospital and community health service providers could inter-work without risk to the patient, their client. There is still some confusion as the management hierarchical structure is yet again changing, with merging Primary Care Trusts, Foundation Hospitals and the inclusion of private care providers into the delivery equation. Questions are being asked as to whether the features of NHS CFH can accommodate the fragmented range of provision.[59] In information terms, much work has been carried out on developing compatible data terming and coding; the introduction of the unique NHS number identifier into wider use; and the establishment of a robust infrastructure to support effective and efficient information flows. Many of the practical processes involved changes of paper documents, and only now with NHS CFH are electronic forms of interaction being introduced across the domain. The reaction of practitioners is not universally welcoming and concerns still unresolved include the following from the *POSITIVE (RADICAL) STRIDES* document in 2005:[58]

- *Care delivery has historically been silo-based and many of the benefits of health informatics will not surface until professional and sectoral inter-working is more widespread.*

- *Strategic specification and development of processes to address future cross-border management should be initiated now, in order to ensure solutions are ready when required.*

- *Initiatives such as the Single Assessment Process (SAP) for children and older people will necessitate much more inter-working between organizations: between health and social care; across home country boundaries; and between public and private sectors. Health care practitioners already recognize that SAP will necessitate flexible, supportive informatics solutions. The experiences of those sites already successfully operating SAP functions can be input to the design of these complex procedures at an early stage of national solutions.*

Converging health and social care

Research carried out in 2001–02 through Warwick University looked at the potential for reuse of ICT across the cusp between health and social care.[60] Issues that are still to be resolved include:

- the procurement methods for equipment;
- attitudinal issues around sharing with trusted third parties or by organizational dictat.

A serious process issue remains where social services staff explicitly record null actions where the generic care pathway indicates an action might be anticipated. In this circumstance the health record only documents actual actions taken. This suggests that, yet again, there are going to be issues of record content transition, primacy of data, opportunities for misunderstanding, incompatible coding schema and usage discrepancies very similar to the problems met as primary and secondary health care records came (or are still coming) together. The lessons and the fixes to solve the problems have not been documented. It is likely that the pitfalls will have to be navigated yet again unless more investment is made retrospectively in capturing experiences from the field.

Many of the issues may possibly be encountered again when the lifestyle (health and fitness) professionals also wish to work together with health and social care. *Radical Steps* goes a very limited way to highlighting issues, but does not document detailed pathways, processes and problem fixes in the management of the change.[58]

These remaining challenges must be reflected against the early 1980s where isolated co-terminous organizations in health and social care (for example in Salford around its PAS) made pragmatic decisions to work together for the local community. In parallel, the ways of working across all the 'welfare' professions in Scotland and Northern Ireland already address many aspects of where England is heading in the mid-term.

In 2002, Sir John Pattison (then head of the NPfIT) first commented at the inaugural *Radical Steps* meeting that health must solve its own problems before it widens its agenda to address those of social care or health and social care collectively. Concerns about professional and inter-sector consistency/incompatibility are therefore not being taken into account (except for being raised in the *Radical Steps* series). These concerns will, however, remain on the BCS HIF agenda for further *Radical Steps* initiatives in future years.

Strategically, the landscape for convergence is still mountainous: 'Information for Health' was rapidly followed by 'Information for Social Care'. These were followed by much angst from the Society of IT Management (SOCITM) as delays to the development of 'Information for Health' solutions jeopardized local government timescales and were beyond their control.

Articles in *BJHC&IM* in 2005 indicated problems in getting front line staff to capture data; systems incompatibilities within a sector; the pace of change being demanded; the lack of basic preparation; and the emotional impact of computerization. Similar challenges revisited indeed! This is all against a background of strategic service revision to support the Single Assessment of Older People and the Integrated Children's System implementation.

Whatever the pressures in the UK, there is a potential for even greater change that could affect the pace and style of convergence between the health and social care sectors. The European Commission has declared that health/welfare topics can now be explicitly addressed at that transnational level. The 'Euro-train' may issue a directive that could have a profound effect on the strategic direction being taken under I4H and I4SC thematic directions. That potential 'time bomb' might be as embedded as messaging standards or could occur at the level of the stimulation of mobility of citizens across the community. I have concerns that UK participation is limited at the Europe-wide level and may not ensure the informatics direction currently being taken in England is preserved. It is therefore very risky not to keep a watching brief on this area. The BCS HIF continues to sponsor involvement of expert members in international standards activity. BCS HIF has also established a task group to firm up links with key decision-influencers across Europe and in the Information Society Directorate-General in Brussels.

Integrating self-management of lifestyles

Another pseudo-convergence of a less formal nature is that of citizen's responsibility for self-management of lifestyles.[46, 61] This may bring additional challenges to the informatics solutions of NHS CFH and the information repositories of clinical, single-issue, bodies like Cancer Backup and Arthritis Care.

The 2004 second Wanless Report (*Wanless2*) creates impetus for the convergence of welfare (health and social care) and lifestyle through raising issues such as:

- *the current absence of Government activity in determining a comprehensive set of objectives for key lifestyle risk factors; albeit it does recognize a '. . . differential impact of interventions across the socio-economic gradient';*

- *a shift of emphasis from interventions at a population level towards targeting individual behaviour change at a population and individual level;*

- *recognizing that individuals may be informed, and still choose unhealthy behaviour because they enjoy it;*

- *the probability that knowledge of genetics and individual risk factors could play an increasing role in effectively delivering individualized health promotion and preventative actions.*

REFERENCE 62

A recent review of *Wanless2* for its potential impact on professionals and the public looked at the potential need for informatics support to both cohorts. It highlights similar issues, first acknowledged in the 1988 Black report,[5] that considered health impacts by the terms 'upstream' (policy interventions) and 'downstream' (individual interventions). A *Wanless2* recommendation about a strategy to increase levels of physical activity and sport in England developed jointly by the DoH and Culture Media and Sport will have been strengthened by the announcement of the 2012 Olympics in the UK.

If a small number of risk factors account for a significant proportion of the burden of disease, there will be a strong demand for evidence and analysis of that hypothesis. Comprehensive community profiles and sophisticated informatics tools will be required to prove the *Wanless2* assertion that the socio-economic gradient in some lifestyle choices can explain a large proportion of health inequalities. The trend in overweight and obesity is inextricably linked to physical activity levels and overall energy expenditure. Wanless also states that physical activity integrated into lifestyle is very important because the increase in sedentary lifestyles in England has lessened the amount of habitual physical activity in which most people engage. The baseline benchmarks are that: just 31 per cent of adults currently meet the recommended level of physical activity; and 38 per cent are sedentary, doing less than 30 minutes of moderate physical exercise per week. About 50 per cent of children are also failing to meet Government recommended levels of physical activity. These figures can be used to audit progress in this area.

Web-based information will support 'structured patient education and informed self-management [that] is a cornerstone of the NHS Plan'.[46] But the information quality caveats should continue to be considered.

Wanless2 challenges the NHS to:

> *... give support to improving the health of its workforce and should undertake pilot exercises to evaluate how to promote the long-term mental and physical good health of their employees. The NHS should become an exemplar for public and private sector employers.*

REFERENCE 62

A comprehensive informatics infrastructure will be necessary to underpin the monitoring of such strategic targets. Unless the citizens' involvement does include the documentation of actions, whether professionally or lay instigated, the veracity and completeness of the holistic patient record will be jeopardized.

The involvement of citizens in their care is changing to be more contributory.[61, 63] This capability is not yet fully embraced by NHS CFH functionality and questions remain unanswered, such as:

- what are the obligations for health care practitioners to take cognizance of any comments made by the patient/client in 'MyHealthSpace'?
- how are alternative therapies and self-administered interventions incorporated into the supposedly holistic record that is a goal of the current strategy?

Initiatives, like NHS Direct Online and NHS Digital TV, plan to involve the public more actively in their health maintenance. Informatics is a catalyst to more open partnerships. Technological developments now support informed participation. Citizens could soon be using ambient (implanted) and assistive (wearable) technologies; carrying smart cards; and consulting with the most appropriate clinical advisors through telehealth media.[64]

NHS CFH has the declared aim of focusing on supporting practitioners in its early phases, leaving open a likelihood that the 'citizen as subject of the record' interface may first become more concrete through the NHS Direct programme. This will require subsequent harmonization and integration with NHS CFH later in the schedule.

Field workers in the domain still have many concerns about whether NHS CFH is going to deliver a service that meets their expectations, either from their existing functionality or their future planned requirements. So far, some of the concerns, such as the issues around the holistic record, have patently not been resolved by NHS CFH.[65] They will therefore continue to be under scrutiny from bodies such as the National Audit Office, the Parliamentary Accounts Committee and the operational practitioner community, stimulated by the *Radical Steps* type of initiative.

There is obviously a continuing agenda to be addressed by the BCS HIF and its constituent groups.

REFERENCES

1   Farrer, J.A., Harvey, P.W. and Roberts, J.M. (1977) Problems arising from the classification in ICD on non-fatal and minor conditions. In Shires, D.B. and Wolf, H. (eds) *MEDINFO 77*, Toronto, pp. 289–92. North-Holland, Amsterdam.

2   Roberts, J.A., Farrer, J.A. and Harvey, P.W. (1977) A progressive study of the Emergency Room demand by the Community. *Medical Informatics*, 2.3, 197–201.

3   Roberts, J.M., Farrar, J.A. and Harvey, P.W. (1977) The use of a computer system in the study of the attendance profile in a district hospital casualty department. *Computers in Biology & Medicine*, 7, 291–9.

4   Farrer, J., Harvey, P., Dyer, J. and Roberts, J. (1978) A study of the demands made on casualty services by the population of socially deprived areas. In Anderson, J. (ed.) *MIE, Cambridge 78: Lecture Notes 1*, pp. 597–603. Springer Verlag, Berlin. ISBN 3 540089 16 0.

5   Whitehead, M., Townsend, P., Davidson, D. and Davidson, N. (1992) *Inequalities in Health: The Black Report and the Social Divide*. Penguin, London. ISBN 0 140172 65 3.

6   Webref (accessed June 2007) www.connectingforhealth.nhs.uk

7   Webref (accessed June 2007) www.dh.gov.uk/PolicyAndGuidance/HealthAndSocialCareTopics/HealthAndSocialCareArticle/fs/en?CONTENT_ID=4070951&chk=W3ar/W

8   James, R., Roberts, J., Harvey, P., Bellis, J. and Cooper, R. (1978) A computerised scheme for the preparation of parenteral nutrition regimes. *Medical Informatics*, 3, 2, 77–86.

9   Shearing, G., Wright, P., James, R.M., Roberts, J.M., Cooper, R.I., Bellis, J.D., Harvey, P.W. and Rich, A.J. (1978) A computer program for calculating intravenous feeding regimes and its application in a teaching hospital and a district general hospital. In Anderson, J. (ed.) *MIE, Cambridge 78: Lecture Notes 1*, pp. 641–50. Springer Verlag, Berlin. ISBN 3 540089 16 0.

10 Roberts, J. and Farrer, J. (1985) EMG processing: an in-service aid. *Medical Informatics*, 10, 4, 319–22.

11 Maguire, G. and Roberts, J. (1982) Nurse allocation with computer assistance – extension to a manpower planner. In O'Moore, R.R., Barber, B., Reichertz, P.L. and Roger, F. (eds) *MIE, Dublin 82: Lecture Notes 16*, pp. 317–21. Springer Verlag, Berlin. ISBN 0 387112 08 1.

12 Mason, A. *et al.* (1983) *Developing a District IT Policy*. King's Fund/Körner for DoH.

13 Roberts, J. and Bryden, J.S. (1989) Towards an index of the community. *BJHC*, March, 34–5.

14   Webref (accessed June 2007) *DTI Foresight 2020 Report.* http://www.foresight.gov.uk/Previous_Rounds/Foresight_1999__2002/Healthcare/index.html

15   Commission of the European Communities (1998) IT EDUCTRA: Awareness, training and education in informatics and telematics, for healthcare practitioners.

16   DoH (1992–96) *Using Information Effectively.* A series of open learning materials. Greenhalgh & Co. Ltd.

17   Roberts, J. (1995) The true cost of informatics procurement. In Richards, B., de Glanville, H. and MacOwan, H. (eds) *Current Perspectives in Healthcare Computing*, pp. 502–5. BJHC for BCS, Weybridge.

18   DoH (2002) *Delivering 21st Century IT Support for the NHS: National Strategic Programme 2002.* DoH.

19   Webref (accessed June 2007) Office of Government Commerce regulations. www.ogc.gov.uk

20   (authors unknown) (2000) Incorporating knowledge into commercially available systems. In Balas, E.A., Boren, S.A. and Brown, G.D. (eds) *Transferring Research to Practice for Health Care Improvement: Information Technology Strategies from the United States and the European Union. Studies in Health Technology and Informatics*, vol. 76. IOS Press, Amsterdam. ISBN 978 1 586030 54 4.

21   Open University (2003) *Knowledge, Information and Care.* Course unit K223. Open University, Milton Keynes.

22   Webref (accessed June 2007) University of Central Lancashire. www.uclan.ac.uk/health informatics

23   Kenny, D. (1978) *Data Protection.* North-Holland, Amsterdam.

24   Webref (accessed June 2007) www.man.ac.uk/idpm

25   BCS HIF Strategic Panel (2006) *The Way Forward for NHS Health Informatics.* www.bcs.org/server.php?show=ConWebDoc.8951

26   Murray, P.J. (2004) *Open Steps Release 1.0.* Report of a think tank meeting on Free/Libre/Open Source Software in the health and health informatics domains: Marwell. IMIA Open Source Working Group and BCS Health Informatics Committee. February.

27   Murray, P.J., Betts, H., Roberts, J., Richardson, G., Ward, R. and Wright, G. (2005) *Report on the Education Steps Project – The Otley Meeting* (full report). BCS HIF, Swindon. www.differance-engine.net/education-steps/documents/otley2005outputs.htm (accessed June 2007)

28   Roberts, J. (1990) The UK has been in IT for years! *Medical Horizons*, April, 18.

29   Webref (accessed June 2007) www.ukbiobank.ac.uk

30   Roberts, J. (1993) Image of a global village. *BJHC*, September, 10–11.

31 Kay, S. (1995) *Comparative Qualities of Academic and Professional Qualifications*. Working document. BCS HISG.

32 Roberts, J. (1998) Health or unidisciplinary informatics initiatives – why are we where we are today? In Cesnik, B., McCray, A.T. and Scherrer, J-R. (eds) *MEDINFO 98*, Seoul, pp. 1184–7. IOS Press, Amsterdam

33 Peel, V. and Roberts, J. (1987) *IT has Finally Arrived*. Poster at IHSM.

34 Hayes, G. (2002) UKCHIP Shadow Steering Group: Notes of Meeting. September.

35 Webref (accessed June 2007) UK Council for Health Informatics Professions. www.ukchip.org

36 Webref (accessed June 2007) www.proms-g.bcs.org.uk

37 Roberts, J. (2000) *Is That a Health Informatician I See Before Me?* Workshop at HC2000.

38 Roberts, J. (2002) Are you an individual professional or a member of a profession – which is the priority for recognition? In Bryant, J. (ed.) *Current Perspectives in Health Computing*, pp. 284–90. BJHC for BCS, Weybridge. CD-Rom.

39 NHS Information Authority (2000–02) *Ways of Working with Information: National Competency Surveys*. CD-Rom.

40 Webref (accessed June 2007) *Agenda for Change.* www.dh.gov.uk/PolicyAndGuidance/HumanResourcesAndTraining/ModernisingPay/AgendaForChange/fs/en

41 Roberts, J. (1994) Medical Records officers debate new roles & electronic patient record. *BJHC*, June, (pages unknown).

42 Roberts, J. (2002) A UK operational practitioner's view: some challenges of health informatics are transnational. *Methods of Information in Medicine*, 41, 55–9.

43 BCS HIC (2004) *(Rough Guide to) Health Informatics*. Promotional material. BCS, Swindon. March.

44 BCS HIF (2006) *Cabinet Office Newsletter: Health Informatics (HI) Professionalism – Similar But Different?* Internal briefing document. BCS.

45 Heeks, R. and Roberts, J. (2003) Learning on health informatics between the UK and developing countries: a one-way or a two-way street? *BJHC&IM*. March.

46 Roberts, J.M. and Copeland, K.L. (2001) Clinical websites are currently dangerous to health. *Int. J. Med. Info.*, 62, 181–7.

47 Roberts, J. (2000) Many audiences, mixed expectations: is enough care taken with health information over the internet? *eHealthcare World – Europe* (September). www.ehealthcareworld.com

48 Roberts, J.M. and Copeland, K.L. (2000) Know your audience: establishing your eHealth site's 'Fitness for Purpose'. *Informedica 2000:*

1st Ibero American Virtual Congress of Medical Informatics. Published online (October 2000).

49 Webref (accessed June 2007) www.hon.ch

50 Rodrigues, R.J., Oliveri, N.C., Monteagudo, J.L., Hernández, A. and Sandor, T. *e-health in Latin America and the Caribbean.* PAHO.

51 Webref (accessed September 2007) www.cfst.dk

52 Webref (accessed June 2007) www.medcom.dk

53 Eysenbach, G., Powell, J., Kuss, O. and Sa, E.R. (2002) Empirical studies assessing the quality of health information for consumers on the World Wide Web: A systematic review. *JAMA*, 287, 2691–700.

54 Crocco, A.G., Villasis-Keever, M. and Jadad, A.R. (2002) Analysis of cases of harm associated with use of health information on the internet. *JAMA*, 287, 21, 2869–71.

55 Webref (accessed June 2007) Asia–Pacific Association for Medical Informatics. www.apami.org/mic2006

56 DoH (1994) *Making IT Work: Implementing the IM&T Strategy.* HMSO. December.

57 DoH (1998) *Information for Health: An Information Strategy for the Modern NHS 1998–2005.* HMSO.

58 Webref (accessed June 2007) BCS HIF *Radical Steps* series of Position Statements (2004-2006). www.bcs.org/BCS/forums/health

59 Collins, T. (2005) Leaked emails emphasize divide between business goals and technology in NHS plan. *Computer Weekly.* 22 November.

60 Roberts, J., Szczepura, A., Bickerstaffe, D. and Fagan, R. (2003) Informatics: an enabler of health and social care interworking. In Bryant, J. (ed.) *Current Perspectives in Healthcare Computing*, pp. 13–19. BJHC for BCS, Weybridge.

61 Roberts, J. (1999) Are healthcare professionals really prepared to deal with information-empowered citizens? In Kokol. P. *et al.* (eds) *MIE, Ljubljana 99*, pp. 523–7. Springer Verlag, Berlin.

62 Wanless, D. (2004) *Securing Good Health for the Whole Population.* HM Treasury.

63 Roberts, J. and Copeland, K. (2002) Stimulating convergence of clinical and lifestyle perspectives on the web: asthma as an example. *Journal of Informatics in Primary Care*, December, 201–4.

64 Roberts, J. (2006) Pervasive health management and health management utilizing pervasive technologies: synergy and issues. *Journal of Universal Computer Science*, 12,1, 6–14.

65 Webref (accessed June 2007) eHealth Insider. www.e-health-insider.com

29 Lessons for the future: the success or failure of health care computing

BARRY BARBER

All introduction of new technology is fraught with problems: the difficulties of handling the technology, the introduction of new concepts and of new methods of working. Projects can be overwhelmed by unexpected problems arising from the technology, by lack of understanding of the new concepts or by the resistance from those expected to adopt new and possibly less favourable working practices. Early users of high-pressure steam engines had to discover the safety requirements necessary to enable those engines to become widely used. The Comet aircraft faced previously unknown problems arising from metal fatigue.

Many early users of computers were scientists who could cope with machine failure in their work, provided that the systems could be got going within the time span of their research activities. The move into operational systems removed this safety margin and organizations became critically reliant on their computer systems in order to handle their routine business. The batch-processing systems in general use had, typically, monthly, weekly or daily cycle times within which problems could be rectified without causing organizational difficulties. However, the real-time systems that were then being designed would have no such luxury. Any failures of the system would become immediately evident to all concerned and would also affect routine activities immediately. Quite apart from their failure rates, these new systems would have a major effect on any business – for good or ill – depending on how well the system had been designed and implemented. Serious failure of systems that are intimately intertwined with the operational activities of an organization can result in devastating financial loss or, equally devastating, a loss of prestige and public confidence. Ivars Peterson, in his book *Fatal Defect: Chasing the Killer Computer Bugs* gives a vivid account of safety problems that have arisen as a result of failures in the design of some computer projects.[1] His most vivid health care example relates to the programming issues associated with the Therac-25 linear accelerator. (See *Why Do Things Go Wrong?*).

WHY COMPUTER PROJECTS FAIL

Brocks, paper in 1969 provides a very early overview of implementation problems that had already emerged.[2] He writes:

> 2 *Various surveys conducted both in this country and in the USA indicate that a high proportion of managements using computers are dissatisfied with their installation in one way or another and that very few indeed can claim that they are obtaining a reasonable return on their investment in computer systems. And yet the sale of computers continues apace and new users are ever hopeful of achieving real benefits from their investment in this equipment. The reason for this continued enthusiasm . . . is probably the underlying conviction that the concept of EDP is good logical common sense and that sooner or later man will triumph over machine and real benefits will accrue to those who are sufficiently enterprising to take the risk.*
>
> 3 *The purpose of this paper is to examine the principal causes of the disappointments which have occurred in the past and to point in a constructive way to means of avoiding the pitfalls, or at least to mitigating the consequences of falling foul of the hazards.*
>
> 4 *In writing the paper, I am conscious that little that I have to say will be new . . .*
>
> REFERENCE 2

On the responsibility of top management he notes that:

> *It is so easy in this highly technical area . . . to blame the equipment, the manufacturer or the EDP staff for lack of success. Each may well have to bear a share of the blame for failure but it must be remembered that it is the responsibility of management to keep abreast of all the relevant factors so as to ensure that things do not go wrong or, if they do go wrong, to minimize the effect.*
>
> REFERENCE 2

Brocks examined the issues in detail and he found the following key problems related to top management:

- Failure to appreciate the need for a feasibility study.
- Failure to appreciate the amount of planning work.
- Failure to appreciate the effect on the organization.
- Failure to appreciate the implications and requirements of the proposed system.
- Failure through lack of understanding of hardware requirements.
- Inadequate communication or inadequate coordination of implementation work.
- Inadequate monitoring of progress.
- Inadequate communications with the data-processing manager.

As time has gone by, the details of the technology have changed and the systems have become immensely more powerful but the underlying top management issues remain the same whether systems are being implemented in a hospital, a community trust or across the whole country. Vital for success are: realistic technical designs; adequate budgeting; user acceptance, if possible enthusiasm; and a thorough understanding of the circumstances in which the systems will be used, both to collect information and to provide information. Vague objectives, changing requirements, budgets that do not match the work to be done and user reluctance or rejection are part of a recipe for failure.

LEARNING THE LESSONS AT THE ROYAL LONDON HOSPITAL

Although some of these issues have been discussed before, it is nevertheless worth recalling some of the key steps that enabled the project to be implemented successfully at a time when computer literacy was rare and computing facilities were very much less powerful than are routinely available now. When The Royal London Hospital embarked on its real-time computer project the chairman, Sir Harry Moore, sought advice from his computer manager at Hill Samuel Ltd. He obtained Brocks' paper and advised the project team that while there might be other reasons for failure, we should study those set out in that paper and certainly avoid them. Many of the issues highlighted by Brocks have echoed down the years as various projects have failed.

Bud Abbott had an overhead slide in his early documents, which summarized his thoughts on project development as follows:

- Prepare a careful, overall, long-range plan.
- Compare tenders of many manufacturers.
- Allow at least two years and possibly longer for the implementation of the process.
- Select a computer manager from among your best managers not necessarily your best technicians.
- Expect everything to cost at least 200 per cent more than estimated by your staff and 500 per cent more than estimated by computer manufacturers.
- Make sure that all your senior and production managers have attended one or more courses on computer planning and systems analysis, and are involved in the design process.
- Avoid dependence on individuals at all costs. At least once a year make sure you have replacements for everyone on the staff.
- Audit the computer's performance against the forecast in your feasibility study.
- Continually educate your staff and management.
- Promote your data-processing management out of data processing at least once every three years.

589

No one would argue with much of this advice and it provided a useful backdrop to the thinking of the Computer Executive.

From the start it must be recognized that there were particular difficulties facing the designers of real-time health information systems in the 1970s. The main problems were:

- health care systems were much more complex than the previous financial and administrative systems;
- health care systems were used by large numbers of people with differing disciplinary backgrounds;
- health care systems required health care professionals and health care support staff, who then had no expertise in using computers, to enter and retrieve information in order to ensure that the information was collected at source and was available where it was needed. Generally, they could not be forced to use systems that were unsatisfactory;
- the funding was generally insufficient for the requirements of the project and ways had to be found of addressing funding gaps so that the deficit was, at least, tolerable.

Many of the solutions that are now generally accepted had to be developed and tested. There was no certainty that non-computer professionals could use computer systems in order to secure the benefits envisaged. 'Tree-branching menu selection' was commonly thought to provide the answer, but it was not clear that the computers could function fast enough to enable health care activities to be handled directly from the computer systems.[3-5]

Project management by a knowledgeable multidisciplinary executive

The Royal London Hospital established a high-level Computer Executive that included a senior administrator, a senior clinician and a senior nurse as well as the directors of the Computing and Operational Research Units. This knowledgeable, multidisciplinary, Executive controlled all aspects of the project and was advised by a Medical Advisory Committee on particular issues such as confidentiality and medical aspects of system usage. The Executive attempted to look at the project from all angles, but especially from the point of view of a system user. Wherever possible it tried to cross-check its decisions with other professional groups and communities, and to use any methods available to assure user satisfaction and project success.

Involvement of key user communities in decision-making

The initial plans were based on the PAS developed in conjunction with the hospital's earlier Elliott 803 computer. However, the Executive took great care to consult with all the relevant 'user communities' about the proposed applications and the ways that the computer system would impact on clinical and administrative activities. These consultations involved all the consultant and senior nurses and heads of departments in a full day's session, which introduced them to the technology and the proposed applications as well as providing opportunities for discussion and modification of the proposals. The concepts of 'menu tree-branching' were tested, as well as the possibility of using touch-wire screens.

Effective education of all users

As the work proceeded, consultation sessions were developed into more obviously training sessions for those who would be the first to use the systems. These sessions included demonstrations on a 'dummy' database so the details of the use of the applications could be understood and explored by those concerned. The training sessions involved peak activity when new applications were being implemented. In addition, regular induction training sessions were developed for those staff who were new to the computer system or to a particular set of applications. The details of these training sessions changed over time as new members of staff were found to be more computer literate. Special arrangements were made so that the hospital could have access to new junior medical staff when the 'house' changed so that they could be advised about the hospital's various systems and arrangements before setting to work in the wards. This 'house officer induction day' obviously included instruction on the relevant uses of the computer system.

Realistic technological plans

Although the Director of Computing had the primary responsibility for the development of the system, the Computer Executive became involved in scrutinizing the technical arrangements and approaches to ensure that they were realistic and compatible with what they knew about the organization and the application area.

Effective monitoring of project progress

The Computer Executive monitored progress with the implementation of new computer applications as well as receiving regular reports on issues affecting the availability of the computer system. Whenever problems arose, steps were taken to rectify the situation and to keep the prospective user community informed.

Learning from others

A great deal of attention was given to the implementation of The London Hospital's computer systems because it was appreciated that it was breaking new ground. Wherever possible the plans were cross-checked with whatever experience could be obtained from other sectors. The most obvious comparison was with airline passenger booking systems and visits were paid to appropriate sites, but it was understood that the resources available to the airlines were far more extensive than those available to the hospital. The American hospital systems tended to focus at that time on billing issues rather than patient care. The Executive was very interested in the system at the small hospital at El Camino, in the USA, and the problems that they had faced,[6-11] although the Executive never managed to make a visit to El Camino to see the system in practice.

At this time some hospitals in Sweden were making great strides in these activities, particularly at the Karolinska Hospital where a pilot medical record system was being developed.[12, 13] At Danderyd and Huddinge Hospitals large computer systems were being installed by Stockholm County to address the computer requirements of a whole hospital.[14-18] The Executive believed that these European activities were closest to the basic requirements of The London Hospital. These sites were studied for any lessons to be learned and were visited.

Finally, and much nearer to home, the King's College Medical Records project was of interest in terms of the amount of equipment available, and the number of wards covered.[19] King's techniques of data capture were of great interest as the system was operational before The London Hospital Project. All this cross-checking helped to keep the project aware of possible problems and their solutions.

THE EXPERIENCE OF MANY NURSING PILOT PROJECTS

It was noted in Chapter 20 that there were a large number of nursing pilot projects that were initiated in hope, but which then fell by the wayside for a variety of reasons – not least because the project funding ran out and did not leave in place a viable operational and maintainable system. There was often the tension between what the nurses needed for the delivery of clinical care and the desire of the management to use systems to achieve reductions in nursing costs. Despite their importance in the delivery of patient care, nursing systems were frequently viewed as a sideshow compared to the implementation of the patient administration and record systems. There was another problem: the lack of an agreed terminology that could be used by the nursing systems. Heather Strachan explored the literature and noted the issues of the management of change and the importance of the 'people issues'. She then explored some of these using Delphi techniques.[20] The issue of health professionals attempting to carry out tasks beyond their competence presented an obvious problem, one that is addressed in the BCS *Code of Good Practice*,[21] and the overarching *Rules of Ethical Conduct for Health Informatics Professionals*.[22]

WHY DO THINGS GO WRONG?

A great deal of effort was put into making the early systems work when the hardware and software available was so limited by comparison with present-day facilities. However, it comes as a shock that subsequently many systems' developments have got into difficulties. As with disease, there are many different causes and, unlike success, which is quickly claimed, failures are ascribed to many and varied causes and people. In retrospect it can often be difficult to discover what the real objectives originally were and where the causes of failure lay or indeed of what failure consisted. Exhaustive investigations into failures suggest that the simple quest for a single cause may not reveal the whole story: much like the historical search for the causes of wars. There may be underlying problems with the project, management problems or immediate technical problems. The organizational or political environment of the project may change and make nonsense of previous decisions.

Ivars Peterson, in *Fatal Defect: Chasing the Killer Computer Bugs*, analysed many of these issues and provided an extensive bibliography for further research.[1] In the health care setting he examined the problems experienced with the Therac-25 linear accelerator and the six overdosed patients and the way in which Nancy Leveson, who was brought in as an expert witness in the case,[23] believed that safety critical software should aim to be 'safe' rather than 'perfect'.[24]

The CCTA Risk Analysis and Management Methodology (CRAMM)[25] provides an approach to risk analysis in information systems that starts with the development of the building of 'worst case scenarios', called the 'data valuation', as a basis for assessing the care required in developing the system (see Chapter 25).[26] On investigation, it becomes clear that errors in medical records or the lack of key information in these records can lead health professionals to initiate contraindicated treatment, or to withhold indicated treatment. In the health care setting information integrity and information availability failures are potentially more serious than confidentiality failures, which have traditionally been of most concern. The CRAMM methodology then goes on to explore the detailed threats and vulnerabilities of the various system components, to focus where effort should be expended in ensuring the security of the system.[27] CRAMM is often criticized for the time-consuming detail of the work entailed: but the issue of the initial data valuation provides a very good overview of the seriousness with which the system should be taken both managerially and technically. 'What could happen if the system goes wrong?', is one of the first questions that should be asked of a project because that should focus the attention of both management and project staff.

Beynon-Davies provided a good review of the issues associated with the initial development of the London Ambulance Service's Despatch System.[28] It was reported that the system became overwhelmed and was late sending ambulances to incidents, and sometimes sent the wrong ambulances at peak times, resulting in patients dying for lack of appropriate care. On such occasions the system appears to have been caught up in a vicious circle of failed allocations, which generated more service calls; and incorrect data in the system arising from crews pressing the wrong buttons, or ambulances being in radio black spots. Any real-time system will fail in some way if it is overloaded, but the key issue is the safety of the ways in which it degrades. In this case it was that system failure could lead to emergency ambulances not arriving at the correct location within the specified times.

Beynon-Davies also discussed the Wessex RHA Regional Information Systems Plan (RISP). There appeared to have been a series of management failures; but the organizational changes over the period of the project meant that the Wessex RHA was no longer able to 'top-slice' funds from DHAs to continue the project.

Smith and Smart[29] and Vic Lane[30] also explored issues of success and failure in NHS information systems. The former provided an indication of the loss from various systems and some analysis of the reasons for the failure.

Projects can be set up with flaws that almost guarantee their failure [such as the following:]

- *Failure to involve users and incorporate organizational needs.*
- *Ill-defined, unrealistic or conflicting objectives.*
- *Suppression or mismanagement of risks or uncertainties.*
- *Ineffective project or resource planning.*
- *Fixing objectives, timeframes or costs where excessive uncertainty still exists.*
- *Insufficient political support.*
- *Over-dependency on other projects, other parts of the organization or suppliers.*
- *Excessive project size and complexity.*
- *Unproven technology or approach.*
- *Lack of project and technical skills, or*
- *Undefined exit conditions.*

REFERENCE 29

Correspondingly:

Even well set up projects can be mismanaged to failure [for example:]

- *Inflexibility, inappropriate aggression and machismo.*
- *Loss of political support.*
- *Resources not properly allocated.*
- *Failure to control suppliers.*
- *Unreliable progress reporting.*
- *Unrealistic timeframes or budgets.*
- *Inattention to quality.*
- *Insoluable technical problems.*

- *Barriers to communication.*
- *Lack of teamwork, and*
- *Fear of failure.*

REFERENCE **29**

Vic Lane outlined various 'Critical Failure Factors', as well as some 'themes which occur in most large failures':

- *Over ambitious.*
- *Technocrats think that they know it all.*
- *Computing must be beneficial.*
- *Management abdicate responsibility.*
- *Credulity – it will turn out alright when needed.*
- *Conflicts may have a conflict of interest.*
- *Custom-built product.*
- *Concealment of bad news by middle managers.*
- *Buck passing, and*
- *Mistaken belief – litigation will solve problems.*

REFERENCE **30**

One project that is not mentioned in these papers, but which received a great deal of national publicity, related to the radiation treatment planning system at the North Staffordshire Royal Infirmary between 1982 and 1991. The error was not discovered until a new radiation treatment planning system was being commissioned. The effect was that some 1,000 patients treated by isocentric rotational therapy were under dosed by 'some 30 per cent'. From a distance it appeared that this might have been a software error in the programming. However, it is clear from the West Midlands RHA reports that it was an error at the interface between the medical physicists and the treating radiographers; an inappropriate inverse square law correction had been applied as a result.[31, 32] There were other treatments for which an inverse square law correction was appropriate, but it is difficult to envisage a computer system design where it would be desirable to make a manual inverse square law correction to the computer output, however good the slide rules or calculators available.

GETTING THINGS RIGHT

The establishment of 'Information for Health' as the basis for the future ensures that health information systems are going to be vital to the development of health care in England.[33, 34] It is therefore imperative that these systems should be developed and implemented effectively. This requires that, as far as possible, these systems should be 'got right' and that the experience of the past should be fully harnessed to support these initiatives. There is a wealth of advice on project management and systems development but the pathology of failed systems has a smell about it. The budgets and project milestones have been set by political rather than technical requirements. The clinical staff who will be using the systems have not been heavily involved in the design processes. The technical staff have not acquired an adequate understanding of the clinical requirements. The organization's management has not been fully involved in the project nor have they grasped the implications of project failure. When Frank Burns talked to the *Health Computing* conference about the lessons learned in setting up the Arrowe Park hospital system in the late 1990s, he was saying in almost exactly the same words the same things that had been said by the project team at The London Hospital some 10 to 15 years before.

The BCS HIF sought to systematize some of these issues, as the NPfIT iniative appeared to be having difficulties in the implementation of its massive programme of expenditure. The result of these exercises has been published as successive *Radical Steps* documents on the BCS HIF website.[35]

Another pressure that is coming is that for safety in health care, which is outlined towards the end of Chapter 25.[36] The European 'safety first principles' for health information systems,[37] Sir Bryan Jarman's Harveian lecture,[38] the *BMJ* issue of April 2000 on safety issues,[39] the establishment of the Patient Safety Agency and the views of its first chairman,[40] and the Royal College of Physicians conference in October 2004[41] are all driving down the same street (see Chapter 25). The long ignored issue of 'adverse incidents' in health care has begun to be taken more seriously as a problem for the NHS. It is to be hoped that health information systems will not be allowed to add additional hazards to the progress of patients through care, but rather will help health professionals avoid adverse incidents in the care of their patients.

REFERENCES

1   Peterson, I. (1996) *Fatal Defect: Chasing the Killer Computer Bugs.* Vintage Books, Random House, New York.

2   Brocks, B.J. (1969) What went wrong? An analysis of mistakes in data processing management. *Accountancy*, September, 666–75.

3   Barber B. (1975) The approach to the evaluation of The London Hospital computer project. In Anderson, J. and Forsythe, J.M. (eds) *MEDINFO 74*, Stockholm, pp. 155–165 and pp. 1011–12. North-Holland, Amsterdam.

4   Barber, B. and Abbott, W.A. (1972) *Computing and Operational Research at The London Hospital*; 'Computers in Medicine' series. Butterworths, London. ISBN 0 407517 00 6.

5   Barber, B., Cohen, R.D. and Scholes, M. (1976) A review of The London Hospital computer project. In Laudet, M., Anderson, J. and Begon, E. (eds) *Proceedings of Medical Data Processing*, Toulouse, pp. 327–38. Taylor & Francis, London. ISBN 0 850661 06 4.

6 Hawkins, R.E. (1975) Introduction of a user-oriented THIS [Total Hospital Information System] into a community hospital setting – introductory agents and their roles. In Anderson, J. and Forsythe, J.M. (eds) *MEDINFO 74*, Stockholm, pp. 75–7 and pp. 155–65. North-Holland, Amsterdam.

7 Gall, J.E. (1975) Introduction of a user-oriented THIS [Total Hospital Information System] into a community hospital setting – tactical management revelations. In Anderson, J. and Forsythe, J.M. (eds) *MEDINFO 74*, Stockholm, pp. 121–6. North-Holland, Amsterdam.

8 Hawkins, R.E. (1975) Introduction of a user-oriented THIS [Total Hospital Information System] into a community hospital setting – introduction and system description. In Anderson, J. and Forsythe, J.M. (eds) *MEDINFO 74*, Stockholm, pp. 295–8. North-Holland, Amsterdam.

9 Cook, M. (1975) Introduction of a user-oriented THIS [Total Hospital Information System] into a community hospital setting – nursing. In Anderson, J. and Forsythe, J.M. (eds) *MEDINFO 74*, Stockholm, pp. 303–4. North-Holland, Amsterdam.

10 Watson, R.J. (1975) Medical staff response to a medical information system with direct physician–computer interface. In Anderson, J. and Forsythe, J.M. (eds) *MEDINFO 74*, Stockholm, pp. 299–302. North-Holland, Amsterdam.

11 Yanez, L.M. (1975) Introduction of a user-oriented THIS [Total Hospital Information System] into a community hospital setting – confidentiality and security. In Anderson, J. and Forsythe, J.M. (eds) *MEDINFO 74*, Stockholm, pp. 201–6. North-Holland, Amsterdam.

12 Hall, P.F.L. (1974) Current status of the Karolinska Hospital computer system (Stockholm). In Collen, M.F. (ed.) *Hospital Computer Systems*, pp. 546–97. John Wiley & Sons, New York.

13 Hallen, B., Mellner, C.H., Selander, H. and Wolodarski, J. (1975) Guidelines for choice of anaesthesia – a result of the computerized patient data. In Anderson, J. and Forsythe, J.M. (eds) *MEDINFO 74*, Stockholm, pp. 443–7. North-Holland, Amsterdam.

14 Abramsson, S., Bergstrom, S., Larsson, K. and Tilman, S. (1970) Danderyd Hospital computer system: a total regional system for medical care. *Computer & Biomedical Research*, 3, 30–46.

15 Abramsson, S. (1971) Danderyd Hospital computer system. *Computer & Biomedical Research*, 4, 126–40.

16 Abramsson, S. (1973) Design of the Stockholm County Regional Medical Computer System. In (eds unknown) *Proceedings of an International Symposium on Medical Information Systems* (MEDIS 73), Osaka, pp. 44–52.

17 Peterson, H. (1975) Training and follow-up of hospital personnel in the use of EDP in Stockholm. In Anderson, J. and Forsythe, J.M. (eds) *MEDINFO 74*, Stockholm, pp. 233–4. North-Holland, Amsterdam.

18 Peterson, H.A. (1975) Password-oriented privacy system for Stockholm County. In Anderson, J. and Forsythe, J.M. (eds) *MEDINFO 74*, Stockholm, pp. 645–8. North-Holland, Amsterdam.

19 Anderson, J. (1974) King's College Hospital computer system. In Collen, M.F. (ed.) *Hospital Computer Systems*, pp. 457–516. Wiley Biomedical Health Publications, John Wiley & Sons, New York.

20 Bottone, H. (neé Strachan) (1994) *The Future Direction of Nursing Informatics in the UK: A Delphi Study.* City University (London) MSc. project. November.

21 British Computer Society (2004) *Code of Good Practice.* BCS, Swindon.

22 Kluge, E.H. (2003) *Rules of Ethical Conduct for Health Informatics Professionals.* BCS HIF, Swindon. Endorsed by IMIA.

23 Leveson, N.G. and Clark, S.T. (1993) An investigation of the Therac-25 accidents. *Computer*, July, 18.

24 Leveson, N.G. (1994) *Software: System Safety in the Computer Age.* Addison-Wesley, Reading, MA.

25 CCTA (1991) *Risk Analysis and Management Methodology (CRAMM) User Manual.* CCTA, IT Security & Privacy Group, 157–61 Millbank, London. (The software has been frequently updated and has now been made available in a commercial context from Insight Consulting at Walton on Thames.)

26 Barber, B., Vincent, R. and Scholes, M. (1992) Worst case scenarios: the legal consequences. In Richards, B. and MacOwan, H. (eds) *Current Perspectives in Healthcare Computing*, pp. 282–8. BJHC for BCS, Weybridge. ISBN 0 948198 12 5.

27 Barber, B. and Davey, J. (1992) The use of the CCTA risk analysis and management methodology (CRAMM) in health information systems. In Lun, K.C., Degoulet, P., Piemme, T.E. and Rienhoff, O. (eds) *MEDINFO 92*, Geneva, pp. 1589–93. North-Holland, Amsterdam.

28 Beynon-Davies, P. (1995) Information systems 'failure': the case of the London Ambulance Service's computer-aided despatch project. *European Journal of Information Systems*, 4, 171–84.

29 Smith, M.F. and Smart, G. (1999) Information technology failures in the NHS. In Bryant, J. (ed.) *Current Perspectives in Healthcare Computing*, pp. 149–55. BJHC for BCS, Weybridge. ISBN 0 948198 11 7.

30 Lane, V. (1999) Information systems projects – are successes or failures congenital or acquired? In Bryant, J. (eds) *Current Perspectives in Healthcare Computing*, pp. 156–64. BJHC for BCS, Weybridge. ISBN 0 948198 11 7.

31 West Midlands RHA (1992) *Report of the Independent Inquiry into the Conduct of Isocentric Radiotherapy at the North Staffordshire Royal Infirmary between 1982 and 1991.* West Midlands RHA.

32 West Midlands RHA (1994) *Second Report of the Independent Inquiry into the Conduct of Isocentric Radiotherapy at the North Staffordshire Royal Infirmary between 1982 and 1991.* West Midlands RHA.

33 Burns, F. (1998) A new information strategy for the NHS. In Bryant, J. (ed.) *Current Perspectives in Healthcare Computing*, pp. 46–53. BJHC for BCS, Weybridge.

34 NHS Executive (1998) *Information for Health: An Information Strategy for the Modern NHS 1998–2005* (The Burns Report). DoH, Ref. A1103.

35 Webref (accessed June 2007) BCS HIF *Radical Steps* series of Position Statements (2004–2006). www.bcs.org/BCS/forums/health

36 Barber, B., Allaert, F-A., Kluge, E.H. (2001) Info-vigilance or safety in health information systems. In Patel, V.L., Rogers, R. and Haux, R. (eds) *MEDINFO 2001*, London, pp. 1229–33. North-Holland, Amsterdam.

37 Barber, B., Jensen, O.A., Lamberts, H., Roger, F., de Schouwer, P. and Zöllner, H. (1991) Six safety first principles of health information systems. In Richards, B. and MacOwan, H. (eds) *Current Perspectives in Healthcare Computing*, pp. 296–314. BJHC for BCS, Weybridge. ISBN 0 948198 11 7.

38 Jarman, B. (2000) The quality of care in hospitals. *Journal Royal College of Physicians*, 34, 1, 75–91.

39 British Medical Journal (2000) Reducing error and improving safety. *BMJ*, 7237, 18 March.

40 Shaw, R. (2004) Patient safety: the need for an open and fair culture. *Clinical Medicine*, March/April, 128–31.

41 Royal College of Physicians (2004) *Making Health Care Safer 2004.* Conference at the Royal College of Physicians, supported by the BMA.

Appendices

Appendix 1: Timeline

Year	Event
1950s	
1956	Ministry of Health working party to consider the use of Automatic Data Processing by the Regional Health Boards
1960s	
1960s	University College Hospital, Royal Infirmary, Edinburgh – pathology systems developments
1961	First in-house hospital finance system, Manchester Regional Hospital Board
1962	Radiotherapy: computers first applied to manage multiple beams, Cardiff
1963	Vaccination & Immunization system produced (West Sussex)
1964	Computer-based immunisation call & recall system West Sussex County Council
1964	Oxford Record Linkage project created person-based data for epidemiological research
1966	BCS London Medical Specialist Group established
1967	Cervical Cytology call & recall, West Sussex
1967	Radiotherapy: stand-alone single task machine operational at Royal Marsden Hospital
1969	Anderson project on lifetime summary and in-depth episode records of hospital care on computer started at King's College Hospital
1968	BCS Northern Medical Specialist Group established
1968	DHSS established Computer Branch and initiated the Experimental Computer Programme
1969	BCS London Medical Specialist Group first conference, Birmingham
1969	International Federation of Information Processing sets up technical committee in health informatics (IFP TC4)

1969   QE Birmingham implemented a ward-based patient administration, drug administration and nursing system

1969   Computer-aided decision support used by de Dombal and colleagues

1969   International Medical Informatics Association started with UK as a member

1970s

1970   Eleven Regional Hospital Board Computer Centres set up, employing 561, total cost of £1.25 million per year

1970   Real-time GP system set up by IBM at Whipton, Exeter

1970   DHSS Experimental Project Programme extended

1970   Computer-aided patient interviewing started at West Middlesex Hospital

1971   DHSS Project No. 2, North Staffordshire Hospital, Stoke-on-Trent

1972   First attempt to link GPs to hospitals by Royal Devon and Exeter

1972   Exeter Community Health Project sponsored by DoH

1972   London Hospital Master Patient Index (MPI/PMI) established

1973   Wythenshawe Hospital started patient monitoring system for nurses, also coronary care and ICU patient management system from Karolinska Hospital, Sweden

1974   International Medical Informatics Association first *MEDINFO* Conference, Stockholm

1976   DELPHI Phoenix CTL (Hammersmith) to Ferranti (Royal Lancaster Infirmary) pathology laboratory system operational

1976   World Health Organization (WHO) supported establishment of European Federation of Medical Informatics

1978   Dental Estimates Board started its computerization

1978   Parliamentary Committee on Accounts (PAC) examination of Experimental Computer Programme

1978   First EFMI Congress, Medical Informatics Europe (MIE) held in Cambridge

1979   IFIP TC4 spun off into International Medical Informatics Association

1980s

1980   *Inequalities in Health*, The Black Report, published

1980   Körner Committee began work. NHS/DHSS Steering Group on Health Services Information

1980   RCGP Report: *Computers in Primary Care*, information systems linked to quality of care

1981	NHS Computer Policy Committee established
1981	BCS Primary Health Specialist Group established
1981	BCS Health Informatics Scottish SG established
1981	Works Information Management (WIMS) system released by DHSS to NHS
1982	National Staff Committee for Nurses and Midwives issue recommendations on Computer Training for Nurses in middle management
1982	First international nursing conference, *The Impact of Computers on Nursing*, held in London and Harrogate
1983	First *Health Computing* conference held in Birmingham
1983	National Poisons Information Service computerized
1983	BCS Nursing Specialist Group formally established
1984	Prescription Pricing Authority established as a Special Health Authority under the NHS Act (1977), computerized
1984	Körner Confidentiality Working Group: *Protection and Maintenance of Confidentiality of Patient and Employee Data*
1984	*British Journal of Healthcare Computing* Issue 1; revised title of *BJHC&IM*
1984	Network of Users of Microcomputers in Nurse Education (NUMINE) formed
1984	Open Software Library for educational software established
1984	Read Code development started
1984	Data Protection Act (1984)
1984	GPASS system accepted by the Scottish Office as standard system
1984	NHS Applications Register set up
1985	Micros for GPs scheme, evaluation
1985	NHS Information Management Group established
1985	Trent Regional Computer Centre entered into commercial agreement with AT&T ISTEL to market Trent PAS
1985	First standards for prescribing by computer
1985	3-D simulation of hip replacement operations, Queen's University Belfast
1985	CAER consortium established
1985	Dental fee claim forms (FP17) amended for computer completion
1986	West Coast consortium (Argyll & Clyde Health Board) took GPASS system
1986	National Strategic Framework for Information Management in Hospital and Community Services

1986	English National Board national survey into Computer-Assisted Learning in nursing
1987	Körner Minimum Data-Sets agreed by Ministers and NHS to supersede existing data
1987	NHS Data Protection Handbook
1987	Community Health Index (CHI) went live in Dundee
1987	GP Free Schemes came to market
1987	Oxford GP links project, connected to John Radcliffe Hospital, Oxford and Wolfson Laboratory, Birmingham for pathology reporting
1987	MACPUFF (physiology of lungs) and MACPEE (body fluid) simulation programmes written by Prof. David Ingram, London
1987	Welsh IT Strategy issued
1987	DoH Health (Scotland) Service Information Systems set up in Common Services Agency
1988	ENB Computer-Assisted Learning Project
1989	A guide to European Legislation for Purchases of IT in the NHS published with establishment of the European single market
1989	BCS NSG official journal *IT in Nursing* first published
1989	GP Minimum System Specification project started
1989	PEN&PAD programme started, MIG Manchester University for clinical workstations and structured data entry

1990s

1990	NHS Common Basic Specification: Generic Model Reference Manual issued
1990	IMIA Workshop on International Primary Care Computing at Brighton
1990	Working for Patients – Framework for Information Systems
1991	HISS set up at Darlington Memorial Hospital
1991	Centre for Coding and Classification set up by DoH Information Management Group
1992	Nursing Terminology Workshop, hosted by BCS NSG in Eastbourne
1992	GALEN clinical terminologies project started
1993	GP/FHSA links for registration/deregistration of patients
1993	First GP systems undergo Requirements for Accreditation (RFA) validation.
1994	*Making IT Work – Implementing the IM&T Strategy'* issued

1994	Electronic Patient Record six-level incremental model structure published
1995	ASSIST established
1995	UK National Standard Clinical EDIFACT messages published
1995	Primary care informatics curriculum developed by JIGSAW group
1995	Promoting Patient Choice, King's Fund for shared clinical decision-making by patient and professional, DoH set up Centre for Health Information Quality
1995	Benefits Realisation Monographs produced by BCS NSG (completed 1996)
1996	National Audit Office report on the HISS Programme
1997	*Report on the Review of Patient Identifiable Information* – Caldicott
1987	PRIMIS pilots started by Nottingham University
1998	*Information for Health: Information Strategy for Modern NHS 1998–2005* published
1998	*Learning to Manage Health Information* published: health informatics for doctors and nurses

2000s

2000	BCS publish *Handbook of Health Informatics*
2000	Regulations allow GPs to 'go paperless'
2001	*MEDINFO* held in London
2002	*Securing Our Future: Taking a Long-Term View* – Wanless first report
2003	ERDIP National Core Evaluation – final report
2004	NHS Institute for Learning, Skills and Innovation established
2005	National Programme for Health renamed *Connecting for Health*

Appendix 2: Lead author biographies

WILLIAM (BUD) ABBOTT, MIHM, MBCS, joined the hospital service in 1947, after war time service in the RAF, taking posts at the Charing Cross Hospital and the Royal Cancer Hospital. He moved to the Finance Department at The London Hospital in 1948, and while a patient he became intrigued with the possibilities of computing to replace machine accounting. He became the Computer Manager when The London established its own computer centre in 1964 and developed a wide range of applications in The London's Experimental Project: the world's first, integrated, real-time hospital systems. In 1974, he set up a region-wide computing bureau for NE Thames RHA. Bud retired from the NHS in 1985, but he continued to be an active member of the BCS and the Institute of Health Management as well as IMIA. Bud has written books, papers and edited proceedings for the international informatics community.

BARRY BARBER, MA, PhD, FBCS, FInstP, ARCP, started as a student mathematician and theoretical physicist at Cambridge (1951–54), going on to be a Medical Physicist, Computer Scientist & Operations Analyst at The London Hospital until 1975 when he was appointed Chief Management Scientist at NE Thames RHA. In 1988, he joined the NHS Executive as Security and Data Protection Manager until he left the NHS in 1997 to set up his own business, Health Data Protection Ltd. He has played very active roles in international informatics including as a Professional Medical Informatician and participant in European Union Projects: AIM Impact and Forecast, SEISMED, ISHTAR, SYNAPSES, EUROMED-ETS, MEDSEC and the standards work of CEN TC251 WGs 6 & III. Barry has also held the roles of IMIA Vice-President (Europe), EFMI Secretary, Vice-President and President, BCS Vice-President (Specialist Groups), Medical Specialist

Groups Coordinating Committee, Medical Specialist Group (London), Hospital Physicists Association Computer Topic Group and Operational Research Society's Health & Welfare Services Study Group.

DENISE BARNETT, MSc, BA(Hons), RN, RCNT, Dip in Nur (Lon), was introduced to computers as a staff nurse at The London Hospital and in 1979 attended the first computer programming workshop for nurses at UMIST. She joined the BCS Nursing Specialist Group, becoming editor of its journal *IT in Nursing* and also an RCN representative at EFMI meetings. After clinical experience in medical and surgical wards, Denise became a nurse manager and later worked in manpower planning and personnel. This was followed by clinical research into the nursing process in Tower Hamlets. She managed the nursing service at the Princess Alexandra Hospital, in Harlow, before becoming the District Adviser for Nursing and Quality Assurance. She left the NHS in 1987 to work on projects and short-term contracts, including a spell as a professional adviser at the National Audit Office and as administrator for the Nursing Professions Information Group. In 1998, she gained a Masters degree in Electronic Publishing from City University and worked as a Web Manager for Citizens Advice, on EEJ Net and as Web Development Officer with the British Association for Adoption and Fostering. Denise has written nursing text books and edited books and booklets about IT for the nursing profession.

ROY BENTLEY, PhD, BSc, took his degree in physics in 1951 and a PhD at the University of Birmingham in 1955, then spent the greater part of his working life at the Institute of Cancer Research and the associated Royal Marsden Hospital. After working on the assessment of radioactive nuclides in the environment, he turned his attention to the application of computers in medicine and in particular to medical physics. This included applications in radiotherapy, measurements of the energy spectra of radiotherapy machines and in nuclear medicine. He spent a year's sabbatical in the biomedical computer department at Washington University, St. Louis, in 1967–68, where he was one of the first people to link a computer online to a gamma camera. With Jo Milan he introduced the RAD-8 system and worked with commercial companies that developed it into various commercial products. He worked on an early system for the checking and recording of parameters in radiotherapy at the point of treatment. Later he was instrumental in the introduction of a computerized information system for the Royal Marsden Hospital. He returned, before retirement in 1995, to an interest in radiotherapy planning including the transferring of a 3-D system from DKFZ in Heidelburg and the networking of planning computers with CT scanners.

JOHN BRYANT, BSc(Hons), MBA, FBCS, CEng, graduated as a physicist and immediately moved into commercial data processing. In 1972, after six years in the commercial sector, he joined the NHS to work in the Experimental Computer Project at Cambridge. John has extensive experience in the use of information and communications technologies. He has directed a wide range of leading edge projects in the health sector including the multimillion pound HISS programme to introduce integrated information systems into acute hospitals, the establishment of the national electronic patient record programme, and the development of benefits management and change management programmes. He also has wide experience of managing information services delivery. During this time he participated actively in the standards field, representing the UK user community at the International Organization for Standardization. He was also active worldwide in the health informatics field, becoming President of the European Federation for Medical Informatics (EFMI) and Vice-President (Europe) of the International Medical Informatics Association (IMIA). He is currently Head of Information Management in the European Institute of Health and Medical Sciences at the University of Surrey, where he has particular interests in strategic management, organizational transformation, the organizational impact of information systems and the prevention of systems failures.

SHEILA BULLAS, BSc, MBCS, CITP, MBA, started her NHS career as a student in medical laboratory technology. She completed a degree in biology at Sussex University specializing in molecular and mathematical biology and worked at Shell Research Laboratories. Sheila returned to the NHS and worked in capital and service planning, information services and computing, heading up RMI systems at the Department of Health. Sheila founded Health Strategies, an independent management and technology consultancy company working in the health care sector on complex, innovative projects, integrating processes, people and information to improve health outcomes. Her international assignments have included strategic advice to the Minister at the Department of Health of Western Australia, Masters in Health Informatics for Erasmus University (The Netherlands) and support to the Swedish Healthcare Planning Institute. Sheila is a fellow of King Alfred's College, Winchester, and a member of CHIRAD, lecturing in strategic management and organizational change to masters students. Currently she is an Associate Director at The Princess Alexandra Hospital NHS Trust, implementing NPfIT and transforming services under difficult circumstances.

ALARIC CUNDY, MSc, BSc, worked in health informatics in the NHS for over 38 years before he retired in 2006. He was a significant contributor to the development of evaluation methodologies for the experimental computer projects that were implemented in around 1970. More recently he has contributed in the field of NHS data standards and he is a Reference Group member of the Information Standards for which he now works on a part-time freelance basis.

MICHAEL FAIREY, CB, MA, is one of the very few general managers who have also held major posts in information management. Michael has had a career-long involvement in the subject. He led the formative stages of The London Hospital Experimental Computer project, was Vice-Chairman and, briefly, Chairman of the Körner Committee, and as the first NHS member of the NHS Management Executive, was for seven years the Director of NHS Information Systems. For 10 years, he was Consulting Editor of the *British Journal of Healthcare Computing and Information Management.*

GLYN M. HAYES, MBChB, DRCOG, FBCS, CITP, qualified at Birmingham Medical School in 1971. He became a principal general practitioner in 1973 and went on to be a founder member of the BCS Primary Health Care Specialist Group, and its Chair in 1985. He is now its President. Glyn designed one of first consulting room GP computer systems. He was Chair of the Hereford and Worcester Local Medical Committee in

1992 and became Medical Director of AAH Meditel & Torex Health until 1999, one of the largest suppliers of IT to the NHS. Glyn is Chair of the BCS Health Informatics Forum. He is also president of the UK Council for Health Informatics Professionals, the registration body for Health Informaticians. He has represented the UK on the International Medical Informatics Association (IMIA) and was the Chair of the Primary Care Working Group of IMIA. He is a respected lecturer, author and keynote speaker at many international conferences.

RAY JONES, FFPH, MBCS, CEng, started his career in health informatics in Nottingham in 1976. His interest in patients' access to their own medical record began in 1979 when, with colleagues, he established the Nottingham Diabetes System, issuing paper copies of medical records to patients. He moved to Glasgow in 1984 where he developed and evaluated the first public access health kiosks (Healthpoint) and carried out research into patients' online access to their own medical record. This included a series of randomized trials, including those of: patient-held records in primary care; a personalised touch-screen system for patients with cancer; an education system for patients with schizophrenia; and a community-based multimedia system for treatment of anxiety. Ray moved to Plymouth in 2002 as Associate Dean for Research in the Faculty of Health and Social Work, and Professor of Health Informatics, at the University of Plymouth. There he has been developing interdisciplinary health and social care research, including e-health. His interest in synchronous technologies started with use of a Plymouth uplink to an ESA satellite. He has been exploring the synergy between e-health and e-learning. Ray is an elected Fellow of the Faculty of Public Health, a member of the British Computer Society and Chartered Engineer.

LLOYD A.W. KEMP, PhD, OBE, studied physics at King's College, London, and went on to obtain a Postgraduate Diploma in Education (1932–36). His first job was at GEC, on the development of television systems, but when war loomed he resigned, and began teaching, having been given exemption from National Service as a conscientious objector. However, impressed by Eve Curie's biography of her mother, he determined to become a medical physicist, and, in 1944, joined Dr John Read at the newly established Medical Physics department at The London Hospital. He followed Dr Read as Head of Department in 1946. In 1956 he was given the Roentgen Award of the British Institute of Radiology in 1956. In 1966 he moved to The National Physical Laboratory, and retired 12 years later.

BERNARD RICHARDS, MSc, PhD, FBCS, FIMA, FIHRIM, CEng, CMath, CSci, CITP, FRAMS, qualified in mathematics, computing and astronomical physics at Manchester University. He became Professor of Computation at UMIST in 1971, and in 1992, concurrently, UK's second Professor of Medical Informatics. In 1997, he was made First President of IHRIM. Bernard was made the BCS 'Fellow of the Year' in 1998 for 'Services to Medical Informatics'. In 1999, he was awarded the Saint Wenceslas Medal of Charles University for 'Services to Medicine in Prague'. Bernard was made an Honorary Fellow of the Medical Computing Societies of the Czech Republic, Hungary, Romania, the Ukraine and Poland, and a Fellow of the Romanian Academy of Medical Science. He was also elected a Council Member of the Ukrainian Association of Computer Medicine. In 2001, Bernard was awarded a plaque by IMIA as the only person to have presented one or more papers at all 10 *MEDINFO*s. He has served on the following committees: NHS Committee on NHS Computer Languages; RCOG Minimum Dataset; Intensive Care Computer Group; British Obstetric Society Computer Group; Occupational Therapy Computer Group; BCS NMS Group (as Chair); the PHCS Group; the LMS Group; BCS Health Informatics Committee (as Chair); and the BCS Manchester Branch.

JEAN ROBERTS, PhD, CEng, FBCS, CITP, MHM, is an internationally known health informatician with extensive experience in the domain: in the NHS, with commercial vendors and consultancies and latterly as an academic and researcher. She has a doctorate mapping health informatics over 30 years. She leads the BCS Health Informatics Forum Policy Group, is a Board member of UKCHIP and the NHS HI Faculty. Jean is a Fellow of the British Computer Society, Chartered Engineer (Information Systems Engineering) and Chartered IT Professional. Her particular research interests are in the maturing of the HI profession; health, social welfare and lifestyle convergence; and effective citizens' information. Jean has played an active role in MIE and IMIA for many years.

MAUREEN SCHOLES, MBCS, CITP, qualified as a nurse at The London Hospital in Whitechapel and held appointments there as a staff nurse, ward sister, night superintendant and assistant matron. In 1967, following her responsibility for nurse allocation, and then with responsibility for the medical side of the hospital, she became the nurse member of the Executive Team steering the introduction of a major computer system at The London. She recognized the importance of nurses meeting together in this new field and became founder Chair of the Computer Projects Nurses Group from 1975 to 1980. Maureen was the Chair of the First International Computer Conference on Nursing and Computing held in London and in Harrogate in 1982. The following year she was a founder member of the BCS Nursing Specialist Group and was the founder Chair of IMIA Working Group 8 (Nursing), later to become SIGNI 1983. Maureen was Director of Nursing Service at The Royal London Hospital until her retirement from the NHS. She has written articles and given international papers at informatics conferences and was a co-author of a history of the first 40 years of international nursing informatics.

HEATHER STRACHAN, MSc, RGN, Dip Nur (Lon), MBCS, trained as a nurse in Glasgow and then worked in Intensive Care Units at a number of London Hospitals including St Mary's Hospital and Westminster Hospital. It was while at Westminster Hospital as leader of the practice development team, in the mid-1980s, that she developed an interest in health informatics while working on a project to develop nurse care planning, scheduling and workload systems. Later, Heather joined the Information Department at Charing Cross Hospital in London. She has held a variety of posts, which have involved acute and community health services management practice development, research, project management, and health care governance. Heather has a Masters in Information Science from London's City University and she is a member of the Centre for Health Informatics Research and Development. She has presented at many health informatics conferences both nationally and internationally and published a number of national reports, book chapters, articles and conference proceedings on a variety of health informatics topics. Heather was a past Chair of the British Computer Society's Health Informatics Group in Scotland and Chair of the International Medical Informatics Association Special Interest Group on Nursing Informatics. Her present post is Nursing, Midwifery and Allied Health Professions eHealth Lead in the Scottish Government.

ANDREW TODD-POKROPEK, PhD, read physics at the University of Oxford. His first post was Assistant Lecturer at the Institute of Nuclear Medicine at The Middlesex Hospital where he undertook postgraduate study, assisting in the development of nuclear medicine instrumentation and one of the first computer systems for use in medical imaging. In 1972, he spent a sabbatical year at the Institut Gustave Roussy in France, returning to take up a post as Senior Lecturer at the University College Hospital. He was appointed Reader in 1989. Andrew became increasingly involved with medical image processing and data handling in nuclear medicine and latterly in radiology in general. His research extends from inverse problems to developing expert systems for medicine. Andrew is Professor and Head of Medical Physics and Bioengineering at University College, London, with a joint appointment in Radiology at Great Ormond Street Hospital for Sick Children. He is also Director of the INSERM Unit in Paris, and of Leapfrog Technology Ltd, which provides consultancy and distributes software. He has published and lectured widely, and participates in standards activities including DICOM, and in the development of Interfile. The honours he has received include the American Society of Nuclear Medicine Annual Lecturership, the Sylvanus Thompson medal and the Normal Veale medal.

ANTHONY R. WILLS, MIET, undertook a student apprenticeship with English Electric Co in Stafford, leading to a BSc(Eng) and research into industrial automation using digital and analogue computers. In 1969, he joined The London Hospital, Whitechapel, at the start of their computer project into hospital-wide information systems, progressing to Applications Development Manager in 1978. He retired in 1996.

ALAN WYNNE THOMAS, BSc(Econ), MTech(Comp Sc), FBCS, CITP, was educated at the London School of Economics and Brunel University. He started 'computing' with the National Coal Board, then ICT, and Rank Xerox/RX Data Systems. He moved to Otago University, New Zealand, to be Head of the new Department of Computer Systems. He was Senior Consultant with ICL Health and Social Security Division and GPSD Italy. Alan was Managing Director of own company, Sifo Ltd, which undertook many health care systems projects in the Middle and Far East, Europe, and the UK. He was a Non-executive Director of NOC NHS Trust in Oxford. He has been Consultant/Project Director with NHSME, NHS in Wales, and NHS Scotland. Alan is also a round-the-world-single-handed sailor.

Index